YOUNG SUPERNOVA REMNANTS

Previous Proceedings
in the Series of Annual Astrophysics Conferences
in College Park, Maryland

	Year	Short Title	Publisher	ISBN
Tenth	1999	Cosmic Explosions	AIP Conference Proceedings 522	1-56396-943-2
Ninth	1998	After the Dark Ages	AIP Conference Proceedings 470	1-56396-855-X
Eighth	1997	Accretion Processes	AIP Conference Proceedings 431	1-56396-767-7
Seventh	1996	Star Formation	AIP Conference Proceedings 393	1-56396-678-6

Other Related Titles from AIP Conference Proceedings

558   High Energy Gamma-Ray Astronomy: International Symposium
Edited by Felix A. Aharonian and Heinz J. Völk, April 2001, 1-56396-990-4

556   Explosive Phenomena in Astrophysical Compact Objects: First KIAS
Astrophysics Workshop
Edited by Heon-Young Chang, Chang-Hwan Lee, Mannque Rho, and Insu Yi,
March 2001, 1-56396-987-4

555   Cosmology and Particle Physics: CAPP 2000
Edited by Ruth Durrer, Juan Garcia-Bellido, and Mikhail Shaposhnikov, March 2001,
1-56396-986-6

540   Particle Physics and Cosmology: Second Tropical Workshop
Edited by José F. Nieves, October 2000, 1-56396-965-3

537   Waves in Dusty, Solar, and Space Plasmas
Edited by F. Verheest, M. Goossens, M. A. Hellberg, and R. Bharuthram, October 2000,
1-56396-962-9

528   Acceleration and Transport of Energetic Particles Observed in the Heliosphere:
ACE 2000 Symposium
Edited by Richard A. Mewaldt, J. R. Jokipii, Martin A. Lee, Eberhard Möbius, and
Thomas H. Zurbuchen, July 2000, 1-56396-951-3

526   Gamma-Ray Bursts: 5th Huntsville Symposium
Edited by R. Marc Kippen, Robert S. Mallozzi, and Gerald J. Fishman, June 2000,
CD-ROM included, 1-56396-947-5

523   Gravitational Waves: Third Edoardo Amaldi Conference
Edited by Sydney Meshkov, June 2000, 1-56396-944-0

To learn more about these titles, or the AIP Conference Proceedings Series, please visit the
webpage **http://www.aip.org/catalog/aboutconf.html**

YOUNG SUPERNOVA REMNANTS

Eleventh Astrophysics Conference

College Park, Maryland   16–18 October 2000

EDITORS
Stephen S. Holt
*Olin College of Engineering
Needham, Massachusetts*

Una Hwang
*NASA/Goddard Space Flight Center
Greenbelt, Maryland*

Melville, New York, 2001
AIP CONFERENCE PROCEEDINGS ■ VOLUME 565

Editors:

Stephen S. Holt
Franklin W. Olin College of Engineering
1735 Great Plain Avenue
Needham, MA 02492-1245
USA

E-mail: stephen.holt@olin.edu

Una Hwang
NASA/Goddard Space Flight Center
Code 662
Greenbelt, MD 20771
USA

E-mail: hwang@orfeo.gsfc.nasa.gov

L.C. Catalog Card No. 2001089085
ISBN 0-7354-0001-6
ISSN 0094-243X
Printed in the United States of America

CONTENTS

Preface xi

INTRODUCTION

Pineapples and Crabs: When Young Supernova Remnants Were
Even Younger 3
    V. Trimble
Young Supernova Remnants: Issues and Prospects 17
    C. F. McKee

STAGE SETTING

Type Ia Supernovae: Toward the Standard Model? 31
    D. Branch
Optical Observations of Core-Collapse Supernovae 40
    A. V. Filippenko
Hydrodynamics of Young Supernova Remnants 59
    J. M. Blondin
Early Optical Spectroscopic Evidence for Conical Ejection
in Supernovae? 69
    E. M. Schlegel
Radio Emission from Supernovae in Binary Systems 73
    F. R. Boffi and N. Panagia
The Fading Radio Emission from Supernova 1961V 77
    C. J. Stockdale, M. P. Rupen, J. J. Cowan, and Y.-H. Chu
Hydrodynamic Instabilities in Young Supernova Remnants 81
    C. L. Hinkle and J. M. Blondin
Modeling of Radiative Blast Waves 85
    K. A. Keilty, E. P. Liang, T. Ditmire, B. A. Remington,
    A. M. Rubenchik, and K. Shigemori
The Evolution of a Supernova Remnant in a Turbulent,
Magnetized Medium 89
    D. Balsara, R. A. Benjamin, and D. P. Cox

SNe TO SNR

SN1987A: The Youngest Supernova Remnant 95
    R. McCray
Supernova Remnants in Molecular Clouds 109
    R. A. Chevalier
The SN−SNR Connection 119
    R. A. Fesen

New Imaging Results in the Formation of SNR 1987A 129
    B. Sugerman, S. Lawrence, A. Crotts, and P. Bouchet
STIS Spectral Imaging of Hot Spots in SNR 1987A 133
    S. Lawrence, B. Sugerman, and A. Crotts
**The Structure of SN 1987A's Outer Circumstellar Envelope
as Probed by Light Echoes** 137
    A. Crotts, B. Sugerman, S. Lawrence, and W. Kunkel

YOUNG SNR I

X-ray Observations of Young Supernova Remnants. 143
    U. Hwang
UV/Optical Observations of Young Supernova Remnants 153
    W. P. Blair
Young SNRs: IR Observations. 163
    R. G. Arendt
Gamma-Ray Line Emission from Supernova Remnants 173
    P. A. Milne
**Chandra Discovery of Ejecta-Dominated X-ray Emission
from the Old SNR Candidate Sgr A East.** 177
    Y. Maeda, F. K. Baganoff, E. D. Feigelson, M. W. Bautz,
    W. N. Brandt, D. N. Burrows, J. P. Doty, G. P. Garmire,
    M. Morris, S. H. Pravdo, G. R. Ricker, and L. K. Townsley
Exploring the Small Scale Structure of N103B 181
    K. T. Lewis, D. N. Burrows, J. A. Nousek, G. P. Garmire,
    P. Slane, and J. P. Hughes
A Look at SNR 0519-69.0 with *Chandra*'s ACIS. 185
    R. Williams, R. Petre, and S. S. Holt
**Far Ultraviolet Spectroscopy of a Nonradiative Shock
in the Cygnus Loop: Implications for the Postshock
Electron-Ion Equilibrium** 189
    P. Ghavamian, J. C. Raymond, and W. P. Blair
**Fabry-Perot [O III] $\lambda5007$ Å Observations of the SMC Oxygen-Rich
SNR 1E 0102-72.9.** 193
    K. A. Eriksen, J. A. Morse, R. P. Kirshner, and P. F. Winkler
**2MASS Near-infrared Images of RCW 103:
A Young SNR in a Dense Medium** 197
    J. Rho, W. T. Reach, B.-C. Koo, and L. Cambresy

YOUNG SNR II

Emission Processes in Young SNRs. 203
    J. C. Raymond

**High Resolution Spectroscopy of Two Oxygen-Rich SNRs
with the Chandra HETG.** 213
    C. R. Canizares, K. A. Flanagan, D. S. Davis, D. Dewey,
    J. C. Houck, and M. L. Schattenburg
An Approximation to the Cooling Coefficient of Trace Elements. 222
    R. A. Benjamin, B. A. Benson, and D. P. Cox
Ionization Structure and the Reverse Shock in E0102-72 226
    K. A. Flanagan, C. R. Canizares, D. S. Davis, D. Dewey,
    J. C. Houck, and M. L. Schattenburg
Spectral Line Imaging Observations of 1E0102.2-7219 230
    D. S. Davis, K. A. Flanagan, J. C. Houck, C. R. Canizares,
    G. E. Allen, N. S. Schulz, D. Dewey, and M. L. Schattenburg

YOUNG SNR III

Radio Emission from SNe and Young SNRs 237
    K. W. Weiler, N. Panagia, M. J. Montes, S. D. Van Dyk,
    R. A. Sramek, and C. K. Lacey
Cas A—A Y2K Status Report 247
    L. Rudnick
Kepler's Supernova Remnant. 257
    A. Decourchelle
Vela Z—So Young and So Exotic 267
    M. D. Filipović, P. A. Jones, and B. Aschenbach
**High Frequency Radio Observations of the Composite Historical
SNR G11.2-0.3.** 271
    R. Kothes and W. Reich
Radio Proper Motions: Cas A and Kepler 275
    B. Koralesky and L. Rudnick
Spectral Index Variations in Kepler's Supernova Remnant 279
    T. DeLaney, B. Koralesky, L. Rudnick, and J. R. Dickel

THE CRAB AND RELATED REMNANTS

The Crab Nebula: The Gift That Keeps on Giving. 285
    J. J. Hester
Plerions and Pulsar-Powered Nebulae. 295
    B. M. Gaensler
Pulsars, AXPs, and SGRs 305
    R. N. Manchester
**NUV and FUV Spectroscopic Timing Observations of the Crab Pulsar
with HST/STIS** 315
    T. R. Gull, J. Sollerman, P. Lundqvist, G. Sonneborn,
    and D. Lindler

**A Brief Update on the First Balloon Flight of a Crystal Diffraction
Lens Telescope, CLAIRE** 317
    R. Smither, P. V. Ballmoos, G. Skinner, and Balloon Launch Crew
    from Toulouse, France

**Young Collapsed Supernova Remnants: Similarities and Differences
in Neutron Stars, Black Holes, and More Exotic Objects** 321
    J. S. Graber

A Zoo of Composite SNRs 325
    J. R. Dickel

Millimetric Observations of Plerionic Supernova Remnants 329
    R. Bandiera, R. Neri, and R. Cesaroni

G291.0−0.1: Powered by a Pulsar? 333
    D. Moffett, B. Gaensler, and A. Green

Chandra Observation of the Hard X-ray Feature in IC 443 337
    C. R. Clearfield, C. M. Olbert, N. E. Williams, J. W. Keohane,
    and D. A. Frail

Evidence of a Pulsar Wind Nebula in Supernova Remnant IC 443 341
    C. M. Olbert, C. R. Clearfield, N. E. Williams, J. W. Keohane,
    and D. A. Frail

**Very Large Array Data Pertaining to the Hard X-ray Source
on the Southern Edge of Supernova Remnant IC 443** 345
    N. E. Williams, C. M. Olbert, C. R. Clearfield, J. W. Keohane,
    and D. A. Frail

COSMIC RAY ACCELERATION AND PULSARS

Pulsar Emission and Central Sources in Young Supernova Remnants 351
    A. K. Harding

**Evidence for Cosmic Ray Acceleration in Supernova Remnants
from X-ray Observations.** 360
    R. Petre

Particle Acceleration in Shock Waves of Young Supernova Remnants 369
    S. P. Reynolds

Nonthermal X-ray Emission from Young Supernova Remnants. 379
    E. van der Swaluw, A. Achterberg, and Y. A. Gallant

Electron Acceleration in Young Supernova Remnants 383
    W. K. Rose

**Maximum Energies of Shock-Accelerated Electrons in Supernova
Remnants in the Large Magellanic Cloud** 387
    S. P. Hendrick, S. P. Reynolds, and K. J. Borkowski

Spatially Varying X-ray Synchrotron Emission in SN 1006 391
    K. K. Dyer, S. P. Reynolds, K. J. Borkowski, and R. Petre

**The Nonthermal X-ray Emission of SN 1006 and the Implications
for Cosmic Rays** 395
    G. E. Allen, R. Petre, and E. V. Gotthelf

X-ray Synchrotron in the Forward Shock of Cassiopeia A 399
    F. Berendse, S. S. Holt, R. Petre, and U. Hwang
Nonthermal X-ray Emission from G266.2−1.2 (RX J0852.0−4622) 403
    P. Slane, J. P. Hughes, R. J. Edgar, P. P. Plucinsky,
    E. Miyata, H. Tsunemi, and B. Aschenbach

SOME SPECIAL EXAMPLES

Circumstellar Nebulae in Young Supernova Remnants 409
    Y.-H. Chu
Young Supernova Remnants in the Magellanic Clouds 419
    J. P. Hughes
An Investigation of SNRs in the Magellanic Clouds—SNR B0450-708
and SNR B0455-687 429
    M. D. Filipović, P. A. Jones, and G. L. White
New Radio Supernova Remnants in the Core of M 31 433
    L. O. Sjouwerman and J. R. Dickel
Assessing the Nature of SNR Candidates in M 101 437
    B. C. Dunne, R. A. Gruendl, Y.-H. Chu, and R. M. Williams
Global VLBI Observations of M82 441
    A. R. McDonald, T. W. B. Muxlow, A. Pedlar,
    M. A. Garrett, K. A. Wills, S. T. Garrington,
    and P. J. Diamond
X-ray and Radio Observations of Supernova Remnants in NGC 300 445
    T. Pannuti, M. D. Filipović, N. Duric, W. Pietsch, and A. Read
X-ray Fluxes of Radio-Selected Remnants 449
    E. A. Miller and L. Rudnick

HYPERNOVAE

Supernovae, Hypernovae, and Gamma Ray Bursts 455
    A. Dar
Hypernovae: Observational Aspects 471
    I. J. Danziger

CONFERENCE SUMMARY

Reflections on Young SNRs: Rapporteur Summary
of the 11th Maryland Astrophysics Conference......................... 481
    P. F. Winkler and K. S. Long

APPENDICES

Conference Program 493
List of Attendees.. 497

Author Index...501
Source Index..505
Subject Index 509

PREFACE

This is the eleventh installment of the current series of annual October Astrophysics Conferences in Maryland. These conferences are organized by astrophysicists at the Goddard Space Flight Center and the University of Maryland. The topic for each conference is selected with the help of an International Advisory Committee, the current membership of which is:

Marek Abramowicz, Gotenborg
Roger Blandford, Pasadena
Claude Canizares, Cambridge, MA
Arnon Dar, Haifa
Alan Dressler, Pasadena
Guenther Hasinger, Potsdam
Steve Holt, Needham
Bob Kirshner, Cambridge, MA
Dick McCray, Boulder

Jim Peebles, Princeton
Sir Martin Rees, Cambridge, UK
Vera Rubin, Washington
Joseph Silk, Oxford
David Spergel, Princeton
Rashid Sunyaev, Moscow
Yasuo Tanaka, Tokyo
Scott Tremaine, Princeton
Simon White, Garching

Each year we attempt to identify a "hot topic" that is of broad interest to the astronomy community, but is not so trendy as to be the subject of lots of other concurrent meetings. The advent of the new X-ray observatories Chandra and XMM-Newton presented us with the opportunity to combine the glorious X-ray imagery and spectroscopy of young supernova remnants that is being obtained from them with work being done from Hubble and ground-based observatories. The programme of the conference was developed by a Scientific Organizing Committee comprised of:

Arnon Dar   *Bob Kirshner*
Alex Filippenko   *Dick McCray*
Steve Holt   *Rob Petre*
Una Hwang   *Virginia Trimble*

As usual, the conference began with introductory invited reviews designed to set the stage for the subsequent business of the conference. Virginia Trimble provided us with a delightful triptych through the history of supernova research, and Chris McKee challenged us with the outstanding questions about young supernovae that that the conference should attempt to address.

The programme then proceeded through the next two days in a series of non-paralleled sessions. Each session was devoted to a specific topic, with two or three invited talks and an extensive discussion period that included the opportunity for each conference attendee to advertise his/her poster paper in one-minute with one viewgraph. The breaks between sessions were extended and drinks and munchies were provided in the same area in which the posters were mounted, so that there was plenty of opportunity for informal dsicussion. We have found that this particular organization, with extended discussion in both the invited oral sessions and the poster sessions, to be very effective for such topical conferences.

The banquet at the conclusion of the second day of these conferences generally features a distinguished speaker who has a clear association with the conference material. This year our speaker was to have been Lo Woltjer, but Lo unfortunately took ill just as he was scheduled to board the plane to Baltimore. Nick White, the new Chief of the Laboratory for High Energy Astrophysics at the Goddard Space Flight Center, was kind enough to step in at the last minute to tell us about NASA's plans for future generations of X-ray astronomical observatories.

The conference traditionally concludes with a rapporteur summarizing some of the more important issues discussed during the previous two days. This year we were lucky enough to have a tandem of rapporteurs, with Knox Long and Frank Winkler challenging each other to be more insightful about the proceedings.

As usual, John Trasco and Susan Lehr made sure that all the logistics were handled flawlessly. Thanks also to Una Hwang, the co-editor of these Proceedings, for doing such an efficient job of handling the details associated with producing this volume.

Steve Holt

January 2001

Introduction

Pineapples and Crabs: When Young Supernova Remnants Were Even Younger

Virginia Trimble

*Physics Department, University of California, Irvine, CA 92697 USA and
Department of Astronomy, University of Maryland, College Park, MD 20742 USA*

Abstract. If a "young" supernova remnant is less than 1000 years old, then all of them have aged at least 10% during the era of photographic astronomy. Nevertheless, historical investigations are, so far, probably telling us more about how astronomers behave than about how SNRs behave. The talk reviewed an assortment of firsts, lasts, and might-have-beens from roughly 5283 BCE to 2000 CE.

INTRODUCTION: NUMBERS OF YOUNG SNR AND METHODS OF RECOGNIZING THEM

The rate of discovery of extragalactic SNe is now in excess of 100 per year. Thus, if a young SNR is less than 1000 years old, there are at least 10^5 examples within telescope range. Another way of looking at the numbers is to note that there are supposed to be 10^{11} galaxies within a Hubble radius. Most of these are dwarfs, and so the rate of one event per 1000 years per galaxy [1] (an old Zwicky estimate) is probably about right, implying 10^{11} young SNRs in the observable universe. Of these, we know something about perhaps 100 and have interpretable data for about 10. Most of these are discussed in the later pages of this volume (and the rest are over-discussed).

The young SNRs about which something can be said are of two types: recoveries (at one or more wavelengths) of events in external galaxies and counterparts, from possible to probable to certain, of events recorded in the Milky Way by pre-telescopic observers. SN 1987A is the first example of what could eventually become a common third class, events that have never been lost (roughly for the same reason that elephants are not often found – they are so big they hardly ever get lost).

Recoveries date back only about a decade, beginning (at least as far as publication goes) with optical emission from 1957D in M83 reported by Long and his colleagues in 1989 [2]. Of the handful known, 1961V was probably not a real supernova; and 1885A (S And) is in absorption (elsewhere in this volume) against the bulge stars of

CP565, *Young Supernova Remnants: Eleventh Astrophysics Conf.*, edited by S. S. Holt and U. Hwang
© 2001 American Institute of Physics 0-7354-0001-6/01/$18.00

the galaxy. SN 1923A, the oldest, is one of a comparable number of radio recoveries [3 and elsewhere in this volume]. X-ray recoveries have lagged a bit (mostly an angular resolution problem one suspects), beginning with 1978K in 1996 [4] and now including about four others plus about eight seen at the time of outburst (E. Schlegel, private communication). All the recovered events seem to be of the core collapse variety. Type I sites have, of course, also been searched at various times. Events whose apparent magnitude at peak luminosity was brighter than m = 10 include 1895B (in NGC 5253), 1937C, 1972E, and 1954A. This is also more or less an exhaustive list, and they are presumably the best bets for further study.

None of the recovered core collapse SNe shows strong evidence for a pulsar. The radio would be very weak, and the X-ray emission is in any case unresolved. Several pulsars have been reported at the site of SN 1987A, most recently one with a period of 2.14 msec [5].

The numbers of historical supernovae (meaning those seen by our pre-telescopic intellectual ancestors) or at least candidates for them are comparable. The standard authority on the subject is the 1977 book by Clark and Stephenson [6] – see [7] and [8] for partial updates. Of the 15 or so candidates, some can be dealt with very quickly. The three events of 1592 were probably sightings of Mira embellished. RCW 103 and MSH 11-54 appear to be supernova remnants not much more than about 1000 years old, but they would have been below the southern horizons of all the chroniclers whose records have been found so far. The event of 1408 is represented by only a single sighting and no indication of duration. Thus it was probably not a supernova, and whether the position was closer to the radio source CTB 80 or the X-ray source Cyg X-1 becomes irrelevant.

Schaefer [9] has presented evidence that the star of 185 CE was really the close juxtaposition of a nova and a comet. Supporting evidence is that the proper motions of the optical filaments at the location of putative radio SNR RCW 186 are less than 10% of what would be required by the supposed age, though one should not forget that much of the Cas A optical remnant consists of quasi-stationary flocculi. Petruk [10], however, still supports a connection and supernova identification. The event of 393 remained visible and, apparently, motionless for eight months. Not surprisingly, the position cannot be reconstructed accurately enough to distinguish between CTB 37 A and B as potential counterpart, and this event did not make the program as either an invited or contributed paper.

The remaining nine candidates are summarized in Table I. It is worth noting that two are quite recent additions to the inventory, SN 837 (regarded as a nova by Clark & Stephenson, [6] = SS 433 (proposed by Wang in 1996, 11, and somewhat earlier at other conferences), and SN 1523 = Kes 25. The three Chinese reports of a "po" star in summer, 1523 are translated and presented for the first time in a preprint by Wang [12]. Properties of the recently-discovered and rapidly slowing pulsar in Kes 75 appear elsewhere in this volume. If the 1523 event was as distant as the SNR appears to be, it may even have been a hypernova!

TABLE I

YEAR	REMNANT	REMARKS
386	G 11.2-0.3	pulsar, P = 0.54 sec
837	SS 433	high galactic latitude for SN (but SS 433 is there!)
1006	PKS 1459-41	VERY bright (record for Milky Way)
1054	Crab Nebula	pulsar, P = 0.033 sec, the "gold standard"
1181	3C 58	first SNR described as Crab-like, in Cas
1523	Kes 75	pulsar, P = 0.324, slowing down time = 700 yr
1572	3C 10	Tycho's event, B Cas
1604	G 4.5 + 68	Kepler's event
1680	Cas A	faint if seen by Flamsteed in 1680; even fainter if not seen at dynamical date of event, 1658 ± 3

Given the uncertainties of the candidates and the certainty that the list must be incomplete (Chinese reports of ominous events cluster near the ends of dynasties, [7]), I am not sure whether any significance at all should be attached to three of the tabulated events/objects being located in Cassiopeia!

The remainder of this introduction is arranged chronologically by events in our study of young SNRs, but divided into Crab Nebula, other historical remnants, and theory.

CHRONOLOGY: THE CRAB NEBULA

The photons (and presumably some neutrinos, gravitinos, etc.) left the progenitor in 5283 BCE (not, of course, known this precisely, but the distance to the nebula is probably within 10% of 2000 kpc, [13]). They arrived some time in 1054 CE, though just who saw them and when is still under discussion. The Chinese sighting on July 4, 1054 is in Biot's 1842 initial list [14], though not as one of the most promising events, and Duyvendak (of Leiden, an Orientalist, not an astronomer) provided a better, confirming translation [15]. A Japanese sighting, of somewhat less certain date, was presented by Iba in 1934 [15a]. The report of Ibn Butlan from Baghdad [16] concurs with first appearance in July. As anyone who has ever observed the nebula optically knows, it is a winter object (on the meridian at midnight on about 10 December), and Fourth of July must have been simply the first evening something "bright like Venus" became obvious against morning twilight. This has, of course, implications for the maximum brightness of the event, since it might well only have been caught past its peak. It has generally been claimed that Europe had particularly bad weather that year, but a recent suggestion is that the bad weather was political-religious, rather than meteorological. Collins et al. [17] have found several apparent European sightings that date from April and May (but nothing new from July onward), and suggest that the supernova was indeed seen, but that the subsequent failure of a last-ditch conference, aimed at holding together the Eastern and Western churches, transformed it into a bad omen, not to be talked about again.

Surprising as it may seem, a number of doubters over the years have suggested that NGC 1952 is not connected with the event of 1054. The basis most often has been either the measured expansion rate (which would place the event in about 1150 in the absence of acceleration) or the peak luminosity (too faint for a typical supernova, whatever that is). The standard responses are acceleration due to the pressure of magnetic field and relativistic particles and that not all supernovae are the same brightness (even if 1054 was caught at maximum).

A more generic response is that one would be surprised if two supernovae had occurred within a few hundred parsecs and a hundred years of each other, one of which was not seen and the other of which left no remnant. In contrast, it is legitimate to doubt that specific South West Indian petroglyphs record a sighting of SN 1054 near a crescent moon. It is true that the moon passed close to the SN position at a time when both were above the horizon at longitude 110° W and not when both were visible from 110° E, but anyone who has seen Venus close to a crescent moon, as is possible every few years, will have an alternative hypothesis.

SN 1054 faded below naked eye visibility in late 1055 and was not heard from again until 1731, when English physician-astronomer John Bevis recorded a fuzzy patch in Taurus. His publisher went bankrupt before his atlas could be published, but at least one copy of the relevant plate survived. His version looks somewhat too large and too circular. The next drawing comes from Charles Messier, who saw the fuzzy patch on 12 September 1758, recorded it as a new discovery, and made a note of its position and appearance so as not to confuse it with Comet Halley, whose return was then expected any month or two, Messier appears in our folklore as a noted hunter for and discoverer of comets. Less widely known is that he became so as a result of being hired as a 21 year old assistant to Delisle of the Marine Observatory on the basis of his neat handwriting and drafting skills [18]. In case you wonder how the Halley story turned out, Messier did indeed recover it in January, 1759, about a month after an accidental discovery by a German amateur. But Messier's drawing was the first to show a nebula roughly the right size and somewhat elongated in the right direction.

Halley of course returned again in 1835, and seems to have left Edouard Biot curious about other perihelion passages. He carried out what he called "Recherches

faites dans le grande collection des historians de la Chine, sur les anciennes appari-
tions de la comete de Halley" in 1842-43 [14], identified the passages in 1222, 1145,
1066 (you know about that one), 989, 837, 684, 451 and -12. He also tabulated an
assortment of guest, new, and broom stars, including that of 1054.

Also in 1842, William Parsons, later the third Earl of Rosse (whose descendent
is a Lord Rosse to this day) cast the first mirror of his "Leviathan", with which, in
due course, he observed M1 [19]. He is said to have assigned the name in 1844, but
has left us two drawings, one nearly indistinguishable from a pineapple standing on
its stem end, the other bearing some remote resemblance to a continuum-emission
photograph of GC 1137 and/or to the larger claw of a Maryland crab. It seems
mostly likely that the former drawing is a mislabeled representation of some other
object (David Dewhirst of Cambridge has been promising for years to tell us which
one), but when, many years ago, I read a Physics Today article by Philip Morrison
on optical imaging which described "resolved objects" as ones that "look like crabs
or pineapples" I knew he had contemplated the same pair of drawings. It is his
impression (personal communication 2000) that the fruit-like image was done by
an assistant, not Rosse himself [20]. Rosse shared the opinion, expressed earlier by
Herschel [21] and others, that the nebula could be resolved into stars, and it was
left to William Lassell, observing under the clearer skies of Malta, to conclude that
the nebula was truly diffuse [22], indeed "very amorphous and indistinct."

Lassell's drawing is not especially Crab-like (put north to the left and it looks
most like Caspar the Friendly Ghost), but his stars are real stars, down to the
central, $m = 15$-16 pair, which is what has led me to wonder whether anyone has
ever scanned drawings by him and his contemporaries of "extragalactic nebulae"
for "guests stars", whose remnants would now be, say, 120-220 years old.

When was the Crab Nebula first photographed? In 1892, with a 20", by someone
named Robertson, so says Shklovsky [23]. Unfortunately, he provides no reference
and cannot now be queried. Casual investigation (that is, I looked in a few early
volumes of Jahresberichte, indices of AJ, AN, and RAS publications from the late
19th century, and asked a few people with historical interests - Gingerich, Hoffleit,
Hazen, Hodge) has not led to an identification for him or his reflector. The first
photograph I have found documented and reproduced dates from 7 February 1896
and was taken on a Guilleminot plate in a 2 hour, 32 minute exposure, as part of
regular astrograph work, by S. K. Kostinsky of Pulkova Observatory [24]. Kostinsky
published his last paper in 1936 in the same issue of Pulkova Observatory Circular
that carried the last paper by Gerasimovich [25]. It was not a good year at Pulkova
for astronomers interested in international collaborations.

Thus the first images were about 20 years old when Vesto Melvin Slipher of
Lowell Observatory began recording spectra [26] in 1913. He correctly described
the split lines, but thought that a strong Stark effect was responsible. A few years
later, Roscoe Sanford held on for a 48-hour exposure, but without doing much more
than confirming some very large velocities by the standards of the times [27].

Carl Lampland, also of Lowell, reported the Crab as the fourth truly variable
nebula in 1921 [28]. He was seeing the complex motions of continuum features (now

associated with the names of Scargle and Hester). But Duncan [29], with access to plates from the 60" at Mt. Wilson (exposed by Ritchey in 1909 and himself in 1920) soon spotted the systematic expansion, and later, its probable acceleration [30].

Also in 1921, Knut Lundmark [31] compared Biot's list of Chinese phenomena with the NGC catalog. His motivation was a firm conviction that novae were associated with nebulosity (not by expelling it, but more like being formed from it, an idea he attributed back to Tycho, whose new star indeed sits right on the edge of the murkiest bit of the Milky Way). Lundmark estimated a distance near 1 kpc from the expansion and spectral data [32] just in time for Hubble to use it in a careful, but non-technical discussion of the 1054-nebula association [33] in 1928. That same year, Adriaan van Maanen [34] first reported a large proper motion for what was then the "south preceeding" star and is now the pulsar. This does not seem to have played a very large part in anybody's thinking, perhaps because he was also the van Maanen of spiral galaxies with rotational motion in the plane of the sky. Indeed when Baade [35] and Minkowski [36] point to that same star it is primarily because it lacks absorption features in its spectrum, though they also report a proper motion based on van Maanen's plates and later ones of their own.

It was from war-torn Holland that Oort and Mayall [37] and Duyvendak [15] submitted their discussions of the association between the guest star of 1054 and the expanding Crab Nebula. Since the technologies of both radio astronomy (via radar dishes and radar engineers) and X-ray astronomy (via captured German V-2 rockets) grew fairly directly out of World War II, this provides a natural break between sections!

THE NON-OPTICAL CRAB

Notoriously, the Crab Nebula was first or nearly first object outside the solar system to be recognized as a source of radio waves, X-rays, gamma rays, and so forth. Radio came first. Bolton and his Australian colleagues employed a virtual interferometer, of which the other arm was reflection from the ocean up to their cliff top to get a good enough position to permit optical identification [38], then as now the sine qua non of feeling that one might have begun to understand a non-optical source. The radio polarization was not measured until 1957 [39], after optical polarization had already been recognized (and this lives in the "theory" section, since it was predicted).

A high-resolution X-ray search of the vicinity was possible early because (and only because) the moon has the dubious taste to pass right across the face of the Crab a time or two every 19 years. Stu Bowyer and his colleagues managed to get a rocket aloft for the critical five minutes in 1963 [40], and their report had already triggered a countably infinite number of calculations of the cooling of neutron stars before a second flight the next year [41] demonstrated that the source was extended and so associated with the nebula as a whole rather than just the central star. To that same year belongs the discovery of the first radio pulsar by Hewish and Okoye [42], only, of course, what they reported was a compact, low frequency radio source, their time resolution having fallen far short (or rather long) of the required 33 msec. Gamma-rays followed within a few years, and later reanalysis of those few photons provided a 1967 period for the pulsar that can never be bested as "first" by any method [43].

Other Crab "firsts" include the pulsar time derivative, a second derivative, optical and X-ray pulsations, and it remained unique until, one fine day [44] 3C 58 was declared to be "Crab-like" and NGC 1952 (3C 144, etc) became merely a prototype.

THEORETICAL CHRONOLOGY

The first theoretical statement that had to be made about supernovae is that they exist as a class distinct from that of the "common" novae. The suggestion came first from Knut Lundmark [45], who had both reconstructed the light curve of SN 1885 (S And) and estimated a distance of 200,000 pc to M31. The implied peak brightness for S And was M = -15, and, in their famous debate, Heber Doust Curtis interpreted this to mean that "a division into two magnitude classes is not impossible" while Harlow Shapley thought it ridiculous and a strong argument against the existence of island universes [46]. Incidentally, Lundmark [47] also coined the name supernova, though credit is generally given to Baade and Zwicky [48] writing two years later, whom you will meet again very shortly. The Gaposchkins, writing in 1938 [49], opined that the name was not very suitable but was also beyond repining. They suggested that the mechanism would prove to be continuous with that of the common novae. By 1957 she, at least, regarded the physics as distinct (partly because the brightest common novae were the fastest, [50]).

Once you accept existence, the next obvious question is energy sources. E. A. Milne provided a "prequel" hypothesis in 1931 [51] when he attributed the energy of common novae to the collapse of normal stars to white dwarfs.

This brings us to the "culture heroes" Walter Baade and Fritz Zwicky. Their 1934 paper [48], and an APS meeting abstract from December, 1933, make several critical points. First, supernovae exist and are different from common novae (with the inventory including S And, the 1054 event, Tycho's new star, and a handful of photographic discoveries between 1890 and 1930). Second, the energy source is the collapse of a normal star to a neutron star (a concept which they also probably invented). And, third, much of the available energy should go into accelerating cosmic rays. They were, of course, telling the truth (if not the whole truth) on all three points. They were not (also of course?) widely believed, and early papers on neutron star structure, like that of Oppenheimer and Volkoff in 1939 do not even cite them.

In fact, there was a giant step backwards, when, at a Paris symposium on novae and white dwarfs, Chadrasekhar associated supernovae with the expulsion of material from massive stars that would otherwise not be able to produce white

9

dwarfs, but then did so. This view was endorsed in print by Chandra himself [52], Minkowski [36], Baade [53], Baum [54], Schwarzschild and Spitzer [55], and various other custodians of received opinion right down to the first, 1964, edition of Abell's [56] pioneering text for non-science students of astronomy. Whichever text was used at UCLA just before Abell published his said the same thing, but the instructor in 1962, Thornton Leigh Page, made sarcastic remarks about mice and elephants' grave yards and was, in retrospect, at least a bit ahead of his time in returning to the neutron star camp, which Zwicky [1, 57] never abandoned. Indeed his last statement on the subject [58] emphasized the three points mentioned above, as well as a rate estimate of one per galaxy per 300 years, large expulsion velocities, and so forth.

After Minkowski separated the supernovae into two classes in 1940 [59], the was room for two mechanisms. Since we have already had core collapse (whether to white dwarf or neutron star), you already know with 20-20 hindsight that the other one is going to be a nuclear explosion of some kind. The standard "first reference" is Hoyle and Fowler in 1960 [60], though precursors in the work of Mestel, Schatzman, and others turn up in reference lists from time to time. Payne- Gaposhkin [50], as meticulous as any secondary source I have found, gives priority to a 1946 Schatzman paper [61], with Gurevich and Lebedinsky, submitted in 1946 but not published until 1947 [62] among the also-rans. One has, however, to suspect that she had not (or not lately) read through these papers. Schatzman's abstract says: "L'energie liberee au cours de l'explosion est due a la transformation de l'energie cinetique d'exclusion des electrons en energie d'agitation thermique." That is, freely translated, he proposes to claw back the energy that has been locked up in electron degeneracy and use it to power the explosion. The paper is explicitly titled "Theorie des Supernovae".

Later papers by Schatzman do, indeed, deal with nuclear burning under degenerate conditions, though it is sometimes a bit difficult to decide whether he is making novae or supernovae at a given time. Meanwhile, however, Gurevich and Lebedinsky [62] have already addressed both novae and supernovae, attributing both to "degenerated" nuclear explosions, with surface ignition (which the star survives) for novae and central ignition (which the star does not survive) for supernovae. No, they had not fully solved the problem, for they consider only hydrogen fusion (the only game in town then) and so decide that a nuclear explosion should occur early in the life of a star and a core-collapse event (for which their source is Gamow and Schoenberg in 1941, [63]) late in stellar lives. It becomes clearer how and why this was an issue if you recall that they were writing during the era in which star formation occurred "in the early universe, when conditions were very different from now," so that age and degree of evolution were not separable parameters.

Even earlier, Zwicky's 1940 review [57] had suggested that the exponentially declining late light curves displayed by many supernovae might be powered by the decay of some radioactive nuclide. Early suggestions included Be^7 (with the right half-life to electron capture but irrelevant if ionized) and Cf^{254} (which would give us a universe drowning in its decay products). The current "best buy" is Co^{56}, which comes from Ni^{56} in about 7 days and goes on to Fe^{56} in 70-some.

A second major theoretical issue was the photon emission process responsible for, initially, the optical and radio Crabs and, later, for other SNRs and X-rays all around. Let's dispose of X-ray bremsstrahlung in blast waves for non-Crabs immediately, with credit, rightly or wrongly to Carl Heiles [64]. And the rest, as no poet has said, is synchrotron. The word had already been whispered (with a slight accent) by Alfven and Herlofson in connection with radio emission from the plane of the galaxy, and the demonstration by Greenstein and Minkowski in 1953 [65] that the radio and optical emission from the Crab Nebula cannot both simultaneously be thermal is clearly part of the story. Several (former) Soviet astronomers have cast doubts on each others' share of the credit. The obvious citation is, however, Shklovski in 1953 [66a] because he explicitly predicted optical polarization as a consequence, and thereby inspired Dombrowski [67a] and Vashakidze [68a] to go out and look for it, successfully. Dombrowski spent most of the rest of his (relatively short) career searching for optical polarization, in other nebulae, typically with much less success. Vashakidze was presumably Georgian.

More detailed study of the polarized optical radiation by Oort and Walraven [69a] and by Woltjer [70a] made it clear that the electrons responsible for the optical synchrotron could have half lives of only about 200 years, implying either that the nebula must have been quite remarkably bright not long ago or that there must be some ongoing electron acceleration.

The solution was, of course, the P = 0.033 sec pulsar, and the measurement of its slowing-down rate and the first generation of models all date from the same year (1968) that I received my PhD with the second thesis ever devoted entirely to the Crab Nebula (Woltjer's was first, and R.E. Williams considered several other nebulae as well). Thus the rest of the story belongs to "current events" rather than to "history".

A different slant on the theoretical chronology can be gained from Shklovsky's book [23]. It is worth noting that (a) he was writing just before the discovery of pulsars and (b) he believed the 1054 event to have been of Type I. His own suggestion for the continuous power source was a neutron (or collapsed) star is an eccentric binary, whose dips into the atmosphere of the companion sent off the moving wisps first reported by Lampland. He favors nuclear over gravitational energy as the main power source, with lots of credit to Hoyle and Fowler [60] and none to Gurevich and Lebedinsky [62]. He accepts the potential existence and importance of neutron stars, but credits them to Oppenheimer and Volkoff [66]. Given that the book as a whole has strong overtones of "The refrigerator was invented by Westinghouski," it is probably significant that he does not attempt to credit neutron stars to Landau.

CHRONOLOGY: OTHER YOUNG SNRS

Supernovae were, in a sense, born with their remnants already attached, because the work of Barnard [67] on Nova Per 1901 and Nova Aql 1918 had firmly established within the astronomical folklore [68] the idea that an old nova ought to be surrounded by an expanding gas cloud, with time scale equal to the time since the outburst [69]. This was the context in which Hubble and Lundmark hunted for the remains of Z Cen = SN 1895B [70]. They were expecting both nebulosity and a hot white dwarf, and found neither. The same apply to Humason and Lundmark's 1924 search for a remnant of Tycho's new star [71] and to Baade's 1938 assault on Tycho [72]. Thus the Crab Nebula was the only "young supernova remnant" for several decades.

This changed in 1943 when Baade and Minkowski [53, 73] imaged a few wisps at the location of Kepler's new star of 1604 (Nova Ophiuchi) and obtained a spectrum. The optical remnant of Tycho's 1572 event, imaged by Baade in 1949, but not published at the time [74] is even less spectacular. Baade and Minkowski [53, 73] remarked upon the absence of a "white dwarf" at the Kepler site, and claimed that this was consistent with the event having been an SNI, in the newly-defined sense.

All the other observations pertain to supernovae whose locations and/or epochs are considerably less well known, and one must begin by asking the extent to which the reality of the events, their coordinates, and, therefore, the connection with assorted nebulosity, radio sources, and X-ray emission are to be trusted. Stephenson [75], shows the part of the sky extending from declination 55° to 75° and R.A. from 22.5 hours around through 0^{hr} to 3.5, on which he has placed the Chinese asterisms that define the location of SN 1181 and its proximity to radio source 3C 58. There are a dozen other 3C and comparable sources in that area (including, as it happens, Cas A and 3C 10, the remnant of Tycho's new star; and I remain uncertain about whether it is of any interest that the three were so close together; remember Tycho's SN was also called B Cas as a star name). Should we, therefore, trust the identification? I naturally think so, based, for instance, on the accuracy with which Venus can be located in Chinese descriptions relative to modern calculations of its motion [7]. Incidentally, the same Stephenson paper [75] lists another candidate historical supernova of 902, that was also in Cassiopeia. Perhaps the Chinese just liked that part of the sky, since it is circumpolar from mid to high northern latitudes.

Radio astronomy enters the picture chronologically with the cataloguing of Cas A by Graham Smith in 1951 [76], but logically not until 1952, when Hanbury Brown and Hazard [77] found emission at the Tycho site, leading to a correction in the position recorded by Brahe and so to increased confidence in the relevance of the optical wisps nearby. SNe 1572 and 1604 were, as should now be clear, along sight lines with much more optical obscuration than SN 1054. The radio counterpart of Kepler, however, was recognized after the optical counterpart by Shakeshaft et al. [78] in 1955, while for Tycho the radio counterpart came first, at least in the published literature. Given that the pioneers of radio astronomy typically did not have backgrounds in traditional astronomy, one might be surprised that the observations and identifications came as quickly as they did. Brown recalled long after the fact [79] that he had been urged to turn his beam in the Tycho direction by Zdenek Kopal, a very traditional astronomer indeed.

With radio supernovae established as "a class" by the second example, it is not surprising that Shklovski almost at once attempted to identify other historical events with the handful of catalogued radio sources [80, 81]. Cas A = SN 369, though now known to be wrong, was perhaps the least improbable suggestion. In the same, 1954, time frame, Bolton et al. [82] reported radio emission from the general direction of Nova Aql 1918, and I have not checked on whether the identification is now generally regarded as correct.

This brings us to the first "large group" of optical identifications of radio sources by Baade and Minkowski in 1954 [93]. This paper is most often cited because the group included Cygnus A, the first recognized radio galaxy (described by them as "colliding galaxies" and so responsible for much theoretical work that should not be described as wasted). But they also found diffuse nebulosities in the (highly absorbed) directions of the radio sources Cas A, Pup A, and Cygnus X (NOT the same as Cygnus X-1, which is a black hole X-ray binary). They regard the first two identifications as reasonably firm (and doubted the third because the filaments show normal HII composition and small velocities). BUT they say with great confidence that "the random velocities are so large that they hide so far any expansion which may be present;" that "the conditions in the radio-source nebulosities are thus diametrically opposed to those in the filaments of supernova remants" (meaning 1054 and 1604); and that "Altogether, no observed fact supports the view that the radio source nebulosities are remnants of super-novae." Baade, at least, had changed his mind within a couple of years [84], and the identification of the radio Cygnus X with the optical Cygnus Loop re-established [85].

SN 1006 (unlike Tycho and Kepler) is still regarded as a Type Ia event. It joined the young SNR class with the discovery of readio emission from the location reported in European and Chinese records in 1965 [86]. The optical nebulosity, for which Minkowski obtained the first image and spectrum [87] is dominated by hydrogen emission. Nevertheless, van den Bergh [88] declared that it had been a type I explosion the same years, 1976, that Winkler and Laird [89] spotted the X-ray emission. It is expanding at the right rate [90] and most of the iron you expect from a nuclear supernova is still (or again!) cold and to be seen only in absorption [91].

Another word should be said about Cas A and S And. That the former acts precisely like a radio, optical, and X-ray supernova is not in doubt (it even, in the Chandra image shown at this meeting, seems to have a neutron star near its center). But we remain uncertain whether the event was witnessed. A star not otherwise to be found in today's skies was recorded on a chart from 1680 by Astronomer Royal John Flamsteed [92]. He seems to have been a fairly reliable chap – another of his phantom stars, from 1690, was a pre-discovery observation, of Uranus. On the other hand, Kamper and van den Bergh [93] have said quite firmly, on the basis of measured proper motions of the nebula filaments, that the explosion must have happened about 20 years earlier. Modest deceleration of the expansion would, of course, reconcile their date of 1658 ± 3 with Flamsteed's star. Astronomers used to say that, when there was another astronomer as great as Kepler, there would be a supernova for him to see. Perhaps it is just that no-one wants to compete for the title of "greatest astronomer since Flamsteed."

For S And (SN 1885 in M31) a major outstanding puzzle has been the absence of expected radio emission, down to 1/15 or less that produced by the Tycho remnant [94]. A poster presentation here may have finally found the radio counterpart, but (like, the optical event that gave rise to Cas A) it is very faint.

And this brings us observationally down essentially to the modern era of optical recoveries, neutrino observations, hypernovae, and the rest. A major feature of that era is, of course, ever more powerful photon collectors, receivers, and processors, which have gradually extended the reach of Crab-like information as far as the LMC, yielding the remnant "more Crable than the Crab" [95, 96]. We can, in due course, look forward to that reach extending to M31 and beyond.

REFERENCES

[1] Zwicky, F. 1938. ApJ 88, 522

[2] Long, D.S. et al. 1989. ApJ 340, L25

[3] Eck, C.R. et al. 1998. ApJ 508, 666

[4] Schlegel, E.M. 1996. ApJ 456, 187

[5] Middleditch, J. et al. 2000. NA 5, 243

[6] Clark, D.H. & F.R. Stephenson 1977. The Historical Supernovae, Oxford, Pergamon

[7] Trimble, V. & D.H. Clark 1985. Bull. Astron. Soc. India 13, 117

[8] Stephenson, F.R. 1988. Int. School of Physics, "Enrico Fermi," Course 95, p. 166

[9] Schaefer, B.E. 1995. AJ 110, 1793

[10] Petruck, O. 1999. A&A 346, 961

[11] Wang, Z-R. 1996. in IAU Colloq. 145, Ed. R. McCray & Z-R. Wang, Cambridge Univ, Press, p. 323

[12] Wang, Z-R. 2000. Preprint

[13] Trimble, V. 1973. PASP 85, 579

[14] Biot, E, 1842. Connaissance de Temps for 1846, Additions

[15] Duyvendak. J.J.L. 1942. PASP 54, 91

[15a] Iba, I. 1934. Pop. Astron. 43, 251

[16] Ibn Butlan 1054. Reproduced in M. Karim & A. Qadir, eds. Experimental Gravitation, Inst. of Physics Publishing (1994)

[17] Collins, G,W. et al. 1999. PASP 111, 871

[18] Yoemans, D.K. 1991. Comets, Wiley, p. 125ff

[19] Hoskin, M.A. (ed.) 1999. The Cambridge Concise History of Astronomy, Cambridge Univ. Press

[20] Rosse, Earl of (Wm. Parsons) 1844. Phil. Trans. Roy Soc. p. 322

[21] Herschel, W. 1818. Phil. Trans. Roy. Soc. p. 435

[22] Lassell, W. 1867. Mem. RAS 36, Plate II, Fig. 6 and p. 41
[23] Shklovsky, I.S. 1968. Supernovae, Wiley
[24] Deutsch, A.N. & V.V. Lavdovsky 1940. Pulk. Obs. Circ. 30, 21
[25] Kostinsky, S.K. 1936. Pulk. Obs. Circ. 20, 23
[26] Slipher, V.M. 1916. PASP 28, 191 and Harv. Bull. #743 (1921)
[27] Sanford, R.F. 1919. PASP 31, 108
[28] Lampland, C.O. 1921. PASP 33, 79
[29] Duncan, J.C. 1921. PNAS 7, 179
[30] Duncan, J.C. 1939. ApJ 89, 482
[31] Lundmark, K. 1921. PASP 33, 227
[32] Lundmark, K. 1926. Pop. Astron. Tidskv. 7, 18
[33] Hubble, E.P. 1929. ASP Leaflet 14
[34] van Maanen, A. 1928. Mt. Wilson Contr. 356
[35] Baade, W. 1942. ApJ 96, 188
[36] Minkowski, R. 1942. ApJ 96, 199
[37] Mayall, N.U. & Oort, J.H. 1942. PASP 54, 95
[38] Bolton, J.G., G.J. Stanley, & O.B. Slee 1949. Nature 164, 101
[39] Mayr, C.H., T.P. McCullough, & R.M. Sloanaker 1957. ApJ 126, 468
[40] Bowyer, C.S. et al. 1964. Science 146, 912 and Nature 201, 1307
[41] Bowyer, C.S. et al. 1965. Science 147, 394
[42] Hewish, A. & S. Okoye 1964. Nature 203, 494
[43] Haymes, R.G. et al. 1968. ApJ 151, L9
[44] Weiler, K.W. & G.A. Seielstad 1971. ApJ 163, 455
[45] Lundmark, K. 1920. Kungl. Svensk. Vetensk Akad. Hand. 60, No. 8
[46] Shapley, H. & Curtis, H.D. 1921. Bull. NRC 2, 171 and 294
[47] Lundmark, K. 1932. Lund Circ. 8, 216
[48] Baade, W. & F. Zwicky 1934. Proc. NAS 20, 254 & 259
[49] Payne-Gaposchkin, C. & S. Gaposchkin 1938. Variable Stars, Harvard
[50] Payne-Gaposchkin, C. 1957. The Galactic Novae. North Holland, p. 307
[51] Milne, E.A. 1931. MNRAS 91, 54
[52] Schoenberg, M. & Chandrasekhar, S. 1942. ApJ 96, 161
[53] Baade, W. 1943. ApJ 97, 119
[54] Baum, W. 1952. AJ 57, 222
[55] Schwarzschild, M. & L, Spitzer 1953. Obs. 73, 77
[56] Abell, G.O. 1964. Exploration of the Universe, Holt Rinehart Winston, p. 445
[57] Zwicky, F. 1940. RMP 12, 72
[58] Zwicky, F. 1974. in C.B. Cosmovici (Ed.) Supernovae and their Remnants, Reidel, p. 1
[59] Minkowski, R.1940. PASP 52, 206
[60] Hoyle, F. & W.A. Fowler 1960. ApJ 132, 565
[61] Schatzman, E. 1946. Ann. d'Ap. 9, 199
[62] Gurevich, L.Z. & A.I. Lebedinsky 1947. Dokl. Akad. Nauk USSR 46, 23
[63] Gamow, G. & M. Schoenberg 1941. RP 59, 539
[64] Heiles, C. 1964, ApJ 140, 477
[65] Greenstein, J.L. & R. Minkowski 1957. ApJ 118, 1

[66] Oppenheimer, J.R & G.K. Volkoff 1939. RP 55, 374

[66a] Shklovsky, I.S. 1953. Dokl. Akad. Nauk USSR 90, 1893

[67] Barnard, E.E. 1917. AJ 30, 16 and ApJ 49, 199 (1919)

[67a] Dombrowsky, V.A. 1954. Dokl. Akad. Nauk USSR 94, 1021

[68] Russell, H.N., R.S. Dugan & J.Q. Stewart 1926. Astronomy, Boston, Ginn & Co. p. 783-87

[68a] Vashakidze, M.A. 1954. Astron. Tsirk. 147, 11

[69] Humason, M.L. 1935. Publ. AAS 8, 115

[69a] Oort, J.H. & T. Walraven 1956. BAN 12, 285

[70] Hubble, E.P. & K. Lundmark 1923. PASP 34, 292

[70a] Woltjer, L. 1957. BAN 13, 301; BAN 14, 39 (1958)

[71] Humason, M.L. & K. Lundmark 1924. PASP 35, 109

[72] Baade, W. 1938. ApJ 88, 285

[73] Minkowski, R. 1943. ApJ 97, 128

[74] Minkowski. R. 1959. IAU Symp, 9, 315

[75] Stephenson, F.R. 1974. in C.B. Cosmovici (ed.) Supernovae and their Remnants, Reidel, p. 75

[76] Smith, F.G, 1951. Nature 168, 555; Nature 170, 1064 (1952)

[77] Brown, R.H. & C. Hazard 1952. Nature 170, 364

[78] Shakeshaft, J.R., et al. 1955. Mem. RAS 67, 106

[79] Brown, R.H., 1984. in W.T. Sullivan (ed.) The Early Years of Radio Astronomy, Cambridge Univ. Press. p. 213

[80] Shklovsky, I.S. 1953. Astron. Zh. 30, 26

[81] Shklovsky, I.S. 1954. Dokl. Akad. Nauk. USSR 94, 417

[82] Bolton., J.G. et al. 1954. Aust. J. Phys. 7, 110

[83] Baade, W. & R., Minkowski 1954. ApJ. 119, 206

[84] Mills, B.Y. et al. 1956. Aust. J. Phys. 9, 84

[85] Walsh, D. & R.H. Brown 1955. Nature 175, 808

[86] Gardner, F.F. & D.K. Milne 1965. AJ 70, 754

[87] Minkowski, R. 1966. AJ 71, 371

[88] van den Bergh, S. 1976. ApJ 208, LI7

[89] Winkler, R.F. & F.N. Laird 1976. ApJ 264, Llll

[90] Hesser, J. 1981. ApJ 251,549

[91] Hamilton, A.J.S. & R.A. Fesen 1988. ApJ 327, 178

[92] Ashworth, W.B. 1980. J. Hist. Astron. 11, 1

[93] Kamper, W. & S. van den Bergh 1983. IAU Symp. 105, 55

[94] Crane, F.C. et al. 1992. ApJ 390, L9

[95] Kirshner, R. et al. 1989. ApJ 342, 260

[96] Kafatos, M.C. & R.B.C. Heny (eds). 1985. The Crab Nebula and Related Supernova Remnants, Cambridge Univ. Press

Young Supernova Remnants: Issues and Prospects

Christopher F. McKee

University of California, Berkeley, CA 94720, USA

Abstract.
    The dynamical evolution of young supernova remnants (YSNRs) is governed by
the density distribution in the ejecta and in the ambient medium. Analytic solutions
are available for spherically symmetric expansion, including the transition from the
ejecta-dominated stage to the Sedov-Taylor stage. YSNRs serve as valuable physics
laboratories, in which we can study nucleosynthesis, the early evolution of compact
objects, pulsar physics, particle acceleration, the formation and destruction of dust,
hydrodynamics at high Reynolds numbers, shock physics at high Mach numbers, and
the effects of thermal conduction in interstellar plasmas. There are several challenges
in YSNR research: (1) Where are the very young remnants in the Galaxy? We expect
5-10 to have occurred since Cas A, but with the possible exception of a remnant
reported at this conference, none have been seen. (2) Can very young SNRs produce
gamma-ray bursts? The acceleration of a shock in the outer layers of a supernova, first
suggested by Colgate, can account for gamma-ray bursts such as that believed to be
associated with SN 1998bw, and more powerful explosions can account for the energies
seen in many cosmological bursts. (3) The Connections Challenge: Can one infer the
nature of the supernova and its progenitor star from observations of the YSNR?

INTRODUCTION

    Young supernova remnants (YSNRs) are fascinating objects. Just as star for-
mation connects the interstellar medium (ISM) with stars through the process
of gravitational contraction, so YSNRs complete the connection back to the ISM
through a titanic explosion driven by the release of either gravitational or nuclear
energy. An SNR is produced by the interaction of the ejecta with the ambient
medium or, alternatively, by the effects of a pulsar left behind by the explosion.
The ambient medium can be circumstellar—i.e., matter ejected by the progenitor
or its companion—or interstellar. I shall use the term "YSNR" to denote an SNR
in which the total mass ejected by the star, both the circumstellar mass, M_{cir*}, and
the mass ejected by the supernova explosion, M_{ej}, exceeds the mass of swept-up
interstellar gas, $M_{sw,is}$—i.e., $M_{cir*} + M_{ej} > M_{sw,is}$.
    In this article, I shall first review the dynamics of YSNRs. I shall then discuss

CP565, *Young Supernova Remnants: Eleventh Astrophysics Conf.*, edited by S. S. Holt and U. Hwang
© 2001 American Institute of Physics 0-7354-0001-6/01/$18.00

YSNRs as physics laboratories that enable us to address problems that are difficult or impossible to address on Earth. Finally, I shall address several challenges posed by YSNRs, including whether they can be the progenitors of gamma-ray bursts.

REVIEW OF YSNR DYNAMICS

The dynamics of a YSNR is driven by the interaction between the ejecta and the surrounding matter. The first step in analyzing the dynamics is therefore to determine the density structure of the ejecta. Chevalier [1] began this effort by working out the dynamics of the blast wave produced by a core-collapse supernova as it propagated in the envelope of a red giant. Subsequently, Chevalier & Soker [2] approximated the envelope of a blue supergiant as having a power-law density distribution with $\rho_0 \propto r^{-17/7}$, which gives rise to a Primakoff blast wave in which the density and velocity are power-laws in radius. A general treatment of the generation of the ejecta density profile (under the assumption of spherical symmetry) was given by Matzner & McKee [3], and we shall briefly review their results here.

They began by developing an analytic expression for the velocity of the supernova shock as it propagates through the progenitor star. In the interior, one expects $v_s \propto [E_{\mathrm{in}}/m(r)]^{1/2}$, where $m(r) \equiv M(r) - M_{\mathrm{rem}}$ is the mass of the ejecta inside r. When the shock reaches the atmosphere, it accelerates down the density gradient according to $v_s \propto [m(r)/\rho_0(r)r^3]^\alpha$, with $\alpha \simeq 0.19$. A general expression for the shock velocity, that is accurate throughout the envelope and atmosphere, is [3]

$$v_s = A \left[\frac{E_{\mathrm{in}}}{m(r)} \right]^{1/2} \left[\frac{m(r)}{\rho_0(r)r^3} \right]^\alpha . \tag{1}$$

It is possible to evaluate the coefficient A analytically as well [4]. This expression is accurate to within 2% in stellar envelopes. It should be contrasted with the result of the Kompaneets approximation, which gives $v_s \propto [E_{\mathrm{in}}/\rho_0(r)r^3]^{1/2}$ and fails in both the interior and the atmosphere.

With this result as a base, Matzner & McKee were able to determine approximate analytic expressions for the density distribution of the ejecta that are quite general. The distribution in the outer ejecta can be approximated as a power-law in velocity,

$$\rho \propto (v_{\mathrm{ej}}/v)^{\ell_\rho}, \tag{2}$$

where v_{ej} is the maximum velocity of the ejecta. In general, ℓ_ρ can depend on position; indeed, by comparing with numerical models, Dwarkadas and Chevalier [5] find that an exponential approximation fits best for Type Ia SNRs. In the outer ejecta, the value of ℓ_ρ is typically greater than 5, which means that the energy as well as the mass of the ejecta are concentrated in the interior.

In the ejecta-dominated stage of evolution of SNRs ($M_{\mathrm{ej}} > M_{\mathrm{sw}}$), there are two shocks: the blast-wave shock that advances into the ambient medium at velocity v_b and a reverse shock that propagates back into the ejecta with a relative velocity v_r

[6]. Between the shocks lie the shocked ambient medium and the shocked ejecta, separated by a contact discontinuity. For the case of a steep density gradient in the ejecta ($\ell_\rho > 5$), the ejecta-dominated stage of evolution can be divided into two parts: First, there is a brief initial stage described by the Hamilton-Sarazin [7] similarity solution, in which the velocity of the blast-wave shock v_b is approximately equal to $v_{\rm ej}$. In a medium of constant density, the radius of the blast wave expands with time as $R_b \simeq 1.10 v_{\rm ej} t(1 - at^{3/2})$, where a is a numerical constant. The velocity of the reverse shock propagating back into the ejecta satisfies $v_r \propto t^{3/2}$. Once the blast-wave velocity has slowed significantly below $v_{\rm ej}$, the evolution enters a second self-similar phase of evolution, which is described by the Chevalier-Nadyozhin solution [8] [9]. In this solution, the blast-wave radius varies as $R_b \propto t^{1-3/\ell_\rho}$; the velocities of the blast-wave shock and the reverse shock both scale as t^{-3/ℓ_ρ}.

When the mass of the swept-up ambient medium exceeds the mass of the ejecta, the dynamics approaches that of an adiabatic blast wave produced by a point explosion. The Sedov-Taylor similarity solution for this problem has $R_b \propto (E_{\rm in} t^2 / \rho_0)^{1/5}$. Many historical remnants are in transition between the ejecta-dominated stage and the Sedov-Taylor stage, and McKee & Truelove [10] proposed an approximate analytic solution that describes the evolution from the Chevalier-Nadyozhin stage through the Sedov-Taylor stage. This solution was developed by Truelove & McKee [11] (note the erratum in [12]), who defined the characteristic quantities

$$R_{\rm ch} \equiv (M_{\rm ej}/\rho_0)^{1/3}, \tag{3}$$

$$v_{\rm ch} \equiv (E_{\rm in}/M_{\rm ej})^{1/2}, \tag{4}$$

$$t_{\rm ch} \equiv R_{\rm ch}/v_{\rm ch}, \tag{5}$$

and then gave approximate expressions for $R_b/R_{\rm ch}$, $v_b/v_{\rm ch}$, and $v_r/v_{\rm ch}$ as functions of $t/t_{\rm ch}$ (or in some cases, $t/t_{\rm ch}$ in terms of $R_b/R_{\rm ch}$, etc). This solution has been successfully compared with the observations of Tycho's SNR by Hughes [13].

Actual YSNRs present a rich range of phenomena that go well beyond the simple dynamics described above. For example, if the YSNR has an embedded pulsar, then the pulsar nebula can be dramatically transformed by the reverse shock [14]. Deviations from spherical symmetry can have major effects. Such deviations can be due to asymmetries in the explosion (e.g., [15]), the ambient medium [16], or due to instabilities (e.g., [17] [18]). Theoretical and observational studies of these effects are given in these conference proceedings.

YSNRS AS PHYSICS LABORATORIES

Young SNRs provide extreme environments in which novel physical effects can be studied. The role of YSNRs in cosmic ray acceleration is reviewed elsewhere in this volume by Steve Reynolds. Some YSNRs harbor pulsars, and these objects are discussed by Manchester and by Gaensler elsewhere in these proceedings. Here I shall briefly comment on how YSNRs can serve as physics laboratories for the study of nucleosynthesis, dust formation and destruction, and interstellar hydrodynamics.

Nucleosynthesis

YSNRs provide laboratories to test both the theory of the formation of the elements and the theory of supernova explosions. X-ray spectroscopy is crucial, since optical spectroscopy is often sensitive to only a small fraction of the mass of the ejecta. Spatially resolved X-ray spectroscopy is now available with *Chandra* and with *XMM-Newton*. The Astronomy and Astrophysics Survey Committee (AASC) [19] has recommended that an even more powerful instrument be built during the coming decade, *Constellation-X*. Consisting of four X-ray telescopes, *Constellation-X* would have an energy resolution $E/\Delta E \sim 300 - 5000$ over the energy range 0.25-40 keV. Its effective area would be about 20-100 times that of existing instruments, and it would have an angular resolution of about 15 arcsec. A particularly strong clue on the nucleosynthesis that occurs in YSNRs is provided by radioactive elements, which can be studied with the INTEGAL spacecraft that is due to be launched in 2001. The Panel on High-Energy Astrophysics from Space of the AASC recommended an Explorer Class mission for nuclear line X-ray spectroscopy.

Dust Formation and Destruction

Most of the refractory elements like silicon and iron in the interstellar medium are contained in dust grains. Observations of isotopic anomalies in meteorites suggest that dust forms in core-collapse supernovae [20]. This idea received observational confirmation when SN1987A showed several signs of dust formation, including dust emission, dust absorption, and a drop in the line intensities of the refractory elements at the time at which the dust emission appeared [21]. Whereas observations of dust emission suggest that only about $(1 - 10) \times 10^{-4} M_\odot$ of dust formed in the supernova, observations of the extinction and of the diminution of the refractory lines—which measure all the dust, not just the hot dust—suggest that most of the refractory elements in the ejecta could have gone into dust [21]. From a theoretical perspective, supernovae have long been thought to be sites of dust formation: Once grains are injected into the ISM they are subject to efficient destruction by SNR shocks, so it is essential that the refractory elements be injected in solid form in order to account for their large depletions (e.g., [22]). Even in this case, significant growth of refractory grains is inferred to occur in the ISM [21] [23].

However, the same shock processes that are effective at destroying grains in the ISM can destroy them in YSNRs. How can the grains survive the reverse shock? How can they survive being embedded in the hot gas in the interior of an SNR? How can they survive being decelerated from velocities in excess of 10^3 km s^{-1}? Sputtering is the dominant process of grain destruction in this case [24] [25]. The effectiveness of sputtering is enhanced by the shattering of the grains that occurs in shocks, since that increases the grain surface area [26]. Furthermore, recent ISO observations have cast doubt on the idea that core collapse supernovae are the dominant source of interstellar grains: Douvion, Lagage, & Cesarsky [27] argue that only a small fraction of the silicon in Cas A is microscopically mixed into the regions necessary to make silicates, whereas Arendt, Dwek, & Mosely [28] find that the spectrum of Cas A indicates that the silicates that do form are not of the type that is typical of interstellar grains.

In order to understand how dust is formed and destroyed in YSNRs, further observational and theoretical work is needed. The *Space Infrared Telescope Facility (SIRTF)* and the *Stratospheric Observatory for Infrared Astronomy (SOFIA)* will provide valuable new data. Further in the future, the *Next Generation Space Telescope (NGST)* offers the possibility of observing the infrared spectra of grain-producing elements and of the grains themselves at high angular resolution.

Interstellar Hydrodynamics

YSNRs offer a fascinating laboratory in which to study hydrodyamic processes in the ISM. As described by McCray elsewhere in these proceedings, observations of the youngest nearby SNR, the remnant of SN1987A, even provide us with an opportunity to observe the *temporal evolution* of these hydrodynamic processes.

YSNRs produce extremely powerful shocks, with shock velocities that can exceed 10^4 km s^{-1} and Mach numbers $\mathcal{M}$ that can exceed 10^3. These shocks may be strongly modified by the acceleration of cosmic rays. Indeed, Hughes, Rakowski, and Decourchelle [29] have analyzed *Chandra* observations of the SMC SNR E0102.2-7219 and have concluded that the most natural interpretation of the low electron temperatures they observe is that most of the shock energy has gone into cosmic rays. A key issue of shock physics that can be addressed through observations of YSNRs is the degree of collisionless heating of electrons behind fast shocks. Ghavamian et al [30] have studied optical emission from nonradiative shocks in several SNRs (the Cygnus Loop, RCW 86, and Tycho), and they infer that the electron heating efficiency falls as the shock velocity increases: they find $T_e/T_i \sim 0.7 - 1$ in the Cygnus Loop ($v_s \sim 300$ km s^{-1}), ~ 0.3 in RCW 86 ($v_s \sim 600$ km s^{-1}), and $\lesssim 0.1$ in Tycho ($v_s \sim 2000$ km s^{-1}). However, they did not allow for the possibility that some of the energy of the shocked gas is in cosmic rays, as found by Hughes et al [29]. Further studies of shock physics using both optical and X-ray observations will be very illuminating.

A problem of fundamental importance in astrophysics is the interaction of a shock wave with a density inhomogeneity (a "cloud"). Astrophysical plasmas are generally inhomogeneous, and shocks are ubiquitous because these plasmas can typically cool to temperatures well below those associated with the violent events that occur in them. Once the shock has passed the cloud, this problem reduces to that of a cloud embedded in a flow, or, equivalently, a clump of ejecta interacting with the ambient medium (an "interstellar bullet"). These problems have been studied both theoretically (e.g., [31], [32], [33]) and in the laboratory [34]. The clouds are subject to both thermal evaporation and hydrodynamic stripping; the first process plays a key role in the three-phase model of the ISM [35]. Magnetic fields have a strong effect on both processes; the effective conductivity in a magnetized, collisionless plasma remains uncertain. The combined effects of hydrodynamic stripping and thermal conduction are invoked in the model of turbulent mixing layers [36] [37], in which it is assumed that conduction is so efficient that the mixed gas starts with an initial temperature that is determined only by mass and energy conservation, but not so efficient that it affects the radiative cooling of the gas. As yet, there is no direct observational evidence that the boundary layers of shocked clouds are described by turbulent mixing layers, but YSNRs provide an ideal laboratory for studying this question, as well as for determining the fate of the "interstellar bullets" seen in the jet in Cas A [38] and for addressing many other problems in interstellar gas dynamics.

CHALLENGE: WHERE ARE THE VERY YOUNG SNRS IN THE GALAXY?

No supernova remnants are known in the Galaxy since Cas A exploded more than 300 years ago, with the possible exception of a very young SNR in the Galactic Center reported at this meeting [39]. Van den Bergh & Tammann [40] cite a range of estimates for the Galactic supernova rate; perhaps the best is from observations of SNe in other galaxies, which gives $8.4h^2$ per century, where h is the Hubble constant normalized to 100 km s^{-1} Mpc^{-1}. Taking $h = 0.63$ from a recent "concordance model" based on observations of the cosmic microwave background and large scale structure [41], we find a Galactic SN rate of 3.3 per century. Subsequently, van den Bergh & McClure [42] estimated a rate of $1.7 - 1.9$ per century for the same value of h. McKee & Williams [43] analyzed data on the OB associations in the Galaxy and estimated that the rate of core-collapse supernovae is 2.0 per century; including Type Ia's, the total SN rate becomes 2.3 per century (note that this estimate is independent of h). In 300 years, we therefore expect that 5-10 SNe have exploded in the Galaxy. Where are their remnants?

Most of these SNe are core-collapse SNe that originated in associations. McKee & Williams [43] worked out the birthrate of associations in the Galaxy based on observations of radio H II regions, radio free-free emission, and N II $\lambda122$ μm emission. They adopted an IMF based on the work of Scalo [44], in which $d\mathcal{N}_*/d\ln m_* \propto m_*^{-1.5}$, slightly steeper that the Salpeter value. If $8M_\odot$ is the lower mass limit for core-collapse SNe, then half the missing SNRs had progenitors less massive than 13 $M_\odot$. Under the assumption that each association has five consecutive generations of star formation (lasting for a total of 18.5 Myr), they estimated that the birthrate of associations that eventually produce at least $\mathcal{N}_{*h}$ stars more massive than 8 $M_\odot$ is

$$\dot{\mathcal{N}}_a(>\mathcal{N}_{*h}) = 0.33 \left(\frac{7200}{\mathcal{N}_{*h}} - 1 \right) \quad \text{Myr}^{-1}, \tag{6}$$

where the largest association in the Galaxy has $\mathcal{N}_{*h} \simeq 7200$ high-mass stars. They extrapolated this distribution down to associations with 100 stars per generation, or 500 stars altogether; for their IMF, this corresponds to a minimum value of $\mathcal{N}_{*h} \simeq 1.3$. From this distribution, one can infer that half of these missing SNRs occurred in associations in which about 35 SNe have already occurred. To find out where in the Galaxy the SNe are likely to have occurred, we use the spatial distribution of associations from reference [43]

$$\frac{d\dot{\mathcal{N}}_a}{dA} = \dot{\mathcal{N}}_a(>\mathcal{N}_{*h}) \left[\frac{\exp(-R/H_R)}{A_{\text{eff}}} \right] \quad (3 \text{ kpc} \lesssim R \lesssim 11 \text{ kpc}), \tag{7}$$

with a radial scalelength $H_R \simeq 3.5$ kpc and an effective area $A_{\text{eff}} = 47$ kpc^2. From this we can infer that half the missing SNRs are between 3 and 6 kpc from the Galactic Center, which includes the highly obscured molecular ring; other SNRs could lie on the far side of the ring, where the obscuration is even greater.

We conclude that many of the missing very young SNRs are either highly obscured or located in regions that have been evacuated by previous SNRs. In this latter case, the SNR is "muffled" by the cavity: the ejecta remain in free expansion out to a radius of about $20[(M_{\text{ej}}/10M_\odot)(10^{-2}\text{cm}^{-3}/n_{\text{H}})]^{1/3}$ pc and radiate their energy less efficiently than in denser environments [32]. YSNRs like Cas A, which is interacting with circumstellar matter, or the Crab, which is a plerion, would be easily seen throughout the Galaxy, however; we conclude that the progenitors of such SNRs are relatively rare.

Two projects recommended by the AASC could dramatically improve the census of YSNRs in the Galaxy: The *Energetic X-ray Imaging Survey Telescope (EXIST)* will survey the sky in the energy range 5-300 keV with 300 arcsecond resolution. It will be able to detect highly obscured SNRs. The *LOw Frequency ARray (LO-FAR)* is a radio telescope with a square kilometer of collecting area at wavelengths between 200 and 1000 cm and an angular resolution of 1 arcsec. With these capabilities, it will be extremely sensitive to the nonthermal radio emission from YSNRs.

CHALLENGE: CAN VERY YOUNG SNRS MAKE GAMMA-RAY BURSTS?

Gamma-ray bursts (GRBs) remain one of the great puzzles in astrophysics more than 30 years after their discovery (see Piran [45] for a review). There are several arguments suggesting that GRBs are associated with the deaths of massive stars: supernova explosions are the most energetic phenomenon known to occur outside galactic nuclei; observations of GRBs with afterglows have suggested that they may be associated with star-forming regions (e.g., [46]); and finally, GRB 980425 appears to be temporally and spatially associated with the unusual supernova SN1998bw [47]. Long before there was any observational evidence for an association of GRBs with massive stars, Colgate [48] suggested that GRBs could be produced by the emission from accelerating shock waves in the outer layers of supernovae in galaxies at distances of 30-50 Mpc. It is now known that most GRBs are at cosmological distances, and Paczynski [49] has suggested that more energetic stellar explosions associated with the collapse of rapidly rotating massive stars ("hypernovae") may underlie GRBs. Some attempts to model the energetics of a gamma-ray burst produced by a spherical explosion of SN1998bw found that the energy in relativistic ejecta was inadequate (e.g., [50]). Extrapolating from a nonrelativistic analysis, however, Matzner & McKee [3] concluded that there was enough energy in relativistic ejecta to produce the observed burst. What was missing from these analyses was the inclusion of the relativistic effects on the hydrodynamics as the shock wave accelerates to velocities near the speed of light. This omission has been rectified by Tan, Matzner, & McKee's analysis of trans-relativistic blastwaves in supernovae [4].

Shock-acceleration models of GRBs have a great advantage in that they naturally deal with the "baryon-loading problem." The observed energies of GRBs are so great that any viable model must be highly efficient, putting a significant fraction of the energy into radiation. Shock acceleration models are intrinsically efficient because they concentrate the energy in the outermost, fastest moving material in an explosion.

The dynamics of a nonrelativistic shock in a stellar envelope are described by equation (1). In the outer layers of the envelope, where the enclosed mass $m(r)$ approaches the total ejected mass M_{ej}, this reduces to

$$v_s \propto \left(\rho_0 r^3 \right)^{-\alpha}, \tag{8}$$

where $\alpha \simeq 0.19$. The shock accelerates if the stellar density ρ_0 falls off faster than $1/r^3$. Gnatyk [51] proposed that a similar relation should hold for trans-relativistic flows, with v_s replaced by $\Gamma_s \beta_s$ and α set equal to 0.2 for accelerating shocks; here $\beta_s = v_s/c$ and $\Gamma_s = (1 - \beta_s^2)^{-1/2}$. Tan et al [4] obtained a more accurate representation of the shock dynamics, which in addition is valid throughout the star:

$$\Gamma_s \beta_s = p(1 + p^2)^{0.12} \tag{9}$$

$$p \equiv A \left[\frac{E_{in}}{m(r)} \right]^{1/2} \left[\frac{m(r)}{\rho_0 r^3} \right]^{0.187}. \tag{10}$$

This reduces to equation (1) in the nonrelativistic limit.

After the passage of the shock through the stellar envelope, the shocked gas is left with an internal energy that is equal to its kinetic energy. As the shocked gas undergoes adiabatic expansion, this internal energy is converted to kinetic energy; in addition, the inner layers do work on the outer layers. Matzner & McKee [3] found that as a result the final velocity in a nonrelativistic, planar flow is about twice the post-shock value. In the relativistic case, Tan et al [4] find

$$\frac{\Gamma_f \beta_f}{\Gamma_s \beta_s} = 2.0 + (\Gamma_s \beta_s)^{\sqrt{3}}, \tag{11}$$

where the scaling in the relativistic limit is the same as that found analytically by Johnson & McKee [52]. The effects of spherical expansion reduce $\Gamma_f \beta_f$ somewhat below the planar value.

Based on their nonrelativistic theory, Matzner & McKee [3] estimated the mass of relativistic ejecta in a supernova; for example, for an $n = 4$ polytrope, the mass with $\Gamma_f \beta_f > 1$ is

$$M_{\rm rel} = 2.7 \times 10^{-6} \left(\frac{E_{\rm in}}{10^{52}~{\rm erg}} \right)^{3.29} \left(\frac{1~M_\odot}{M_{\rm ej}} \right)^{2.29} \quad M_\odot. \tag{12}$$

The corresponding energy is

$$M_{\rm rel} c^2 = 3.1 \times 10^{48} \left(\frac{E_{\rm in}}{10^{52}~{\rm erg}} \right)^{3.29} \left(\frac{1~M_\odot}{M_{\rm ej}} \right)^{2.29} \quad {\rm erg}. \tag{13}$$

Tan et al [4] show that this is a reasonably good estimate for $M_{\rm rel}$, and furthermore, that the kinetic energy of the ejecta above some value of $\Gamma_f \beta_f$ is

$$E_k \propto F(\Gamma_f \beta_f) M_{\rm rel} c^2, \tag{14}$$

where (for $n = 4$) $F \propto (\Gamma_f \beta_f)^{-4.69}$ in the nonrelativistic limit and $F \propto (\Gamma_f \beta_f)^{-0.98}$ in the relativistic limit.

To determine if the explosion of SN1998bw had enough energy to power GRB980425, Tan et al used a model of the supernova kindly provided by Stan Woosley. This model, which is consistent with the observed light curve of the supernova, had a Wolf-Rayet progenitor with a mass of $6.55 M_\odot$, an ejecta mass of $4.77 M_\odot$, and an energy of 2.8×10^{52} erg. Both the analytic theory and numerical simulation yield about 2×10^{48} erg in ejecta with $\Gamma_f \beta_f > 1$; by comparison, the observed energy of the burst was $(8.1 \pm 1.0) \times 10^{47}$ erg, under the assumption that it was in fact associated with the supernova [47]. The energy content of the ejecta therefore appears sufficient to power the burst. In order to release the energy, however, it must first be randomized. This can be done through interaction with the dense stellar wind expected for a Wolf-Rayet star [50]; a mass loss rate of a few times $10^{-4} M_\odot~{\rm yr}^{-1}$ is required. Because the emission is due to the interaction with the ambient medium, this model for GRBs is in fact a YSNR model. Tan et al did not attempt to carry out a calculation of the emitted spectrum, so it remains to be seen whether an approximately spherical shock-acceleration model can account for the observed burst. However, they did show that the characteristic synchrotron energy from such a model is compatible with the observations and that this model can satisfy the constraints imposed by observations of the radio emission [53].

The inferred isotropic energies of GRBs range up to in excess of 10^{54} erg, so if GRB 980425 was in fact associated with SN1998bw, it was a very puny burst. Can shock acceleration models account for these much more luminous cosmological bursts? To address this question, Tan et al considered an extreme hypernova model in which $E_{in} = 5 \times 10^{54}$ erg is released into about $5M_{\odot}$ of ejecta. Whether such an energetic explosion is possible is not known; whereas most of the energy in a conventional core-collapse supernova is released in the form of neutrinos and possibly gravitational radiation, much of the energy in a hypernova is assumed to go into kinetic form as the result of a rapidly rotating, magnetized black hole at the center [49]. Tan et al find that in this model almost 1% of the explosion energy goes into material with $\Gamma_f \beta_f > 10$. The principal observational limits on the minimum value of $\Gamma_f \beta_f$ come from the constraints that the opacity due to photon-photon interactions and to the external medium be small. These limits in turn depend upon the size of the burst and therefore the timescale of the variations. While some GRBs show very rapid variability, most of the bursts with observed afterglows (and therefore measured energies) are relatively smooth. By assuming that the variations observed in these bursts could be explained by fluctuations in the ambient medium or by instabilities in the ejecta, Tan et al inferred that the minimum values of $\Gamma_f \beta_f$ were of order 10, almost an order of magnitude smaller than the limits inferred by some other workers (see [45]). As a result, they concluded that some of the cosmological GRBs with measured afterglows could be accounted for by an extreme hypernova model with a spherical explosion. However, a number of the bursts were too energetic to be accounted for by such a model. The most likely conclusion is that the cosmological bursts are aspherical, so that the emission is beamed towards us and the actual explosion energy is much smaller than the isotropic value (see [45]). Furthermore, beaming provides the most natural explanation for the optical and X-ray light curves of some bursts, such as GRB 990510 [54] [55].

Tan et al conclude that very young SNRs (including very young hypernova remnants) might indeed be the engines that underlie GRBs. Whereas the explosion that produced the GRB that is believed to be associated with SN1998bw could have been approximately spherical, the explosions associated with the much more energetic cosmological bursts are most likely quite aspherical. The recently launched *HETE-2* spacecraft, the proposed *Swift* mission, and the AASC-recommended *Gamma-ray Large Area Space Telescope (GLAST)* and *EXIST* missions should provide the data that will enable us to finally determine the nature of the enigmatic GRBs.

CONCLUSION: THE CONNECTIONS CHALLENGE

In this brief review, I have tried to show how YSNRs are at the nexus of many important problems in contemporary astrophysics: the late stages of stellar evolution, supernovae, the formation of compact objects, nucleosynthesis, the formation and destruction of interstellar dust, astrophysical gas dyanamics, and possibly even gamma-ray bursts. In order for the study of YSNRs to reach its potential, however, we must overcome a major challenge, which I call the *"Connections Challenge:"* How can one infer the nature of the supernova and its progenitor from observations of the YSNR? Some steps have been taken in this direction for the best-studied YSNRs such as Cas A (e.g., [56]), but much more needs to be done. A key step in this process is to identify the compact object that remains from the explosion, when it exists. In some cases this can be done through observations of radio pulsars. However, the recent discovery of an X-ray point source in Cas A [57] shows that other approaches to this problem are possible and opens up a new window on understanding supernovae and the formation of compact objects [58].

The results presented at this conference show that we are making progress in addressing the challenges I have presented. If the ambitious program for new instruments and theory recommended by the AASC is carried out, the coming decade could well see the resolution of most of these challenges.

Acknowledgments. I wish to thank the organizers for inviting me to an extremely valuable conference. Comments on a draft of this paper by Roger Chevalier, Chris Matzner, and Jonathan Tan are gratefully acknowledged. My research is supported in part by the National Science Foundation under grant AST 95-30480.

REFERENCES

1. Chevalier, R.A. 1976, ApJ, 207, 872
2. Chevalier, R.A., & Soker, N. 1989, ApJ, 341, 867
3. Matzner, C.D., & McKee, C.F. 1999, ApJ, 510, 379
4. Tan, J.C., Matzner, C.D., & McKee, C.F. 2000, ApJ (in press), astro-ph 0012003
5. Dwarkadas, V.V., & Chevalier, R.A. 1998, ApJ, 497, 807
6. McKee, C. F. 1974, ApJ, 188, 335
7. Hamilton, A.J.S., & Sarazin, C.L. 1984, ApJ, 281, 682
8. Chevalier, R.A. 1982, ApJ, 258, 790
9. Nadyozhin, D.K. 1985, Ap&SS, 112, 225
10. McKee, C.F., & Truelove, J.K. 1995, Phys. Rep., 256, 157
11. Truelove, J.K., & McKee, C.F. 1999, ApJS, 120, 299
12. Truelove, J.K., & McKee, C.F. 2000, ApJS, 128, 403
13. Hughes, J.P. 2000, ApJ (submitted), astro-ph 0010122
14. Chevalier, R.A. 1998, Memorie della Societa Astronomia Italiana, 69, 977
15. Blondin, J.M., Borkowski, K., & Reynolds, S.P. 2000, ApJ (submitted), astro-ph 0010285
16. Maciejewski, W., & Cox, D.P. 1998, ApJ, 511, 792
17. Gull, S.F. 1973, MNRAS, 161, 47
18. Chevalier, R.A., Blondin, J.M., & Emmering, R.T. 1992, ApJ, 392, 118
19. McKee, C.F., & Taylor, J.H. 2001, *Astronomy and Astrophysics in the New Millennium*, National Academy Press (in press).
20. Clayton, D.D. 1975, ApJ, 199, 765
21. Dwek, E. 1998, ApJ, 501, 643
22. Draine, B.T., & Salpeter, E.E. 1979, ApJ, 231, 438

23. McKee, C.F. 1989, in *IAU Symp. 135 Interstellar Dust*, ed. L.J. Allamandola & A.G.G.M. Tielens, Dordrecht: Kluwer, 431

24. Tielens, A.G.G.M., McKee, C.F., Seab, C.G., and Hollenbach, D.J., 1994, ApJ, 431, 321

25. Dwek, E., Foster, S.M., & Vancura, O. 1996, ApJ, 457, 244

26. Jones, A.P., Tielens, A.G.G.M., & Hollenbach, D.J. 1996, ApJ, 469, 740

27. Douvion, T., Lagage, P.O., & Cesarsky, C.J. 1999, A&A, 352, L111

28. Arendt, R.G., Dwek, E., & Mosely, S.H. 1999, ApJ, 521, 234

29. Hughes, J.P., Rakowski, C.E., & Decourchelle, A. 2000, ApJ, 543, L61

30. Ghavamian, P., Raymond, J., Hartigan, P., & Blair, W.P. 2001, ApJ (submitted), astro-ph 0010496

31. McKee, C.F. & Cowie, L.L. 1975, ApJ, 195, 715

32. McKee, C.F. 1988, in *IAU Colloq. 101, Supernova Remnants and the Interstellar Medium*, ed. R.S. Roger and T.L. Landecker, Cambridge: Cambridge University Press, p. 205

33. Klein, R.I., McKee, C.F., & Colella, P. 1994, ApJ, 420, 213

34. Klein, R.I., Budil, K.S., Perry, T.S., & Bach, D.R. 2000, ApJS, 127, 379

35. McKee, C.F., & Ostriker, J.P. 1977, ApJ, 218, 148

36. Begelman, M.C., & Fabian, A.C. 1990, MNRAS, 244, 26P

37. Slavin, J.D., Shull, J.M., & Begelman, M.C. 1993, ApJ, 407, 83

38. Fesen, R.A., & Gunderson, K.S. 1996, ApJ, 470, 967

39. Senda, A., et al 2001, poster 4.03 at this conference

40. van den Bergh, S., & Tammann, G.A. 1991, *Annual Reviews of Astronomy and Astrophysics*, 29, 363

41. Tegmark, M., Zaldarriaga, M., & Hamilton, A.J.S. 2000, astro-ph 0008167

42. van den Bergh, S., & McClure, R.D. 1994, ApJ, 425, 205

43. McKee, C.F., & Williams, J.P. 1997, ApJ, 476, 144

44. Scalo, J. 1986, Fund. Cosmic Phys., 11, 1

45. Piran, T. 1999, Phys. Reports, 314, 575

46. Kulkarni, S.R., et al 2000, in *Gamma-Ray Bursts: 5th Huntsville Symposium*, eds, Kippen, R.M., Malozzi, R.S., & Fishman, G.J., New York, AIP, p. 277

47. Galama, T.J., et al 1998, Nature, 395, 670

48. Colgate, S.A. 1974, ApJ, 187, 333

49. Paczynski, B.P. 1998, in *Gamma-Ray Bursts: 4th Huntsville Symposium*, eds, C.A. Meegan, R.D. Preece, & T.M. Koshut, New York, AIP

50. Woosley, S.E., Eastman, R.G., & Schmidt, B.P. 1999, ApJ, 516, 788

51. Gnatyk, B.I. 1985, Sov. Astron. Lett., 11(5), 331

52. Johnson, M.H., & McKee, C.F. 1971, PRD, 3, 858

53. Kulkarni, S.R., et al 1998, Nature, 395, 663

54. Stanek, K.Z., et al. 1999, ApJ, 522, L39

55. Pian, E., et al 2000, A&A (submitted), astro-ph 0012107

56. Fesen, R.A., Becker, R.A., & Goodrich, R.W. 1988, ApJ, 329, L89

57. Tananbaum, H. 1999, IAU Circular 7246

58. Umeda, H., Nomoto, K., Tsuruta, S., & Mineshige, S. 2000, ApJ, 534, L193

Stage Setting

Type Ia Supernovae: Toward the Standard Model?

David Branch

Department of Physics and Astronomy, University of Oklahoma, Norman, OK 73019, USA

Abstract.
In this short review I suggest that recent developments support the conjecture that Type Ia supernovae (SNe Ia) are the complete disruptions of Chandrasekhar–mass carbon–oxygen white dwarfs in single–degenerate binary systems. The causes of the observational diversity of SNe Ia within the context of this standard model, and the implications of the model for young remnants of SNe Ia, are briefly discussed.

INTRODUCTION

The intense current interest in Type Ia supernovae (SNe Ia) is mainly due to their value as distance indicators for cosmology. At a conference on young supernova remnants, however, we're more concerned with what SNe Ia eject, and what, if anything, they leave behind. Because what they leave behind depends on where they come from, we're also concerned with the nature of the progenitors of SNe Ia. Those who are interested in SNe Ia as distance indicators are concerned with these issues too.

We have a pretty good idea of what a typical SN Ia ejects. The composition structure of the venerable carbon–deflagration model W7 [1] has passed the test of time as a first approximation to what is ejected: a Chandrasekhar mass, half in the form of low–velocity iron–peak isotopes (initially mostly radioactive ^{56}Ni); some high–velocity unburned carbon and oxygen; and plenty of intermediate–mass elements such as silicon, sulfur, and calcium in between, around 10,000 km s^{-1}. This model, and others that are not too unlike it, give a good account of the spectra of typical SNe Ia [2].

W7 is a specific example of what can be considered to be the standard model for SNe Ia: Chandrasekhar mass ejection following the ignition of carbon at or near the center of a carbon–oxygen (C–O) white dwarf that has been accreting matter from a nondegenerate companion star. In this "single–degenerate" (SD) scenario the donor star survives the explosion of the white dwarf. The Tokyo supernova group has been advocating the standard model for years (see [3] for their most recent review), and recently some others have been moving, more cautiously, toward the

CP565, *Young Supernova Remnants: Eleventh Astrophysics Conf.,* edited by S. S. Holt and U. Hwang
© 2001 American Institute of Physics 0-7354-0001-6/01/$18.00

same conclusion (see, e.g., [4] for another recent review). I would like to climb onto this bandwagon because recent developments make it seem to be headed in the right direction. In this short review the emphasis will be on these recent developments; most of the references will be to papers that have been written within the last few years.

OPTIONS

Sub–Chandrasekhar Helium Ignitors

During the 1990s there has been some interest in sub–Chandrasekhar off–center helium–ignitor models, in which the first nuclear ignition occurs at or near the bottom of an accumulated helium layer surrounding a C–O core. Those who have calculated the spectra and light curves of such models have found that they do not match the observations as well as carbon–ignitor models do [5,6,7]. It is unlikely that SNe Ia come from helium–ignitors. If the models change sufficiently, this conclusion can be reexamined.

Double Degenerates

In the "double–degenerate" (DD) scenario, two white dwarfs merge following orbital decay caused by gravitational wave radiation. The mass ejection can be super–Chandrasekhar (up to double–Chandrasekhar, in principle), and no donor star is left behind. The DD scenario has been taken seriously, not so much because of strong evidence in its favor, but because of perceived deficiencies in the rival SD scenario. The main worry about the SD scenario has been that because there are so many ways for the accreting white dwarf to lose mass (e.g., nova explosions, common envelope formation), it may be difficult or impossible for it to reach the Chandrasekhar mass. The DD scenario provides a natural way to assemble a super–Chandrasekhar mass. Another possible problem with the SD scenario is that some circumstellar matter is expected to be present, while no convincing observational evidence for circumstellar matter associated with any SN Ia has been found. Circumstellar matter is not expected in the DD scenario. On the other hand, the main worry about the DD scenario has been that most of those who have tackled the difficult problem of modeling a white–dwarf merger have ended up being pessimistic about getting a SN Ia out of it, rather than a collapse to a neutron star. The problem is that nuclear burning begins in the outer layers and propagates inward, converting the C–O composition to an O–Ne–Mg mixture that eventually must collapse [8,9]. Another worry about the DD scenario is that it may be unable to account for the rate at which SN Ia are observed to occur [10].

One particular SN Ia for which a super–Chandrasekhar mass ejection has been suspected is the well observed, peculiar SN 1991T. Its light curve was broader than the light curves of most SNe Ia, and until the time of maximum brightness its spectra showed strong lines of Fe III instead of the usual lower–excitation lines of Si II, S II, Ca II, and O I. Both of these properties suggest an unusually large ejected mass of ^{56}Ni. If the distance to NGC 4527, the parent galaxy of SN 1991T, is 16.4 Mpc — the distance found from Cepheids for NGC 4536 [11] and NGC 4496A [12] which have been thought to be co–members with NGC 4527 of a small group of galaxies on the near side of the Virgo cluster — then SN 1991T appears to have been too luminous for a Chandrasekhar–mass explosion, so it must have come from a DD progenitor [13]. But on the basis of a simple model of the light echo of SN 1991T from interstellar dust in NGC 4527, the maximum distance has been estimated to be 15 Mpc [14], and recent studies of Cepheids in NGC 4527 have given $14.1 \pm 0.8 \pm 0.8$ Mpc [15] and $13.0 \pm 0.5 \pm 1.2$ Mpc [16]. (An even shorter distance of 11.3 ± 0.4 Mpc has been obtained from an application of the surface–brightness–fluctuations technique to NGC 4527 [17].) It now appears that as expected from its spectrum and light curve, SN 1991T was somewhat overluminous for a SN Ia, but not so luminous that super–Chandrasekhar mass ejection and a DD progenitor are required.

It has been inferred from the spectra of SN 1991T that unburned carbon and freshly synthesized nickel coexisted in velocity space [13], which is not the case in any published one–dimensional hydrodynamical model for SNe Ia. This characteristic of SN 1991T, although not directly indicating a DD progenitor, has seemed to provide further evidence that SN 1991T was different in some distinct way from normal SNe Ia. Recently, however, the first successful *three*–dimensional numerical simulation of a deflagration explosion of a (non–rotating) Chandrasekhar–mass white dwarf has been carried out [18], and in the resulting model, unburned carbon and freshly synthesized nickel do coexist in velocity space. It will be very interesting to see whether careful analysis of the spectra of normal SNe Ia reveals signs (more subtle than in SN 1991T) that the coexistence of carbon and nickel in velocity space is a general characteristic of SNe Ia.

An important observational development has been the discovery of events such as SN 1999aa that have normal spectral features near the time of maximum brightness, but spectra that are intermediate between those of SN 1991T (strong Fe III) and normal SNe Ia (strong Ca II) a week before maximum [19]. This seems to me [20] (but see [19]) to make SN 1991T less likely to be physically distinct from spectroscopically normal SNe Ia.

All of the above results make SN 1991T begin to look more like an event near one end of a continuum rather than like a physically distinct phenomenon. This is important, because if the DD scenario is not needed for SN 1991T, then we are left with little or no evidence that it is needed at all. From time to time other arguments for a range in the ejected mass of SNe Ia have been advanced (e.g., [21,22]); in particular it has often been suggested that subluminous SNe Ia like SN 1991bg eject *less* than a Chandrasekhar mass, but even for such extremely peculiar events the arguments for a non–Chandrasekhar mass are not compelling [23]. Given the absence of positive evidence for DD progenitors, the lack of persuasive evidence for a range in ejected masses among SNe Ia, and recent developments indicating that the SD scenario is likely to be able to produce SNe Ia (see below), it is tempting to think that those who have concluded that DD mergers make neutron stars rather than SNe Ia may have been correct. (The recent suggestion that a merger makes a neutron star *and* a SN Ia [24] does not seem promising because in most cases the ejected mass would be too much less than a Chandrasekhar mass.) Of course, if a single SN Ia proves to be too luminous for Chandrasekhar mass ejection (for some candidates see [15]), this conclusion will need to be revised.

SINGLE DEGENERATES

A very attractive feature of the SD scenario is that if the accreting white dwarf does manage to approach the Chandrasekhar mass, it is more likely than a merging pair of white dwarfs to actually explode, because unlike in the DD case burning will begin in the dense inner layers and propagate outward. On the issue of whether the white dwarf can get to the Chandrasekhar mass, there have been important recent developments. It has been argued [25] that at the fairly high desired accretion rates (10^{-7} to 10^{-6} $M_\odot$ y^{-1}) the white dwarf develops a strong fast ($\lesssim 1000$ km s^{-1}) wind that stabilizes the mass–transfer process and enables common–envelope evolution and ruinous mass loss from the system to be avoided. Both main–sequence donors and red–giant donors of appropriate masses and initial separations from the white dwarf now appear to be able to drive it to the Chandrasekhar mass [26,27,28]. According to recent more detailed and selfconsistent calculations, which take into account the evolution of the structure of the main sequence donor, the mass transfer rate, and the orbit, it turns out that even with less optimistic assumptions about the role of the fast wind it is possible for the white dwarf to reach the Chandrasekhar mass for a variety of initial parameters [29]. For some systems it is not even a very close call, i.e., if the white dwarf did not explode at 1.4 $M_\odot$ it could reach as much as 2 $M_\odot$. Given these recent results, the tables seem to have been turned: it now appears that SD systems with main sequence donors should not fail to produce Chandrasekhar–mass SNe Ia. (And because the accretion rate does not remain low enough long enough to build up a dangerously massive helium layer, SD systems with main–sequence donors do not produce sub–Chandrasekhar helium–ignitors.)

Observational support for the SD scenario with main–sequence donors is provided by the existence of supersoft X–ray sources [30,31] and recurrent novae [32,33] in which a massive white dwarf is accreting at a suitably high rate and appears to be on its way to the Chandrasekhar mass.

SD systems with main sequence donors cannot produce SNe Ia in stellar populations much older than a Gyr. It will be interesting to see the approach of [29] applied to SD systems with red giant donors, which are needed in the standard model to produce the SNe Ia that are observed to occur in older populations.

As for the doubt about the SD scenario because of the lack of evidence for circumstellar matter — the existing upper limits on the amount of circumstellar matter from optical, radio, and x–ray observations are interesting but not sufficiently stringent to rule out the SD scenario [34,35]. I'm not aware of any important observational developments on this issue during the last few years. The detection of circumstellar matter associated with SNe Ia should be a high priority for observers, but if there is a fast wind from the white dwarf, the circumstellar density will be lower than expected previously, and detection may be very difficult.

DIVERSITY

If we suppose that the standard model is correct for all SNe Ia, so the ejected mass is always 1.4 $M_\odot$, then what causes the observational diversity? The primary physical variable almost certainly is M_{Ni}, the ejected mass of ^{56}Ni. The higher the value of M_{Ni} the higher the peak luminosity; also, because radioactivity is responsible for heating the ejecta, the higher the temperature and the opacity [36] and the broader the light curve [37,38]. Thus it is likely that the observed correlation between light–curve width and peak luminosity is mainly due to a range in M_{Ni}. The temperature also determines which features appear in the early–time spectra [6,39,40]. If M_{Ni} is sufficiently high ($\gtrsim 0.8\ M_\odot$?) then Fe III lines, instead of the usual lower–excitation lines, become prominent. If M_{Ni} is sufficiently low ($\lesssim 0.4\ M_\odot$?), Ti II lines accompany the usual lines (and because a strong Ti II absorption blend falls right in the middle of the B band, the $B - V$ color becomes quite red).

What causes the range in M_{Ni}? Perhaps the most likely cause is a range in the carbon–to–oxygen ratios of the white dwarfs, since fusing carbon releases more energy than fusing oxygen. The C/O ratio is expected to be determined primarily by the main–sequence mass and the metallicity of the progenitor of the white dwarf [41]. In the three–dimensional modeling cited above [18], the amount of nuclear burning does appear to depend strongly on the C/O ratio.

Not all of the observational diversity can be attributed to M_{Ni}. Photometrically, it appears that a *two–parameter* empirical correction (using both the light–curve decline parameter, Δm_{15}, and the maximum–light color, $B-V$) is both necessary and (so far) sufficient to standardize the SN Ia peak luminosities [42,43]. Spectroscopically, a clear indication that SNe Ia cannot be fully described by a one–parameter sequence is that some events such as SN 1984A have only the usual spectral features but they are unusually broad and blueshifted, indicating that the ejecta are not unusually hot as in the case of SN 1991T but that an unusually large amount of mass has been ejected at high velocity ($\sim 20,000$ km s^{-1}) [44,45]. This may mean that most SNe Ia are pure subsonic deflagrations while the high–velocity events like SN 1984A are delayed detonations, in which the deflagration makes a transition to a supersonic detonation with the result that the density of the high–velocity layers is much higher than in a pure deflagration. (For a recent review of the propagation of nuclear burning fronts in SNe Ia, see [46].) Why two SNe Ia that produce similar amounts of M_{Ni} would have different modes of burning propagation remains to be understood; or perhaps not understood — it could be just a matter of chance [47]. (The observed correlation between SN Ia properties and the characteristics of the stellar population at the SN site [47,48] at least proves that the outcome is not *entirely* unrelated to the initial conditions.)

To summarize, it seems likely that within the SD scenario the observational diversity can be attributed mainly to differences in M_{Ni}, and secondarily to differences in the amount of mass ejected at high velocity (whether both deflagrations and delayed detonations are actually involved or not). There are plenty of other things that might contribute to the diversity, probably to a lesser extent: local composition asymmetries in the ejecta of deflagrations (e.g., clumps of iron–peak elements surrounded by intermediate–mass elements, embedded in unburned C–O [18]); global shape asymmetries of the ejecta in delayed detonations caused by the transition taking place at a point (or points) rather than simultaneously everywhere on a spherical shell [50]; a range of metallicities of the white dwarf [51,52]; the orientation of the donor star with respect to the line of sight of the observer [53]; effects associated with differences in the white–dwarf rotation speeds which, owing to angular momentum acquired during the accretion process, are expected to range up to a significant fraction of the break–up speed [29]; and global shape asymmetries caused by a strong magnetic field of the progenitor white dwarf [54].

IMPLICATIONS FOR YOUNG SUPERNOVA REMNANTS

According to the standard model SN Ia ejecta should consist of 1.4 $M_\odot$ of heavy elements from the white dwarf, accompanied by some hydrogen–rich mass knocked off the donor star: about 0.15 $M_\odot$ from a main–sequence star and about 0.5 $M_\odot$ from a red giant [53]. Most of this mass should be deep inside the supernova ejecta, expanding at low velocity, less than 1000 km s^{-1} [53,55]. The surviving donor star probably has peculiar surface abundance ratios: no lithium, beryllium, or boron; enhanced carbon; and perhaps enhanced iron–group elements caused by fallback of some of the lowest–velocity supernova ejecta [29,53]. A main sequence donor will have a space velocity (determined primarily by its pre–explosion orbital velocity) of more than about 450 km s^{-1} [56] and, having been pretty badly shaken up by its encounter with the supernova ejecta, it will be quite overluminous, as high as 5000 L$_\odot$ [53,56] before its swollen envelope relaxes back into equilibrium on a thermal timescale. A red giant donor will be moving faster than about 100 km s^{-1} [56], and because it will have been stripped of most but not quite all of its envelope it will continue to look like a red giant throughout the SNR phase [53]. Both kinds of donor stars will remain inside their associated SNRs during the lifetime of the remnants.

Were any of the historical Galactic supernovae produced by SNe Ia? SNe 1006 and 1572 (Tycho's supernova) are often cited as possibilities, but the arguments are not completely convincing. The heliacal risings and settings of SN 1006 have been used to make a much improved estimate of its peak apparent magnitude, which leads to a peak absolute visual magnitude, M_V, in the range -15.9 to -17.4 [57]. Normal SNe Ia have $M_V \simeq -19.4$, so if SN 1006 was a SN Ia it would seem to have been a peculiar one. If the luminosity was low then M_{Ni} also was low, and not much iron ($\lesssim 0.1\ M_\odot$) should be found in the remnant. (This is true whether SN 1006 was a peculiar weak SN Ia, a SN Ib/c, or some other kind of peculiar event.) If SN 1572 was a SN Ia then it may also have been a peculiar one, because its light curve appears to have been very fast for a SN Ia and its peak absolute visual magnitude has been estimated to be -18.64 ± 0.31 [57], a bit subluminous for a SN Ia and implying a reduced iron content in the Tycho remnant, too. The discovery in SN 1006 or 1572 of a star having the expected properties of a SN Ia donor would be exciting because it would confirm both that the event really was a SN Ia and that it was produced by a SD progenitor system; on the other hand, establishing the *absence* of a donor star would not rule out the SD scenario if we're not completely sure that these events were SNe Ia. To the extent that the Balmer-dominated remnants in the LMC are thought to have been SNe Ia because they resemble the remnants of SNe 1006 and 1572, then the case that the LMC events were SNe Ia may also not be compelling.

I am grateful to the members of the University of Oklahoma supernova group for discussions. This work was supported by NSF grant AST 9986965 and NASA grant NAG5–3505.

REFERENCES

1. Nomoto K., Thielemann F.—K., and Yokoi K. 1984, ApJ, 286, 644
2. Lentz E. J., Baron E., Branch D., and Hauschildt P. H. 2001, ApJ, in press
3. Nomoto K., Umeda H., Kobayashi C., Hachisu I., Kato M., and Tsujimoto T. 2000, in Cosmic Explosions, ed. S. S. Holt and W. W. Zhang, American Institute of Physics, in press
4. Livio M. 2000, in The Greatest Explosions Since the Big Bang: Supernovae and Gamma–Ray Bursts, ed. M. Livio, N. Panagia, and K. Sahu, Cambridge University Press, in press
5. Höflich P., Khokhlov A., Wheeler J. C., Nomoto K., and Thielemann F.-K. 1997, in Thermonuclear Supernovae, ed. P. Ruiz–Lapuente, R. Canal, and J. Isern, Kluwer, p. 659
6. Nugent P., Baron E., Branch D., Fisher A., and Hauschildt P. H. 1997, ApJ, 485, 812
7. Höflich P., and Khokhlov A. 1996, ApJ, 457, 500
8. Segretain L., Chabrier G., and Mochkovitch R. 1997, ApJ, 481, 355
9. Saio H., and Nomoto K. 1998, ApJ, 500, 388
10. Ruiz–Lapuente P., and Canal R. 2001, ApJ, in press
11. Saha A., Labhardt L., Schwengeler H., Macchetto F. D., Panagia N., Sandage A., and Tammann G. A. 1994, ApJ, 425, 14

12. Saha A., Sandage A., Labhardt L., Tammann G. A., Macchetto F. D., and Panagia, N. 1996, ApJS, 107, 693
13. Fisher F., Branch D., Hatano K., and Baron E. 1999, MNRAS, 304, 67
14. Sparks W. B., Macchetto F., Panagia N., Boffi F. R., Branch D., Hazen M., and Della Valle M. 1999, ApJ, 523, 585
15. Saha A., Sandage A., Thim F., Tammann G. A., Labhardt L., Christensen J., Maccheto F. D., and Panagia N. 2001, ApJ, in press
16. Gibson B. K., and Stetson P. B. 2001, ApJ, in press
17. Richtler T., Jensen J. B., Tonry J., Barris B., and Drenkahn G. 2001, A&A, in press
18. Khokhlov A. 2001, ApJ, in press
19. Li W., Filippenko A. V., Treffers R. R., Riess A. G., Hu J., and Qiu Y. 2001, ApJ, in press
20. Branch D. 2001, PASP, in press
21. Leibundgut B. 2001, Astr. and Astrophys. Rev., in press
22. Contardo G., Leibundgut B., & Vacca W. D. 2000, A&A, 359, 876
23. Modjaz M., Li W., Filippenko A. V., King J. Y., Leonard D. C., Matheson T., and Treffers R. T. 2001, PASP, in press
24. King A. R., Pringle J. E., and Wickramasinghe D. T. 2001, MNRAS, in press
25. Hachisu I., Kato M., and Nomoto K. 1996, ApJ, 470, L97
26. Li X.–D., and van den Heuvel E. P. J. 1997, A&A, 322, L9
27. Hachisu I., Kato M., Nomoto K., and Umeda H. 1999, ApJ, 519, 314
28. Hachisu I., Kato M., and Nomoto K. 1999, ApJ, 522, 487
29. Langer N., Deutschmann A., Wellstein S., and Höflich P. 2000, A&A, 362, 1046
30. Kahabka P., and van den Heuvel E. P. J. 1997, ARAA, 35, 69
31. Kahabka P., Puzia T. H., and Pietsch W. 1999, A&A, 347, L43
32. Hachisu I., Kato M., Kato T., and Matsumoto K. 2000, 528, L97
33. Ergma E., Gerskevits J., and Sarna S. 2001, A&A, in press
34. Cumming R., Lundqvist P., Smith L. J., Pettini M., and King D. L. 1996, MNRAS, 283, 1355
35. Cumming R., and Lundqvist P. 1996, in Advances in Stellar Evolution, ed. R. T. Rood, Cambridge University Press, p. 297
36. Khokhlov A., Müller E., and Höflich P. 1993, A&A, 270, 223
37. Hoflich P., Khokhlov A., Wheeler J. C., Phillips M. M., Suntzeff N. B., and Hamuy M. 1996, ApJ, 472, L81
38. Mazzali P., Nomoto K., Cappellaro E., Nakamura T., Umeda H., Iwamoto K 2001, ApJ, in press
39. Mazzali P., Lucy L. B., Danziger I. J., Guiffes C., Cappellaro E., and Turatto M. 1993, A&A, 279, 447
40. Hatano K., Branch D., Fisher A., Millard J., and Baron E. 1999, ApJS, 121, 233
41. Umeda H., Nomoto K., Kobayashi C., Hachisu I., and Kato M. 1999, ApJ, 522, L43
42. Tripp R., and Branch D. 1999, ApJ, 525, 209
43. Parodi B. R., Saha A., Sandage A., and Tammann G. A. 2000, ApJ, 540, 634
44. Hatano K., Branch D., Lentz E. J., Baron E., Filippenko A. V., and Garnavich P. M. 2000, ApJ, 543, L49
45. Lentz E. J., Baron E., Branch D., and Hauschildt P. H. 2001, ApJ, 547, in press

46. Hillebrandt W., and Niemeyer J. C. 2000, ARAA, 38, 191
47. Sorokina E. I., and Blinnikov S. I. 2000, Astronomy Letters, 26, 67
48. Hamuy M., Trager S. C., Pinto P. A., Phillips M. M., Schommer R. A., Ivanov V., and Suntzeff N. B. 2000, AJ, 120, 1479
49. Ivanov V. D., Hamuy M., and Pinto P. A. 2000, ApJ, 542, 5881
50. Livne E. 1999, ApJ, 527, L97
51. Höflich P., Wheeler J. C., and Thielemann F.–K. 1998, ApJ, 495, 617
52. Lentz E. J., Baron E., Branch D., Hauschildt P. H., and Nugent P. E. 2000, ApJ, 530, 966
53. Marietta E., Burrows A., and Fryxell B. 2000, ApJS, 128, 615
54. Ghezzi C. R., de Gouveia Dal Pinto E. M., & Horvath J. E. 2001, ApJ, in press
55. Chugai N. N. 1986, Soviet Astronomy, 30, 563
56. Canal, R., Mendez J., and Ruiz–Lapuente P. 2001, ApJ, in press
57. Schaefer B. E. 1996, ApJ, 459, 438

Optical Observations of Core-Collapse Supernovae

Alexei V. Filippenko

Department of Astronomy, University of California, Berkeley, CA 94720-3411

Abstract. I present an overview of optical observations (mostly spectra) of Type II, Ib, and Ic supernovae (SNe). SNe II are defined by the presence of hydrogen, and exhibit a very wide variety of properties. SNe II-L tend to show evidence of late-time interaction with circumstellar material. SNe IIn are distinguished by relatively narrow emission lines with little or no P-Cygni absorption component and (quite often) slowly declining light curves; they probably have unusually dense circumstellar gas with which the ejecta interact. Some SNe IIn, however, might not be genuine SNe, but rather are "impostors" — specifically, super-outbursts of luminous blue variables. SNe Ib do not exhibit the deep 6150 Å absorption characteristic of "classical" SNe Ia; instead, their early-time spectra have He I absorption lines. SNe Ic appear similar to SNe Ib, but lack the helium lines as well. Spectra of SNe IIb initially exhibit hydrogen, yet gradually evolve to resemble those of SNe Ib; their progenitors seem to contain only a low-mass skin of hydrogen. Spectropolarimetry thus far indicates large asymmetries in the ejecta of SNe IIn, but much smaller ones in SNe II-P. As one peers deeper into the ejecta of core-collapse SNe, the asymmetry (indicated by the amount of polarization) seems to increase. There is intriguing, but inconclusive, evidence that some peculiar SNe IIn might be associated with gamma-ray bursts. The rates of different kinds of SNe as a function of Hubble type are still relatively poorly known, although there are good prospects for future improvement.

INTRODUCTION

Supernovae (SNe) occur in several spectroscopically distinct varieties; see reference [1], for example. Type I SNe are defined by the absence of obvious hydrogen in their optical spectra, except for possible contamination from superposed H II regions. SNe II all prominently exhibit hydrogen in their spectra, yet the strength and profile of the Hα line vary widely among these objects.

The early-time ($t \approx 1$ week past maximum brightness) spectra of SNe are illustrated in Figure 1. [Unless otherwise noted, the optical spectra illustrated here were obtained by my group, primarily with the 3-m Shane reflector at Lick Observatory. When referring to phase of evolution, the variables t and τ denote time since *maximum brightness* (usually in the B passband) and time since *explosion*, respectively.] The lines are broad due to the high velocities of the ejecta, and most of

CP565, *Young Supernova Remnants: Eleventh Astrophysics Conf.,* edited by S. S. Holt and U. Hwang
© 2001 American Institute of Physics 0-7354-0001-6/01/$18.00

them have P-Cygni profiles formed by resonant scattering above the photosphere. SNe Ia are characterized by a deep absorption trough around 6150 Å produced by blueshifted Si II λ6355. Members of the Ib and Ic subclasses do not show this line. The presence of moderately strong optical He I lines, especially He I λ5876, distinguishes SNe Ib from SNe Ic.

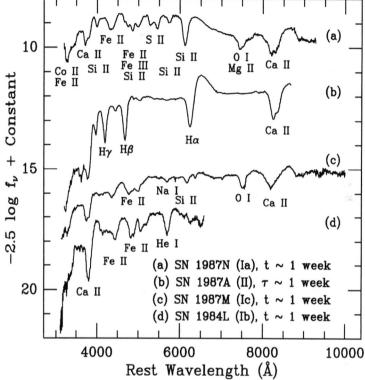

Figure 1: Early-time spectra of SNe, showing the main subtypes.

The late-time ($t \gtrsim 4$ months) optical spectra of SNe provide additional constraints on the classification scheme (Figure 2). SNe Ia show blends of dozens of Fe emission lines, mixed with some Co lines. SNe Ib and Ic, on the other hand, have relatively unblended emission lines of intermediate-mass elements such as O and Ca. At this phase, SNe II are dominated by the strong Hα emission line; in other respects, most of them spectroscopically resemble SNe Ib and Ic, but with narrower emission lines. The late-time spectra of SNe II show substantial heterogeneity, as do the early-time spectra.

To a first approximation, the light curves of SNe I are all broadly similar [2]. SNe Ib usually have slower decline rates than SNe Ic; however, SNe Ic may come in "slow" and "fast" varieties [3,4]. The light curves of SNe II exhibit much dispersion [5], though it is useful to subdivide the majority of them into two relatively distinct subclasses [6,7]. The light curves of SNe II-L ("linear") generally resemble those of SNe I, with a steep decline after maximum brightness followed by a slower exponential tail. In contrast, SNe II-P ("plateau") remain within ~ 1 mag of maximum brightness for an extended period. The light curve of SN 1987A, albeit atypical, was generically related to those of SNe II-P.

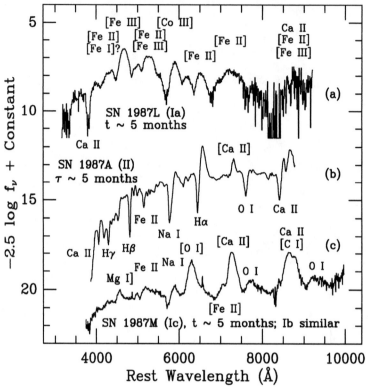

Figure 2: Late-time spectra of SNe. At even later phases, SN 1987A was dominated by strong emission lines of Hα, [O I], [Ca II], and the Ca II near-infrared triplet.

The locations at which SNe occur provide important clues to their nature, and to the mass of their progenitor stars. SNe II, Ib, and Ic have *never* been seen in elliptical galaxies, and rarely if ever in S0 galaxies. They are generally in or near spiral arms and H II regions [8], implying that their progenitors must have started their lives as massive stars ($\gtrsim 10\ M_\odot$). The progenitors of SNe II are thought to suffer core collapse and subsequently "rebound" with help from neutrinos [9,10], leaving a neutron star or perhaps in some cases a black hole [11]. Most workers now believe that SNe Ib/Ic are produced by the same mechanism as SNe II, except that the progenitors were stripped of their hydrogen (SN Ib) and possibly helium (SN Ic) envelopes prior to exploding, either via mass transfer to companion stars [12,13] or through winds (e.g., [14,15]). White dwarf models have been discussed [16] but are unlikely.

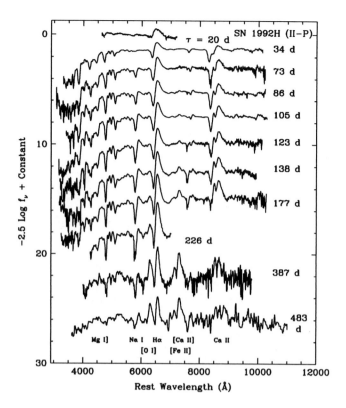

Figure 3: Montage of spectra of SN 1992H in NGC 5377. Epochs (days) are given relative to the estimated time of explosion, February 8, 1992.

SUBCLASSES OF TYPE II SUPERNOVAE

Most SNe II-P seem to have a relatively well-defined spectral development, as shown in Figure 3 for SN 1992H (see also reference [17]). At early times the spectrum is nearly featureless and very blue, indicating a high color temperature ($\gtrsim$ 10,000 K). He I λ5876 with a P-Cygni profile is sometimes visible. The temperature rapidly decreases with time, reaching $\sim$ 5000 K after a few weeks, as expected from the adiabatic expansion and associated cooling of the ejecta. It remains roughly constant at this value during the plateau (the photospheric phase), while the hydrogen recombination wave moves through the massive ($\sim$ 10 $M_\odot$) hydrogen ejecta and releases the energy deposited by the shock. At this stage strong Balmer lines and Ca II H&K with well-developed P-Cygni profiles appear, as do weaker lines of Fe II, Sc II, and other iron-group elements. The spectrum gradually takes on a nebular appearance as the light curve drops to the late-time tail; the continuum fades, but Hα becomes very strong, and prominent emission lines of [O I], [Ca II], and Ca II also appear.

Few SNe II-L have been observed in as much detail as SNe II-P. Figure 4 shows the spectral development of SN 1979C [18], an unusually luminous member of this subclass. Near maximum brightness the spectrum is very blue and almost featureless, with a slight hint of Hα emission. A week later, Hα emission is more easily discernible, and low-contrast P-Cygni profiles of Na I, Hβ, and Fe II have appeared. By $t \approx 1$ month, the Hα emission line is very strong but still devoid of an absorption component, while the other features clearly have P-Cygni profiles. Strong, broad Hα emission dominates the spectrum at $t \approx 7$ months, and [O I] $\lambda\lambda 6300$, 6364 emission is also present. Several authors [19–21] have speculated that the absence of Hα absorption spectroscopically differentiates SNe II-L from SNe II-P, but the small size of the sample of well-observed objects precluded definitive conclusions.

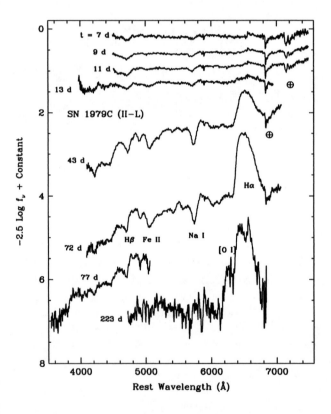

Figure 4: Montage of spectra of SN 1979C in NGC 4321, from reference [18]; reproduced with permission. Epochs (days) are given relative to the date of maximum brightness, April 15, 1979.

The progenitors of SNe II-L are generally believed to have relatively low-mass hydrogen envelopes (a few $M_\odot$); otherwise, they would exhibit distinct plateaus, as do SNe II-P. On the other hand, they may have more circumstellar gas than do SNe II-P, and this could give rise to the emission-line dominated spectra. They are often radio sources [22]; moreover, the ultraviolet excess (at $\lambda \lesssim 1600$ Å) seen in SNe 1979C and 1980K may be produced by inverse Compton scattering of photospheric radiation by high-speed electrons in shock-heated ($T \approx 10^9$ K) circumstellar material [23,24]. Finally, the light curves of some SNe II-L reveal an extra source of energy: after declining exponentially for several years, the Hα flux of SN 1980K reached a steady level, showing little if any decline thereafter [25,26]. The excess almost certainly comes from the kinetic energy of the ejecta being thermalized and radiated due to an interaction with circumstellar matter [27,28].

The very late-time optical recovery of SNe 1979C and 1980K [26,29,30] and other SNe II-L supports the idea of ejecta interacting with circumstellar material. The spectra consist of a few strong, broad emission lines such as Hα, [O I] $\lambda\lambda6300, 6364$, and [O III] $\lambda\lambda4959, 5007$. A *Hubble Space Telescope (HST)* ultraviolet spectrum of SN 1979C reveals some prominent, double-peaked emission lines with the blue peak substantially stronger than the red, suggesting dust extinction within the expanding ejecta [30]. The data show general agreement with the emission lines expected from circumstellar interaction [31], but the specific models that are available show several differences with the observations. For example, we find higher electron densities (10^5 to 10^7 cm^{-3}), resulting in stronger collisional de-excitation than assumed in the models. These differences can be used to further constrain the nature of the progenitor star. Note that based on photometry of the stellar populations in the environment of SN 1979C (from *HST* images), the progenitor of the SN was at most 10 million years years old, so its initial mass was probably 17–18 $M_\odot$ [32].

During the past decade, there has been the gradual emergence of a new, distinct subclass of SNe II [20,33,34,28] whose ejecta are believed to be *strongly* interacting with dense circumstellar gas, even at early times (unlike SNe II-L). The derived mass-loss rates for the progenitors can exceed $10^{-4} M_\odot$ yr^{-1} [35]. In these objects, the broad absorption components of all lines are weak or absent throughout their evolution. Instead, their spectra are dominated by strong emission lines, most notably Hα, having a complex but relatively narrow profile. Although the details differ among objects, Hα typically exhibits a very narrow component (FWHM $\lesssim 200$ km s^{-1}) superposed on a base of intermediate width (FWHM ≈ 1000–2000 km s^{-1}; sometimes a very broad component (FWHM ≈ 5000–10,000 km s^{-1}) is also present. This subclass was christened "Type IIn" [34], the "n" denoting "narrow" to emphasize the presence of the intermediate-width or very narrow emission components. Representative spectra of five SNe IIn are shown in Figure 5, with two epochs for SN 1994Y.

The early-time continua of SNe IIn tend to be bluer than normal. Occasionally He I emission lines are present in the first few spectra (e.g., SN 1994Y in Figure 5). Very narrow Balmer absorption lines are visible in the early-time spectra of some of these objects, often with corresponding Fe II, Ca II, O I, or Na I absorption as well (e.g., SNe 1994W and 1994ak in Figure 5). Some of them are unusually luminous at maximum brightness, and they generally fade quite slowly, at least at early times. The equivalent width of the intermediate Hα component can grow to astoundingly high values at late times. The great diversity in the observed characteristics of SNe IIn provides clues to the various degrees and forms of mass loss late in the lives of massive stars.

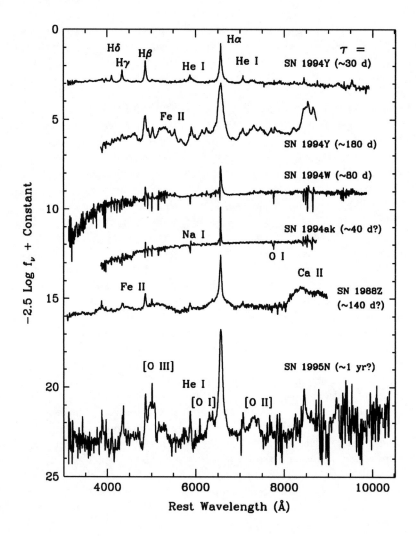

Figure 5: Montage of spectra of SNe IIn. Epochs are given relative to the estimated dates of explosion.

TYPE II SUPERNOVA IMPOSTORS?

The peculiar SN IIn 1961V ("Type V" according to Zwicky [36]) had probably the most bizarre light curve ever recorded. (SN 1954J, also known as "Variable 12" in NGC 2403, was similar [37].) Its progenitor was a very luminous star, visible in many photographs of the host galaxy (NGC 1058) prior to the explosion. Perhaps SN 1961V was not a genuine supernova (defined to be the violent destruction of a star at the end of its life), but rather the super-outburst of a luminous blue variable such as η Carinae [38,39].

A related object may have been SN IIn 1997bs, the first SN discovered in the Lick Observatory Supernova Search (LOSS, described later in this review). Its spectrum was peculiar (Figure 6), consisting of narrow Balmer and Fe II emission lines superposed on a featureless continuum. Its progenitor was discovered in an *HST* archival image of the host galaxy [40]. It is a very luminous star ($M_V \approx -7.4$ mag), and it didn't brighten as much as expected for a SN explosion ($M_V \approx -13$ at maximum). These data suggest that SN 1997bs may have been like SN 1961V — that is, a supernova impostor. The real test will be whether the star is still visible in future *HST* images obtained years after the outburst.

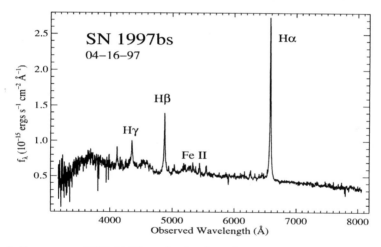

Figure 6: Spectrum of SN 1997bs, obtained on April 16, 1997 UT.

TYPE Ib AND Ic SUPERNOVAE

In the 1960s, Bertola and collaborators [41,42] recognized that not all SNe I are of the "classical" variety (now known as SNe Ia) with a strong absorption trough near 6150 Å. By the mid-1980s these objects came to be known as SNe Ib [43], and He I lines were identified in their early-time spectra [44]. Gradually it became clear that SNe Ib constitute a heterogeneous subclass, with substantial variations in the observed He I strengths in spectra obtained around maximum brightness. Wheeler & Harkness ([45]; see also [44]) suggested that SNe Ib should actually be divided into two separate categories: SNe Ib are those showing strong He I absorption lines (especially He I λ5876) in their early-time photospheric spectra, while SNe Ic are those in which He I is not easily discernible. However, they modeled SNe Ic in the same physical way as SNe Ib [46], but with different relative concentrations of He and O in the envelope.

A large, comprehensive study of SNe Ib and SNe Ic was recently completed by my group [47]. The relative depths of the helium absorption lines in the spectra of the SNe Ib appear to provide a measurement of the temporal evolution of the supernova, with He I λ5876 and He I λ7065 growing in strength relative to He I λ6678 over time. Some SNe Ic show evidence for weak He I absorption, but most do not. Aside from the presence or absence of the helium lines, there are other spectroscopic differences between SNe Ib and SNe Ic. On average, the O I λ7774 line is stronger in SNe Ic than in SNe Ib. In addition, the SNe Ic have distinctly broader emission lines at late times, indicating either a consistently larger explosion energy and/or a lower envelope mass for SNe Ic than for SNe Ib. These results are consistent with the idea that the progenitors of SNe Ic are massive stars that have lost more of their envelope (i.e., much of the helium layer) than the progenitors of SNe Ib. The general hypothesis that SNe Ib/Ic have "stripped" progenitors is greatly supported by the discovery of links between SNe II and SNe Ib/Ic, as I discuss next.

LINKS BETWEEN TYPE II AND TYPE Ib/Ic SUPERNOVAE

Filippenko [48] presented the case of SN 1987K, which appeared to be a link between SNe II and SNe Ib. Near maximum brightness, it was undoubtedly a SN II, but with rather weak photospheric Balmer and Ca II lines. Many months after maximum brightness, its spectrum was essentially that of a SN Ib. The simplest interpretation is that SN 1987K had a meager hydrogen atmosphere at the time it exploded; it would naturally masquerade as a SN II for a while, and as the expanding ejecta thinned out the spectrum would become dominated by emission from deeper and denser layers. The progenitor was probably a star that, prior to exploding via iron core collapse, lost almost all of its hydrogen envelope either through mass transfer onto a companion or as a result of stellar winds. Such SNe were dubbed "SNe IIb" by Woosley et al. [49], who had proposed a similar preliminary model for SN 1987A before it was known to have a massive hydrogen envelope.

The data for SN 1987K (especially its light curve) were rather sparse, making it difficult to model in detail. Fortunately, the Type II SN 1993J in NGC 3031 (M81) came to the rescue, and was studied in greater detail than any supernova since SN 1987A [50]. Its light curves [51] and spectra [52–55] amply supported the hypothesis that the progenitor of SN 1993J probably had a low-mass (0.1–0.6 $M_\odot$) hydrogen envelope above a $\sim 4\ M_\odot$ He core [56–58]. Figure 7 shows several early-time spectra of SN 1993J, showing the emergence of He I features typical of SNe Ib. Considerably later (Figure 8), the Hα emission nearly disappeared, and the spectral resemblance to SNe Ib was strong. The general consensus is that its initial mass was $\sim 15\ M_\odot$. A star of such low mass cannot shed nearly its entire hydrogen envelope without the assistance of a companion star. Thus, the progenitor of SN 1993J probably lost most of its hydrogen through mass transfer to a bound companion 3–20 AU away. In addition, part of the gas may have been lost from the system. Had the progenitor lost essentially *all* of its hydrogen prior to exploding, it would have had the optical characteristics of SNe Ib. There is now little doubt that most SNe Ib, and probably SNe Ic as well, result from core collapse in stripped, massive stars, rather than from the thermonuclear runaway of white dwarfs.

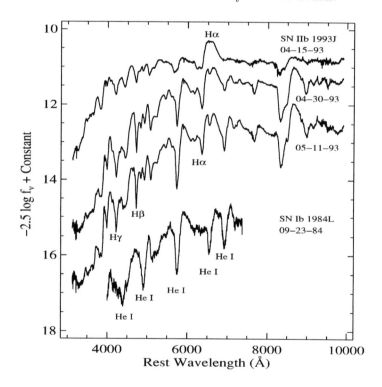

Figure 7: Early-time spectral evolution of SN 1993J. A comparison with the Type Ib SN 1984L is shown at bottom, demonstrating the presence of He I lines in SN 1993J. The explosion date was March 27.5, 1993.

SN 1993J held several more surprises. Observations at radio [59] and X-ray [60] wavelengths revealed that the ejecta are interacting with relatively dense circumstellar material [61], probably ejected from the system during the course of its pre-SN evolution. Optical evidence for this interaction also began emerging at $\tau \gtrsim 10$ months: the Hα emission line grew in relative prominence, and by $\tau \approx 14$ months it had become the dominant line in the spectrum [53,62,63], consistent with models [31]. Its profile was very broad (FWHM $\approx$ 17,000 km s^{-1}; Figure 8) and had a relatively flat top, but with prominent peaks and valleys whose likely origin is Rayleigh-Taylor instabilities in the cool, dense shell of gas behind the reverse shock [64]. Radio VLBI measurements show that the ejecta are circularly symmetric, but with significant emission asymmetries [65], possibly consistent with the asymmetric Hα profile seen in some of the spectra [53].

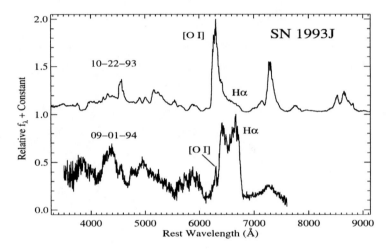

Figure 8: In the *top* spectrum, which shows SN 1993J about 7 months after the explosion, Hα emission is very weak; the resemblance to spectra of SNe Ib is striking. A year later (*bottom*), however, Hα was once again the dominant feature in the spectrum (which was scaled for display purposes).

SPECTROPOLARIMETRY OF SUPERNOVAE

Spectropolarimetry of SNe can be used to probe their geometry. The basic question is whether SNe are round, and the idea is simple: A hot young SN atmosphere is dominated by electron scattering, which by its nature is highly polarizing. For an unresolved, spherical source, however, the directional components of the electric vectors cancel exactly, yielding zero net linear polarization. Conversely, if the source is aspherical, incomplete cancellation occurs and a net linear polarization results, with typical continuum polarizations of $\sim 1\%$ expected for objects with moderate ($\sim 20\%$) asphericity (e.g., [66]), although the exact amount of polarization observed is also sensitive to the viewing angle. Spectropolarimetry is important for a full understanding of the physics of SN explosions and can also provide information on the circumstellar environment of SNe.

My group obtained spectropolarimetry of one object from each of the major SN types and subtypes [67], generally with the Keck-II 10-m telescope (but in a few important cases with the Lick 3-m Shane reflector). Most of the objects exhibit a change in both the magnitude and direction of the polarization across strong lines, especially the absorption troughs of the strongest P-Cygni lines. This may result from global asymmetry of the electron-scattering atmosphere and/or the underlying continuum region [68,66].

We have studied SN IIn 1998S in some detail [69] (see also [70]); its optical spectrum is dominated by strong, multi-component emission lines, thought to be produced by an intense interaction between the supernova and its dense circumstellar environment. We combined our early-time (3 days after discovery) spectropolarimetric observation with total flux spectra spanning nearly 500 days. We measure an intrinsic continuum polarization of $p \approx 3\%$ (one of the highest yet found for a SN), suggesting a global asphericity of $\gtrsim 45\%$ from the models of Höflich [66]. The line profiles favor a ring-like geometry for the circumstellar gas, generically similar to what is seen directly in SN 1987A, although much denser and closer to the progenitor in SN 1998S.

We also found that one month after exploding, the Type IIb SN 1993J had a polarized flux spectrum resembling spectra of SNe Ib, with prominent He I lines [71]. The data are consistent with models in which the polarization is produced by an asymmetric He core configuration of material. It is interesting that the percent polarization may increase with decreasing envelope mass, along the sequence Type II, IIb, Ib, and Ic [72,70,73], suggesting that asymmetries in massive stars become more pronounced as one probes deeper into the core.

SN 1999em, an extremely bright ($m_V \approx 13.5$) SN II-P, provided the rare opportunity to study the geometry of a "normal" core-collapse event at multiple epochs [74]. We obtained spectropolarimetry at 3 plateau-phase epochs, and then one final epoch long after it had dropped off the plateau (Fig. 9). A very low but temporally increasing polarization level suggests a substantially spherical geometry at early times that becomes more aspherical at late times as ever-deeper layers of the ejecta are revealed. We speculate that the thick hydrogen envelope intact at the time of explosion in SNe II-P might serve to dampen the effects of an intrinsically aspherical explosion. The increase in asphericity seen at later times is consistent with the trend identified above among stripped-envelope core-collapse SNe: the deeper we peer, the more evidence we find for asphericity. The natural conclusion that it is *explosion* asymmetry that is responsible for the polarization has fueled the idea that some core-collapse SNe produce gamma-ray bursts (GRB; e.g., [75]) through the action of a jet of material aimed fortuitously at the observer, the result of a "bipolar," jet-induced, SN explosion [76,72].

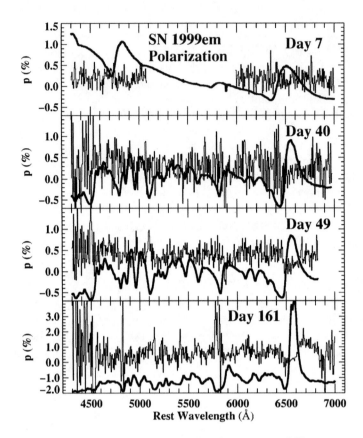

Figure 9: Montage of the observed optical polarization of SN 1999em, with arbitrarily scaled total flux spectra *(thick lines)* overplotted for comparison of features. The polarization spectra are binned 5 Å bin^{-1} to improve the S/N ratio. See reference [73] for details.

SUPERNOVAE ASSOCIATED WITH GAMMA-RAY BURSTS?

At least a small fraction of gamma-ray bursts (GRBs) may be associated with nearby SNe. Probably the most compelling example thus far is that of SN 1998bw and GRB 980425 (e.g., [77–80]), which were temporally and spatially coincident. SN 1998bw was, in many ways, an extraordinary SN; it was very luminous at optical and radio wavelengths, and it showed evidence for relativistic outflow. Its bizarre optical spectrum is often classified as that of a SN Ic, but the object should be called a "peculiar SN Ic" if not a subclass of its own; the spectrum was distinctly different from that of a normal SN Ic.

Models suggest that SNe associated with GRBs are highly asymmetric; thus, spectropolarimetry should provide some useful tests. In particular, perhaps objects such as SN 1998S, mentioned above, would have been seen as GRBs had their rotation axis been pointed in our direction. That of SN 1998S was almost certainly *not* aligned with us [69]; both the spectropolarimetry and the appearance of double-peaked Hα emission suggest an inclined view, rather than pole-on.

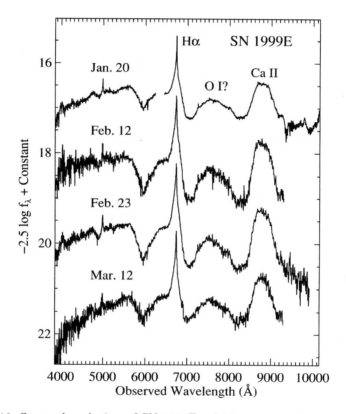

Figure 10: Spectral evolution of SN 1999E, which may have been associated with GRB 980910.

The case of GRB 970514 and the very luminous SN IIn 1997cy is also interesting [81,82]; there is a reasonable possibility that the two objects were associated. The optical spectrum of SN 1997cy was highly unusual, and bore some resemblance to that of SN 1998bw, though there were some differences as well. SN 1999E, which might be linked with GRB 980910 but with large uncertainties [83], also had an optical spectrum similar to that of SN 1997cy [84,85]; see Figure 10. The undulations are very broad, indicating high ejection velocities. Besides Hα, secure line identifications are difficult, though some of the emission features seem to be associated with oxygen and calcium. Perhaps SN 1999E was produced by the highly asymmetric collapse of a carbon-oxygen core.

SUPERNOVA RATES

I conclude with a brief summary of supernova rates. The derived rates of various types of SNe as a function of Hubble type are still quite uncertain; no single search has found a sufficiently large sample of SNe, and combining different searches can be misleading due to differences in the dominant selection effects. Probably the best existing determination of supernova rates is that of Capellaro et al. [86], who combined the results of five searches while being attentive to various selection effects. Four of the searches had been done with photographic plates, while the fifth was a visual search by the Rev. Robert Evans. Given that photographic searches can be rather insensitive to SNe near the central regions of galaxies, and that the Evans visual search did not go very deep, we must avoid the temptation to take the overall results (see table below) too literally. Nevertheless, they can be used as a rough guideline.

$$\text{SNu } [\#(100 \text{ yr})^{-1}(10^{10}L_\odot^B)^{-1}]$$

Galaxy type	Ia	Ib/c	II	All
E-S0	0.18 ± 0.06	< 0.01	< 0.02	0.18 ± 0.06
S0a-Sb	0.18 ± 0.07	0.11 ± 0.06	0.42 ± 0.19	0.72 ± 0.21
Sbc-Sd	0.21 ± 0.08	0.14 ± 0.07	0.86 ± 0.35	1.21 ± 0.37
Others*	0.40 ± 0.16	0.22 ± 0.16	0.65 ± 0.39	1.26 ± 0.45
All	0.20 ± 0.06	0.08 ± 0.04	0.40 ± 0.19	0.68 ± 0.20

Note: $H_0 = 75$ km s^{-1} Mpc^{-1}; scale by $(H_0/75)^2$.
*Others include Sm, Irr, Pec.

My group is trying to remedy the situation by conducting a long-term search for nearby SNe (with redshifts generally less than 5000 km s^{-1}) in a uniform manner [87,88]. Special emphasis is placed on finding them well before maximum brightness. We are using the Katzman Automatic Imaging Telescope (KAIT) at Lick Observatory, a fully robotic 0.75-m reflector equipped with a CCD imaging camera. Its telescope control system checks the weather, opens the dome, points to the desired objects, finds and acquires guide stars (only for long integrations), exposes, stores the data, and manipulates the data without human intervention. There is a filter wheel with 20 slots, including UBVRI. Five-minute guided exposures yield $R \approx 20$ mag.

A limit of about 19 mag (4σ) is reached in the 25-second unfiltered, unguided exposures of our Lick Observatory Supernova Search (LOSS). We observe up to 1200 galaxies per night, and try to cycle back to the same galaxies after 3–4 nights. Our software automatically subtracts new images from old ones and identifies supernova candidates that are subsequently examined by undergraduate research assistants. LOSS found 20 SNe in 1998, 40 in 1999, and 36 in 2000, making KAIT the world's most successful search engine for nearby SNe. (Note that we found SN 2000A and SN 2001A — and hence the first supernova of the new millennium regardless of one's definition of the turn of the millennium!) After a few more years, we hope to have found enough SNe to provide a meaningful update on the supernova rates given above.

Multi-filter follow-up photometry is conducted of the most important SNe, and all objects are monitored in unfiltered mode. A Web page describing LOSS is at http://astro.berkeley.edu/~bait/kait.html . As a byproduct of LOSS, we also find novae in the Local Group, comets, asteroids, and cataclysmic variables.

ACKNOWLEDGMENTS

My recent research on SNe has been financed by the NSF, most recently through grant AST-9987438, as well as by NASA grants GO-7821, GO-8243, GO-8648, AR-6371, and AR-8006 from the Space Telescope Science Institute, which is operated by AURA, Inc., under NASA Contract NAS5-26555. Additional funding was provided by NASA/Chandra grant GO-0-1009C. KAIT and its associated science have been made possible with funding or donations from NSF, NASA, the Sylvia and Jim Katzman Foundation, Sun Microsystems Inc., Lick Observatory, the Hewlett-Packard Company, Photometrics Ltd., AutoScope Corporation, and the University of California. I am grateful to the students and postdocs who have worked with me on SNe over the past 15 years for their assistance and discussions.

REFERENCES

1. Filippenko, A. V., *ARAA*, **35**, 309 (1997).
2. Leibundgut, B., Tammann, G. A., Cadonau, R., & Cerrito, D., *A&AS*, **89**, 537 (1991a).
3. Clocchiatti, A., et al., *ApJ*, **483**, 675 (1997).
4. Clocchiatti, A., & Wheeler, J. C., *ApJ*, **491**, 375 (1997).
5. Patat, F., Barbon, R., Cappellaro, E., & Turatto, M., *A&AS*, **98**, 443 (1993).
6. Barbon, R., Ciatti, F., & Rosino, L., *A&A*, **72**, 287 (1979).
7. Doggett, J. B., & Branch, D., *AJ*, **90**, 2303 (1985).
8. Van Dyk, S. D., Hamuy, M., & Filippenko, A. V., *AJ*, **111**, 2017 (1996).
9. Arnett, W. D., Bahcall, J. N., Kirshner, R. P., & Woosley, S. E., *ARAA*, **27**, 629 (1989).
10. Burrows, A., & Young, T., *Phys Rep*, **333**, 63 (2000).
11. Brown, G. E., & Bethe, H. A., *ApJ*, **423**, 659 (1994).
12. Nomoto, K., et al., *Nature*, **371**, 227 (1994).
13. Woosley, S. E., Langer, N., & Weaver, T. A., *ApJ*, **448**, 315 (1995).

14. Woosley, S. E., Langer, N., & Weaver, T. A., *ApJ*, **411**, 823 (1993).

15. Swartz, D. A., Filippenko, A. V., Nomoto, K., & Wheeler, J. C., *ApJ*, **411**, 313 (1993).

16. Branch, D., & Nomoto, K., *A&A*, **164**, L13 (1986).

17. Clocchiatti, A., et al., *AJ*, **111**, 1286 (1996).

18. Branch, D., Falk, S. W., McCall, M. L., Rybski, P., Uomoto, A. K., & Wills, B. J., *ApJ*, **224**, 780 (1981).

19. Wheeler, J. C., & Harkness, R. P., *Rep. Prog. Phys.*, **53**, 1467 (1990).

20. Filippenko, A. V., in *Supernovae*, ed. S. E. Woosley (New York: Springer), 467 (1991a).

21. Schlegel, E. M., *AJ*, **111**, 1660 (1996).

22. Sramek, R. A., & Weiler, K. W., in *Supernovae*, ed. A. G. Petschek (New York: Springer), p. 76 (1990).

23. Fransson, C., *A&A*, **111**, 140 (1982).

24. Fransson, C., *A&A*, **133**, 264 (1984).

25. Uomoto, A., & Kirshner, R. P., *ApJ*, **308**, 685 (1986).

26. Leibundgut, B., et al., *ApJ*, **372**, 531 (1991b).

27. Chevalier, R. A., in *Supernovae*, ed. A. G. Petschek (New York: Springer), p. 91 (1990).

28. Leibundgut, B., in *Circumstellar Media in the Late Stages of Stellar Evolution*, eds. R. E. S. Clegg, I. R. Stevens, & W. P. S. Meikle (Cambridge: Cambridge Univ. Press), p. 100 (1994).

29. Fesen, R. A., Hurford, A. P., & Matonick, D. M., *AJ*, **109**, 2608 (1995).

30. Fesen, R. A., et al., *AJ*, **117**, 725 (1999).

31. Chevalier, R. A., & Fransson, C., *ApJ*, **420**, 268 (1994).

32. Van Dyk, S. D., et al., *PASP*, **111**, 315 (1999).

33. Filippenko, A. V., in *Supernova 1987A and Other Supernovae*, eds. I. J. Danziger & K. Kjär (Garching: ESO), p. 343 (1991b).

34. Schlegel, E. M., *MNRAS*, **244**, 269 (1990).

35. Chugai, N. N., in *Circumstellar Media in the Late Stages of Stellar Evolution*, eds. R. E. S. Clegg, I. R. Stevens, & W. P. S. Meikle (Cambridge: Cambridge Univ. Press), p. 148 (1994).

36. Zwicky, F, in *Stars and Stellar Systems*, Vol. 8, ed. L. H. Aller & D. B. McLaughlin, pp. 367. Chicago: Univ. Chicago Press (1965).

37. Humphreys, R. M., & Davidson, K., *PASP*, **106**, 1025 (1994).

38. Goodrich, R. W., Stringfellow, G. S., Penrod, G. D., & Filippenko, A. V., *ApJ*, **342**, 908 (1989).

39. Filippenko, A. V., et al., *AJ*, **110**, 2261 (1995) [Erratum: **112**, 806].

40. Van Dyk, S. D., Peng, C. Y., King, J. Y., Filippenko, A. V., Treffers, R. R., Li, W., & Richmond, M. W., *PASP*, **112**, 1532 (2000).

41. Bertola, F., *Ann Ap*, **27**, 319 (1964).

42. Bertola, F., Mammano, A., & Perinotto, M., *Asiago Contr*, **174**, 51 (1965).

43. Elias, J. H., Matthews, K., Neugebauer, G., & Persson, S. E., *ApJ*, **296**, 379 (1985).

44. Harkness, R. P., et al., *ApJ*, **317**, 355 (1987).

45. Wheeler, J. C., & Harkness, R. P., 1986, in *Galaxy Distances and Deviations from*

Universal Expansion, ed. B. F. Madore & R. B. Tully (Dordrecht: Reidel), p. 45.

46. Wheeler, J. C., et al., *ApJ*, **313**, L69 (1987).

47. Matheson, T., Filippenko, A. V., Li, W., Leonard, D. C., & Shields, J. C., *AJ*, in press (2001).

48. Filippenko, A. V., *AJ*, **96**, 1941 (1988).

49. Woosley, S. E., Pinto, P. A., Martin, P. G., & Weaver, T. A., *ApJ*, **318**, 664 (1987).

50. Wheeler, J. C., & Filippenko, A. V., in *Supernovae and Supernova Remnants*, eds. R. McCray & Z. Wang (Cambridge: Cambridge Univ. Press), p. 241 (1996).

51. Richmond, M. W., Treffers, R. R., Filippenko, A. V., & Paik, Y., *AJ*, **112**, 732 (1996).

52. Filippenko, A. V., Matheson, T., & Ho, L. C., *ApJ*, **415**, L103 (1993).

53. Filippenko, A. V., Matheson, T., & Barth, A. J., *AJ*, **108**, 2220 (1994).

54. Matheson, T., et al., *AJ*, **120**, 1487 (2000a).

55. Matheson, T., et al., *AJ*, **120**, 1499 (2000b).

56. Nomoto, K., Suzuki, T., Shigeyama, T., Kumagai, S., Yamaoka, H., & Saio, H., *Nature*, **364**, 507 (1993).

57. Podsiadlowski, P., Hsu, J. J. L., Joss, P. C., & Ross, R. R., *Nature*, **364**, 509 (1993).

58. Woosley, S. E., Eastman, R. G., Weaver, T. A., & Pinto, P. A., *ApJ*, **429**, 300 (1994).

59. Van Dyk, S. D., Weiler, K. W., Sramek, R. A., Rupen, M. P., & Panagia, N., *ApJ*, **432**, L115 (1994).

60. Suzuki, T., & Nomoto, K., *ApJ*, **455**, 658 (1995).

61. Fransson, C., Lundqvist, P., & Chevalier, R. A., *ApJ*, **461**, 993 (1996).

62. Patat, F., Chugai, N., & Mazzali, P. A., *A&A*, **299**, 715 (1995).

63. Finn, R. A., Fesen, R. A., Darling, G. W., & Thorstensen, J. R., *AJ*, **110**, 300 (1995).

64. Chevalier, R. A., Blondin, J. M., & Emmering, R. T., *ApJ*, **392**, 118 (1992).

65. Marcaide, J. M., et al., *Science*, **270**, 1475 (1995).

66. Höflich, P., *A&A*, **246**, 481 (1991).

67. Leonard, D. C., Filippenko, A. V., & Matheson, T., in *Cosmic Explosions*, ed. S. S. Holt & W. W. Zhang (New York: AIP), p. 165 (2000a).

68. McCall, M. L., *MNRAS*, **210**, 829 (1984).

69. Leonard, D. C., Filippenko, A. V., Barth, A. J., & Matheson, T., *ApJ*, **536**, 239 (2000b).

70. Wang, L., Howell, D. A., Höflich, P., & Wheeler, J. C., *ApJ*, in press (2001).

71. Tran, H. D., et al., *PASP*, **109**, 489 (1997).

72. Wheeler, J. C., in *Cosmic Explosions*, ed. S. S. Holt & W. W. Zhang (New York: AIP), p. 445 (2000).

73. Leonard, D. C., et al., in preparation (2001b).

74. Leonard, D. C., Filippenko, A. V., Ardila, D. A., & Brotherton, M. S., *ApJ*, in press (2001a).

75. Bloom, J. S., et al., *Nature*, **401**, 453 (1999).

76. Khokhlov, A. M., et al., *ApJ*, **524**, L107 (1999).

77. Galama, T. J., et al., *Nature*, **395**, 670 (1998).

78. Iwamoto, K., et al., *Nature*, **395**, 672 (1998).

79. Woosley, S. E., Eastman, R. G., & Schmidt, B. P., *ApJ*, **516**, 788 (1999).

80. Stathakis, R. A., et al., *MNRAS*, **314**, 807 (2000).

81. Germany, L. M., Reiss, D. J., Sadler, E. M., & Schmidt, B. P., *ApJ*, **533**, 320 (2000).

82. Turatto, M., et al., *ApJ*, **534**, L57 (2000).

83. Thorsett, S. E., & Hogg, D. W., *GCN Circ.* 197 (1999).

84. Filippenko, A. V., Leonard, D. C., & Riess, A. G., *IAU Circ.* 7091 (1999).

85. Cappellaro, E., Turatto, M., & Mazzali, P., *IAU Circ.* 7091 (1999a).

86. Cappellaro, E., Evans, R., & Turatto, M., *A&A*, **351**, 459 (1999b).

87. Li, W. D., et al., in *Cosmic Explosions*, ed. S. S. Holt & W. W. Zhang (New York: AIP), p. 103 (2000).

88. Filippenko, A. V., et al., in preparation (2001).

Hydrodynamics of Young Supernova Remnants

John M. Blondin

North Carolina State University, Raleigh, NC 27695, USA

Abstract.
    The structure and evolution of young supernova remnants are determined, to a large extent, by using the equations of ideal, compressible gas dynamics to evolve a hypersonically expanding ejecta propagating into an ambient gas. Even within this limited domain of ideal gas dynamics, this simple problem can become extremely complex. Here we review some of the hydrodynamic solutions describing young supernova remnants, from one-dimensional self-similar models to three-dimensional simulations incorporating nonuniform ejecta. An online version of the conference presentation is available at
                http://wonka.physics.ncsu.edu/~blondin/october2000

INTRODUCTION

A detailed understanding of the dynamics of supernova remnants (SNRs) is essential for progress in interpreting new observations. Out of necessity, one typically uses the simplest dynamical model that can successfully explain the data, and astrophysicists have relied heavily on two self-similar, spherically symmetric solutions: the self-similar driven wave (SSDW) describing the early two-shock phase of SNR evolution [1,2], and the Sedov-Taylor blastwave [3] describing the later single shock wave evolution. Not unexpectedly, these simple dynamical models rarely match a given SNR. One must therefore develop more sophisticated dynamical models, both to understand the limitations of the simple models and to provide more detailed models capable of matching the observations.

Examples of when self-similar spherically-symmetric models will not work, forcing one to use more sophisticated multidimensional hydrodynamic models, include the presence of circumstellar features like shells and rings that break the scale-free character of the problem, hydrodynamic instabilities that alter the structure of the remnant, the transition epochs between different self-similar phases, and non-uniformity in either the ejecta or the circumstellar medium (CSM). The challenge is to understand which of these processes affect the observational properties of SNRs.

In keeping with the focus on young supernova remnants, this review will focus primarily on the ejecta-dominated phase of the evolution of SNRs. We begin with a

CP565, *Young Supernova Remnants: Eleventh Astrophysics Conf.*, edited by S. S. Holt and U. Hwang

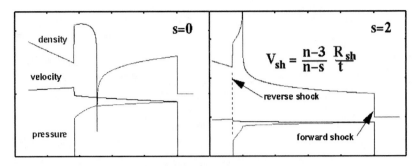

FIGURE 1. Radial profiles for the SSDW solutions corresponding to $s = 0$ (a uniform ambient density) and $s = 2$ (a steady progenitor wind), both with $n = 7$.

discussion of one-dimensional models, building from the SSDW model. From those humble beginnings we will explore the hydrodynamics of young SNRs as various assumptions are removed, with an emphasis on the growing level of complexity as we approach more realistic dynamical models.

SPHERICALLY SYMMETRIC MODELS

The basic dynamics of young SNRs is relatively simple; cold and dense supernova (SN) ejecta expanding with highly supersonic velocities, $v = r/t$ into an ambient medium. A shockwave is driven out in front of the expanding ejecta. As this shockwave sweeps up ambient gas it slows down, and before long it has slowed down to velocities lower than the fastest ejecta, such that the ejecta is running into the backside of the shockwave. However, because the ejecta is very cold (from rapid adiabatic spherical expansion over many orders of magnitude in radius) its velocity relative to the forward shock is supersonic, and a reverse shock is formed to decelerate the ejecta.

Chevalier [1] pointed out that if the expanding SN ejecta is described by a power-law in radius, $\rho_{sn} = At^{-3}v^{-n}$, and the SNR is propagating into an ambient medium that can also be described by a powerlaw (e.g. an exponent of zero for a constant density), there is no natural length scale in the problem and one can find a self-similar solution to describe the propagation of the forward/reverse shock pair. Two example solutions of this SSDW are shown in Figure 1.

Dwarkadas & Chevalier [4] present an alternative solution in which the SN ejecta is described by an exponential distribution in radius, $\rho_{sn} = At^{-3}\exp(-v/v_o)$. Based on radial profiles of ejecta calculated from hydrodynamic simulations of SN explosions, they argue that an exponential profile is a better fit than a power law for Type Ia SNe. The radial profiles of the interaction region are qualitatively similar to the SSDW solutions, but with the key difference that the exponential solution contains a characteristic length scale.

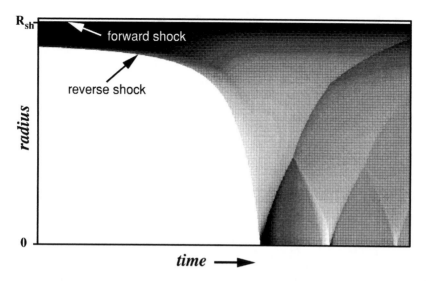

FIGURE 2. A space time diagram showing the crash of the reverse shock as illustrated with the exponential ejecta solution. The grey scale depicts gas pressure, with black the maximum and white the minimum. In the early two-shock phase the high pressure is confined to a small region near the forward shock. At late times the pressure profile is similar to the Sedov-Taylor solution, with a high pressure at the shock front tapering off to a constant interior value.

The exponential solution is thus not self-similar, and gradually evolves from something that resembles the SSDW model into something that resembles a Sedov-Taylor blastwave. This evolution is shown in Figure 2. As the density at the reverse shock decreases, so too does the ram pressure driving the shock interaction region outwards. Eventually the ejecta density and corresponding ram pressure drop so quickly that the reverse shock crashes into the center of the SNR. Following the arrival of the reverse shock at the center of the SNR, pressure waves propagate back and forth until the solution gradually settles into the self-similar Sedov-Taylor blastwave.

The same process must eventually occur in the SSDW model, once the swept up mass is greater than the mass ejected in the SN explosion. The powerlaw distribution of ejecta must in fact be truncated at some small radius in order to avoid an infinite ejecta mass. One typically defines some ejecta velocity, inside of which the ejecta density is constant (in radius, not time). This velocity is a function of the ejecta mass, powerlaw exponent, and SN energy:

$$v_e = \left(\frac{10n - 50}{3n - 9} \frac{E_{sn}}{M_{ej}} \right)^{1/2} \qquad (1)$$

When the edge of this plateau region reaches the reverse shock of the SSDW solution, the self-similarity is lost and the reverse shock crashes into the interior, much like the exponential model. Truelove & McKee [5] studied this transition between the ejecta-dominated two-shock phase and the Sedov-Taylor phase in one dimension, presenting useful analytic approximations to match the two phases. Note that for typical parameters, this event will occur around $\sim 10^3$ years.

Running out of ejecta is not the only possible fate for the two-shock model of the ejecta-dominated phase of a SNR. This simple picture may be broken if the SNR hits something; The forward shock may run into a circumstellar shell left behind by changes in the mass loss of the progenitor star, as in models for Cas A [6], or the reverse shock may be impacted by an expanding pulsar nebula.

SMOOTH, BUT NOT SYMMETRIC

Even if the ejecta and CSM are spherically symmetric, 1D models may be insufficient to describe SNRs due to the presence of hydrodynamic instabilities. While the Sedov-Taylor solution describing older remnants is dynamically stable, Gull [7] showed that young SNRs can be subject to convective instability. Chevalier, Blondin & Emmering [8] analyzed the stability of SSDWs with analytic and numerical techniques. An example of this instability, for a SSDW with $s = 2$, is shown in Figure 3. Here the layer of dense shocked ejecta is decelerated by the lower density shocked ambient gas (see Fig. 1): the classical Rayleigh-Taylor instability. Provided there are significant inhomogeneities to seed this instability, the growth is expected to saturate relatively quickly (after several doublings of the SNR radius), with dense Rayleigh-Taylor fingers reaching roughly half the distance to the shock front. The instability is slightly different for $s = 0$, but the end result of mixing shocked ejecta and CSM is the same. Wang & Chevalier [9] studied this same phenomena in the exponential ejecta solution.

Alas, the real world is in fact three-dimensional, so how good are the myriad of 2D simulations reported in the literature? Jun & Norman [10] reported on 3D simulations of the SSDW and found relatively little difference between 2D and 3D. Hinkle (these proceedings) has presented a systematic investigation of the SSDW in 3D.

In addition to the convective, or Rayleigh-Taylor instabilities in SSDWs, one also finds many instances of impulsive, or Richtmeyer-Meshkov instabilities in young SNRs. These might arise, for example, when the forward blastwave strikes a shell of swept-up circumstellar gas, or when the reverse shock impacts a dense shell swept up by a pulsar wind. Borkowski, Blondin & Sarazin [11] modeled Kepler's SNR in terms of a blastwave striking a circumstellar shell. Their 2D simulations clearly show the instability of the inside edge of the circumstellar shell after it has been impacted by the blastwave.

Another example of the shock/shell interaction is shown in Figure 4, where we show some preliminary results from 2D simulations of the interaction between a reverse shock and a pulsar nebula.

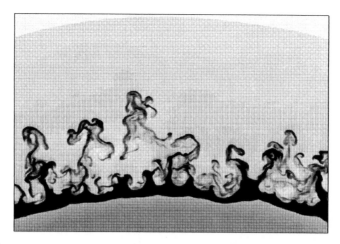

FIGURE 3. The end result of the convective instability in a SSDW with $n = 7$ and $s = 2$. This image shows the gas density (black is high density) from a 2D simulation on a grid of 1000^2 zones. The curved surface separating white from grey at the top of the plot is the forward shock, propagating upwards. The black regions represent shocked ejecta, including the fingers generated by the Rayleigh-Taylor instability. The surface separating black from grey along the bottom of the plot is the reverse shock.

FIGURE 4. The result of a 2D simulation illustrating the crash of the reverse shock into a pulsar nebula. The long fingers of dense gas (black) are produced by a combination of the convective instability of the SSDW and the Richtmeyer-Meshkov instability from the reverse shock impacting a shell swept up by the pulsar wind.

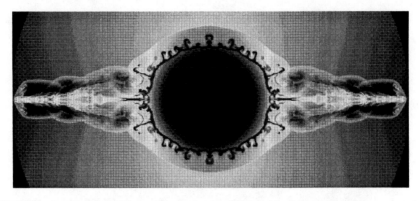

FIGURE 5. A relatively mild asymmetry in the CSM can lead to extreme asymmetry in the SNR. Here the equatorial region of the circumstellar medium was 5 times denser than the poles. (Don't try to make sense of the grey scale!)

ASYMMETRIC EJECTA/CSM

So far we have maintained the assumption that both the ejecta and the ambient gas are smooth and spherically symmetric. The weakest break from this symmetry is to assume a smooth CSM but with some gradual global variation. For example, a stellar wind with a polar variation in the wind density or asymmetric ejecta propagating out through a spherical star [12]. Blondin, Lundqvist & Chevalier [13] explored the effects of an asymmetry on the shape of young SNRs following the SSDW model. For the most part the asymmetry of the SNR followed the asymmetry of the CSM in that one could estimate the shape of the SNR by comparing the radii of the pole and equator as if they evolved independently. However, for relatively mild asymmetries that were sufficiently peaked at the pole, they found a very nonlinear effect that produced such large asymmetries some people might start using the word "jet" (see Figure 5).

Alternatively, the asymmetry may be local rather than global. There is in fact relatively little evidence for assuming a smooth CSM or ejecta. Instabilities are expected both in the fast winds from hot stars and in the slow winds from cool red supergiant progenitors. Jun, Jones & Norman [14] considered small cloudlets in the CSM. They found that these small-scale (compared to the radius of the remnant) anisotropies enhanced the vorticity within the SSDW, and could effectively lengthen the long fingers produced by the convective instability discussed earlier. It was not clear, however, if the fingers of dense, shocked ejecta could reach all the way to the shock front. Furthermore, this simulation was two-dimensional, making these small infinite cylinders rather than small clouds.

Jun & Jones [15] modeled the hydrodynamic interaction of a larger cloud, one with a radius comparable to the separation between the forward and reverse shocks. As expected, this had a more dramatic effect on the SNR, but this was a single

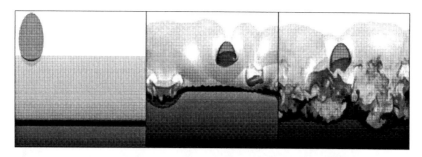

FIGURE 6. Propagation of a SSDW through a cloudy CSM. The first frame (left) shows a spherical SSDW beginning to overrun a cloud. The dark band near the bottom is the layer of dense shocked ejecta. The middle frame is from a short time later, after a few clouds have passed through. The last frame (right) is from a late time in the simulation, after dozens of clouds have been swept up. Note the partially shocked cloud in the middle of this image.

cloud and it was still restricted to two dimensions. In Figure 6 we show three frames from a 3D simulation of many small clouds. The clouds have a density contrast of 10, and a volume filling fraction of ~ 0.1. The general structure of the SSDW is still present, although the width of the interaction region has grown by $\sim 50\%$ and both the forward and reverse shocks are significantly distorted.

There is also little reason to believe that the ejecta is smooth. The presence of large-scale mixing of ejecta in core-collapse supernovae has been well established on both observational and theoretical grounds. The early, unexpected emergence of X-rays and γ-rays shortly after the explosion of SN 1987A provided dramatic evidence of mixing of radioactive Ni throughout its He core and its H-rich envelope. Extensive mixing of Ni is also required to explain the light curve and spectral observations of SN 1987A (see [16] for a review of SN 1987A). From studies of light curves and spectra of numerous core-collapse SNe, it is now known that such large-scale mixing is nearly always present.

Large-scale turbulence generated during the explosion appears not to be sufficient to mix SN ejecta completely, but only macroscopically. In Cas A, optical observations revealed the presence of ejecta with very different chemical abundances, such as O-, S-, Ar-, and Ca-rich ejecta knots, which were macroscopically mixed during the SN explosion as evidenced by the lack of spatial stratification expected in the absence of mixing [17]. The most recent X-ray observations of Cas A provide further evidence for this macroscopic but not microscopic mixing[18].

In SN 1987A, the inhomogeneities take a particularly interesting form: Fe-Ni bubbles [19], inferred from observations of Fe, Co, and Ni lines. Ni-rich ejecta clumps, transported outwards by turbulent motions, are heated by their own radioactive energy input, and expand in the ambient substrate of other heavy elements, forming low-density Fe bubbles. The SN structure is then expected to be like a Swiss cheese, with Fe bubbles occupying a substantial (~ 0.5) fraction of the ejecta volume [19].

Blondin, Borkowski & Reynolds [20] presented 3D hydrodynamic simulations

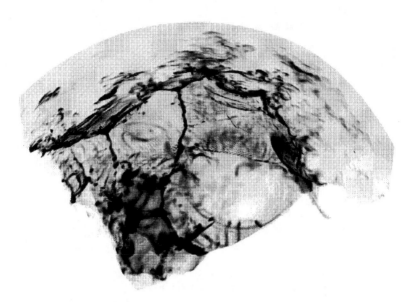

FIGURE 7. A volume rendering of the shocked gas from the 3D simulation of a SSDW with FeNi bubbles in the ejecta.

illustrating the dynamical effects of this FeNi bubble scenario. They modeled the SN ejecta with a density power law with $n = 9$, but filled with low density bubbles occupying a volume filling fraction of roughly 0.5. They find that the presence of FeNi bubbles leads to (1) vigorous turbulence and mixing of Fe with other heavy elements and with the ambient normal-abundance gas, (2) a ten-fold increase in the turbulent energy density within the interaction region, and (3) a strong clumping of the shocked ejecta, with some narrow filaments and clumps moving with radial velocities larger than the velocity of the forward shock. To get a sense of how FeNi bubbles effect the morphology of a SNR, we show a volume rendering of the gas density in Figure 7.

Perhaps the most dramatic consequence of FeNi bubbles is the severe deformation of the reverse shock. Any deviation from a spherical shock will decrease the shock velocity and hence the postshock temperature. Furthermore, as the reverse shock moves backward through a bubble (in the expanding frame), the shock velocity can be higher than in the spherical case, resulting in higher postshock temperatures (but with small emission measure because of the low density).

To estimate these effects without undertaking the complicated effort of calculating X-ray spectra, we have summed up the emission measure (EM) for each zone in the simulation, and plotted this as a function of temperature and radial velocity (scaled to the postshock temperature and shock velocity). (This is only a very approximate procedure as we neglected variations in the mean molecular weight between Fe bubbles and the ambient ejecta. For detailed comparisons with X-ray

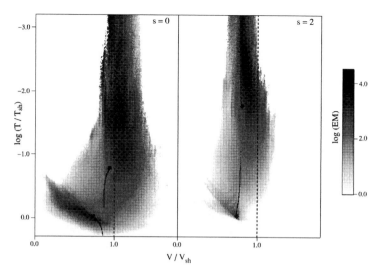

FIGURE 8. A map of the emission measure as a function of temperature and radial velocity for 3D simulations with FeNi bubbles in the ejecta. The temperature is scaled to the postshock temperature. The velocity is scaled to the velocity of the forward shock. The black lines map out the region of emission for the corresponding spherical SSDW. The stars mark the forward and reverse shocks.

observations one also needs to consider electron temperature which is generally lower than the mean gas temperature discussed here.)

The results are shown in Figure 8. Superimposed on these plots are the curves corresponding to emission from spherically symmetric SSDWs. In both cases these curves are composed of two pieces, one for the shocked ejecta and one for the shocked ambient gas. Shocked gas immediately behind the forward shock is located at $\log(T) = 0$ and $V = 0.75$. For the $s = 2$ case the gas temperature in the SSDW decreases with increasing distance from the forward shock, while the opposite holds for $s = 0$. Shocked ejecta immediately behind the reverse shock are located at $\log(T) = -1.76$ and $V = 0.82$ and decrease in temperature away from the shock in the case of $s = 2$. For $s = 0$, the shocked ejecta start off at $\log(T) = -0.75$ and $V = 0.95$ and increase in temperature away from the shock.

In both cases X-ray emitting gas is significantly spread out in temperature. This is most dramatic in the $s = 0$ case, for which there is no cool gas in the spherical solution. There is also a large spread in the radial velocity of the X-ray emitting gas, particularly for the case of $s = 0$. In both simulations the radial velocity of the shocked ambient gas is spread to lower velocities as a result of protrusions in the forward shock, while the radial velocity of the shocked ejecta is spread to higher velocities because the ejecta are not fully decelerated by the often oblique reverse shock. In the case of $s = 0$, the bulk of the X-ray emitting ejecta is actually traveling faster than the forward shock.

CONCLUSION

Although the generic picture of a two-shock structure for the dynamics of young SNRs is relatively simple and robust, in practice the details may be very complex and object-dependent. Obvious questions such as "where is the reverse shock" may not be so simple. There is clearly much work to be done in exploring the complexities of SNR dynamics at the level accessible to the observers. To that end, the hydrodynamics code used in all of the simulations shown here is available to any interested person at http://wonka.physics.ncsu.edu/pub/VH-1/

REFERENCES

1. Chevalier, R. A. 1982, ApJ, 258, 790
2. Nadyozhin, D. K. 1985, Ap&SS, 112, 225
3. Sedov, L. I. 1959, (New York: Academic Press) *Similarity and dimensional methods in mechanics*
4. Dwarkadas, V.V. & Chevalier, R.A. 1998, ApJ, 497, 807
5. Truelove, J. K. & McKee, C. F. 1999, ApJS, 120, 299
6. Borkowski, K. J., Symkowiak, A. E., Blondin, J. M., & Sarazin, C. L. 1996, ApJ, 466, 866
7. Gull, S. F. 1973, MNRAS, 161, 47
8. Chevalier, R. A., Blondin, J. M., & Emmering, R. T. 1992, ApJ, 392, 118
9. Wang, C.-Y. & Chevalier, R. A. 2000, ApJ, submitted
10. Jun, B.-I. & Norman, M. L. 1996, ApJ, 465, 800
11. Borkowski, K. J., Blondin, J. M., & Sarazin, C. L. 1992, ApJ, 400, 222
12. Chevalier, R. A. & Soker, N. 1989, ApJ, 341, 867
13. Blondin, J. M., Lundqvist, P., & Chevalier, R. A. 1996, ApJ, 472, 257
14. Jun, B.-I., Jones, T. W. & Norman, M. L. 1996, ApJ, 468, L59
15. Jun, B.-I. & Jones, T. W. 1999, ApJ, 511, 774
16. McCray, R. 1993, ARAA, 31, 175
17. Fesen, R. A., & Gunderson, K. S. 1996, ApJ, 470, 967
18. Hughes, J. P., Rakowski, C. E., Burrows, D. N., & Slane, P. O. 2000, ApJ, 528, L109
19. Li, H., McCray, R., & Sunyaev, R. A. 1993, ApJ, 419, 824
20. Blondin, J. M., Borkowski, K. J. & Reynolds, S. P. 2001, ApJ, submitted

Early Optical Spectroscopic Evidence for Conical Ejection in Supernovae?

Eric M. Schlegel

Harvard-Smithsonian Center for Astrophysics,
60 Garden Street, Cambridge, MA 02138, USA

Abstract.
We suggest that a parabolic emission feature observed in the optical spectra of several recent supernovae indicates the conical ejection of matter.

INTRODUCTION

Supernovae are assigned a type based upon spectral features [1], with the presence or absence of hydrogen lines as the dominant criterion. Subtypes historically recognized have been based on light curve shape (II-Plateau, II-Linear; e.g., [2]). Recently, the shape of the Hα profile has led to a subclass (the IIn's; [3],[4]) although the class is not yet as cleanly defined as the others.

Overall, the line shapes are produced by the expanding "atmosphere" of the supernova which produces a P Cygni profile. The absorption component is produced by the portion of the atmosphere between the observer and the deeper, more optically thick layers (the "photosphere"). Light resonantly scattered into the line of sight from the entire, un-occulted atmosphere produces the emission component. The basic line shape was worked out by [5] for an expanding atmosphere of a spherical star.

An expanding supernova atmosphere can be considered as an extreme and catastrophic case of a stellar wind. Adopting that viewpoint means that we can use the development of stellar wind theory (e.g., [6]), namely, that the line shapes carry information about the structure of the atmosphere. This paper discusses an unidentified emission line that can be understood from the stellar wind viewpoint.

SPECTRA

Figure 1 presents several optical spectra of Type IIn supernovae as well as a spectrum of SN1987A for comparison. The unidentified spectral feature is most easily visible in the spectrum of SN1997cy [7] at $\sim\lambda 5600$Å. That feature is parabolic in

CP565, *Young Supernova Remnants: Eleventh Astrophysics Conf.*, edited by S. S. Holt and U. Hwang
© 2001 American Institute of Physics 0-7354-0001-6/01/$18.00

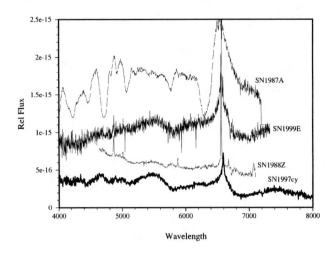

FIGURE 1. Optical spectra of several Type II supernovae. The parabolic spectral feature is most visibly present in the Type IIn SN1997cy spectrum at ∼5500Å. Note that the feature is also present, but weaker, in the Type IIn SN1999E and possibly in the Type IIn SN1988Z, but not in the Type IIpec SN1987A spectrum.

shape and approximately symmetric about the peak. Note the lack of damping wings for the emission feature, in contrast to the shape of Hα, for example, particularly in the spectrum of SN1987A (Figure 2). The parabolic feature is present, but weaker, in the spectrum of SN1999E [8] and possibly in SN1988Z [9]. There is no evidence for such a feature in the spectrum of SN1987A.

Previously, these lines have been explained as blends of, for example, emission lines of iron (e.g., [7]). We judge the parabolic feature to be too symmetric to be produced by a blend of lines. For the feature to be a blend, the lines must necessarily be blended identically in several different supernovae even though the Hα lines in these supernovae differ noticeably in shape.

THE MODEL

We suggest the parabolic spectral features can be explained as evidence of non-spherical ejection of matter, specifically, conical ejection. Spectral features with similar shapes are observed in the stellar wind spectra of T Tau stars (e.g., [10]), novae, and Wolf-Rayet stars [11].

If we assume conical ejection, the resulting shape of the line profiles will be dictated by the line optical depth, the ejection direction, and the viewing angle, among other variables. Two angles must be considered: the opening angle of the cone and the direction of the center of emission of the cone. For a optically thick cone pointed directly towards the observer, we expect to see blue-shifted emission

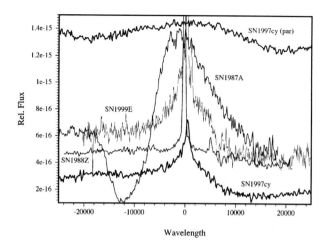

FIGURE 2. Expanded comparison of the Hα profiles with the parabolic feature of SN1997cy.

(and, perhaps, red-shifted emission from the reverse cone if it is not occulted). For optically thick ejection across our line of sight, the emission will symmetric about the Hubble-redshifted line center. The profile will fall smoothly to either side because of the decreasing emitting area of the cone. Other angles will produce intermediate profiles.

DISCUSSION

We turn now to addressing some of the implications. Why have these features not been modeled? How widespread among the various supernovae subclasses are these features? What predictions follow if the hypothesis is firmly established?

A survey of the available literature on models of supernovae spectra (see [12] for a list) shows the consistent adoption of the assumption of spherical symmetry for line transfer and spectrum calculations. In the absence of compelling observations, that assumption is certainly reasonable. Assuming the identification made here is supported by additional observations, that assumption will have to be relaxed.

How many supernovae have been observed to show parabolic spectral features? From a list of recent supernovae, a brief, and incomplete, survey of their spectra was made. The survey was constructed with the following criteria: (i) all Type II supernovae detected more recently than 1995 January 1; (ii) brighter than mag 15.5 in *any* band at discovery; (iii) for which at least one optical spectrum was readily available; and (iv) for which the spectrum's age was less than ∼20 days. Our imposition of a short time limit will be explained shortly. The numbers of supernovae that meet criteria (i) and (ii) are: 2000: 8; 1999: 18; 1998: 11; 1997:

9; 1996: 9; and 1995: 7. The number of supernovae for which "young" spectra were readily available[1] is considerably smaller. The table is presented in [12]. The clearest examples are provided by some of the Type IIn SNe.

If we assume all core collapse SNe produce a γ-ray Burst (GRB), and if we assume that GRBs signal pole-on beaming (see [12] for citations), then the presence of parabolic spectral features should be anti-correlated with the SNe-GRB association. In contradiction to this supposition, SN1997cy has been associated with GRB970514 [7].

Supernovae with parabolic features should show the strongest polarization, assuming all other parameters are constant. A variation in illumination can also lead to non-zero polarization, rendering the interpretation difficult.

Finally, we note recent papers (e.g., [13], [14], [15]) that discuss models of jet-like supernova explosions, some arguing that jet-induced explosions can explain the apparent high luminosities of the "hypernovae" [16]. Jet-induced explosions presumably will affect the ejection dynamics and be detectable in the line profiles of moderate- to high-resolution spectroscopy. Detailed spectral integration of these models has not yet been done, but we expect the jet to impress the spectrum with a signature.

A more complete discussion has been submitted for publication.

REFERENCES

1. Zwicky, F. 1965 in Stellar Structure, ed L. Aller & D. McLaughlin (University of Chicago Press), pp367-423
2. Kirshner, R. P. 1990, in Supernovae, ed. A. Petschek, (Springer-Verlag, Inc.), p59
3. Schlegel, E. M. 1990, MNRAS, 244, 269
4. Filippenko, A. 1997, ARAA, 35, 309
5. Sobolev, V. V. 1960, Moving Envelopes of Stars, (Cambridge: Harvard University Press)
6. Lamers, H. J. G. & Cassinelli, J. 1999, Introduction to Stellar Winds, Cambridge University Press
7. Turatto, M. et al. 2000, ApJ, 534, L57.
8. Silotti, S. Z., Schlegel, E. M., et al. 2000, in preparation
9. Turatto, M. et al. 1993, MNRAS, 262, 128.
10. Appenzeller, I., Jankovics, I., & Östreicher, R. 1984, A&A, 141, 108
11. Beals, C. 1931, MNRAS, 91, 966
12. Schlegel, Eric M. et al. 2000, in preparation
13. Nagataki, S. 2000, ApJS, 127, 141
14. Höflich, P., Wheeler, J. C., & Wang, Lifan 2000, ApJ, 521, 179
15. Khokhlov, A. M., Höflich, P. A., Oran, E. S., Wheeler, J. C., Wang, L., & Chtchelkanova, A. Yu. 2000, ApJ, 524, L107
16. Iwamoto, K. et al. 1998, Nature, 395, 672

---

[1] Largely via spectra in the CfA SNe database

Radio Emission from Supernovae in Binary Systems

Francesca R. Boffi[1] and Nino Panagia[1,2]

[1] *Space Telescope Science Institute, Baltimore, MD 21218, USA*
[2] *On Assignment from the Astrophysics Division, Space Science Department of ESA*

Abstract.
We have started a study of the radio emission from supernovae (SNe) exploding in binary systems. Here, we present our results for the case of a Type II SN in a wide binary system, which is viewed pole–on, and in which the companion star is a massive Main Sequence star. *At early times* the light curves rise more slowly than in the single–star case and are sensitive functions of the separation of the two stars. *At late times* the light curves are strongly affected by the regions of highly compressed material and the fluxes are higher than in the single–star case. For a given mass loss rate of the SN progenitor star the radio emission of a binary system SNII mimicks the emission that would be produced by a single–star SNII with a higher mass loss rate.

INTRODUCTION

We have started a study of the interaction of stellar winds in binary systems which may produce a supernova explosion, of its effects on the circumstellar environments and on the resulting SN radio emission. Here we present the first results on the radio emission from Type II supernovae exploding in binary systems.

When the primary reaches the red supergiant phase, its slow wind will be constrained within a bi–conical structure, a *double-funnel*, by the action of the hot companion fast wind (see Figure 1) (*e.g.* [1],[2])

Inside the funnel, the density is that of the RSG unperturbed wind whereas the layer at the interface between the two winds will be constituted by RSG material that has been compressed by a factor of, say, 2 to 10, depending on the parameters of the system (*e.g.* [3]). The main parameter determining the properties of the funnel is the ratio of the momentum carried by the companion wind to the primary star wind momentum.

In particular, the higher the wind momentum of the companion, the smaller the opening angle of the funnel (*e.g.* [4]). This is illustrated in Figure 2 in which we plot the expected opening angle for the case of a 15 $M_\odot$ Red Supergiant (RSG) primary

CP565, Young Supernova Remnants: Eleventh Astrophysics Conf., edited by S. S. Holt and U. Hwang
© 2001 American Institute of Physics 0-7354-0001-6/01/$18.00

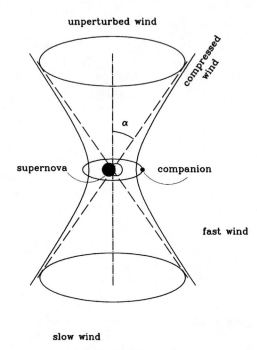

FIGURE 1. The *double funnel* structure produced by the interaction of the slow red supergiant wind of the primary with the fast wind of the companion star.

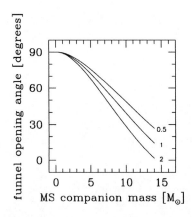

FIGURE 2. The expected funnel opening angle for the case of a 15 $M_\odot$ Red Supergiant (RSG) primary as a function of the Main Sequence companion mass, adopting three efficiencies of the MS mass loss.

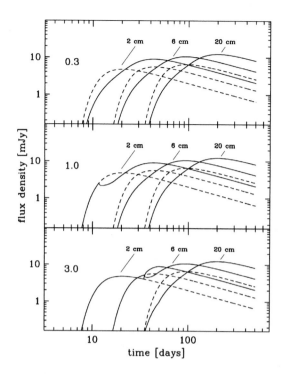

FIGURE 3. Light curves for three values of the separation (0.3, 1 and 3 $\times 10^{15}$ cm) and three wavelengths (2, 6 and 20 cm). The solid-line curves are the results for the *binary system* supernova and the dashed-line curves correspond to the case of a *single* supernova.

with a Main Sequence companion of mass in the range 1-14 $M_\odot$ and adopting three efficiencies of the MS mass loss (*e.g.* [5],[6]).

Here we present and discuss our calculations of the radio emission of a Supernova (SN) with RSG progenitor mass ($M_{primary} \sim 15\ M_\odot$) which had a companion with mass approximately half that of the primary ($M_{secondary} \sim 8\ M_\odot$, which is approximately a B5V type star on the main sequence).

MODEL CALCULATIONS AND RESULTS

The model assumptions are:

- The isotropic mass loss rate and wind velocity from the pre-supernova RSG are $\dot{M} = 6 \times 10^{-6}\ M_\odot\ yr^{-1}$ and $w = 10\ km\ s^{-1}$, respectively.

- With an 8 $M_\odot$ MS companion, the opening angle of the funnel is $\alpha = 45°$.

- The asymptotic angular width of the interface region is assumed to be $10°$.

- The intrinsic radio emission is assumed to be a power law of frequency and time $S \propto \nu^{-0.7} t^{-0.7}$.

- Three values of the separation between the two stars are considered, 3×10^{14}, 10^{15}, and 3×10^{15} cm (i.e. 20, 67, and 200 AU).

- The system is viewed pole-on.

The radio light curves for three wavelengths, 2, 6 and 20 cm, which were computed following the prescriptions of [7] and [8], are shown in Figure 3. The solid curves represent the emission from the binary system supernova, whereas the dashed curves are the light curves predicted in the case of a *single* supernova (*i.e.* a supernova with the same progenitor as considered here but that had no companion to affect its wind).

The main results can be summarized as follows:

- At early times the light curve evolution is a strong function of the system separation. This is because at the beginning the SN shock front interacts with circumstellar material which is made entirely of unperturbed RSG wind, whereas at later times both the unperturbed and the compressed winds become important.

- The rising branches of the light curves (solid lines) are more gradual than in the cases of a purely unperturbed wind (dashed lines). At higher frequencies, where the evolution is faster, the rise depends more appreciably on the separation of the system.

- Since both the emission and the absorption are proportional to powers of density which are greater than unity, at late times the emission originating from the compressed wind dominates the light curves, at all frequencies and all separations. In particular, in the case presented here, the emission at late times is a factor of 3 higher than it is predicted for a single supernova with equal progenitor.

REFERENCES

1. Girard T., and Willson L.A. 1987, A&A, 183, 247
2. Podsiadlowksi Ph., Fabian A.C., and Stevens I.R. 1991, Nature 354, 43
3. Lloyd H.M., O'Brien T.J., and Kahn F.D. 1995, MNRAS 273, L19
4. Eichler D., and Usov V. 1993, ApJ 402, 271
5. Panagia N., and Macchetto F.D. 1982, A&A, 106, 226
6. Scuderi S., Panagia N., Stanghellini C., Trigilio C., and Umana G. 1998, A&A, 332, 251
7. Weiler K., Sramek R.A., Panagia N., van der Hulst J.M., and Salvati M. 1986, ApJ 301, 790
8. Weiler K., Panagia N., and Sramek R.A. 1990, ApJ 364, 611

The Fading Radio Emission from Supernova 1961V

Christopher J. Stockdale[1], Michael P. Rupen[2], John J. Cowan[1], and
You-Hua Chu[3]

[1] *Department of Physics and Astronomy, 440 West Brooks Room 131, University of Oklahoma,
Norman, OK 73019, USA; stockdal@mail.nhn.ou.edu, cowan@mail.nhn.ou.edu*
[2] *National Radio Astronomical Observatories, PO Box 0, 1003 Lopezville Road, Socorro, NM
87801, USA; mrupen@nrao.edu*
[3] *Astronomy Department, University of Illinois, 1002 West Green Street, Urbana, IL 61801,
USA; chu@astro.uiuc.edu*

Abstract. Using the Very Large Array (VLA), we have detected radio emission from
the site of SN 1961V in the Sc galaxy NGC 1058. With a peak flux density of $0.084 \pm
0.011$ mJy/beam at 6 cm and 0.145 ± 0.028 mJy/beam at 18 cm, the source is non-
thermal, with a spectral index of -0.51 ± 0.21. Within errors, this index is the same
value reported for previous VLA observations taken in 1984 and 1986. The radio
emission at both wavelengths has decayed since the mid 1980's observations with an
averaged power-law index of $\beta_{20cm\&6cm} = -0.98 \pm 0.13$. We discuss the radio properties
of this source and compare them with those of Type II radio supernovae and luminous
blue variables.

INTRODUCTION

Supernova (SN) 1961V, identified as the prototype of Zwicky's Type V SNe (now
classified as Type II Peculiar) [1], is unique in several respects. Its progenitor was
visible as an 18th magnitude star from 1937 to 1960. It is the first SN, prior to
SN 1987A, whose parent star was identified before it exploded. The bolometric
correction, the exact distance, and the extinction are all uncertain, but its pre-
outburst luminosity apparently exceeded 10^{41} ergs s^{-1}, which is the Eddington limit
for a 240 M$_\odot$ star. After the explosion in late 1961, the initial peak of the optical
light curve was more complex and much broader than for any other supernova ever
observed. Optical spectra taken during this extended bright phase showed that the
characteristic expansion velocity was 2,000 km s^{-1}, which differs from the typical
value of 10,000 km s^{-1} for most SNe. This velocity is similar with nova expansion
velocities and with the expansion of SN 1986J (the only other Type II Peculiar SN)
which had an expansion velocity (taken well after maximum optical brightness) of
1,000 km s^{-1} [21,16,22]. Subsequently, the optical light curve of SN 1961V decayed

CP565, *Young Supernova Remnants: Eleventh Astrophysics Conf.*, edited by S. S. Holt and U. Hwang
© 2001 American Institute of Physics 0-7354-0001-6/01/$18.00

TABLE 1. Radio Observations of SN 1961V

Right Ascension (J2000)	$02^h43^m36\overset{s}{.}44 \pm 0\overset{s}{.}01$
Declination (J2000)	$+37°20'43\overset{''}{.}6 \pm 0\overset{''}{.}3$
18 cm[a] (14 Sept. 1999)	0.145 ± 0.028
6 cm[a] (21 & 25 Jan. 2000)	0.084 ± 0.008
Spectral Index[b] α_{6cm}^{18cm}	-0.51 ± 0.21
20 cm[a] (15 Nov. 1984)	0.258 ± 0.020
6 cm[a] (13 Aug. 1986)	0.126 ± 0.013
Spectral Index[b] α_{6cm}^{20cm}	-0.60 ± 0.13

[a] peak flux density (mJy/beam)
[b] $S \propto \nu^\alpha$

very slowly, by about 5 magnitudes in 8 years. Few SNe have been followed optically for more than 2 years. Goodrich *et al.* (1989) suggest instead that this object was a luminous blue variable (LBV) similar to η Carinae, and that the supposed supernova was an outburst of the variable star. Subsequently, Filippenko *et al.* (1995) observed SN 1961V using the Hubble Space Telescope (HST), although at that time the HST had not been refurbished. Those observations seemed to suggest a (very faint) star is still present at the site, which might or might not argue against a supernova origin. Among the brightest LBVs (e.g. η Car, P Cygni, and V 12 in NGC 2403), η Car is reported to have been the most luminous, reaching $M_{Bol} \simeq -14$. In comparison, SN 1961V was reported to have peaked at $M_{Bol} \simeq -17$ [11]. This peak estimate for SN 1961V is likely underestimated by 1.2 magnitudes if one accounts for the more recently derived Cepheid distance [18]. This would make SN 1961V nearly 100× brighter in the optical than η Car at maximum brightness.

RADIO PROPERTIES OF SN 1961V

Using the Very Large Array (VLA)[1], observations of SN 1961V were made in the mid 1980s, with the most definitive search in 1986 [4]. The source was reobserved in late 1999 and early 2000 to detect variability in its radio flux density and spectral index [19]. The results of these two searches are presented in Table 1.

These radio observations identify an object with radio luminosities at 6 cm and 20 cm similar to those found for many intermediate-age Type II SNe, including SNe 1950B, 1957D, 1970G, and 1980K (see Figure 1). Our measured flux densities at both wavelengths indicate a clear decline in the radio emissions from SN 1961V. The recently measured 18 cm flux, when scaled to 20 cm using the newly determined spectral index, indicates a reduction in the 20 cm peak flux intensity by 41% from 1984 to 1999. The 6 cm peak flux intensity has also dropped by 41% in the interval from 1986 to 2000. The decline in the radio flux density is consistent with models for radio emission from SNe [2]. As indicated in Table 1, our new observations

---

[1] The National Radio Astronomy Observatory is a facility of the National Science Foundation operated under cooperative agreement by Associated Universities, Inc.

indicate that SN 1961V remains a non-thermal radio source. The spectral index, α, is relatively unchanged although the error bars are rather large. The current and previous values of α for SN 1961V are still consistent with spectral indices of other intermediate-age RSNe at similar wavelengths [19]. The rate of decline of radio emission for SN 1961V, as measured by a power-law index ($S \propto t^\beta$), is similar to the decline rates of known Type II intermediate-age RSNe (see Figure 1). The power-law indices for SN 1961V were determined from the peak intensities to be $\beta_{20cm} = -1.03 \pm 0.21$ and $\beta_{6cm} = -0.93 \pm 0.14$. SN 1961V's rate of decay is not as steep as than the rates of of similarly aged RSNe. This may indicate that there is more circumstellar material around SN 1961V than around SNe 1970G or 1959D, or that SN 1961V's shock is traveling at a slower speed than found in other SNe.

Radio comparisons between η Car, the super-luminous LBV, and SN 1961V are more problematic since the first radio observations of η Car were made 100 years after its eruption. η Car, with a 20 cm flux density of 0.9 ± 0.3 Jy [15], is in fact not a strong radio source when compared to SN 1961V. And when its luminosity is extrapolated to the age and distance of SN 1961V, it is at least $1,000\times$ weaker than SN 1961V [19]. η Car's spectral index between 3 cm and 6 cm appears to peak at $+1.8$ at the position of η Car and then drops radially toward an index of 0 [6]. The thermal nature of η Car is consistent with most LBVs and in contrast

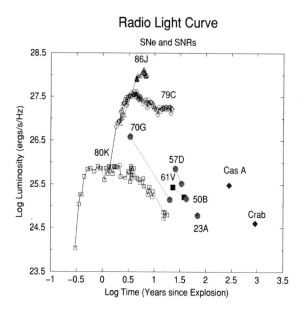

FIGURE 1. Radio light curve for SN 1961V at 20 cm compared to several RSNe and SNRs. Data, fits and distances for SN 1923A [7,17]; for SNe 1950B & 1957D [5,17]; for SN 1961V [19,18]; for SN 1970G [3,12]; for SN 1979C [23,26,14,8]; for SN 1980K [23,25,13,20]; and for SN 1986J [16,24,18]. Luminosities for Cas A and the Crab [7].

with the non-thermal nature of SN 1961V and RSNe.

While the present radio observations seem to be consistent with other well-known radio RSNe and thus support a supernova interpretation, the lack of radio observations of similarly-aged bright LBVs prevents a definitive identification. Additional multiwavelength observations of SN 1961V, as the object evolves, will be needed to make a final judgment about the nature of this object.

REFERENCES

1. Branch, D., & Greenstein, J. L., *ApJ*, **167**, 89, (1971).
2. Chevalier, R. A., *ApJ*, **285**, L63 (1984).
3. Cowan J. J., Goss, W. M., & Sramek, R. A., *ApJ* **379**, L49 (1991).
4. Cowan, J. J., Henry, R. B. C., & Branch, D., *ApJ* **329**, 116 (1988).
5. Cowan, J. J., Roberts, D. A., & Branch, D., *ApJ* **434**, 128 (1994).
6. Duncan, R. A., White, S. M., & Lim, J., *MNRAS* **290**, 680 (1997).
7. Eck, C. R., Cowan, J. J., & Branch, D., *ApJ* **508**, 664 (1998).
8. Ferrarese, L. *et al.*, *ApJ* **464**, 568 (1996).
9. Filippenko, A. V., Barth, A. J., Bower, G. C., Ho, L. C., Stringfellow, G. S., Goodrich, R. W., & Porter, A. C., *AJ* **110**, 2261 (1995).
10. Goodrich, R. W., Stringfellow, G. S., Penrod, G. D., & Filippenko, A. V., *ApJ* **342**, 908 (1989).
11. Humphries, R. M., & Davidson, K., *PASP* **106**, 1025 (1994).
12. Kelson, D. *et al.*, *ApJ* **463**, 26 (1996).
13. Montes, M. J., Van Dyk, S. D., Weiler, K. W., Sramek, R. A., & Panagia, N., *ApJ* **506**, 874 (1998).
14. Montes, M. J., Weiler, K. W., Van Dyk, S. D., Panagia, N., Lacey, C. K., Sramek, R. A., & Park, R., *ApJ* **532**, 1124 (2000).
15. Retallack, D. S., *MNRAS* **204**, 669 (1983)
16. Rupen, M. P., van Gorkom, J. H., Knapp, G. R., & Gunn, J. E., *AJ* **94**, 61 (1987).
17. Saha, A., Sandage, A., Labhardt, L., Schwengler, H., Tammann, G. A., Panagia, N., & Macchetto, F. D., *ApJ* **438**, 8 (1995).
18. Silbermann, N. A. *et al.*, *ApJ* **470**, 1 (1988).
19. Stockdale, C. J., Rupen, M., Cowan, J. J., Jones, S. & Chu, Y.-H., in preparation (2000).
20. Tully, R. B., *Nearby Galaxies Catalog*, Cambridge: Cambridge Univ. Press (1988).
21. van Gorkom, J. H., Rupen, M. P., Knapp, G. R., & Gunn, J. E., *IAU Circ.* **No. 4248** (1986).
22. Weiler, K. W., & Sramek, R. A., *Ann. Rev. Astr. Ap.* **26**, 295 (1988).
23. Weiler, K. W., Sramek, R. A., Panagia, N., van der Hulst, J. M., & Salvati, M., *ApJ* **301**, 790 (1986).
24. Weiler, K. W., Panagia, N., & Sramek, R. A., *ApJ* **364**, 611 (1990).
25. Weiler, K. W., Van Dyk, S. D., Panagia, N., & Sramek, R. A., *ApJ* **398**, 248 (1992).
26. Weiler, K. W., Van Dyk, S. D., Panagia, N., Sramek, R. A., & Discenna, J. L., *ApJ* **380**, 161 (1991).

Hydrodynamic Instabilities in Young Supernova Remnants

Christopher L. Hinkle and John M. Blondin

North Carolina State University, Raleigh, North Carolina 27607, USA

Abstract. The interaction of a young supernova remnant and its surroundings can be described by a self-similar solution if the expanding supernova ejecta can be described by a power law density profile of $\rho \propto r^{-n}$ with n>5, and the surrounding ambient medium has a density power law of $\rho \propto r^{-s}$ with s<3 [1]. It has been shown in two dimensions that initializing a hydrodynamics code with the self-similar solution for the interaction region and perturbing the density randomly causes Rayleigh-Taylor instabilities [2]. We present hydrodynamic simulations of self-similar driven waves, with different parameters, for the purpose of comparing the growth and saturation of the convective instability in three dimensions compared to two dimensions. The contact discontinuity in all simulations produced Rayleigh-Taylor fingers with predictable morphologies; long, continually growing spikes for s=2, and shorter, mushroom capped blobs for s=0. A fourier spectrum was obtained and compared in both two and three dimensions to analyze the evolution of the instability's growth. Turbulent energy density was measured and compared for different initial conditions. These simulations and diagnostics show how closely two-dimension codes can approximate three-dimensions and also provide information on expansion and mixing of young supernova remnants.

INTRODUCTION

Young supernova remnants have been observed to display Rayleigh Taylor instabilities during interaction with their surrounding medium. These instabilities likely play a role in mixing the metal-enriched ejecta with the ambient medium, clump formation in young remnants, and in amplifying any trace magnetic field. The interaction region of a supernova with a power law density profile has been studied in one and two dimensions. Using a parallel code and supercomputer, we have completed three-dimensional simulations of the interaction region.

PROCEDURE

The hydrodynamics code Virginia-Hydrodynamics 1 (VH1) was used to reproduce Chevalier's 1982 findings on self-similar solutions for power law density profile supernova models. The results, done in one-dimension, were exact and show the

CP565, *Young Supernova Remnants: Eleventh Astrophysics Conf.,* edited by S. S. Holt and U. Hwang
© 2001 American Institute of Physics 0-7354-0001-6/01/$18.00

density, pressure, and velocity signatures expected depending on the power law. The supernova gas was initialized with a profile of $\rho \propto r^{-n} (n>5)$, and the surrounding medium was $\rho \propto r^{-s}$ (s<3). VH1 was then equipped with a "moving grid" code to follow the interaction region as it progressed. The 1-D results were then used to initialize a two-dimension moving grid VH1 code. To start some instability, zones near the contact discontinuity were given slight density perturbations of <20%. This perturbed code was run and reproduced the results of Chevalier, Blondin, and Emmering (1992). Diagnostics were taken to compare the different initial conditions and the results of the three-dimension simulations. Turbulent energy density was tracked and the growth rate of the instabilities was found by a fourier spectrum. A computer code of VH1, using MPI parallel programming, was used for the three dimensional simulations. The initial conditions were the results of the 1-D runs along with the same perturbations as used in two-dimensions. The parallel code was then run on an IBM SP computer for faster computations. Again the turbulent energy density and a fourier spectrum were gathered for analysis.

COMPUTATIONS

Numerical hydrodynamics provides a useful way to quantitatively and qualitatively evaluate models for supernova remnants. For this project, the hydrodynamics code VH1 (http://wonka.physics.ncsu.edu/pub/VH-1) was used. VH1 can be modified to suit a particular simulation, and in this case was equipped with a moving spherical grid to follow the expansion of the SNR, and coded in parallel structure to take advantage of computing facilities. These facilities included the IBM SP machine located at the North Carolina Supercomputing Center (NCSC) in the Research Triangle Park. The code was run on 32 nodes with each node using four processors. This greatly increases the speed of computations and allows for more intensive diagnostics to be run. These simulations ran at a speed of approximately 45000 zones per second on a grid of 600x256x256 evolved for 20,000 time steps. This meant less than 48 hours of computing time for a full three dimensional simulation.

RESULTS

The evolution of the instabilities in three-dimensions proceed in the same manner as the well studied two-dimensional Rayleigh-Tayor instabilities [2,3]. In the s=0 simulations, the density perturbations produce Rayleigh-Taylor fingers slicing through the medium into the high entropy region. The dense gas is then turned back on itself, forming a mushroom shaped cap that begins a fountain-like convection cycle. The s=2 simulations begin in the same manner as the s=0 simulations with the perturbations producing instabilities that push into the region between the contact discontinuity and the forward shock. However, in the s=2 runs, the fingers continue further into the medium forming much smaller caps, and growing at a faster rate than the s=0 models. A look at the diagnostics provides more

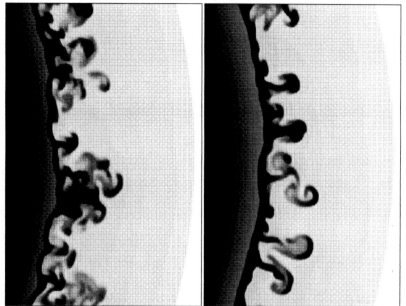

FIGURE 1. Density in the interaction region for n=7, s=2 case, showing mushroom capped Rayleigh-Taylor fingers. This image shows the "choppy" appearance of the three-dimesional simulation (left) as compared to the two-dimensional simulation (right).

FIGURE 2. Volume rendering of a three-dimensional simulation. This image shows the instabilities as if they were approaching the observer.

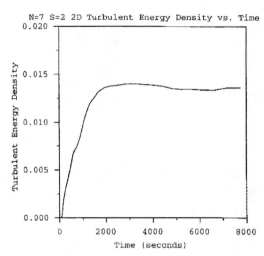

FIGURE 3. This graph shows the time to saturation of turbulent energy density in a two-dimensional simulation. Saturation occurs approximately two times later in three-dimensions.

comparisons between the two types of simulations. The turbulent energy density, an indicator of the amount of metal mixing and magnetic field amplification, was gathered. Plots of turbulent energy density versus time show different times to saturation depending on the number of dimensions. The three-dimensional simulations take longer for the saturation presumably because the extra dimension allows more room for the convection to evolve. Saturation occurred approximately twice as fast in two dimensions as compared to three dimensions. Looking at the fourier spectrum of the simulations gives information about the growth rate and peak wavelengths of the instabilities. All simulations had a region of power law growth which was analyzed to find the slope. The three-dimensional simulations all show a steeper power law growth rate than the two-dimensional simulations. It is reasonable to expect the fingers to be able to push the surrounding medium away more easily in three dimensions allowing for more rapid growth of the instabilities. The power law was approximately 1.2-1.4 times greater in three-dimensions. The peak wavelength of a late time power spectrum is shifted to a higher wavenumber/shorter wavelength in the three-dimensional simulations. This too agrees with expectations since the three-dimensional simulations appear more "choppy" than the smooth two-dimensional simulations.

REFERENCES

1. Chevalier,R. A., *ApJ* **258**, 790 (1982).
2. Chevalier,R. A., Blondin, J. M., Emmering, R. T., *ApJ* **392**, 118 (1991).
3. Jun, B. -I., Norman, M. L., *ApJ* **472**, 245 (1996).

Modeling of Radiative Blast Waves

K.A. Keilty[1], E.P. Liang[1], T. Ditmire[2,3], B.A. Remington[2],
A.M. Rubenchik[4] and K. Shigemori[5]

[1] *Rice University, Houston, TX 77005, USA*
[2] *Lawrence Livermore National Laboratory, Livermore, CA 94550, USA*
[3] *University of Texas, Austin, TX 78712, USA*
[4] *University of California, Davis, CA 95616, USA*
[5] *Osaka University, Osaka, Japan*

Abstract.
We simulate experiments performed with the Falcon laser at Lawrence Livermore
National Laboratory to generate strong blast waves expanding in cylindrical geometry
of relevance to astrophysics. In particular, we are interested in producing and modeling
radiative shocks. Our goal is to develop a laboratory setting for studying radiative
shocks of relevance to supernova remnants (SNR). Although late-term supernovae are
known for exhibiting radiative shocks, it is also likely that some young SNR are also
radiative when they expand into a dense interstellar medium (ISM). In previous work
we have demonstrated that it is possible to generate radiative shocks in the laboratory.
In addition, we have shown how we can determine the energy-loss rate of the shock
from the blast wave evolution using a simple analytic method that is independent of
the details of radiative cooling, and is scalable to both the laboratory and astrophysical
blast waves. Our future work deals with instabilities associated with radiative blast
waves and their application to the laboratory and astrophysics. This paper examines
some of the previous work done in the area of radiative instabilities and discusses the
challenges in adapting this work to the laboratory setting.

INTRODUCTION

It has long been understood that SNR eventually enter a "pressure driven snowplow" (PDS) phase in which the remnant forms a thin, dense shell and the expansion is driven by the pressure from the hot, tenuous interior. During this radiative
phase, the energy lost due to radiative cooling becomes significant compared to
the energy flux entering the shock front [1–3]. We expect that the radius R of the
blast wave is a power-law in time with $R \propto t^n$. For the early adiabatic phase of
expansion, $n = 2/5$; this changes to $n = 2/7$ at the beginning of the PDS phase [3].
We will describe the cooling function during the PDS phase with $\Lambda \propto T^\alpha$ [ergs
cm^{-3} sec^{-1}] where T is temperature. The rate of radiative energy loss is therefore
proportional to Λn^2, where n is the hydrogen density of the preshock region. Linear

CP565, *Young Supernova Remnants: Eleventh Astrophysics Conf.*, edited by S. S. Holt and U. Hwang
© 2001 American Institute of Physics 0-7354-0001-6/01/$18.00

perturbative analysis and simulations show that there is a one-dimensional cooling instability in radiative blast waves [4,5]. For a spherically symmetric blast wave with adiabatic index $\gamma = 5/3$, the shock becomes unstable when $\alpha \leq 0.8$ [4].

Previous work has demonstrated our ability to generate cylindrically symmetric radiative blast waves in the laboratory using tabletop fs pulse lasers and xenon gas [6–8]. We use a gas jet of ~ 1 atm, room temperature xenon, irradiated from the side by a ~ 15 mJ laser with a pulse length of 30 fs, to create an expanding cylindrical blast wave. In addition, we have developed a simple analytic method of determining the energy-loss rate from the blast wave evolution [9,6]. This method is independent of the method of cooling, and is scalable to both the laboratory and astrophysical environments. We are interested in exploring the possibility of studying the one-dimensional instability in a laboratory setting that is relevant to SNR. The rest of this paper will describe the benefits and challenges of using the materials and geometry appropriate to the laboratory setting.

USE OF MATERIAL

Xenon has been demonstrated to have a strong laser absorption rate [10], which makes it an ideal candidate for generating high temperature, low density plasmas. In addition, we expect that the importance of radiative cooling in the blast wave and ionization as an energy sink will increase with high-Z elements.

Although there are still many questions dealing with the possibility of scaling radiative phenomena to the laboratory, it is interesting to compare the cooling functions for xenon and for cosmic abundances (see Figure 1). The average slopes of the areas where the $\alpha \leq 0.8$ are very similar. In addition, the cooling rate for xenon is higher by a factor of $10^4 - 10^5$. This indicates some promise in using high-Z materials to study radiative phenomena.

Xenon absorbs at a lower energy and radiates more efficiently than lower-Z elements at comparable temperatures. This will, among other things, allow us to see the same effects in a shorter amount of time compared to using materials that are perhaps more realistic from an astrophysical point of view. For a SNR, the time of shell formation (when radiative losses begin to be important) is: $t_{sf} = 3.61 \times 10^{-4} E_{51}^{3/14} n^{-4/7}$ yr [3], where E_{51} is the energy of the initial blast in 10^{51} erg and n is the number density of the ISM. Under laboratory conditions (assuming spherical geometry, $E = 10^5$ erg and $n = $ a few times 10^{18} cm^{-3}), we find that $t_{cool} \simeq 5 \times 10^{-11}$ sec. We should therefore be able to see radiative effects from the very beginning of the formation of the blast wave; this is indeed what we have observed [6]. Xenon appears to go straight into a radiative phase without passing through a lengthy adiabatic phase first. If we are looking to study the radiative phase of a blast wave in particular, this is a significant improvement in terms of time over traditional SNR simulations [3] which indicate a significant amount of time before the onset of the PDS.

There are certain particular issues that must also be addressed if we wish to

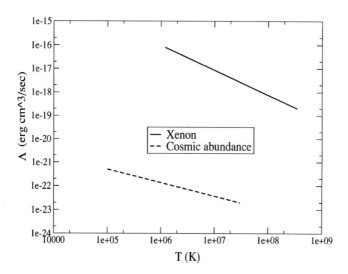

FIGURE 1. Comparison of cooling rates for cosmic abundances [taken from Raymond et al., *ApJ* **204**, 290 (1976)] and xenon [taken from Post et al., *At. Data Nucl. Tables* **20**, 5 (1977)]. These straight lines represent the similar area in the cooling curves which are of interest; i.e., the general region where ($\alpha \leq 0.8$).

continue to use high-Z elements in the laboratory. Previous experiments have shown that partially ionized xenon gas has an adiabatic index (γ) of much closer to 1.1 than to 5/3 (the adiabatic case) when it is in a radiative blast wave stage [11,6,7]. Since most of the original equations dealing with radiative cooling instablities in SNR [4] were not meant to take into account the particular laboratory conditions with which we deal, it will be necessary to reexamine the equations closely for their γ dependence.

USE OF GEOMETRY

When the well-known Sedov-Taylor equation is modified for cylindrical geometry, we see that $R \propto t^{1/2}$ rather than $R \propto t^{2/5}$. One of the benefits of using cylindrical geometry is that the blast wave moves faster that in spherical geometry. The growing interest in astrophysical regimes where cylindrical geometry may be important (including systems with jets) makes this geometry increasingly relevant.

We are currently conducting experiments using cylindrically symmetric geometry. While many of the parameters (including the inital analytic treatment of instabilities in radiative blast waves [4]) are geometry-independent, there are other factors to consider. In particular, as noted, we expect different times for the development of the blast wave and t_{cool} depending on the geometry used. There is also the logistics of the laboratory setup. Further experiments have indicated that at long times (after 50 ns), the xenon blast wave may actually gain back some of the energy it has lost as it sweeps up ambient material that has been preheated [8]. This is one of the challenges of using such a high-density gas that must be addressed if we wish to design meaningful, astrophysically relevant experiments.

CONCLUSIONS

Our past work has demonstrated that it is indeed possible to generate radiative blast waves in the laboratory using short-pulse lasers. In this paper we have mentioned the one-dimensional cooling length instability associated with radiative blast waves in general and with radiative shocks in SNR in particular. We have also mentioned both the difficulties of and benefits associated with working with a high-Z element in cylindrical geometry in the laboratory.

Future work in this area will include a detailed analysis of the equations governing the action of radiative SNR with respect to the importance of γ. We are performing further 1-D and 2-D simulations of high-Z element blast waves with an emphasis on observing instabilities associated with the cooling length. Finally, we will be using these simulations and our current approximations to design further laboratory experiments in multiple geometries. One of the possibilities involves adding a magnetic field to the problem.

Although there is still much work to do, the field of laser astrophysics has shown great promise in a short amount of time. We believe that the physics of radiative shocks is a good example of an area of overlapping interest in the plasma physics and astrophysics communities, and we intend to continue with laboratory experiments intended to relate to astrophysical phenomena.

REFERENCES

1. Cox, D.P., *ApJ* **178**, 159 (1972).
2. Dopita, M.A., *ApJ* **209**, 395 (1976).
3. Cioffi, D.F., McKee, C.F. & Bertschinger, E., *ApJ* **334**, 252 (1988).
4. Chevalier, R.A. & Imamura, J.N., *ApJ* **261**, 543 (1982).
5. Kimoto, P.A. & Chernoff, D.F., *ApJ* **485**, 274 (1997).
6. Keilty, K.A. et al., *ApJ* **538**, 645 (2000).
7. Shigemori, K. et al., *ApJ* **533**, L159 (2000).
8. Edwards, M.J. et al., *Phys. Rev. Let. submitted* (2001).
9. Liang, E.P. & Keilty, K.A., *ApJ* **533**, 890 (2000).
10. Ditmire, T. et al., *Phys. Rev. Let.* **78**, 3121 (1997).
11. Grun, J. et al., *Phys. Rev. Let.* **66**, 2738 (1991).
12. Ryutov, D. et al., *ApJ* **518**, 821 (1999).

The Evolution of a Supernova Remnant in a Turbulent, Magnetized Medium

Dinshaw Balsara*, Robert A. Benjamin†, and Donald P. Cox†

*NCSA, Univesity of Illinois at Urbana-Champaign, Urbana, IL 61820, USA
†Department of Physics, University of Wisconsin-Madison, Madison, WI 53706, USA

Abstract. We discuss the results of three dimensional calculations for the MHD evolution of an adiabatic supernova remnant in both a uniform and turbulent interstellar medium using the RIEMANN framework of Balsara. In the uniform case, which contains an initially uniform magnetic field, the density structure of the shell remains largely spherical, while the magnetic pressure and synchrotron emissivity are enhanced along the plane perpendicular to the field direction. This produces a bilateral or barrel-type morphology in synchrotron emission for certain viewing angles. We then consider a case with a turbulent external medium characterized by $v_A(rms)/c_s = 2$. Several important changes are found. First, despite the presence of a uniform field, the overall synchrotron emissivity becomes approximately spherically symmetric, on the whole, but is extremely patchy and time-variable, with flickering on the order of a few computational time steps, reminiscent of the Cas A observations of Anderson & Rudnick. We suggest that the time and spatial variability of emission in early phase SNR evolution provides information on the turbulent medium surrounding the remnant.

The morphology and emission of SNR is in large part determined by the nature of the surrounding circumstellar and interstellar medium: SNR in the halo of the Galaxy will have low X-ray surface brightness. SNR in dense molecular clouds will be elongated in the direction of the density gradient. SNR at the edges of clouds will have offsets between the morphological and kinematic centers. Barrel-shaped remnants may probe the global magnetic field structure, and so on. There have been several calls by observers to introduce a realistic representation of interstellar turbulence in SNR simulations. We have carried out the first 3D MHD simulations of SNR evolution in both a uniform and turbulent, magnetized ISM.

The simulations presented here are calculated using the RIEMANN framework for parallel, self-adaptive computational astrophysics described in [1], [2], and references therein. This framework uses some of the most accurate higher order Godunov methods for non-relativistic and relativistic MHD and radiative transfer that have been devised. The advantages of this method in modelling turbulent flows is described in the above references. The RIEMANN framework has been applied and tested on numerous computational astrophysics problems that require

CP565, *Young Supernova Remnants: Eleventh Astrophysics Conf.*, edited by S. S. Holt and U. Hwang
© 2001 American Institute of Physics 0-7354-0001-6/01/$18.00

numerical 3D-MHD.

We present the results of two high-resolution $(256)^3$ zone runs ($L = 40$ pc), comparing the results of evolution in a medium with a uniform and turbulent medium. The initial uniform hydrogen particle density is 1 cm^{-3}; the initial temperature is 8000 K ($c_s = 9.8$ km s^{-1}), the initial velocities are zero and there is a uniform $2\mu G$ magnetic field parallel to the x-axis. We then introduce 10^{51} ergs of thermal energy within 2 pc of the computation grid center. The simulations are evolved to the point when the shock reaches the computational boundary. These parameters may be rescaled without requiring additional simulations.

In the case of the turbulent ISM, we introduce uncorrelated Gaussian random fluctuations in the velocity and magnetic fields, corresponding $v_A(rms)/c_s = 2$ [$B(rms) = 10\mu G$] and $v(rms)/c_s = 2$. As the blast wave sweeps outward, there is some evolution in the preshock medium, with growing density fluctuations and decaying magnetic and velocity fluctuations. These fluctuations quickly establish a power spectrum of $E(k) \propto k^{-5/3}$ in the surrounding medium well before the blast wave propagates a large distance.

RESULTS

Despite the presence of a magnetic field, the thermal pressure of the explosion is so large that the evolution agrees very well with the analytical Sedov solution. The shock position is given by $R_{Sedov} \propto t^{-2/5}$. The density profile is self-similar and nearly agrees with the Sedov solution. Limited resolution prevents us from getting the full factor of four density jump right at the shock front; the postshock density is 2.8 times greater than the preshock density. The most notable feature of the uniform ISM simulation is that the magnetic field, initially parallel to the x-axis, is swept up into a compressed shell, yielding a barrel morphology.

Several differences result for the case of a remnant evolving in a turbulent medium. The azimuthally averaged density profile still evolves self-similarly and agrees with the Sedov solution. However, in each radial shell, there are azimuthal variations in density just behind the blast wave. The ambient medium is characterized by fluctuations which grow from $\delta\rho/\rho \approx 0.1 - 0.4$ over the course of the simulation. At the shock front, the density fluctuations jumps by a factor of 4 at early times. This decreases to 1.2 at late times, since the effect of individual lumps becomes less important as the surface area of the remnant increases. Behind the shock front, $\delta\rho/\rho$ decreases as the increased thermal pressure works to smooth out the density fluctuations.

The variation in the magnetic field strength is also noteworthy. The overall shape of the remnant is again spherical, but with large azimuthal variations. The mean magnetic field jumps by a factor of 2.4 across the shock front. This value is about what one would expect with a factor $\sim$ four jump in the two components of field perpendicular to the shock front, i.e. $B_{postshock} = \sqrt{(16B_{x,o}^2 + 16B_{y,o}^2 + B_{z,o}^2)/3} = 3.3$. The random component of the magnetic field in the ambient medium is much larger than the uniform component, so the barrel-like morphology noted in the uniform case is destroyed.

The resulting synchrotron shell is very patchy and shows no evidence of the initial uniform magnetic field orientation, although perhaps polarization of the synchrotron emisison might still retain evidence of this configuration. While the magnetic pressure dominates the pressure in the ambient medium, the pressure behind the blast wave is dominated by thermal pressure for most of the evolution of the model remnant. At late times, there are some spots in which the pressure magnetic regions actually dominate over the thermal pressure. In these regions, the remnant can be synchrotron bright and relatively X-ray faint.

Our simulations also show rapid "flickering" of the estimated synchrotron emissivity. Over just a few timesteps ($\Delta t=300$ years), regions can brighten or fade by

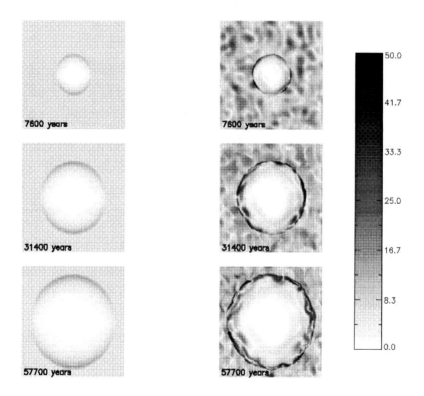

FIGURE 1. Two dimensional slices through our computational grid showing the evolution of the total magnetic field strength (μG) for a uniform ISM (left column) and a turbulent ISM (right column). The introduction of strong MHD turbulence erases the bilateral morphology present in the uniform case, and results in extremely strong patches of magnetic field. These patches appear and dissipate over only a few computational timesteps, suggestive of the synchrotron "flickering" observed in Cas A by [3].

up to a factor of 2. The timescale of this flickering is limited by the lower size scale of the input spectrum of velocity and density fluctuations, which is 10 grid cells. Higher resolution simulations would yield much shorter timescale flickering like the varations that are observed in Cas A over 5-10 year time scales by [3]. We expect that the intensity fluctuations in the image should correlate with the time scales of the intensity fluctuations.

DISCUSSION

Supernova remnants serve as useful "test explosions" that can be used to probe the parameters characterizing the surrounding circumstellar and interstellar medium. In a quiescent medium, the remnant sweeps up the uniform field, and yields a bilateral morphology. However the introduction of realistic, strongly turbulent fluctuations erases this asymmetry, and produces remnants that are (on average) spherical. The resulting projected X-ray and synchrotron maps show a range of fluctuation in the emission characteristics of the remnant. These calculations inject an element of realism into the modelling of the pre-SNR environment that has heretofore been lacking.

In our simulations, the temporal and angular variations in emission are associated with the nature of the magnetic fluctuations in the unshocked interstellar medium. It therefore seems that variations in emission characteristics of SNR may be a valuable probe of the turbulence in the pre-SNR environment. In the future, we intend to run a grid of such models to determine how the nature of the turbulence is related to the emission characteristics of the remnant.

In addition, the interaction of SNR blastwave with a region of supersonic MHD turbulence provides an opportunity to investigate the degree to which magnetic fluctuations affect the effective equation of state of a magnetized plasma. The importance of understanding the nature of this behavior in interstellar conditions has been emphasized by [4]. We will also examine how this effective equation of state varies over time with the parameters characterizing the MHD turbulence.

This work was supported by NASA Astrophysical Theory Grant NAG5-8417. We also thank NCSA and SDSC for use of their supercomputers.

REFERENCES

1. Roe, P. and Balsara, D., 1996, SIAM J. Appl. Math., 56, 57
2. Balsara, D. and Norton, C.,2000, J. Parallel Comp. , in press
3. Anderson, M.C. and Rudnick, L., 1996, ApJ 456, 234
4. McKee, C.F. and Zweibel, E. G., 1995, ApJ, 440, 686

SNe to SNR

SN1987A: the Youngest Supernova Remnant

Richard McCray

JILA, University of Colorado, Boulder, CO 80309-0440, USA

Abstract.

Supernova 1987A became a supernova remnant about 10 years after outburst, when the luminosity resulting from the impact of its debris with its circumstellar matter dominated the luminosity due to internal radioactivity. Today, with the Hubble Space Telescope, we clearly see optical and ultraviolet emission from the reverse shock and from several "hot spots," where the blast wave is beginning to strike the inner circumstellar ring. With the *Chandra* Observatory, we also see X-ray emission from the hot shocked gas between the reverse shock and the blast wave. Here I review the observations of this remarkable event and forecast what we can expect to see during the coming decade.

INTRODUCTION

One of the most exciting mysteries of SN1987A is its remarkable system of circumstellar rings. Evidently, the rings were ejected by the supernova progenitor some 20,000 years before it exploded. But how do we account for their morphology? At the moment, there is no satisfactory explanation. But, as I shall describe, events beginning now will give us a new window on the supernova's past.

The supernova blast wave is now beginning to strike the inner ring. This impact marks the birth of a supernova remnant, defined as the stage when the supernova light is dominated by the impact of the supernova debris with circumstellar matter. It will be a spectacular event. During the coming decade, the remnant SNR1987A will brighten by orders of magnitude at wavelengths ranging from radio to X-ray. With the Hubble Space Telescope (*HST*), we have obtained exquisite optical and ultraviolet images and spectra of this unique event. With the *Chandra* observatory, we have already obtained our first images and spectra of the X-ray source. Large ground-based telescopes equipped with adaptive optics have already begun to provide excellent images and spectra at optical and near-infrared wavelengths. Future observatories, such as the Space Infrared Telescope Facility and the Atacama Large Millimeter Array, will provide data at other wavelength bands to complement these observations. The combination of these data will give us a unique opportunity to

CP565, *Young Supernova Remnants: Eleventh Astrophysics Conf.*, edited by S. S. Holt and U. Hwang

probe the rich range of physical phenomena associated with astrophysical shocks and to learn about the death throes of a supernova progenitor.

Before describing the observations of SN1987A, it might be useful to review its energy sources. SN1987A has three different sources of energy, each of which emerges as a different kind of radiation and with a different timescale. The greatest is the collapse energy itself ($\sim 10^{53}$ ergs), which emerges as a neutrino burst lasting a few seconds. The energy provided by radioactive decay of newly synthesized elements ($\sim 10^{49}$ ergs) is primarily responsible for the optical display. Most of this energy emerged within the first year after outburst, primarily in optical and infrared emission lines and continuum from relatively cool ($T \lesssim 5,000$ K) gas. Note that the radioactive energy is relatively small, $\sim 10^{-4}$ of the collapse energy.

The kinetic energy of the expanding debris ($\sim 10^{51}$ ergs) can be inferred from observations of the spectrum during the first three months after explosion. Astronomers infer the density and velocity of gas crossing the photosphere from the strengths and widths of hydrogen lines in the photospheric spectrum. By tracking the development of the spectrum as the photosphere moved to the center of the debris, astronomers can measure the integral defining the kinetic energy. Doing so, they find that $\sim 10^{-2}$ of the collapse energy has been converted into kinetic energy of the expanding debris.

This kinetic energy will be converted into radiation when the supernova debris strikes circumstellar matter. When this happens, two shocks always develop: the blast wave, which overtakes the circumstellar matter; and the reverse shock, which is driven inwards (in a Lagrangean sense) through the expanding debris. The gas trapped between these two shocks is typically raised to temperatures in the range $10^6 - 10^8$ K and will radiate most of its thermal energy as X-rays with a spectrum dominated by emission lines in the range $0.3 - 10$ keV.

Most of the kinetic energy of the debris will not be converted into thermal energy of shocked gas until the blast wave has overtaken a circumstellar mass comparable to that of the debris itself, $\sim 10 - 20 M_\odot$. Typically, that takes many centuries, and as a result, most galactic supernova remnants (*e.g.*, Cas A) reach their peak X-ray luminosities after a few centuries and fade thereafter.

As I discuss below, we believe that SN1987A is surrounded by a few $M_\odot$ of circumstellar matter within a distance of parsec or two. Thus, a significant fraction of the kinetic energy of the debris will be converted into thermal energy within a few decades as the supernova blast wave overtakes this matter.

THE CIRCUMSTELLAR RINGS

The first evidence for circumstellar matter around SN1987A appeared a few months after outburst in the form of narrow optical and ultraviolet emission lines seen with the *International Ultraviolet Explorer* [1]. Even before astronomers could image this matter, they could infer that:

- the gas was nearly stationary (from the linewidths);

- it was probably ejected by the supernova progenitor (because the abundance of nitrogen was elevated);

- it was ionized by soft X-rays from the supernova flash (from emission lines of N v $\lambda\lambda 1239, 1243$ and other highly ionized elements in the spectrum);

- it was located at a distance of about a light year from the supernova (from the rise time of the light curve of these lines); and

- the gas had atomic density $\sim 3 \times 10^3 - 3 \times 10^4$ cm^{-3} (from the fading timescale of the narrow lines).

The triple ring system was first seen in images obtained by the ESO *NTT* telescope [2], but the evidence of the outer loops was not compelling until astronomers obtained an image with the *HST* WFPC-2 [3]. By measuring the Doppler shifts of the emission lines, Crotts & Heathcote [4] found that the inner ring is expanding with a radial velocity ≈ 10 km s^{-1}. Dividing the radius of the inner ring (0.67 lt-year) by this velocity gives a kinematic timescale $\approx 20,000$ years since the gas in the ring was ejected, assuming constant velocity expansion. The more distant outer loops are expanding more rapidly, consistent with the notion that they were ejected at the same time as the inner ring.

The rings observed by *HST* may be only the tip of the iceberg. They are glowing by virtue of the ionization and heating caused by the flash of EUV and soft X-rays emitted by the supernova during the first few hours after outburst. But calculations [5] show that this flash was a feeble one. The glowing gas that we see in the triple ring system is probably only the ionized inner skin of a much greater mass of unseen gas that the supernova flash failed to ionize. For example, the inner ring has a glowing mass of only about $\sim 0.04 M_\odot$, just about what one would expect such a flash to produce.

In fact, ground-based observations of optical light echoes during the first few years after outburst provided clear evidence of a much greater mass of circumstellar gas within several light years of the supernova that did not become ionized [2,6]. The echoes were caused by scattering of the optical light from the supernova by dust grains in this gas. They became invisible about five years after outburst.

What accounts for this circumstellar matter and the morphology of the rings? My hunch is that the supernova progenitor was originally a close binary system, and that the two stars merged some 20,000 years ago. The inner ring might be the inner rim of a circumstellar disk that was expelled during the merger, perhaps as a stream of gas that spiraled out from the outer Lagrangean (L2) point of the binary system. Then, during the subsequent 20,000 years before the supernova event, ionizing photons and stellar wind from the merged blue giant star eroded a huge hole in the disk. Finally, the supernova flash ionized the inner rim of the disk, creating the inner ring that we see today.

The binary hypothesis provides a natural explanation of the bipolar symmetry of the system, and may also explain why the progenitor of SN1987A was a blue giant rather than a red giant [7]. But we still lack a satisfactory explanation for the

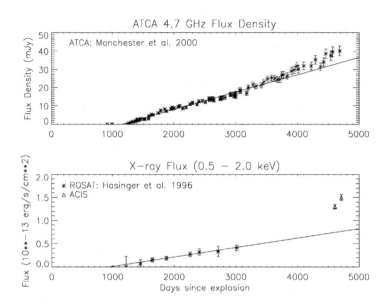

FIGURE 1. Radio (upper) and X-ray (lower) light curves.

outer loops. If we could only see the invisible circumstellar matter that lies beyond the loops, we might have a chance of reconstructing the mass ejection episode.

Fortunately, SN1987A will give us another chance. When the supernova blast wave hits the inner ring, the ensuing radiation will cast a new light on the circumstellar matter. As I describe below, this event is now underway.

THE CRASH BEGINS

The first evidence that the supernova debris was beginning to interact with circumstellar matter came from radio and X-ray observations. As Figure 1 shows, SN1987A became a detectable source of radio [8] and soft X-ray [9] emission about 1200 days after the explosion and has been brightening steadily in both bands ever since. Shortly afterwards, astronomers imaged the radio source with the Australia Telescope Compact Array (*ATCA*) and found that the radio source was an elliptical annulus inside the inner circumstellar ring observed by *HST* (Fig. 5). From subsequent observations, they found that the annulus was expanding with a velocity $\sim 3,500$ km s^{-1}.

Chevalier [10] recognized that the radio emission most likely arose from relativistic electrons accelerated by shocks formed inside the inner ring where the supernova debris struck relatively low density ($n \sim 100$ cm^{-3}) circumstellar matter, and that the X-ray emission probably came from the shocked circumstellar matter and supernova debris. Subsequently, Chevalier & Dwarkadas [11] suggested a model for the circumstellar matter, in which the inner circumstellar ring is the waist of an hourglass-shaped bipolar nebula. The low density circumstellar matter is a thick layer of photoionized gas that lines the interior of the bipolar nebula. The inner boundary of this layer is determined by balance of the pressure of the hot bubble of shocked stellar wind gas and that of the photoionized layer. In the equatorial plane, the inner boundary of this layer is located at about half the radius of the inner ring. According to this model, the appearance of X-ray and radio emission at ~ 1200 days marks the time when the blast wave first enters the photoionized layer.

THE REVERSE SHOCK

Following Chevalier & Dwarkadas [11], Borkowski et al. [12] developed a more detailed model to account for the X-ray emission observed from SN1987A. They used a 2-D hydro code to simulate the impact of the outer atmosphere of the supernova with an idealized model for the photoionized layer. They found a good fit to the *ROSAT* observations with a model in which the thickness of the photoionized layer was about half the radius of the inner ring and the layer had atomic density $n_0 \approx 150$ cm^{-3}.

With the same model, we found to our delight that Lyα and Hα emitted by hydrogen atoms crossing the reverse shock should be detectable with the STIS. Then, in May 1997, only three months after our predictions [10] appeared, the first STIS observations of SN1987A were made, and broad ($\Delta V \approx \pm 12,000$ km s^{-1}) Lyα emission lines were detected [13]. Within the observational uncertainties, the flux was exactly as predicted.

One might at first be surprised that such a theoretical prediction of the Lyα flux would be on the mark, given that it was derived from a hydrodynamical model based on very uncertain assumptions about the density distribution of circumstellar gas. But, on further reflection it is not so surprising because the key parameter of the hydrodynamical model, the density of the circumstellar gas, was adjusted to fit the observed X-ray flux. Since the intensity of Lyα is derived from the same hydrodynamical model, the ratio of Lyα to the X-ray flux is determined by the ratio of cross sections for atomic processes, independent of the details of the hydrodynamics.

The broad Lyα and Hα emission lines are not produced by recombination. (The emission measure of the shocked gas is far too low to produce detectable Lyα and Hα by recombination.) Instead, the lines are produced by neutral hydrogen atoms in the supernova debris as they cross the reverse shock and are excited by collisions with electrons and protons in the shocked gas. Since the cross sections for excitation of the $n \geq 2$ levels of hydrogen are nearly equal to the cross sections for impact ionization, about one Lyα photon is produced for each hydrogen atom that crosses the shock. Thus, the observed flux of broad Lyα is a direct measure of the flux of hydrogen atoms that cross the shock. Moreover, since the outer supernova envelope

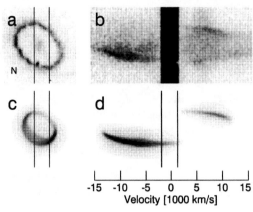

FIGURE 2. STIS spectrum of Lyα emission from the reverse shock [14].

is expected to be nearly neutral, the observed flux is a measure of the mass flux across the shock.

The fact that the Lyα and Hα lines are produced by excitation at the reverse shock gives us a powerful tool to map this shock. Since any hydrogen in the supernova debris is freely expanding, its line-of-sight velocity, $V_{\parallel} = z/t$, where z is its depth measured from the mid-plane of the debris and t is the time since the supernova explosion. Therefore, the Doppler shift of the Lyα line will be directly proportional to the depth of the reverse shock: $\Delta\lambda/\lambda_0 = z/ct$. Thus, by mapping the Lyα or Hα emission with STIS, we can generate a 3-dimensional image of the reverse shock.

Figure 2 illustrates this procedure. Panel **a** shows the location of the slit superposed on an image of the inner circumstellar ring, with the near (**N**) side of the tilted ring on the lower left. Panel **b** shows the actual STIS spectrum of Lyα from this observation. The slit is black due to geocoronal Lyα emission. The bright blue-shifted streak of Lyα extending to the left of the lower end of the slit comes from hydrogen atoms crossing the near side of the reverse shock, while the fainter red-shifted streak at the upper end of the slit comes from the far side of the reverse shock.

From this and similar observations with other slit locations we have constructed a map of the reverse shock surface, shown in panel **c**. Note that the emitting surface is an annulus that lies inside the inner circumstellar ring. Presumably, the reverse shock in the polar directions lies at a greater distance from the supernova, where the flux of atoms in the supernova debris is too low to produce detectable emission. Panel **d** is a model of the STIS Lyα spectrum that would be expected from hydrogen atoms crossing the shock surface illustrated in panel **c**. By comparing such model spectra with the actual spectra (e.g., panel **b**), we may refine our model of the shock surface.

Note that the broad Lyα emission is much brighter on the near (blue-shifted) side of the debris than on the far side, and so is the reconstructed shock surface. There is one obvious reason why this should be so: the blue-shifted side of the reverse shock is nearer to us by several light-months, and so we see the emission from the near side as it was several months later than that from the far side. Since the flux of atoms across the reverse shock is increasing, the near side should be brighter. But this explanation fails quantitatively. The observed asymmetry is several times greater than can be explained by light-travel time delays, and must be attributed to real asymmetry in the supernova debris. As we shall see, observations at radio and X-ray wavelengths also provide compelling evidence for asymmetry of the supernova debris.

THE HOT SPOTS

One can estimate the time that the blast wave should strike the inner circumstellar ring from Chevalier's self-similar solutions [15] for the hydrodynamics of a freely expanding stellar atmosphere striking a circumstellar medium. Stellar atmosphere models give a good fit to the spectrum of SN1987A during the early photospheric phase with a stellar atmosphere having a power-law density law $\rho(r,t) = At^{-3}(r/t)^{-9}$ [16,17]. If this atmosphere strikes a circumstellar medium having a uniform density n_0, the blast wave will propagate according to the law $R_B(t) \propto A^{1/9}n_0^{-1/9}t^{2/3}$. For such a model, the time of impact can be estimated from the equation $t \propto A^{-1/6}n_0^{1/6}$. The coefficient, A, of the supernova atmosphere density profile can determined from the fit to the photospheric spectrum, leaving the density, n_0, of the gas between the supernova and the circumstellar ring as the main source of uncertainty. With various assumptions about the density distribution of this gas, predictions of the time of first contact ranged from 1999 ± 3 to 2005 ± 3 [11,18,19].

In April 1997, Sonneborn et al. [13] obtained the first STIS spectrum of SN1987A with the 2×2 arcsecond aperture. Images of the circumstellar ring were seen in several optical emission lines. No Doppler velocity spreading was evident in the ring images except at one point, located at $P.A. = 29°$ (E of N), which we now call "Spot 1," where a Doppler-broadened streak was seen in Hα and other optical lines.

Figure 3 shows a portion of a more recent (March 1998) STIS spectrum of Spot 1 [20], where one can see vertical pairs of bright spots corresponding to emission from the stationary ring at Hα and [NII]$\lambda\lambda6548, 6584$ (one also sees three more fainter spots at each wavelength where the outer loops cross the slit, and a broad horizontal streak at the center due to the Hα emission from the rapidly expanding inner debris). The emission lines are broadened (with $FWHM \approx 250$ km s^{-1}) and blue-shifted (with $\Delta V \approx -80$ km s^{-1}) at the location of Spot 1, which is located slightly inside the stationary ring.

Spot 1 evidently marks the location where the supernova blast wave first touches the dense circumstellar ring. When a blast wave propagating with velocity $V_b \approx$

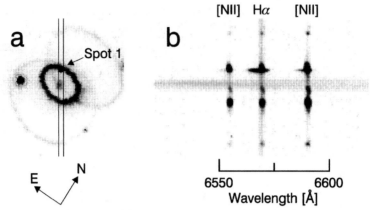

FIGURE 3. Spectrum of Spot 1. (a) Slit orientation on image of SN1987A's triple ring system; (b) Section of STIS G750M spectrum.

$4,000$ km s^{-1} through circumstellar matter with density $n_0 \approx 150$ cm^{-3} encounters the ring, having density $n_r \approx 10^4$ cm^{-3}, one would expect the transmitted shock to propagate into the ring with $V_r \approx (n_0/n_r)^{1/2}V_b \approx 500$ km s^{-1} if it enters at normal incidence, and more slowly if it enters at oblique incidence. Since Spot 1 is evidently a protrusion, a range of incidence angles, and hence of transmitted shock velocities and directions, can be expected. Obviously, the line profiles will be sensitive to the geometry of the protrusion. Since the protrusion is on the near side of the ring and is being crushed by the entering shocks, most of the emission will be blue-shifted, as is observed. But part of the emission is red-shifted because it comes from oblique shocks entering the far side of the protrusion.

Looking back to previous WFPC images, Garnavich et al. [21] found that Spot 1 had begun to brighten as early as 1996. It has continued to brighten steadily, with a current doubling time scale of about one year. As of February 2000, Spot 1 had a flux $\approx 7\%$ of the rest of the inner circumstellar ring.

Since the detection of Spot 1, several new spots have appeared, of which four (at $P.A. \approx 91°, 106°, 123°$ and $230°$, are evident in Figure 4 [21,22]. Clearly, the blast wave is beginning to overtake the inner circumstellar ring in several places.

The emission line spectrum of Spot 1 resembles that of a radiative shock, in which the shocked gas has had time to cool from its post-shock temperature $T_1 \approx 1.6 \times 10^5[V_r/(100$ km s$^{-1})]^2$ K to a final temperature $T_f \approx 10^4$ K or less. As the shocked gas cools, it is compressed by a density ratio $n_f/n_r \approx (T_1/T_f) \approx 160[V_r/(100$ km s$^{-1})]^2 \, [T_f/(10^4\text{K})]^{-1}$. We see evidence of this compression in the observed ratios of forbidden lines, such as [NII]$\lambda\lambda6548, 6584$ and [SII]$\lambda\lambda6717, 6731$, from which we infer electron densities in the range $n_e \sim 10^6$ cm^{-3} using standard nebular diagnostics.

The fact that the shocked gas in spot 1 was able to cool and form a radiative layer

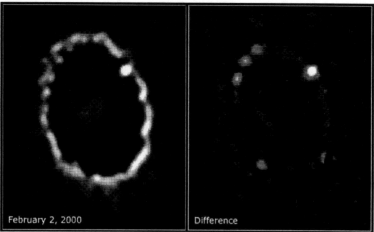

February 2, 2000 Difference

FIGURE 4. Hot spots on the inner ring. The right panel shows the difference between WFPC observations taken on 2 February 2000 and 21 February 1996, showing that the hot spots have brightened while the rest of the ring has faded.

within a few years sets a lower limit, $n_r \gtrsim 10^4$ cm^{-3}, on the density of unshocked gas in the protrusion. Given that limit, we can estimate an upper limit on the emitting surface area of Spot 1, from which we infer that Spot 1 should have an actual size no greater than about one pixel on WFPC2. This result is consistent with the imaging observations.

The cooling timescale of shocked gas is sensitive to the postshock temperature, hence shock velocity. For $n_r = 10^4$ cm^{-3}, shocks faster than 250 km s^{-1} will not be able to radiate and form a cooling layer within a few years. It is quite possible that such fast non-radiative shocks are present in the protrusions but are invisible in optical and UV line emission. For example, I estimated above that a blast wave entering the protrusion at normal incidence might have velocity ~ 500 km s^{-1}. We would still see the line emission from the slower oblique shocks on the sides of the protrusion, however.

We have attempted to model the observed emission line spectrum of Spot 1 with a radiative shock code kindly provided by John Raymond. Up to now, our efforts have met with only partial success. This is perhaps not surprising, given the complexity of the hydrodynamics. It is known, for example, that radiative shocks are subject to violent thermal instabilities [23], which we have not included in our initial attempts to model the shock emission.

THE X-RAY SOURCE

As I have already mentioned above, we believe that the X-ray emission from SNR1987A seen by *ROSAT* (Fig. 1) comes from the hot shocked gas trapped

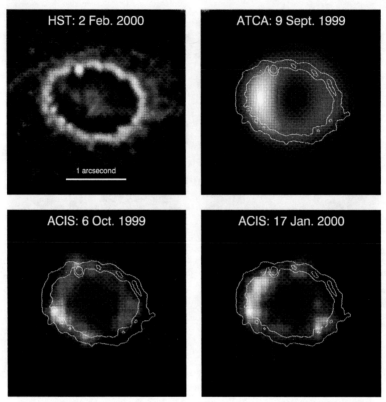

FIGURE 5. Optical, radio and X-ray images of SNR1987A.

between the supernova blast wave and the reverse shock. But, with its 10" angular resolution, *ROSAT* was unable to image this emission; nor was *ROSAT* able to obtain a spectrum.

Very recently, Burrows et al. [24] used the new *Chandra* observatory to advance our knowledge of the image and spectrum of X-rays from SNR1987A. Figure 5 is a montage of images of SNR1987A, showing: the *HST* optical image (a); the *ATCA* radio image from the (b); and the X-ray images observed by *Chandra* on 6 October 1999 (c) and 17 January 2000 (d). The optical image (a) is replicated as contour lines on images (b), (c), and (d). We see immediately that the *ATCA* radio source (b) is an annulus that is somewhat smaller than the optical ring and is much brighter on the **E** side than on the **W**.

The two *Chandra* images appear different, but we are not sure whether this difference represents an actual change of the X-ray source or is an artifact of the limited photon statistics (only about 600 counts per image) and the deconvolution procedure we used to achieve the maximum possible angular resolution. We can be sure, however, that the X-ray source, like the radio source, is an annulus that lies mostly within the optical ring, and that it is brighter on the **E** side than on the **W**.

The X-ray images of SNR1987A are consistent with the model by Borkowski et al. [12], except that they obviously do not have the cylindrical symmetry assumed in that model. The fact that the X-ray and radio images have roughly the same morphologies suggests that the relativistic electrons presumed responsible for the non-thermal radio emission have energy density proportional to that in the X-ray emitting gas and reside in roughly the same volume.

The fact that the X-ray and radio images are both brighter on the **E** side than on the **W** could be explained by a model in which either: (a) the circumstellar gas inside the inner ring had greater density toward the **E**; or (b) the outer supernova debris had greater density toward the **E**. But the fact that most of the hot spots are found on the **E** side favors the latter hypothesis. If the circumstellar gas had greater density toward the **E** side and the supernova debris were symmetric, the blast wave would have propagated further toward the **W** side, and the hot spots would have appeared there first.

This conclusion is also supported by observations of Hα and Lyα emission from the reverse shock (see below), which show that the flux of mass across the reverse shock is greater on the **W** side.

These observations highlight a new puzzle about SN1987A: why was the explosion so asymmetric? We might explain a lack of spherical symmetry by rapid rotation of the progenitor, but how do we explain a lack of azimuthal symmetry?

With the grating spectrometer on *Chandra*, Burrows et al. [24] also obtained a spectrum of the X-rays from SNR1987A, shown in Figure 6. It is dominated by emission lines from helium- and hydrogen-like ions of O, Ne, Mg, and Si, as well as a complex of Fe-L lines near 1 keV, as predicted by Borkowski et al. [12]. The characteristic electron temperature inferred from the spectrum, $kT_e \sim 3$ keV, is much less than the proton temperature, $kT_p \sim 30$ keV for a blast wave propagating with $V_b \approx 4,000$ km s^{-1}. This result was expected because Coulomb collisions are too slow to raise the electron temperature to temperature equilibrium with the ions.

The *Chandra* observations show that the current X-ray flux from SNR1987A is about twice the value that would be estimated by extrapolating the *ROSAT* light curve to January 2000 (Figure 1). The X-ray flux is expected to increase by another factor $\sim 10^2$ during the coming decade as the blast wave overtakes the inner circumstellar ring [25]. Are we already beginning to see the X-ray emission from the shocked ring? Further imaging (with *Chandra*) and spectroscopic (with *XMM*) observations will tell.

THE FUTURE

SNR1987A has been tremendous fun so far, but the best is yet to come. During the next ten years, the blast wave will overtake the entire circumstellar ring. More hot spots will appear, brighten, and eventually merge until the entire ring is blazing

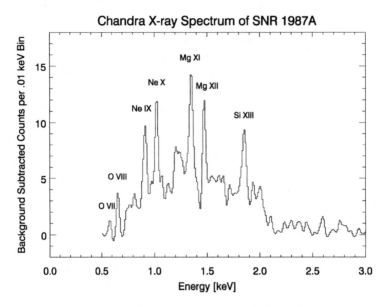

FIGURE 6. X-ray spectrum of SNR1987A.

brighter than Spot 1. We expect that the Hα flux from the entire ring will increase to $F_{H\alpha} \gtrsim 3 \times 10^{-12}$ ergs cm^{-2} s^{-1}, or $\gtrsim 30$ times brighter than it is today and that the flux of ultraviolet lines will be even greater [19].

As we have already begun to see, observations at many wavelength bands are needed to tell the entire story of the birth of SNR1987A. Fortunately, powerful new telescopes and technologies are becoming available just in time to witness this event.

Large ground-based telescopes equipped with adaptive optics will provide excellent optical and infrared spectra of the hot spots. We need to observe profiles of several emission lines at high resolution in order to unravel the complex hydrodynamics of the hot spots. These telescopes also offer the exciting possibility to image the source in infrared coronal lines of highly ionized elements (e.g., [Si IX] 2.58, 3.92μm, [Si X] 1.43μm) that may be too faint to see with *HST*. Observations in such lines will complement X-ray observations to measure the physical conditions in the very hot shocked gas.

The circumstellar rings of SN1987A almost certainly contain dust, and so does the envelope of SN1987A [26]. When dusty gas is shocked to temperatures $\gtrsim 10^6$ K, its emissivity at far infrared wavelengths ($\gtrsim 10\mu$m) exceeds its X-ray emissivity by factors $\sim 10^2 - 10^3$ [27]. Therefore, we expect that SNR1987A will be brightest at far infrared wavelengths, with luminosity $\sim 10^2 - 10^3$ times its X-ray luminosity. This makes SNR1987A a prime target for *SIRTF*. Together with the *Chandra* and *XMM* observations of the X-ray spectra, the *SIRTF* observations will give us a unique opportunity to investigate the destruction of dust grains in shocked gas.

The observations with the *ATCA* have given us our first glimpse of shock acceleration of relativistic electrons in real time, but the angular resolution of *ATCA* is not quite good enough to allow a detailed correlation of the radio image with the optical and X-ray images. This will become possible several years from now when the Atacama Large Millimeter Array (*ALMA*) is completed. Such observations will give us a unique opportunity to test our theories of relativistic particle acceleration by shocks.

Of course, we should continue to map the emission of fast Lyα and Hα from the reverse shock with *STIS*. Such observations give us a three-dimensional image of the flow of the supernova debris across the reverse shock, providing the highest resolution map of the asymmetric supernova debris. We expect this emission to brighten rapidly, doubling on a timescale ~ 1 year. Most exciting, such observations will give us an opportunity to map the distribution of nucleosynthesis products in the supernova debris. We know that the debris has a heterogeneous composition. The early emergence of gamma rays from SN1987A showed that some of the newly synthesized ^{56}Co (and probably also clumps of oxygen and other elements) were mixed fairly far out into the supernova envelope by instabilities following the explosion [26]. When such clumps cross the reverse shock, the fast Hα and Lyα lines will vanish at those locations, to be replaced by lines of other elements. If we keep watching with *STIS*, we should see this happen during the coming decade.

Likewise, we should continue to monitor the development of the hot spots with *HST*, using both *WFPC2* to observe images and *STIS* to observe spectra. Optical and infrared spectra obtained with ground-based telescopes will be of limited value unless they are complemented by *HST* observations that tell us which spots are producing which lines. Moreover, only *HST* can observe UV lines such as N IV] $\lambda\lambda1483, 1486$ and N V $\lambda\lambda1239, 1243$, the ratio of which are sensitive functions of shock velocity. We need to measure these ratios as a function of Doppler shift to untangle the complex hydrodynamics of the shocks entering the spots. Spot 1 is now becoming bright enough to do that with *HST*.

A deeper *Chandra* image will tell us whether we are seeing X-ray emission from the hot spots.

The shocks in the hot spots are surely producing ionizing radiation, roughly half of which will propagate ahead of the shock and ionize heretofore invisible material in the rings. The effects of this precursor ionization will soon become evident in the form of narrow cores in the emission lines from the hot spots.

The circumstellar rings of SN1987A represent only the inner skin of a much greater mass of circumstellar matter, and we only obtained a fleeting glimpse of this matter through ground-based observations of light echoes. The clues to the origin of the circumstellar ring system lie in the distribution and velocity of this matter, if we could only see it clearly. Fortunately, SNR1987A will give us another chance. Although it will take several decades before the blast wave extends to the outer rings, the impact with the inner ring will eventually produce enough ionizing radiation to cause the unseen matter to become an emission nebula. Luo et al. [19]

have estimated that the fluence of ionizing radiation from the impact will equal the initial ionizing flash of the supernova within a few years after the ring reaches maximum brightness. I expect that the circumstellar nebula of SNR1987A will be in full flower within a decade. In this way, SN1987A will be illuminating its own past.

Many of the results reported here are the fruits of my happy collaborations with John Blondin, Kazik Borkowski, Dave Burrows, Steve Holt, Una Hwang, Eli Michael, Jason Pun, George Sonneborn, Svet Zhekov, and the Supernova INtensive Study team led by Bob Kirshner. My own work on SN1987A is currently supported by grants from NASA (ATP NGT 5-80) and the Space Telescope Science Institute (GO-08243.03-97A).

REFERENCES

1. Fransson, C., et al. 1989, ApJ, 336, 429–441
2. Wampler, J., et al. 1990, ApJ, 362, L13–16
3. Burrows, C. J. et al. 1995, ApJ, 452, 680–684
4. Crotts, A., & Heathcote, S. R. 1991, Nature, 350, 683–685
5. Ensman, L., & Burrows, A. 1992, ApJ, 393, 742–755
6. Crotts, A. P. S., Kunkel, W. E., & Heathcote, S. R. 1995, ApJ, 38, 724–734
7. Podsiadlowski, P. 1992, Publ. Astr. Soc. Pacific, 104, 717–729
8. Manchester, R. N., et al. 2000, Publ. Astr. Soc. Australia, in press.
9. Hasinger, G., Aschenbach, B., & Trümper, J. 1996, Astr. & Ap., 312, L9–12
10. Chevalier, R. A. 1992, Nature, 355, 617–618
11. Chevalier, R. A., & Dwarkadas, V. I. 1995, ApJ, 452, L45–48
12. Borkowski, K. J., Blondin, J. M., & McCray, R. 1997, ApJ, 476, L31–34
13. Sonneborn, G., et al. 1998, ApJ, 492, L139–142
14. Michael, E., et al. 1998, ApJ, 509, L117–120
15. Chevalier, R. A. 1982, ApJ, 258, 790–797
16. Eastman, R. G., & Kirshner, R. P. 1989, ApJ, 347, 771–793
17. Schmütz, W. et al. 1990, ApJ, 355, 255–270
18. Luo, D., & McCray, R. 1991, ApJ, 372, 194–198.
19. Luo, D., McCray, R., & Slavin, J. 1994, ApJ, 430, 264–276
20. Michael, E., et al. 2000, ApJ, 543, L53–56
21. Garnavich, P., Kirshner, R., & Challis, P. 2000a, *IAU Circ. 7360*
22. Lawrence, S. S., et al. 2000, ApJ, 537, L123–126
23. Innes, D. E., Giddings, J. R., & Falle, S. A. E. G. 1987, MNRAS, 226, 67–93
24. Burrows, D. N., et al. 2000, ApJ, 543, L149–152
25. Borkowski, K. J., Blondin, J. M., & McCray, R. 1997, ApJ, 477, 281–293
26. Dwek, E., & Arendt, R. G. 1992, Ann. Rev. Astr. & Ap., 30, 11–50
27. McCray, R. 1993, Ann. Rev. Astr. & Ap., 31, 175–216

Supernova Remnants in Molecular Clouds

Roger A. Chevalier

Department of Astronomy, University of Virginia
P.O. Box 3818, Charlottesville, VA 22903, USA

Abstract. Supernovae are expected to occur near the molecular material in which the massive progenitor star was born, except in cases where the photoionizing radiation and winds from the progenitor star and its neighbors have cleared out a region. The clumpy structure in molecular clouds is crucial for the remnant evolution; the supernova shock front can become radiative in the interclump medium and the radiative shell then collides with molecular clumps. The interaction is relevant to a number of phenomena: the hydrodynamics of a magnetically supported dense shell interacting with molecular clumps; the molecular emission from shock waves, including the production of the OH 1720 MHz maser line; the relativistic particle emission, including radio synchrotron and γ-ray emission, from the dense radiative shell; and the possible gravitational instability of a compressed clump.

INTRODUCTION

Massive stars are born in molecular clouds and those with masses $\gtrsim 8$ $M_\odot$ end their lives as supernovae after $\lesssim 3 \times 10^7$ years. This age is less than the typical age of molecular clouds so that the supernovae may explode in the vicinity of molecular cloud material. Early studies of the interaction assumed that the supernovae interacted directly with molecular cloud material, with density $n_H = 10^4 - 10^5$ cm^{-3} [1], [2]. At these high densities, the supernova remnant evolves on a timescale of 10's of years during which it is a luminous infrared source.

Such sources have not been clearly observed. However, supernova remnant interaction with molecular gas has been observed, starting with the remnant IC 443 [3]. Continued observations have shown IC 443 to be one of the best cases of molecular cloud line emission [4]. The emission appears to be associated with shock interaction with dense clumps, as expected in a molecular cloud; the point of view adopted here involves the expansion of the supernova remnant shell in the interclump medium of a cloud with some collisions with clumps [5].

CP565, *Young Supernova Remnants: Eleventh Astrophysics Conf.,* edited by S. S. Holt and U. Hwang
© 2001 American Institute of Physics 0-7354-0001-6/01/$18.00

THE SUPERNOVA SURROUNDINGS

A crucial point for the evolution of supernova remnants in molecular clouds is the clumpy structure of these clouds. The basic picture that has emerged is of dense clumps with most of the mass embedded in a lower density interclump medium. Blitz [6] notes some properties of typical Giant Molecular Clouds: total mass $\sim 10^5 \ M_\odot$, diameter ~ 45 pc, H_2 density in clumps of 10^3 cm^{-3}, interclump gas density in the range $5-25$ H atoms cm^{-3}, and clump filling factor of $2-8\%$. The clumps have an approximate power law mass spectrum, such that there are more small clumps, but most of the mass is in the large clumps. There have been recent attempts to model the clump properties in terms of MHD (magnetohydrodynamic) turbulence [7,8]. In this case, the clumps may be transitory features in the molecular cloud.

If the dense clumps are confined by the interclump pressure, a pressure $p/k \approx 10^5$ K cm^{-3} is needed [6]. Even if the clumps are features in a fluctuating velocity field and are not necessarily confined, a comparable pressure is needed to support the clouds against gravitational collapse. The thermal pressure in the interclump gas is clearly too small and magnetic fields are a plausible pressure source. Studies of the polarization of the light from stars behind molecular clouds implies that the interclump magnetic field does have a uniform component [9]. Setting $B_o^2/8\pi$ equal to the above pressure yields $B_o = 19p_5^{1/2} \ \mu$G, where B_o is the uniform field component and p_5 is p/k in units of 10^5 K cm^{-3}. However, a uniform magnetic field would not provide support along the magnetic field and would not explain the large line widths observed in clouds. An analysis of the polarization and Zeeman effect in the dark cloud L204 shows consistency with equal contributions to the pressure from a uniform field and a fluctuating, nonuniform component [9]. The uniform component is then $B_o = 13p_5^{1/2} \ \mu$G. The magnetic field strength deduced in L204 is consistent with this value.

The molecular cloud can be influenced by stellar mass loss and photoionization by the progenitor star before the supernova. The effect of these processes can be to remove molecular gas from the vicinity of the central star. Draine and Woods [10] found that stars with initial mass $\lesssim 20 \ M_\odot$ in clouds of density $n_H \gtrsim 10^2$ cm^{-3} would have such a small effect that the surroundings could be treated as homogeneous in discussing the blast wave evolution. For a 25 $M_\odot$ star, they found the presupernova effects to be small provided $n_H \gtrsim 10^3$ cm^{-3}. In an interclump medium with $n_H \approx 10$ cm^{-3}, Chevalier [5] found that stars with mass $\lesssim 12 \ M_\odot$ would clear a region with radius $\lesssim 5$ pc around the star because of the effects of photoionization and winds. The high pressure of the molecular cloud plays a role in limiting the effects of the progenitor star. If the massive star has a significant velocity (~ 5 km s^{-1}), the star can move into a region of undisturbed cloud material and the progenitor star effects are lessened [10]. On the other hand, massive stars tend to form in clusters and the immediate environment of a star could be affected by nearby, more massive stars. In a sparse cluster, it is still possible for a massive star to interact with its natal cloud material.

THE CASE OF IC 443

The remnant IC 443 shows many of the features that characterize molecular cloud interaction. The remnant appears to be interacting with a relatively low mass molecular cloud ($\lesssim 10^4\ M_\odot$) that is primarily in front of the remnant [11]. This low mass suggests that there were few or no massive stars which would be likely to disrupt the cloud with their photoionizing fluxes and winds. A lower mass Type II supernova could thus interact directly with the molecular cloud. I take the distance to the remnant to be 1.5 kpc [12].

The morphology of the remnant is a shell in the molecular cloud region, with an apparent "break-out" region to the southwest (see [13] for an X-ray image and [14] for a radio image). I consider the shell part of the remnant to be interacting with the molecular cloud. Fesen and Kirshner [12] found that the spectra of the optical filaments in this region imply an electron density for the [S II] emitting region of $\lesssim 100$ to 500 cm^{-3} and shock velocities in the range $65 - 100$ km s^{-1}. If the magnetic field does not limit the compression at a temperature $\sim 10^4$ K, the preshock density implied by the higher density filaments is $10 - 20$ cm^{-3}. The filaments with a lower density may have their compression limited by the magnetic field. These observations point to a radiative phase of evolution for the remnant in the interclump medium of the molecular cloud. The cooling shock front is expected to build up a shell of HI [15]. For an ambient density of 15 cm^{-3} and a radius of 7.4 pc, the total swept-up mass is 1000 $M_\odot$ for a half-sphere. HI observations suggest that there is $\sim 1000\ M_\odot$ of shocked HI in the northeast shell part of the remnant, with velocities up to ~ 110 km s^{-1} [16]. These results are in quantitative accord with expectations for a 10^{51} erg supernova in a medium with $n_o = 15$ cm^{-3} [5,15], if account is taken of the fact that some energy has been released in the blow-out region.

In a restricted region across the central, eastern part of IC 443, a rich spectrum of molecular emission has been observed [4,17]. A set of spectroscopically distinct CO clumps has been labelled A–G [3,18], although the H$_2$ emission indicates that the clumps are associated through a filamentary molecular structure [19]. These clumps, with sizes ~ 1 pc, have masses of $3.9 - 41.6\ M_\odot$ as deduced from ^{12}CO lines [20]. The corresponding densities, $n_H \lesssim 500$ cm^{-3}, are low but should be regarded as lower limits because the ^{12}CO emission is assumed to be optically thin. Also, there is likely to be density structure within individual clumps. Analysis of absorption in one place yielded a preshock density of $n(H_2) \approx 3,000$ cm^{-3} [4]. The properties of these clumps are consistent with those seen in quiescent molecular clouds. High spatial resolution is possible in the H$_2$ infrared lines, which show clumps down to a scale of $1''$ (2×10^{16} cm$= 0.007$ pc at a distance of 1.5 kpc) but not smaller [17]. The emission has a knotty appearance that is unlike the filamentary appearance of the optical emission in IC 443 and the H$_2$ emission in the Cygnus Loop [17].

This type of structure is consistent with the possible fractal structure of molecular clumps [8].

The 2MASS survey has provided a K_s band image of IC 443 that is likely to be dominated by shocked H_2 emission [21]. The image clearly shows a ridge of emission passing from the south to the central part of the remnant. The J and H band images with the 2MASS survey show emission from the northeast part of the remnant, which is likely to be [Fe II] emission from the ~ 100 km s^{-1} shock wave. Observations of the [O I] 63 μm line with *ISO* (Infrared Space Observatory) show that it is particularly strong in the northeast shell region [21], which implies that it is also from the ~ 100 km s^{-1} shock. This is consistent with expectations that most of the [O I] 63 μm line luminosity should be from the shock front in the interclump medium [5].

The interpretation of the molecular line strengths is still controversial, but the emission is most consistent with that from a partially dissociating J(jump)-type shock [4,17,19,22]. The required shock velocity, $25 - 30$ km s^{-1}, is consistent with the velocities observed in the strongest emission. The dissociated H_2 can be observed as a high column density of shocked HI at the positions of the clumps [23]. Following [5], I interpret the clump shock as being driven by the radiative shell in the interclump region (Fig. 1); this can generate a pressure in the interaction region

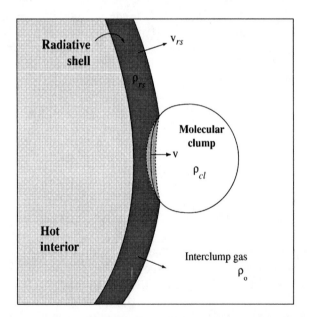

FIGURE 1. Interaction of a radiative shell, moving at velocity v_{rs}, with a molecular clump. The interaction generates a dense slab bounded by shock waves (dashed lines) and moving at velocity v. The densities of the molecular clump, ρ_{cl}, the radiative shell, ρ_{rs}, and the interclump medium, ρ_o, are indicated. From [5].

significantly above that expected for ram pressure equilibrium. For a clump shock velocity $v = 25 - 30$ km s^{-1} and a radiative shell velocity $v_{rs} = 100$ km s^{-1}, the ratio of clump density to shell density ρ_{cl}/ρ_{rs} is in the range $5.4 - 9.0$. The density ratio can vary significantly with relatively little effect on the shock velocity. The clump shock velocity is consistent with shell density $n_{rs} = 500$ cm^{-3} and preshock clump density $n_{cl} = 3000$ cm^{-3}, which are plausible values. The ratio of postshock pressure in the clump to that in the interclump medium (with $n_o = 15$ cm^{-3}) is 18.

The column density through the radiative shell is $N_H \approx n_o R/3 \approx 10^{20}$ cm^{-2}. For $n_{cl} = 3000$ cm^{-3}, the column density through the clumps is $10^{22}\ell_{\rm pc}$ cm^{-2}, where $\ell_{\rm pc}$ is the path length through the clump in pc. The larger observed clumps are thus expected to be passed over by the radiative shell and left in the interior of the remnant. For clumps with a size $\lesssim 0.02$ pc, the shock front breaks out of the clump first and there is the possibility of further acceleration by the radiative shell. The initial shock front through the clump may not dissociate molecules, which are then accelerated by the shell. Low column densities of high velocity molecular gas can be produced in this way. High velocity molecular gas has been observed in IC 443 at the edge of one of the clumps [24].

Cesarsky et al. [25] have observed H$_2$ rotational lines lines from clump G with *ISO*. They find that the line fluxes are consistent with a shock velocity ~ 30 km s^{-1}, a preshock density $\sim 10^4$ cm^{-3}, and an evolutionary time of $1,000 - 2,000$ years. At constant velocity, the shock has penetrated $\sim 0.03 - 0.06$ pc into the clump. This result is consistent with the clump interaction scenario.

Another important diagnostic of the molecular interaction is the OH 1720 MHz maser line, which Claussen et al. [14] find to be associated with clump G in IC 443. Lockett et al. [26] examined the pumping of the maser and found that the emission implies temperatures of $50 - 125$ K and densities $\sim 10^5$ cm^{-3}; the shock must be C-type. An important aspect of the OH maser observations is that it is possible to estimate the line-of sight magnetic field from the Zeeman effect. Although they do not have a result for IC 443, Claussen et al. [14] find $B_{\parallel} \approx 0.2$ mG for the maser spots in W28 and W44. This field strength is compatible with the field strength expected in the radiative shells of the supernova remnants [5].

In addition to the molecular line emission, IC 443 is a source of nonthermal continuum emission. At low frequencies, the power law radio spectrum is likely to be explained by synchrotron emission. The observed spectral index, $F_\nu \propto \nu^{-0.36}$ [27], implies a particle energy spectrum of the form $N(E) \propto E^{-1.72}$. Duin and van der Laan [28] explained the radio emission by the shock compression of the ambient magnetic field and relativistic electrons. First order Fermi accleration should be included and is consistent with the observations if ambient cosmic rays are accelerated [5,29]. In the shock acceleration of ambient particles, the particle spectrum is maintained in the postshock region if the particle spectrum is flatter that $N(E) \propto E^{-2}$. Chevalier [5] argued that the relatively flat spectrum may be consistent with the Galactic cosmic ray electron spectrum, which is observed to be flat at low energies, perhaps because of Coulomb losses in the interstellar medium.

If particles are injected into first order shock acceleration at a low energy, the

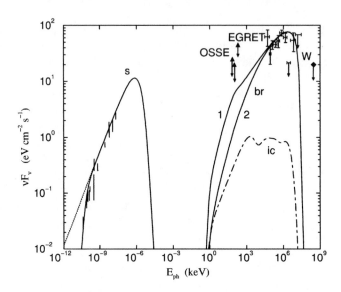

FIGURE 2. Broadband νF_ν spectrum of the shell of IC 443 calculated from a model of nonthermal electron production by a radiative shock with direct injection of electrons from the thermal pool. The shock velocity is 150 km s^{-1}, the interclump number density is 25 cm^{-3}, and the interclump magnetic field is 1.1×10^{-5} G. The two solid curves labeled by numbers 1 and 2 correspond to two limiting values of an MHD turbulence parameter. The extended inverse Compton emission from the whole remnant is shown as a dashed line (ic). The observational data points are from [32] for the *EGRET* source 2EG J0618 +2234 and [27] for the IC 443 radio spectrum. Upper limits for MeV emission are from *OSSE* observations [34]. Upper limits for γ-rays $\geq$ 300 GeV are from *Whipple* observations of IC 443 [36]. From [31].

spectrum tends to $N(E) \propto E^{-2}$ in the postshock region in the test particle limit. However, Ostrowski [30] found that second order Fermi acceleration in the turbulent medium near the shock could give rise to a relatively flat spectrum, as observed in IC 443. In another calculation including shock acceleration of electrons from a thermal pool, Bykov et al. [31] found that the combination of nonlinear effects on the shock compression and free-free absorption of the radiation could give rise to the observed spectrum (Fig. 2).

An additional constraint on nonthermal particles comes from the γ-ray detection of IC 443 by the *EGRET* experiment on *CGRO* (Compton Gamma-Ray Observatory) [32], although the poor spatial resolution of the *CGRO* still leaves some doubt about the detection. Assuming that the detection is correct, the emission mechanisms that might contribute to the radiation are electron bremsstrahlung, pion decays, and inverse Compton radiation [33]. A number of studies have fit the γ-ray spectrum of IC 443 with these processes [34,35], but they have assumed adiabatic shock wave dynamics. In view of the strong evidence for radiative shocks in IC 443, Bykov et al. [31] studied electron injection and acceleration in the context of radiative shock waves in a relatively dense medium. A model with injection of electrons from the thermal pool could approximately fit the observations, with bremsstrahlung being the dominant radiation mechanism (see Fig. 2). As can be seen in Fig. 2, the upper limit on TeV γ-rays from the Whipple Observatory [36] places an important constraint on the high energy part of the particle spectrum.

The continuum radio image of IC 443 [14] shows that although there is some correlation with the molecular emission [21], the shocked molecular clumps do not stand out as sources of radio synchrotron emission. In hard X-ray observations with *ASCA*, Keohane et al. [37] found two emission regions in IC 443, which have also been found in *BeppoSAX* observations [38]. Recent observations with *Chandra* show that the brighter source is likely to be a pulsar wind nebula [39], as indicated by earlier observations [5]. The other source may be associated with the molecular interaction along the southeast rim of the remnant. Keohane et al. [37] suggested that the emission could be synchrotron emission from shock accelerated electrons, but the slow shock velocity in the radiative shock model makes such an interpretation unlikely [5]. Bykov et al. [31] suggested that the emission could be bremsstrahlung from electrons accelerated in ionizing shock waves in relatively dense gas. The evidence for C-shocks giving rise to OH maser emission implies non-ionizing shock waves, but different conditions could prevail in different clumps. The prediction is that the shock waves in the interclump gas can produce bremsstrahlung γ-ray emission, while the shock waves in the clump gas can in some cases produce hard X-ray emission.

The radio emission from IC 443 is relatively highly polarized for a supernova remnant, indicating a preferred direction for the magnetic field in the shell [40]. The field direction agrees with that in the ambient cloud, as determined by the polarization of starlight. This is consistent with an ordered magnetic field in the cloud, as mentioned for molecular cloud support in the previous section.

The ordered field observed in radio emission refers to the field in the swept up radiative shell. However, there may also be a uniform component to the magnetic field in the hot interior of the radiative remnant, which would allow heat conduction to operate in the interior. IC 443 is observed to have a relatively isothermal interior with $T \approx 10^7$ K [41]. An isothermal interior with enhanced central X-ray emission is a characteristic of a number of remnants that are interacting with molecular gas. Shelton et al. [42] have discussed a conduction model for the remnant W44, which is a member of this class. They find that the model X-ray emission distribution profile is close to that observed, although it is not as centrally concentrated.

DISCUSSION AND FUTURE PROSPECTS

After a period of neglect, the study of supernova remnants interacting with molecular material has entered a phase of more intense study. In addition to IC 443, a number of remnants have come in for study. One reason for the increase is the re-discovery of the OH 1720 MHz maser line emission as a signpost of molecular clump interaction. Although initial observations were made more than 30 years ago [43], it was not until after the recent observations by Frail et al. [44] that the importance of the emission for remnant studies was realized. Surveys of remnants [45,46] have yielded 17 remnants with the OH maser line emission out of 160 observed. In addition, Koralesky et al. [47] made VLA observations searching for maser emission toward 20 remnants and found shock excited emission in 3 of them. When remnants with OH maser emission are observed in the CO line, molecular clumps have been observed in a number of cases. Frail and Mitchell [48] found CO clumps associated with the OH emission in W28, W44, 3C 391 and Reynoso and Mangum [49] found such clumps in three other remnants.

Another source of progress has been the availability of infrared observations, especially recently with *ISO*. *ISO* has been instrumental in detecting atomic fine structure lines, molecular lines, and dust continuum emission from remnants that appear to be interacting with molecular gas [50–52]. Of particular interest is the study by Reach and Rho [52] of emission from W28, W44, and 3C 391. They found evidence for interaction with a clumpy medium in which higher pressures were attained at higher densities, as discussed for IC 443. This situation can be explained by a dense radiative shell interacting with molecular clumps. More recently, the 2MASS survey has provided near-infrared imaging of the sky. The 2MASS K_s band image of IC 443 is likely to be dominated by shocked H_2 emission [21]. The detection of supernova remnants in this band thus provides a good indication that molecular interaction is taking place. The advent of the future NASA infrared observatories *SOFIA* and *SIRTF* will allow more complete studies of the density structure in shocked molecular clouds.

Reach and Rho [53] suggest that the mass of compressed gas in the shocked clump in 3C 391 is sufficiently high that self-gravity is significant and that it could eventually form one or more stars. Star formation in shocked supernova clumps cannot be directly verified because the clump collapse time is considerably longer than the age of the supernova remnants. By the time stars have formed, it is difficult to unambiguously discern a supernova trigger. One reason for interest in supernova triggered star formation has been the evidence for extinct radioactivities in the early solar system and this has stimulated computer simulations of the process [54,55]. The observations of supernova remnants in molecular clouds can be useful in providing realistic initial conditions for such simulations.

In cases where massive stars form in a dense cluster, the combined effects of the photoionizing radiation and winds can effectively clear molecular cloud material from the immediate vicinity of the stars. However, observations of the Trapezium cluster in Orion have shown the presence of small, dense gaseous regions, which O'Dell et al. [56] have identified as proplyds (protoplanetary disks). The eventual supernova explosion of one of the massive stars interacts with the disk material; nearby disks are disrupted, but ones farther out can survive [57]. These interactions may have observable consequences in supernova remnants.

The study of supernova remnants in molecular clouds has attracted renewed attention in the past few years and can be expected to blossom with the advent of new observatories and space missions, such as the *ALMA* millimeter array, the *SOFIA* and *SIRTF* infrared observatories, and the *GLAST* γ-ray mission. These observatories will make possible the detailed study of molecular shock waves and γ-ray emission from relativistic particles. The shock waves in molecular clouds provide an important probe of molecular cloud structure, which remains uncertain. A combination of hydrodynamic studies with emission calculations will be useful in elucidating this area.

ACKNOWLEDGMENTS

This work was supported in part by NASA grants NAG5-8232 and NAG5-8088.

REFERENCES

1. Shull, J. M., *ApJ* **237**, 769 (1980)
2. Wheeler, J. C., Mazurek, T. J., and Sivaramakrishnan, A., *ApJ* **237**, 781 (1980)
3. DeNoyer, L., *ApJ* **232**, L165 (1979)
4. van Dishoeck, E. F., Jansen, D. J., and Phillips, T. G., *A&A* **279**, 541 (1993)
5. Chevalier, R. A., *ApJ* **511**, 798 (1999)
6. Blitz, L., in *Protostars and Planets III*, Tucson: Univ. of Arizona, 1993, p. 125
7. Vázquez-Semadeni, E., Ostriker, E. C., Passot, T., and Gammie, C. F., in *Protostars and Planets IV*, Tucson: Univ. of Arizona, 2000, p. 3
8. Williams, J. P., Blitz, L., and McKee, C. F., in *Protostars and Planets IV*, Tucson: Univ. of Arizona, 2000, p. 97
9. Heiles, C., Goodman, A. A. McKee, C. F., and Zweibel, E. G., in *Protostars and Planets III*, Tucson: Univ. of Arizona, 1993, p. 279
10. Draine, B. T., and Woods, D. T., *ApJ* **383**, 621 (1991)
11. Cornett, R. H., Chin, G., and Knapp, G. R., *A&A* **54**, 889 (1977)
12. Fesen, R. A., and Kirshner, R. P., *ApJ* **242**, 1023 (1980)
13. Asaoka, I., and Aschenbach, B., *A&A* **284**, 573 (1994)
14. Claussen, M. J., Frail, D. A., Goss, W. M., and Gaume, R. A., *ApJ* **489**, 143 (1997)
15. Chevalier, R. A., *ApJ* **188**, 501 (1974)
16. Giovanelli, R., and Haynes, M. P., *ApJ* **230**, 404 (1979)
17. Richter, M. J., Graham, J. R., and Wright, G. S., *ApJ* **454**, 277 (1995)
18. Huang, Y.-L., Dickman, R. L., and Snell, R. L., *ApJ* **302**, L63 (1986)
19. Burton, M. G., Hollenbach, D. J., Haas, M. R., and Erickson, E. F., *ApJ* **355**, 197 (1990)
20. Dickman, R. L., Snell, R. L., Ziurys, L. M., and Huang, Y.-L., *ApJ* **400**, 203 (1992)
21. Rho, J., Jarrett, T. H., Cutri, R. M., and Reach, W. T., *ApJ* in press (astro-ph/0010551) (2000)
22. Richter, M. J., Graham, J. R., Wright, G. S., Kelly, D. M., and Lacy, J. H., *ApJ* **449**, L83 (1995)

23. Braun, R., and Strom, R. G., *A&A* **164**, 193 (1986)
24. Tauber, J. A., Snell, R. L., Dickman, R. L., and Ziurys, L. M., *ApJ* **421**, 570 (1994)
25. Cesarsky, D., Cox, P., Pineau des Forets, G., van Dishoeck, E. F., Boulanger, F., and Wright, C. M., *A&A* **348**, 945 (1999)
26. Lockett, P., Gauthier, E., and Elitzur, M., *ApJ* **511**, 235 (1999)
27. Erickson, W. C., and Mahoney, M. J., *ApJ* **290**, 596 (1985)
28. Duin, R. M., and van der Laan, H., *A&A* **40**, 111 (1975)
29. Blandford, R. D., and Cowie, L. L., *ApJ* **260**, 625 (1982)
30. Ostrowski, M., *A&A* **345**, 256 (1999)
31. Bykov, A. M., Chevalier, R. A., Ellison, D. C., and Uvarov, Yu. A., *ApJ* **538**, 203 (2000)
32. Esposito, J. A., Hunter, S. D., Kanbach, G., and Sreekumar, P., *ApJ* **461**, 820 (1996)
33. Gaisser, T. K., Protheroe, R. J., and Stanev, T., *ApJ* **492**, 219 (1998)
34. Sturner, S. J., Skibo, J. G., Dermer, C. D., and Mattox, J. R., *ApJ* **490**, 619 (1997)
35. Baring, M. G., Ellison, D. C., Reynolds, S. P., Grenier, I. A., and Goret, P., *ApJ* **513**, 311 (1999)
36. Buckley, J. H., et al., *A&A* **329**, 639 (1998)
37. Keohane, J. W., Petre, R., Gotthelf, E. V., Ozaki, M., and Koyama, K., *ApJ* **484**, 350 (1997)
38. Bocchino, F., and Bykov, A. M., *A&A* **362**, L29 (2000)
39. Clearfield, C. R., et al., these proceedings
40. Kundu, M. R., and Velusamy, T., *A&A* **20**, 237 (1972)
41. Petre, R., Szymkowiak, A. E., Seward, F. D., and Willingale, R., *ApJ* **335**, 215 (1988)
42. Shelton, R. L., Cox, D. P., Maciejewski, W., Smith, R. K., Plewa, T., Pawl, A., and Różycska, M., *ApJ* **524**, 192 (1999)
43. Goss, W. M., *ApJS* **15**, 131 (1968)
44. Frail, D. A., Goss, W. M., and Slysh, V. I., *ApJ* **424**, L111 (1994)
45. Frail, D. A., Goss, W. M., Reynoso, E. M., Giacani, E. B., Green, A. J., and Otrupcek, R., *AJ* **111**, 1651 (1996)
46. Green, A. J., Frail, D. A., Goss, W. M., and Otrupcek, R., *AJ* **114**, 2058 (1997)
47. Koralesky, B., Frail, D. A., Goss, W. M., Claussen, M. J., and Green, A. J., *AJ* **116**, 1323 (1998)
48. Frail, D. A., and Mitchell, G. F., *ApJ* **508**, 690 (1998)
49. Reynoso, E. M., and Mangum, J. G., *ApJ* **545**, 874 (2000)
50. Reach, W. T., and Rho, J., *A&A* **315**, L277 (1996)
51. Reach, W. T., and Rho, J., *ApJ* **507**, L93 (1998)
52. Reach, W. T., and Rho, J., *ApJ* **544**, 843 (2000)
53. Reach, W. T., and Rho, J., *ApJ* **511**, 836 (1999)
54. Foster, P. N., and Boss, A. P., *ApJ* **468**, 784 (1996)
55. Vanhala, H. A. T., and Cameron, A. G. W., *ApJ* **508**, 291 (1999)
56. O'Dell, C. R., Wen, Z., and Hu, X., *ApJ* **410**, 696 (1993)
57. Chevalier, R. A., *ApJ* **538**, L151 (2000)

The SN - SNR Connection

Robert A. Fesen

Department of Physics and Astronomy
Dartmouth College, Hanover NH 03755, USA

Abstract.
One of the longstanding goals of SN/SNR research is the establishment of firm observational links between the different types of extragalactic SNe and the various sorts of young remnants seen in our galaxy and the LMC/SMC. Here I briefly review some recent observational research on late-time SN emissions and links between the properties of young supernova remnants and SN subtypes.

INTRODUCTION

The phase transition when a supernova (SN) becomes a supernova remnant (SNR) is both poorly defined and relatively unexplored. Although past and on-going studies of SN 1987A have been helpful in allowing us to witness certain aspects of this identity change for this one object, its particular properties do not offer broad insights about the formation and evolution of young remnants across the whole spectrum of SN subclasses.

It should be noted at the outset, that there is no generally accepted definition for the point when a supernova turns into a supernova remnant. Supernova researchers often refer to even decades old objects as still SNe whereas SNR investigators tend to look upon such objects as young remnants. One suggested operational definition is that the remnant phase begins when a SN begins to strongly interact with its surrounding interstellar medium (ISM). However, such a definition would seem to fail for even the least controversial cases. For example, by this definition the nearly 1000 yr old Crab Nebula would technically still be termed a supernova since there is little in the way of X-ray, radio, or optical data suggesting its interaction with the local ISM [1–3]. In fact, its outward expansion has not decreased but rather accelerated due to spin-down energy off its pulsar. This definition might also require describing the 115 yr old remains of SN 1885 (S And) in M31 as still a SN rather than a SNR. It lies in a low density ISM region in the bulge of M31 and is still in nearly free expansion [4].

A somewhat better definition for the start of a SN's remnant phase would be perhaps when UV/optical line and continuum emissions from SN ejecta fall below that generated by interaction with either surrounding circumstellar/interstellar material

CP565, *Young Supernova Remnants: Eleventh Astrophysics Conf.,* edited by S. S. Holt and U. Hwang
© 2001 American Institute of Physics 0-7354-0001-6/01/$18.00

or the emission from a compact stellar remnant. By this definition, SN 1987A is only now beginning its remnant phase [5].

There are two major divisions of SNe: thermonuclear explosions of white dwarfs (Type Ia) and core collapse explosions of massive stars (Type IIs and Ib/c). The range of observed properties of core collapse events is broader (II-L, II-P, IIn, and IIpec) than for SN Ia (slow-to-fast) with Ib/c objects thought related to SN II through greater pre-SN mass loss histories resulting in smaller envelope masses at time of outburst [6]. This SN diversity is in sharp contrast to just three types of young SNRs known: O-rich remnants (e.g. Cas A), pulsar or plerion remnants (e.g. Crab), and collisionless shock remnants (e.g. Tycho's SNR).

Past attempts at relating SN events with SN remnants has met with only limited success. van den Bergh suggested SN Ib have O star or WR star progenitors and leave O-type SNRs, whereas B type stars were connected to ordinary Type II SNe and leave plerionic remnants [7]. However, most SNRs have strong individual characteristics which are not always shared by other young SNRs or easily connected to SN subclasses. For example, Cas A shows optical properties unlike any of the other O-rich SNRs for which it is suppose to be the prototype, while the galactic remnant W50/SS433 and the O-rich ejecta + 50 msec pulsar remnant 0540-69 in the LMC defy simple cataloging into these three classes. An important underlying complication in relating core collapse SNe to specific remnant classes is the role CSM plays in affecting a remnant's observed properties and evolution.

Attempts at identifying young remnants of SN Ia have been a bit more successful (but see [8] for an alternative view). Tycho's SN of 1572 has long been regarded as a probable Type Ia based first on Baade's reconstructed light curve [9] but more recently on its optical and x-ray properties [10–12]. However, it may have been significantly subluminous [13,14]. There is stronger evidence for the Type Ia nature of the SN 1006 remnant based on X-ray data [15], UV absorptions [16–19] and historical observations [14]. Lastly, the remnant of SN 1885 has been recently detected with properties consistent with a Type Ia nature (see discussion below).

Despite these difficulties, significant benefits would result for both the SN and SNR communities if we could confidently connect the various SN types seen in extragalactic systems with nearby, young galactic and LMC/SMC supernova remnants. Young SNRs, like the Crab and Cas A, offer explosion dynamical data (e.g. expansion asymmetries) and ejecta composition and knot dimensions that are lost when viewing many extragalactic events. Studies of young SNRs can also provide detailed information about the progenitor's local CSM/ISM environment which in turn give us insights into its past evolutional history as well as predict future emission levels. Finally, SNR studies can offer kinematic information of central compact sources when present.

On the other side, SN studies offer data on diversity and rates of occurrence regarding specific SN types and subclasses. SN studies can also provide peak mean ejection velocities of chemically distinct layers and information about the presence of CSM. Finally, detailed SN modeling gives us the foundations upon which to interpret chemical and kinematic observational data on young remnants.

TABLE 1. Possible SN Progenitor Types for Young (age < 5000 yr) SNRs

SN Type	Galactic SNRs	LMC/SMC SNRs	Extragalactic SNRs
Ia	Tycho's SNR, SN 1006	0509-67.5, 0519-69.0	SN 1885 (S And)
		0509-67.5, N103B	
Ib/c	Cas A, Puppis A, G292.0+1.8	N132D, 1E0102-72	NGC 4449
IIpec	Crab, 3C58	SNR 1987A	?
II-L, II-P	Kepler's SNR ?	0540-69 ?	?
IIn	?	?	?

Currently, there are less than two dozen young SNRs where we have decent clues as to the likely SN type progenitor. Table 1 lists some of these SN/SNR connections.

LATE-TIME EMISSIONS FROM SUPERNOVAE

There are too few extragalactic SNe which have been detected decades after outburst to firmly establish clear and unambiguous empirical SN-SNR links. The basic problem is simple. Extragalactic SNe typically fade below readily detectable X-ray, UV/optical, and IR emission levels within 2-3 years after maximum light. Those few objects for which emissions have been seen for over 2 yrs or so are mainly objects having strong circumstellar (CSM) or interstellar (ISM) shock and ejecta interactions.

This situation has prevented significant progress to be made in empirically relating SNe across all subclasses with the various types of young SNRs seen both locally in our galaxy and in the LMC and SMC. Still, in cases where late-time emission has been observed several years or even decades after maximum light, the source powering these late-time emissions vary among SN types. A sample of late-time, optical SN detections are listed below.

SNe with Late-Time Emission via Strong Interactions with CSM/ISM

- Type II-L: SN 1985L, SN 1980K, SN 1979C, SN 1970G

- Type IIn: SN 1998S, SN 1997ab, SN 1995N, SN 1988Z, SN 1986J, SN 1978K

- Type II-unkn: SN 1957D

SNe with Late-Time Emission via Radioactive Heating

- Type IIpec: SN 1987A

SNe detected via Photoionization of Ejecta

- Type Ia: SN 1885 (S And)

Despite the observational limitations, late-time emissions from some SN II events provide powerful diagnostics on the physical conditions of the SN ejecta and its

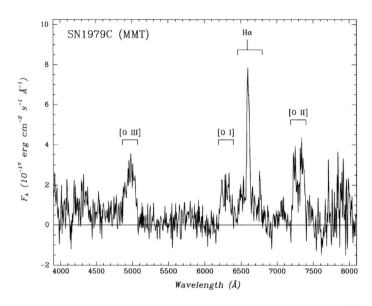

FIGURE 1. A May 1993 optical spectrum of SN 1979C in M100 [23].

CSM environment. SN 1979C in M100 is one of the best studied of the small group of relatively nearby SN II-L objects which have been followed decades after maximum light. Strong, late-time emissions have been seen from 79C in the X-rays (1.0×10^{39} erg s^{-1} from 0.1-2.4 keV), radio (2.6×10^{28} erg s^{-1} Hz^{-1} at 6 cm) and UV/opt (Hα luminosity: 15×10^{37} erg s^{-1}) due to surrounding CSM [20–23]. Variations in the late-time radio flux have been interpreted as due to modulation of the CSM density by a binary companion with a recent halt in the radio flux interpreted as due to the shock entering a denser region [24]. Several broad, 6000 km s^{-1} optical and UV line emissions are observed (Fig. 1) with relative fluxes that match reasonably well shock CSM models [23,25]. Only lines with critical electron densities above 10^5 cm^{-3} are seen here and in other late-time SN II-L emissions such as SN 1980K. HST imaging of SN 1979C and its environment suggest that the progenitor may have had a mass around 17-18 solar masses and lies within an OB cluster off the southern arm of M100 [26].

While the late-time luminosities of SN II-L objects are considerable, they can be dwarfed by those of SN IIn. SN 1988Z in MCG +03-28-022 is a good example. It like other IIn objects was characterized by narrow and intermediate emission line widths on top of a broad (15 000 km s^{-1}) component and an unusually slow decaying light curve, making it some 5 mag brighter in V band at day $\simeq$ 600 than ordinary SN II-L or SN II-P. Late-time X-ray & Hα emissions exceeded 10^{41} erg s^{-1} with an 2-20 cm radio flux some 3000 times stronger than that of Cas A [27,28].

Other SN IIn examples include SN 1978K [29] and SN 1995N [30]. Both the

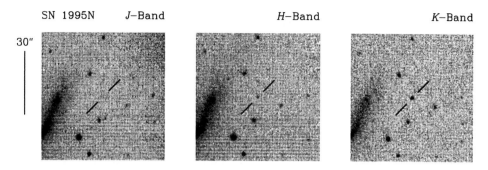

FIGURE 2. Near infrared images of SN 1995N taken in May 2000 [32].

variety of line widths plus the enormous late-time luminosities are well explained by a clumpy wind model [31]. In this picture, the broad component arises from shocked SN ejecta expanding into a relatively rarefied wind (supported by the presence of several coronal lines), while the intermediate width lines come from a shocked, clumpy wind component. Unshocked, progenitor wind material is the source of the narrow lines seen at early times. The density of this CSM probably exceeds that present in SN II-L objects and may contain considerable dust possibly explaining the extraordinary late-time IR luminosities seen in SN 1995N (see Fig. 2).

Young SNRs thought associated with SN Ia events include SN 1006, Tycho's SNR, and several Balmer-dominated LMC remnants (see Table 1). Unlike the case for remnant's of massive stars, ejecta clumps have not been observed, with metal-rich ejecta usually only reported via X-ray data. However in the case of SN 1006, a fortuitous placement of a background sdO star has permitted detailed study of the remnant's ejecta via absorption lines [16,18]. This powerful technique has recently be employed in the study of the young SN Ia remnant of SN 1885 in M31 [4,36]. HST WFPC2 images show the remnant as a 0.7 arcsec diameter absorption disk silhouetted against M31's central bulge, corresponding to a diameter of 2.5 ±0.4 pc at M31's distance. This size, in turn, implies an average expansion velocity of 11 000 ±2000 km s^{-1} since its outburst in 1885. Figure 3 shows the remnant's near UV absorption spectrum revealing smooth, broad Ca I λ4227 and Ca II $\lambda\lambda$3934,3968 K and H line absorptions. The absorption profiles suggest a spherically symmetric distribution of SN ejecta freely expanding at up to 13 000 km s^{-1}. Little deceleration via interaction with local ISM is consistent with the remnant's exceedingly faint radio emission recently detected [37].

If the ionization state of the remnant's iron is similar to the observed ionization state of calcium, then the estimated mass of Fe II is $\simeq$ 0.21 solar masses, consistent with SN 1885's subluminous classification [33–35]. However, large measurement errors permit either subluminous or normal SN Ia predicted iron masses to be present. In fact, these data together with an ultraviolet image taken with HST at resonance Fe I and Fe II lines are consistent with a broad Fe mass range of 0.1–1.0

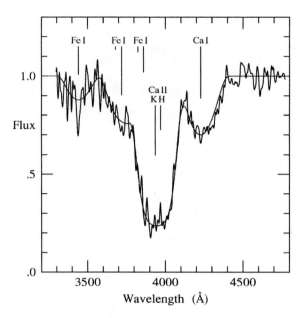

FIGURE 3. HST spectrum of SN 1885 (S And) in M31 [36].

solar masses [36]. Nonetheless, its chemical and kinematic properties strongly point to a SN Ia progenitor, even if a peculiar subluminous event.

CAS A - THE YOUNGEST GALACTIC SNR

As an example of the sort of SN - SNR connections that can result from studies of young SNRs, I briefly discuss below some recent work on the youngest and best studied nearby SNR, namely Cas A.

Likely sighted in August of 1680 by Rev. J. Flamsteed, the Cas A SN was probably subluminous since no other historical sightings of it have been discovered. With a 4′ diameter and a distance of 3.4 kpc, it has a main shell radius of 1.5 pc which is expanding at $\simeq$ 6000 km s^{-1}. This shell contains numerous optical/IR knots of O, S, Ar rich ejecta [38–40] along with a few dozen CSM clumps rich in He and N. The remnant also shows a flare or jet of O, S, Ar ejecta off to the northeast with knot velocities about twice that of the main shell [41]. HST images of the main shell and jet [42] show the ejecta to consist of clusters of tiny knots 0.1-0.3 arcsec in size (see Fig. 4).

Investigations of Cas A over the last decade have successfully identified knots of ejecta from a variety of individual layers of the progenitor star. Optical images and spectra have recently shown additional high-velocity, outlying ejecta in more than just the NE jet region. Most of these knots show strong N lines but no hydrogen or other lines [43], and space velocities around 10,000 km s^{-1} indicating this as a minimum expansion velocity of the progenitor's outermost layers capable of generating observable optical knots. The chemistry of these N-rich outer knots indicates a likely SN Ib/c and WN Wolf-Rayet progenitor [44].

In the main shell, knots of nearly pure O have been seen [39]. These are in stark contrast to some S, Ca, Ar-rich fragments which have nearly no O, representative of Si-group material due to explosive O-burning in the layer just above the Ni-rich core [45]. New ISO infrared observations have also detected a 22 μm emission feature indicate of newly formed silicate dust [40].

Clear evidence for considerable mixing in the remnant, however, has also emerged. X-ray observations made with the Chandra Observatory have shown iron-rich ejecta outside of Si-rich material in some parts of the main shell [46]. This abundance "inversion" has been interpreted as an indication that energetic, bulk motions occurred during the SN explosion involving significant portions of the remnant. Such an inversion is likely to be the result of plumes of Fe-rich ejecta from a neutrino-driven convection layer and/or bubbles of radioactive decay ^{56}Ni clumps.

Supporting evidence for a turbulent expansion comes from optical velocity studies that have revealed a pronounced, multi-ring like main shell structure [47,48], with one of the most prominent rings located at the base of the NE jet of high velocity ejecta. Furthermore, several outlying optical knots have been shown to have a mixed chemistry. Namely, they show lines of nitrogen like the N-rich knots but also have strong emissions of S, O, and Ar like the main shell knots. Such "mixed" ejecta have only been found in the NE jet and along the western rim – regions where the remnant's highest velocity outer ejecta have been detected [43].

First light Chandra images discovered a compact x-ray point source very near the estimated expansion center whose nature is currently unknown [49,50]. A recent and more accurate explosion center determination, making use of several recently discovered outlying knots which show up faintly on historical 5 m plates, shows that this x-ray point source lies 7″ away from the remnant's explosion point [51]. This implies a transverse velocity of 360 km s^{-1} at d = 3.4 kpc. These data also indicate an explosion date of 1671 assuming no deceleration, lending support to Flamsteed's 1680 sighting.

When taken all together, the Cas A observations paint a persuasive story where a WR star asymmetrically exploded probably in a SN Ib/c type event. This resulted in a remnant where some of the highest velocity ejecta fragments originated from the deepest layers of the progenitor star. Such observations have given us fundamental information about how the SN exploded, its surrounding CSM, and other properties such as the kick velocity of the resulting stellar remnant some 300 yrs after outburst. Similar data on other young SNRs represent a seemingly under utilized resource for testing and refining SN models.

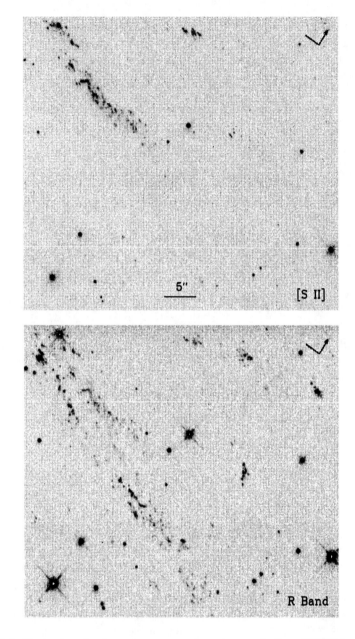

FIGURE 4. Ejecta clumping in a young remnant can be seen in these HST R band and [S II] images of the lower section Cas A's NE jet. Differences between the images are mainly due to different radial velocities covered by the two filters [42].

SUMMARY

There has been great progress over the last decade in following old SNe turn into young remnants. Whereas in 1990, we had only fragmentary information about the very late-time emission properties of SNe, today we have partial X-ray, UV/optical/IR, and radio data on over two dozen objects. However, investigating the transition of SNe into SNRs is observationally difficult except in cases where there is a dense CSM through which to convert the SN's kinetic energy quickly into radiation. Consequently, data on the young developing SN/SNR have been heavily biased towards SN IIn and SN II-L events over the time frame of 1-100 yrs.

For objects 100-300 yrs old, there's currently only object to study: SN 1885 (S And). Though a subluminous SN Ia event, it offers wonderfully detailed kinematic and abundance information despite a weak UV background by which to probe its inner structure. Further study of its properties may give additional insights into the properties of SN Ia events in general.

The majority of known "young SNRs" lie in the age range of 300-3000 yr. These have yielded a wealth of detailed data not possible from extragalactic SNe studies. However, there are too few currently known upon which we can build clear remnant class associations and divisions or establish links to SN progenitor types. So far, reported searches for young SNRs in M31 and M33 have failed to turn up Cas A or Crab like young objects and to date the NGC 4449 SNR is only young remnant outside of our galaxy and the LMC/SMC. Until a much larger set of youthful remnants can be identified either from the ranks of detectable late-time SNe or in nearby galaxies, firm SN-SNR connections will remain limited.

REFERENCES

1. Predehl, P., and Schmidt, J. H. 1995, A&A, 293, 887
2. Frail, D. A., Kassim, N. E., Cornwell, T. J., and Goss, W. M. 1995, ApJ, 454, L129
3. Fesen, R. A., Shull, J. M., and Hurford, A. P. 1997, AJ, 113, 354
4. Fesen, R. A., Gerardy, C. L., McLin, K. M., Hamilton, A. J. S. 1999, ApJ, 514, 195
5. McCray, R. 2001, this conference
6. Filippenko, A. V. 1997, Ann. Rev. A&A, 35, 309
7. van den Bergh, S. 1988, ApJ, 327, 156
8. Branch, D. 2001, this conference
9. Baade, W. 1945, ApJ, 102, 309
10. Chevalier, R. A., and Raymond, J. C. 1978, ApJ, 225, L27
11. Hamilton, A. J. S., Sarazin, C. L., and Szymkowiak, A. E. 1986, ApJ, 300, 713
12. Hwang, U., and Gotthelf, E. V. 1997, ApJ, 475, 665
13. van den Bergh, S. 1993, ApJ, 413, 67
14. Schaefer, B. E. 1996, ApJ, 459, 438
15. Vancura, O., Gorenstein, P., and Hughes, J. P. 1995, ApJ, 441, 680
16. Fesen, R. A., Wu, C.-C., Leventhal, M., and Hamilton, A. J. S., 1988, ApJ, 327, 164

17. Wu, C.-C., Crenshaw, D. M., Fesen, R. A., Hamilton, A. J. S., and Sarazin, C. L., 1993, ApJ, 416, 247

18. Wu, C.-C., Crenshaw, D. M., Fesen, R. A., Hamilton, A. J. S., Leventhal, M., and Sarazin, C. L., 1997, ApJ, 447, L53

19. Hamilton, A. J.S., Fesen, R. A., Wu, C.-C., Crenshaw, D. M., and Sarazin, C. L., 1997, ApJ, 482, 838

20. Weiler, K. W., Van Dyk, S. D., Pringle, J., and Panagia, N. 1992, ApJ, 399, 672

21. Schwartz, D. H., and Pringle, J. E. 1996, MNRAS, 282, 1018

22. Immler, S., Pietsch, W., and Aschenbach, B. 1998, A&A, 331, 601

23. Fesen, R. A. et al. 1999, AJ, 117, 725

24. Montes, M. J., Weiler, K. W., Van Dyk, S. D., Panagia, N., Lacey, C. K., Sramek, R. A., and Park, R., 2001, ApJ, 532, 1124

25. Chevalier, R. A., and Fransson, C. 1994, ApJ, 420, 268

26. Van Dyk, S. D., Peng, C. Y., Barth, A. J., and Filippenko, A. V., 1999, AJ, 118, 2331

27. Turatto, M., Cappellaro, E., Danziger, I. J., Benetti, S., Gouiffes, C., Della Valle, M. 1993, MNRAS, 262, 128

28. Aretxage, I. et al. 1999, MNRAS, 309, 343

29. Schlegel, E. M., Ryder, S., Staveley-Smith, L., Petre, R., Colbert, E. Dopita, M., and Campbell-Wilson, D. 1999, AJ, 118, 2689

30. Filippenko, A. V., et al. in preparation

31. Chugai, N. N., and Danziger, I. J. 1994, MNRAS, 268, 173

32. Gerardy, C. L., et al. 2001, in preparation

33. de Vaucouleurs, G. and Corwin, H. G. 1985, ApJ, 295, 287

34. Chevalier, R. A., and Plait, P. C. 1988, ApJ, 331, L109

35. van den Bergh, S. 1994, ApJ, 424, 345

36. Hamilton, A. J. S., and Fesen, R. A., ApJ, 542, 779

37. Sjouwerman, L. O., and Dickel, J. R. 2001, this conference

38. Chevalier, R. A., and Kirshner, R. P. 1978, ApJ, 219, 931

39. Chevalier, R. A., and Kirshner, R. P. 1979, ApJ, 233, 154

40. Arendt, R. G., Dwek, E., and Moseley, S. H. 1999, ApJ, 521, 234

41. Fesen, R. A., and Gunderson, K. S. 1996, ApJ, 470, 967

42. Fesen, R. A., et al. 2001, in preparation

43. Fesen, R. A., 2001, ApJS, 132, in press

44. Fesen, R. A., Becker, R. H., and Blair, W. P. 1987, ApJ, 313, 378

45. Fesen, R. A. and Hurford, A. P. 1996, ApJ, 469, 246

46. Hughes, J. P., Radowski, C. E., Burrows, D. N., and Slane, P. O. 2000, ApJ, 528, L109

47. Reed, J. E., Hester, J. J., Fabian, A. C., and Winkler, P. F. 1991, BAAS, 23, 826

48. Lawrence, S. S., et al. 1995, AJ, 109, 2635

49. Pavlov, G. G., Zavlin, V. E., Ascenbach, B., Trumpler, J., and Sanwal, D. 2000, ApJ, 531, L53

50. Umeda, H., Nomoto, K., Tsuruta, S., and Mineshige, S. 2000, ApJ, 534, L193

51. Thorstensen, J. R., Fesen, R. A., and van den Bergh, S. 2001, AJ, submitted

New Imaging Results in the Formation of SNR 1987A

Ben Sugerman[1], Stephen Lawrence[1], Arlin Crotts[1], Patrice Bouchet[2]

[1]*Dept. of Astronomy, Columbia University, New York, NY 10027, USA*
[2]*CTIO, Casilla 603, La Serena, Chile*

Abstract. We report *HST* imaging results of the equatorial circumstellar ring (ER) and its interaction with the ejecta (i.e. hot spots) of SN 1987A. Using archival and DD WFPC2 and STIS imaging, we study the ER morphology, produce multi-wavelength light curves of the first hot spot and spatially-resolved light curves for the ER in broad and narrow bands. The first hot spot grows with a power law $t^{\sim 3}$ until late 1999, where it appears to be turning over. Spatially-resolved light curves for the ER reveal that the flux is brighter, and fades more rapidly to the North (nearer to Earth) than to the South, indicating a possible density gradient in the ring.

INTRODUCTION

The development of SNR 1987A provides an unprecedented opportunity to observe, at high spatial, spectral, and temporal resolution, the birth of a supernova remnant (SNR). Observations of the SNR might reveal the angular and velocity distribution of the ejecta by highlighting where it impacts a *presumably* well-known nebular structure. As the interaction between the SN ejecta and the circumstellar equatorial ring (ER) intensifies, this region will undergo radical changes, leaving increasingly-less pristine ER with which to reconstruct the mass-loss history of the progenitor and the ring formation mechanism. As such, the early discovery of these interaction regions (hot spots) and a complete understanding of the nebular structure are crucial.

We use archival HST images taken in the WFPC2 F255W, F336W, F439W, F502N, F555W, F656N, F658N, F675W, and F814W filters between 1994 Feb and 1999 Apr, and STIS 28x50LP images from 1997 Dec through Nov 2000 [days 2537–5000], and reduce the data as explained in [1,2]. As in [1], we study variability using the non-iterative, non-invasive procedure of PSF-matched difference imaging. With this dataset, we study the ER morphology, produce multi-wavelength light curves of the first hot spot and spatially-resolved light curves for the ER in broad and narrow bands.

CP565, *Young Supernova Remnants: Eleventh Astrophysics Conf.*, edited by S. S. Holt and U. Hwang
© 2001 American Institute of Physics 0-7354-0001-6/01/$18.00

THE EVOLUTION OF HS 1-029

As reported in [1], the first hot spot, HS 1-029[1], began developing on or before 1995 Mar. We present in Fig. 1 lightcurves of HS 1-029, averaged together from both direct and difference-image measurements. Error bars are formally-propagated photon-count noise only. We assume the flux of HS 1-029 was negligible in 1994.

In all WFPC2 filters, HS 1-029 grows (after day~2400) with a power-law of $t^{\sim 3}$. We note the apparent break in flux from 1999 Jan (day 4335) to 1999 Apr (day 4439) in 439W, 656N and 675W. In the first case, a bad pixel is coincident with the hot spot, hence the flux at this epoch is uncertain. Comparison of these curves with the STIS F28x50LP filter data (which covers F675W and F814W) reveals that this was a real short-term break in the light-curve, as seen from data between 1999 Aug – 1999 Oct. Following 1999 Oct, the light curve may be slowly turning over. This appears consistent with a simple toy model in which a shock impacts and refracts around a protrusion from the ER, with the gas cooling time $\gtrsim$ the time for the refracted shock to move through the protrusion [2]. The appearance of a new spot (HS 10-040) just east of HS 1-029 by 1999 Oct [3], together with the aforementioned break in the curve around the same epoch, is suggestive that the forward SN blast wave swept through the HS 1-029 material by late 1999.

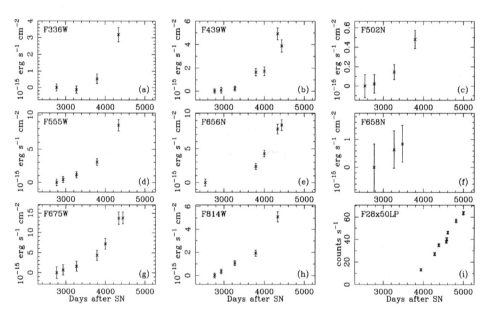

FIGURE 1. Lightcurves of HS 1-029 made through WFPC2 (panels [a]-[h]) and STIS F28x50LP (panel [i]) filters. The hot spot is assumed to have zero flux in 1994.

---

[1] Hot spot ID "HS I-NNN": I gives discovery order and NNN is the first reported imaging P.A.

THE EQUATORIAL RING

The most comprehensive study to date of the ER was performed by [4], using pre-SV1 WFPC and FOC images of SN 1987A predominantly in [OIII], but also including [NII] and Hβ. Using the five-year database of WFPC2 images, we perform a similar analysis on the physical parameters of the ER. We generate high signal-to-noise images of the 3-ring system in [OIII] and the [NII]+Hα complex by median-combining flux-calibrated images from 1994-1996 (to avoid significant emission from HS 1-029). We study the physical parameters of the ER using a flux-weighted moment-of-inertia analysis (see [5]). The results from this analysis are shown in Table 1; errors listed do not include systematics. Our results are in good agreement with previous values, as expected.

We measure the radial profile of the ER using an elliptical aperture of width $1/2$ PC pixel, whose center and eccentricity are defined in Table 1, and find the [OIII] emission is offset from the [NII]+Hα by $\sim 0''032$, as expected from analyses by [6]. The width of the ER is measured as the FWHM of the radial profile (c.f. [4]) and also as the width of an elliptical aperture containing 80% of the flux within $1''8$ (the edge of the diffuse ER emission). Again, our results ($\pm 0''023$) are within the uncertainties of previous work. The 80%-width of the ER in [OIII] is larger than that in [NII]+Hα, which supports the model of [6] in which [OIII] originates from less-dense gas located above and below the [NII]+Hα.

Ground-based monitoring of SNR 1987A provides good temporal coverage in imaging and high-resolution spectra, but with limited spatial information. The installation of STIS in 1997 introduced the possibility of taking detailed spectra *with* high-spatial resolution, with which to monitor all aspects of the SN and SNR evolution. However, to date only 4 spectra have been taken with the $2''$ slit (1997 Apr, 1997 Dec, 2000 May and 2000 Oct) which fully images the ER. Prior to 1997, and in the absence of such coverage in 1998 and 1999, a confluence of WFPC2 imaging, ground-based spectra and narrow-slit STIS spectra must reconstruct the spatially-resolved monochromatic evolution of the ER. As such, characterizing the spatial evolution of the ER is a crucial step in constraining its density, composition, temperature, morphology and formation models.

TABLE 1. Physical Parameters of the ER

| | This Work | | Plait et al. (1995) | |
Parameter	Hα+[NII]	[OIII]	[NII]	[OIII]
Eccentricity	$0.707 \pm .02$	$0.707 \pm .02$	0.711	0.724
Semi-Major Axis (mas)	831	863	845	858
FWHM Width (mas)	192	200	229	173
80%-Flux Width (mas)	478	554	...	...
PA of major axis (° E of N)	84.1 ± 2.4	87.1 ± 2.4	84.1 ± 3	88.6 ± 3
Inclination Angle	45.0 ± 1.8	45.0 ± 1.9	44 ± 2	43.7 ± 1
Offset from SN (mas N)	-54 ± 21	-33 ± 22	-31 ± 44	7 ± 11
Offset from SN (mas E)	-72 ± 44	0 ± 22	-13 ± 44	4 ± 11

We present our first results in studying the ER evolution. We measure the surface brightness of the ER in elliptical arcs defined by Table 1. Lightcurves for the ER were made through two arcs containing the northern half (PA 270–90) while excluding the HS 1-029 region (PA 10–40), and the southern half (PA 90–270) excluding Star 5 (PA 220–240), thereby roughly dividing the ER into the near and far halves, respectively. In general, the northern half is brighter than the southern half, consistent with previous studies: the region from PA 300–360 is ~2.5 times brighter in all filters redward of F439W than from PA 140–200, with the minimum flux around PA 150. The northern half also fades more rapidly than the southern half, in contrast to [4], who found the [OIII] far side to fade ~ 5% faster than the near side. In our data, the [OIII] fading rate is identical between the two halves, while in other filters the far side fades ~6 – 15% more rapidly than its northern counterpart. While this may result from column-depth differences looking into the different sides of the ring, we speculate that the southern limb is denser, and thus is cooling more rapidly from the initial SN UV outburst. The fading rate has also slowed in [OIII], since [4] found the ER to fade by a factor of 3.7 in ~3.3 years before 1994, while we find it has faded by only 1.8 in ~3.5 years after 1994. The [NII]+Hα flux has faded by 1.5 in ~4.1 years since 1994. Comparison of the wide-band fluxes with dominant emission lines from STIS and high-resolution ground-based spectra (in which we can separate the SNR from the ER emission) should yield information on the spatial and temporal variation of other line emission as well.

CONCLUSIONS

HS 1-029 emission is coincident with an inward-protrusion on the ER 1.5–2 PC pixels, or $0\rlap{.}''68 - 0\rlap{.}''91$ in radial size. If the forward shock impacted the hot spot in early 1995 and propagated through by late 1999, the radial shock velocity would be ~3000–4500 km s^{-1} (assuming $d_{SN} = 50$ kpc). While the SN ejecta has expanded linearly since day ~1400 at 3500 km s^{-1} [7], the forward shock could have propagated more rapidly from the SN to the ER. Provided $v \gtrsim 35000$ km s^{-1} before day 1400, the shock could have reached HS 1-029 by early 1995 at $v \gtrsim 4000$ km s^{-1}, while impacting spots 2–6 (see [1]) by early–mid 1998. This offers one resolution to the 4-year turn-on delay between HS 1-029 and subsequent hot spots.

REFERENCES

1. Lawrence, S. *et al.,* 2000, *Ap. J.,* 537, L123
2. Sugerman, B., Lawrence, S., Crotts, A., 2001, in preparation
3. Sugerman, B., Lawrence, S., Crotts, A., 2000, *IAUC* 7520
4. Plait, P., Lundqvist, P., Chevalier, R., Kirshner, R., 1995, *Ap. J.,* 439, 730
5. Sugerman, B., Summers, F.J., Kamionkowski, M., 2000, *MNRAS*, 311, 762
6. Lundqvist, P., & Sonneborn, G., asrto-ph/9707144
7. Manchester, R. *et al.,* this proceedings

STIS Spectral Imaging of Hot Spots in SNR 1987A [1]

Stephen Lawrence, Ben Sugerman, and Arlin Crotts

Department of Astronomy, Columbia University, New York, NY 10027, USA

Abstract. We report on Space Telescope Imaging Spectrograph (STIS) 52X2 G750M spectra of SNR 1987A obtained in 2000 May, and compared with nearly-identical observations taken in 1997 April. Such carefully-matched observations allow for a clean subtraction of the inner equatorial ring (ER) emission and an improved measurement of the hot spot fluxes. We present Hα fluxes and line widths for the seven hot spots clearly detected in the 2000 May data. We also provide the first description of spatially-compact emission associated with HS 1-029 which spans wide wavelength ranges of the G140L, G230L, G430L and G750L spectra taken in 1999 September and October. We suggest that this apparent continuum most likely arises from free-bound, free-free, and two-photon decay transitions in H I.

INTRODUCTION

The conversion of SN 1987A into supernova remnant SNR 1987A is now unambiguously underway. The fastest-moving ejecta have reached the inner edge of the circumstellar equatorial ring (ER), generating large arcs of radio and X-ray emission and localized "hot spots" of UV-optical-IR emission lines. Following nearly three years after the discovery of the first hot spot, ground-based and HST observations in early 2000 have now detected at least seven hot spots, scattered around the circumference of the ER [1–3]. In this article we use the hot spot naming convention of [4]

All observations reported here were taken with the Space Telescope Imaging Spectrograph (STIS) between 1997 April 26 and 2000 May 01. The G750M, G140L, G230L, G430L, and G750L gratings were used with a variety of slit widths and orientations. Raw data files were reduced using the "On-The-Fly-Calibration" option, although some additional cosmic-ray rejection has been performed. Spectral extractions and fits used standard IRAF routines.

---

[1] This research supported by NASA grant NAG5-3502 and STScI grants GO 8806 and 8872.

CP565, *Young Supernova Remnants: Eleventh Astrophysics Conf.*, edited by S. S. Holt and U. Hwang
© 2001 American Institute of Physics 0-7354-0001-6/01/$18.00

G750M SPECTRAL IMAGING

By carefully matching the P.A. of observation of 2000 May to that of 1997 Apr, we were able to record nearly identical monochromatic images of the ER in each line with little spatial or spectral confusion. The new hot spots are localized areas of brightening on the inner edge of the ER, which is large, spatially-complex, and generally fading with different time scales in each line. Carefully planned and matched observations such as these allow us to register, scale, and subtract each emission line individually, revealing subtle flux differences. As the hot spots also have significantly higher velocity widths than the ER emission, their emission tends to be dispersed away from the ER. Given that the hot spots are brightest in Hα, and that the WFPC2 656N filter is contaminated by [N II] λ6548, spectrally-dispersed imaging through the G750M grating is perhaps the most sensitive means to detect faint new hotspot activity. It also gives us the ability to monitor the whole ER efficiently with a single pointing.

In Figure 1 we present spectral difference images of the ER in the Hα + [N II] λλ6548,6583 region. The 1997 Apr image has been carefully registered, scaled downward by a factor of 1.15 (to account for fading of the ER in Hα), and then subtracted from the 2000 May image. By scaling to Hα, the [N II] monochromatic images have been strongly oversubtracted. Hot spots HS 5-139 and 7-289 (at the bottom of the ER) are not directly apparent in the 2000 May image, but stand out significantly in the difference image. Also note the extension to the east (top) of the core of HS 1-029, which G750M data from 2000 Oct reveal to be a newly-forming spot near the base of HS 1-029 [5].

TABLE 1. Hα Emission from the Hot Spots in 2000 May.

Spot	Flux 10^{16} erg s^{-1} cm^{-2}	FWHM km s^{-1}
HS 1-029	148 (20)	215 (10)
HS 2-106	39 (7)	190 (10)
HS 3-126	29 (5)	240 (15)
HS 4-091	5 (2)	100 (70)
HS 5-139	10 (2)	165 (20)
HS 6-229	15 (3)	180 (15)
HS 7-289	6 (3)	220 (70)

Once the ER flux has been scaled down and removed in this manner, spectral extractions can be performed on the difference images, one for each emission line, in order to obtain their fluxes and velocity dispersions. This provides a valuable complement to measurements from narrow slit spectra. There is little ER flux remaining to skew the fit, and no need to assume a Gaussian form to the complex ER spatial profile in the slit. Slit losses and centering issues are also minimized. Table 1 presents the Hα fluxes and velocity widths as determined from the difference image. [A ~7% correction has been added to the flux of HS 1-029 to replace the oversubtraction from its flux in the 1997 Apr data.]

CONTINUUM EMISSION FROM HS 1-029

In Figure 2 we present low resolution spectra taken with the the G140L, G230L, G430L, and G750L gratings in September and October of 1999. The wavelength ranges displayed are indicated at the ends of each spectrum and the spatial position of HS 1-029 along the slit is located by the horizontal arrows. Note that there is faint, spatially-unresolved continuous emission that spans the 9000 Å wavelength range sampled by these gratings. (The similar band of continuous emission in the upper portion of the G230L spectrum comes from a star superposed on the SW side of the ER, which was included in this pointing only.) This continuous emission is spatially coincident with the discrete emission lines from HS 1-029 in all four spectra, despite the G140L aperture having had a significantly different position angle on the sky. This argues against the emission coming from a background star, unless it is extremely serendipitously aligned with HS 1-029, and no evidence for such a star is seen in early WFPC2 imaging. The emission disappears in G140L spectra taken 10 days earlier through the same position angle but offset 0″.2 to the SE, so it is unlikely that it originates in the large-scale reverse shock.

Possible sources for this continuous emission include free-bound (recombination) and free-free (bremsstrahlung) transitions, two-photon decays from 2s ^{2}S in H I, the superposition of many faint emission lines from HS 1-029 (with typical widths of ~200 km s^{-1}), emission from strong lines in extremely high-velocity gas, and optical synchrotron continuum. The variations in intensity that appear to occur across a single grating seem inconsistent with the smooth power law expected from synchrotron radiation, nor is HS 1-029 clearly detected in *Chandra* X-ray images [6] or in radio synchrotron maps [7]. There does not appear to be a clear correlation of the continuous emission with strong H I or He I lines, which argues against high-velocity gas. Nor do there appear to be sufficient lines of astrophysically abundant elements to blend together at low velocity dispersion to form this continuum. Free-bound, free-free, and two-photon decay continuum transitions seem to be the most likely explanation. The apparent strengthening of the emission blueward of [O II] $\lambda3727$ in the G430L spectrum is certainly suggestive of Balmer recombination. Using the treatment of Osterbrock [8], for temperatures from 10,000—20,000 K and densities from 10—10^6 cm^{-3}, the two-photon decay continuum dominates free-bound and free-free emission near the higher-order Balmer lines and across most of the G140L spectrum. We plan a careful analysis of this emission with respect to wavelength, which should shed further light on its origin.

REFERENCES

1. Bouchet, P. *et al.*, *IAU Circ.* **7354** (2000).

2. Maran, S, Pun, C. S. J., and Sonneborn, G., *IAU Circ.* **7359** (2000).

3. Garanavich, P, Kirshner, R., and Challis, P., *IAU Circ.* **7360** (2000).

4. Lawrence, S. *et al.*, *ApJ* **537**, L123 (2000).

5. Sugerman, B., Lawrence, S., and Crotts, A., *IAU Circ.* **7520** (2000).

6. Burrows, D. *et al.*, *ApJ* **543**, L149 (2000).

7. Gaensler, B. *et al.*, *ApJ* **479**, 849 (1997).

8. Osterbrock, D., *Astrophysics of Gaseous Nebulae and Active Galactic Nuclei*, Mill Valley, CA: University Science Books, 1989, ch. 4, pp. 86-95.

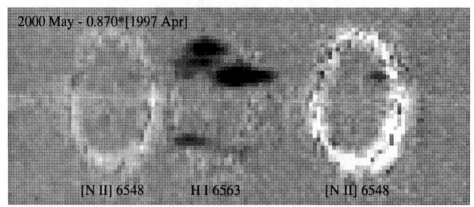

FIGURE 1. A G750M spectral difference image highlighting the brightening of hot spots 1-029 through 7-289 between 1997 Apr and 2000 May. North is to the right, east is up.

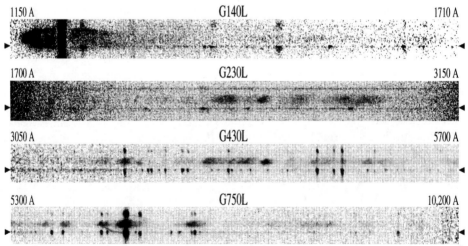

FIGURE 2. Continuum emission from the spatial location of HS 1-029 is seen across ∼9000 Å sampled by the STIS low resolution gratings. The spectra are ∼2".5 high, with slit P.A.s between -125 and -150 degrees E of N. Arrows indicate the spatial location of HS 1-029.

The Structure of SN 1987A's Outer Circumstellar Envelope as Probed by Light Echoes

Arlin Crotts*, Ben Sugerman*, Stephen Lawrence* and William Kunkel[†]

*Dept. of Astronomy, Columbia University, 550 W. 120th St., New York, NY 10027 U.S.A.
[†]Carnegie Inst. of Washington, Las Campanas, Obs., Casilla 601, La Serena, Chile

Abstract. We present ground-based and *HST* images processed by image subtraction to highlight transient reflection nebulae or "light echoes" of the maximum light pulse of the explosion of SN 1987A from surrounding material. Along with numerous structures already discussed elsewhere, we have found (in multiple epochs of data) a new feature opposite the SN from the mysterious "Napoleon's Hat" which indicates a symmetric structure due to shocks internal to the SN's red supergiant wind and probably caused by the pile-up of gas due to differential velocities within the outflow. We also show how echoes betray the ram pressure distribution of the progenitor mass loss flow.

OUTER ENVELOPE'S SPATIAL MORPHOLOGY

The image subtraction technique for light echoes producing the three-dimensional maps of circumstellar structure [1] and interstellar structure [2] in the vicinity of SN 1987A have been applied to the region just outside the bipolar/triple-ring nebula, revealing the evolution of four features: 1) a ring at radii of 9-15 arcsec (depending on the epoch of observation), discovered by Bond et al. [3] and modeled [4] as a contact discontinuity between the red supergiant (RSG) wind and the surrounding bubble blown by the main-sequence (MS) progenitor wind. For the sake of discussion, we will use the more general term "circumstellar boundary" (CB); 2) a sheet of material laying along the equatorial plane defined by the inner, equatorial ring (ER) and bisecting the bipolar nebula (c.f. [1], Fig. 21); 3) a diffuse echo [5] extending from the bipolar nebula outward to feature #1, and 4) "Napoleon's Hat" [6], apparently a discontinuity in the gradient of feature #3. We have followed these features since 1988 or 1989 until their disappearance at various times since.

Beyond these structures is a large volume, devoid of structure to a radius of 130 pc from the SN. This presumably corresponds to the bubble blown by the main sequence wind from the progenitor, at velocities of several thousand km s^{-1} over about 10 My.

CP565, *Young Supernova Remnants: Eleventh Astrophysics Conf.*, edited by S. S. Holt and U. Hwang
© 2001 American Institute of Physics 0-7354-0001-6/01/$18.00

3-D GEOMETRY OF CIRCUMSTELLAR BOUNDARY

While the echo from the CB was flocculent in appearance, its individual components tended to lie in well-defined ellipses, until the feature began fading drastically in 1991-2. Beyond these epochs we find the locus of the overall echo difficult to trace. (We are trying to improve the signal-to-noise of the residual images by using PSF-matching techniques e.g. [7], but cannot report these results yet.) In the meantime we simply centroid on individual echo patches and display them in three dimensions (x and y on the plane of the sky, and line-of-sight depth $z = (x^2 + y^2)/2ct - ct/2$, where t is time since maximum light). Light echoes allow one to trade information from the delay time t for the third spatial dimension z.

Since the echo at a given z expands in x and y rapidly with time, especially for large z values, one can collect a time series of images of the region, then isolate the rapidly moving echoes within the image by subtracting images from different epochs. Results from this process are shown in the Figures 1-3.

The implications of these structures are significant. If the CB is a contact discontinuity [4] between the RSG wind and the ISM bubble blown by the MS blue supergiant (BSG) progenitor, the pressure in this bubble should be nearly uniform over a few parsecs (since the reverse shock long ago back-filled the bubble), and the shape of the CB is due primarily to the ram pressure ρv^2 against the nearly

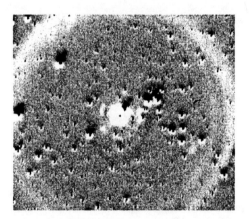

FIGURE 1. A residual image, once the average from a sequence of images taken at different times in the same band is subtracted, showing, in order of increasing radius, echoes from the diffuse circumstellar medium (bright central blob), circumstellar boundary (faint, broken ellipse) and ISM 130 pc away from the SN (circle ringing picture). The SN, and companion Stars 2 and 3 are contained within the diffuse echo spot. With permanent nebulosity and the net flux from stars removed (but some stellar residuals remaining - including from Star 2), the echoes become more visible. Contained within the diffuse echo is the bipolar/triple-ring nebula. The field of view is 85 arcsec in width, with north up, east left. The image was obtained on day 750 after core collapse at the LCO 2.5-meter in a band centered on 6023Å.

uniform bubble pressure. If the CB is aspherical, this is due to anisotropy in ρv^2. Given the equatorial overdensity (feature #2 mentioned above) in the RSG wind, a polar bulge or even a spherical CB indicates a RSG wind velocity at the poles higher than at the equator. This is a wind configuration that has not yet been considered in interacting wind models for the SN progenitor, and might lead to qualitatively different results than those published e.g. [8,9,10]. The 3-D geometry of the CB and its relation to other structures is shown in Figure 2.

The presence of the CB and the diffuse material inside it contradicts some complicated geometries proposed for the circumstellar environment of SN 1987A (c.f. [11]); we see a continuously distributed, higher density region (in terms of its echo reflectivity) bounded on its interior and exterior by low densities. While this corresponds to the predicted morphology of a BSG-RSG-BSG interacting wind model, the shape of the inner bipolar nebula and rings, and now the outer CB, have yet to be successfully modeled. Still, the presence and shape of this diffuse medium presents a strong challenge to models outside the BSG-RSG-BSG evolutionary picture.

NEW ECHO FEATURE AND NAPOLEON'S HAT

We recently discovered a new light echo feature (Fig. 3) which appeared and then disappeared suddenly in the mid-1990s. This corresponds to a reflecting sheet of material behind SN 1987A, nearly tangent to the echo paraboloids of constant delay time, as in Fig. 2. In Figure 2 the new feature is nearly symmetric with respect to the nebula's axis of symmetry opposite Napoleon's Hat. Contrary to some former hypotheses, then, these features appear to reside in the mass outflow from the SN progenitor, and might be explained by a pile-up of material as ejection from the star accelerated due an episode that persisted from ~ 50000y to $\gtrsim 10^5$y before explosion, gleaned from the distance and outflow velocity from the SN.

REFERENCES

1. Crotts, A.P.S., Kunkel, W.E. & Heathcote, S.R., *Ap. J. Let.*, **438**, 724 (1995).
2. Xu, J., Crotts, A.P.S. & Kunkel, W.E., *Ap. J.*, **451**, 806 (1995); Erratum: **463**, 391.
3. Bond, H., Gilmozzi, R., Meakes, M., & Panagia, N., *Ap. J. Let.*, **354**, L49 (1990).
4. Chevalier, R.A. & Emmering, R.T., *Ap. J. Let.*, **342**, L75 (1989).
5. Crotts, A.P.S. & Kunkel, W.E., *Ap. J. Let.*, **366**, L73 (1991).
6. Wampler, E.J., et al., *Ap. J. Let.*, **362**, L13 (1990).
7. Tomaney, A.B. & Crotts, A.P.S., *A. J.*, **112**, 2872 (1996).
8. Blondin, J.M. & Lundqvist, P., *Ap. J.*, **405**, 337 (1993).
9. Blondin, J.M., Lundqvist, P. & Chevalier, R.A., *Ap. J.*, **472**, 257 (1996).
10. Martin, C. & Arnett, D., *Ap. J.*, **447**, 378 (1995).
11. Podsiadlowski, Ph., Fabian, A.C. & Stevens, I.R., *Nature*, **354**, 43 (1991).

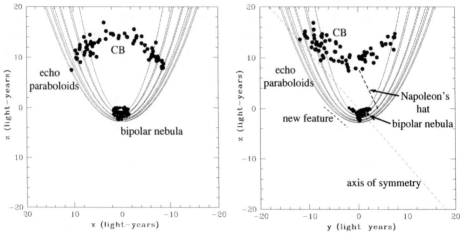

FIGURE 2. The 3-D geometry of the circumstellar boundary, as viewed from (left) far to the north, and (right) far to the east, and place relative to the bipolar nebula (near the center). The CB wraps around the SN at a radius of about 5 pc. The only points sampled must be contained within the light-echo parabolae for the epochs of observation 1989-1995. Points far from the plane of the sky containing the SN appear interior to the parabolae. in the left panel, it is evident that the CB viewed from the east shows a larger radius of curvature than when viewed far to the north, as to the left. Possibly the CB is elongated, in the sense that the region radially outward from the southern pole of the bipolar nebula, perpendicular to the ER plane, is farther from the SN than points near the ER plane. The dashed line extending from the northern end of the bipolar nebula to the CB at the same y as the SN shows the position of Napoleon's Hat, derived consistently from several epochs of data. It does not extend beyond the CB.

Difference Images in WFPC2 F675W Filter

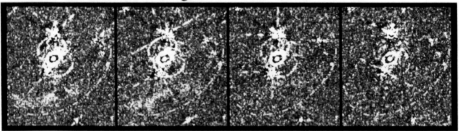

Sep 1994 - Jan 1999   Mar 1995 - Jan 1999   Feb 1996 - Jan 1999   Jul 1997 - Jan 1999

FIGURE 3. The newly discovered echo feature as it moves inwards (and disappears) over several epochs during 1994-1997. An excess of flux in WFPC2 at each epoch over that in Jan 1999 is portrayed as white. The light curve below the echo is meant to serve as a positional reference only.

140

Young SNR I

X-ray Observations of Young Supernova Remnants

Una Hwang

*NASA Goddard Space Flight Center, Greenbelt, MD 20771, and
University of Maryland, College Park, MD 20742, USA*

Abstract.
This brief review of recent X-ray observations of young supernova remnants highlights results obtained by the Chandra and XMM Newton Observatories since their launch last year. Their impressive capabilities are illustrated by results for spectral imaging, and for spatially resolved spectroscopy that isolates emission from individual ejecta knots and from the forward shock. I also review X-ray dynamical studies of supernova remnants, which should undergo significant advances during this new era.

INTRODUCTION

Young supernova remnants are expected to emit brightly in X-rays, as the forward and reverse shocks in remnants typically heat the ambient gas and ejecta to X-ray emitting temperatures. Moreover, supernova remnants are expected to be rich X-ray emission line sources: the gas ionizes slowly after being suddenly heated behind the shock, so that the populations of certain stable ions can be significantly higher than would be expected at equilibrium for a given electron temperature. The result is correspondingly enhanced line emission from those ions. For remnants whose X-ray emission is dominated by the reverse-shocked ejecta, the line emission will be further enhanced by the high element abundances in the ejecta that result from nuclear processing, both before and during the supernova explosion.

It has been twenty years since early X-ray spectral observations at both moderate and high resolution first verified the rich X-ray emission line spectra of supernova remnants (e.g., Becker et al. [1], Winkler et al. [2]); meanwhile, ever improving X-ray images have revealed their beautiful and complex morphologies (e.g., see Seward [3] and references therein, Aschenbach et al. [4], Levenson et al. [5], Williams et al. [6]). Opportunities for simultaneous imaging and spectroscopy, however, have been limited until relatively recently. The launch of the ASCA satellite in 1993 [7] marked a significant advance, in that 1' optics were coupled with CCD imaging spectrometers of moderate spectral resolution that were capable of isolating certain emission lines and blends. ASCA thus provided the first X-ray spectral images [8,9],

CP565, *Young Supernova Remnants: Eleventh Astrophysics Conf.,* edited by S. S. Holt and U. Hwang

as well as important spatially resolved spectral results for remnants of larger angular size.

In 1999, the launch of Chandra and XMM Newton provided an enormous further advance in that the CCD spectrometers were placed at the focus of far better mirrors—XMM Newton provides a point source image with a roughly 6″ core, while Chandra provides a 0.5″ angular resolution that is the best ever realized for X-rays. The two missions are complementary, with Chandra's strength being its superior mirrors, and XMM's strength being its higher collecting area, particularly at energies above 5 keV (for example, see the impressive very high energy X-ray image of Cas A taken by XMM [10]). Both missions also include dispersive high resolution spectrometers, with complementary characteristics, that have already provided exciting results for supernova remnants in the Magellanic Clouds (see the review by Canizares [11]). Unfortunately, such observations are difficult for remnants of any significant angular size because of the complicated blending of highly detailed spatial and spectral information.

This review highlights a few of the many results obtained by Chandra and XMM during their first year, with the goal of illustrating the power of simultaneous imaging and spectroscopy with moderate resolution CCD spectrometers. It is restricted to a few illustrative examples of spectral imaging and of detailed spectroscopy of compact regions of thermal X-ray emission from shocked ejecta and ambient gas. Numerous other such examples have been presented at this conference. I conclude by looking forward to the coming of age of X-ray dynamical studies of supernova remnants.

X-RAY SPECTRAL IMAGING

The capability to image a source in specific spectral features from a single element is particularly important for young supernova remnants whose emission is dominated by their ejecta, as such images can shed light on the different physical conditions and spatial distribution of various elements in the ejecta. Use of such techniques has been commonplace at optical wavelengths, but optically emitting ejecta generally represent a relatively small fraction of the total shocked ejecta mass; few remnants even have significant optical emission from their ejecta to begin with, excepting Cas A and other so-called "oxygen-rich" remnants. The wide bandpass of the ASCA, Chandra, and XMM CCD spectrometers also allows the comparison of relatively hard (4–6 keV) continuum images with softer ($\lesssim$ 2 keV) line images, which has been useful in identifying emission from the forward shock, as discussed below.

Later in this paper, we consider spectral imaging results from Chandra in conjunction with spectroscopic results. Here we consider the recent XMM Newton observation of Tycho's SNR.

Tycho's SNR with XMM

Tycho's supernova remnant (SN 1572), with its 8' diameter on the sky, is ideally suited for XMM, with its 37' field of view and 6" PSF core. Tycho's X-ray spectrum is notable for the richness of its line emission, from Fe L transitions (to the n=2 level) with energies near 1 keV, and an exceptionally prominent Si He α blend near 1.86 keV, to Fe K transitions (to the n=1 level) with energies near 6.4 keV. An imaging study with the ASCA satellite by Hwang & Gotthelf [12] suggested that the Fe L and Fe K emission are not spatially coincident, with the Fe K emission at a smaller average radius than other emission lines. That earlier study is limited by both photon statistics and angular resolution, but new XMM images presented by Decourchelle et al. [13] clearly show that the azimuthally averaged radial distribution of Fe K emission has a peak that lies well inside that of the Fe L emission. By contrast, the Fe L and Si He α emission are consistent with having the same average radial distribution.

These differences in the radial profiles may be qualitatively understood as arising from the temperature and ionization structure of the ejecta behind the reverse shock. The ejecta of Type Ia supernovae have a steep outer density profile surrounding an inner plateau of more constant density. As the reverse shock reaches the inner plateau in the ejecta distribution, the temperature of the shocked gas should increase and the density decrease. The more recently reverse-shocked gas is therefore hotter and has a lower ionization age, as would be required to explain the presence of Fe K emission at smaller radii, as pointed out by Decourchelle et al. [13]. The low density, hot Fe emission has also been suggested to be associated with radioactively heated bubbles of Fe ejecta by Wang & Chevalier [14]. Detailed further analysis of observations with Chandra and XMM should shed further light on these issues.

SPATIALLY RESOLVED SPECTROSCOPY

For young remnants whose emission is dominated by their reverse-shocked ejecta, the capability to do detailed spatially resolved X-ray spectroscopy with Chandra and XMM is a wonderful opportunity to study the X-ray emitting ejecta. The XMM observation of Tycho's SNR, for example, straighforwardly confirms the very different compositions of Si and Fe in the two bright knots at the eastern periphery of the remnant [13], a result indicated previously by Vancura et al. [15], but only with considerable data manipulation and luck using the limited instruments then available. XMM isolates the spectra of these knots for the first time. The superior angular resolution of the Chandra mirrors allows one to probe the spectra of even more compact regions, and for this illustration we turn to the brightest Galactic remnant, Cas A.

Ejecta Knots in Cas A

Cas A has long been considered the undisputed product of the explosion of a massive star, probably a Wolf-Rayet star [16,17], though the compact stellar remnant that then must have been produced remained elusive until Chandra's first light revealed it in X-rays [18–20]. This brief initial observation with the Advanced CCD Imaging Spectrometer (ACIS) also showed that the individual ejecta knots in Cas A have a range of spectral characteristics (Hughes et al. [21]), some of which are qualitatively consistent with their origin in different explosive nucleosynthesis burning zones of a massive star. For instance, the spectrum of a typical knot with emission lines of Si, S, Ar, and Ca was shown to be consistent with the element abundances expected for the zone of explosive O burning, whereas another knot also showing Fe emission is consistent with the zone of incomplete Si burning. A region in the outer southeast region of the remnant that was examined by Hughes et al. required the presence of still more Fe in the ejecta, suggesting a macroscopic mixing and inversion of the Si rich ejecta layer and the innermost layer of Fe.

An observation of Cas A twenty times deeper than the first light observation was also obtained by Chandra. With this observation, the distributions of the Si, S, Ar, Ca, and Fe-emitting ejecta were mapped through the equivalent widths of their emission lines by Hwang, Holt, & Petre [22]. The location of Fe-emitting ejecta mapped in this way is exterior to that of the Si-emitting ejecta throughout the southeast region, but not elsewhere. Because of this singularity of the southeast region, we turn to a more detailed spectral analysis using the deep Chandra observation.

Figure 1 shows the southeastern part of Cas A and indicates the regions studied; the spectral extraction regions were as small as $3''$ in diameter. The spectra were analyzed using models for plane parallel shocks in XSPEC version 11.0 [23] with a single electron temperature, and including a range of ionization ages from zero up to a fitted maximum value. The ionization age $n_e t$ is defined as the product of the ambient electron density and the time since the gas was heated by the shock, with equilibrium corresponding to roughly 10^{12} cm^{-3} s or higher. Also indicated in the figure are the fitted Fe/Si abundance ratios for the various knots, classified broadly as dominated by Si (Fe/Si < 0.1 Fe$_\odot$/Si$_\odot$), dominated by Fe (Fe/Si > 1.5 Fe$_\odot$/Si$_\odot$) and with a mixture of Fe and Si (0.1 Fe$_\odot$/Si$_\odot$ $<$ Fe/Si < 1.5 Fe$_\odot$/Si$_\odot$). The highest Fe/Si abundance ratios systematically occur in the outermost regions in the southeast, with fitted ratios as high as $4-5$ Fe$_\odot$/Si$_\odot$ in the regions that we examined. An intermediate region with a more solar mixture of Si and Fe gives way to a Si-dominated region coincident with the bright ring evident in the broadband images.

Two extreme examples of the spectra for a *single* highly Fe-rich knot and a more typical S-rich knot are shown in Figure 2. Both have a fitted electron temperature of kT ~ 2 keV, that is typical of the ejecta knots. What is obviously very different between the two is the relative abundance of Fe to Si (roughly a factor of 200 higher in the spectrum on the left), that manifests itself in the shape of the continuum

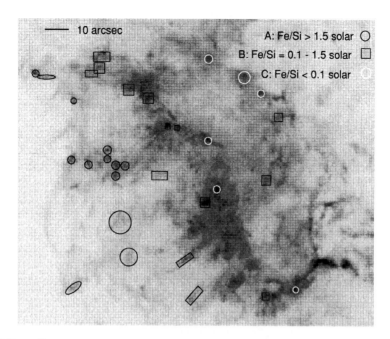

FIGURE 1. Closeup view of the southeastern section of Cas A with Chandra ACIS in a deep 50 ks observation (region is 2.5′ across). Spectral extraction regions are overlaid and coded according to the fitted Fe/Si element abundance ratio (relative to solar): A (high Fe) circular and elliptical regions, B (mixture) square and rectangular regions, C (high Si) white circular regions.

below the Si blends. Because of the high Fe abundance in the spectrum on the left, this continuum comes predominantly from Fe rather than from H and He. The ionization age $n_e t$ is also some 50 times higher for the Fe-rich knot than it is for the Si-rich knot. The high ionization age is evident in the shape of the Fe L emission in the spectrum on the left, and is also reflected in the increased strength of the Si Ly α (2.006 keV) line relative to the Si He α ($\sim$ 1.85 keV) blend. Assuming that, to zeroth order, the Fe and Si knots were shocked at roughly the same time, this indicates much higher densities in the Fe-rich knots, consistent with their origin deep within the inner layers of the supernova ejecta.

The region of the northeast jet, the largest radial extension seen in Cas A, is well-known from detailed optical observations to be enriched by ejecta emitting in O, S, and Ar [16,17]. The Chandra data show that the X-ray emission arises largely from ejecta highly enriched in Si, Si, Ar, and Ca. Fits to their X-ray spectra also show higher ionization ages than in the bright ring, suggesting that these are also high density clumps flung outward at high velocity during the explosion.

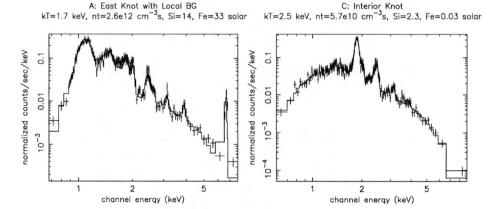

FIGURE 2. Representative spectra of extremely Fe- (left) and Si-dominated (right) knots in the southeast region of Cas A. The the best-fit plane-parallel shock model with a single electron temperature is overlaid on the data, and fitted parameter values are indicated.

ISOLATION OF EMISSION FROM THE FORWARD SHOCK

It was evident from the earliest nonimaging spectral observations that the reverse-shocked ejecta in Tycho and Cas A are the dominant contributors to the soft X-ray emission at energies below about 5 keV [24,1]. The reverse shock cannot exist, however, without the forward shock. For Cas A and Tycho, spatial X-ray components corresponding to the forward shock were suggested from early Einstein observations, for example by Fabian et al. [25] and Seward, Harnden & Tucker [26], and hard components in the spectra were attributed early on to the forward shock (e.g. Pravdo & Smith [27]. The true situation is more complex because theory also predicts nonthermal X-ray emission from particle acceleration at the forward shock and such emission is now observed, with the most famous example being SN1006 [28]. To unambiguously identify the location of the forward shock and to isolate its spectrum, one requires the capability to disentangle spatial and spectral information that is now available with Chandra and XMM.

For Cas A, the Chandra X-ray continuum image at energies between $4 - 6$ keV shows a "rim" surrounding the bright ring of emission from the ejecta, as shown by Gotthelf et al. [29]. As discussed elsewhere in these proceedings by Rudnick [30], this is persuasively identified as the location of the forward shock. The location of the forward shock in Tycho is similarly identified by XMM from the good correlation of this X-ray continuum emission with radio emission [13].

At distances of a few kpc within the Galaxy, Cas A and Tycho are relatively nearby, but Chandra's spatial resolution is good enough to isolate the forward shock in much more distant remnants. The bright ejecta-dominated ring of E0102-72, the brightest remnant in the Small Magellanic Cloud, has been previously imaged [31] ; Chandra reveals a faint shelf of emission exterior to the ejecta ring that is identified as the forward shock in the observation by Gaetz et al. [32]. The spectrum of this outer region is qualitatively different from that of the ring, as demonstrated by Hughes, Rakowski, & Decourchelle [33], with line strengths that are well-explained by the ambient abundances in the SMC, in contrast to the prominent O and Ne emission from the ejecta-enriched ring. Importantly, the measured temperature of the blast-shocked gas is barely 1 keV. Together with the shock velocity inferred from the measured X-ray expansion of the forward shock, this has the interesting implication that the acceleration of cosmic rays at the forward shock is highly efficient, as the shock should have heated the electrons to substantially higher temperature, even if the electrons equilibrate to the much higher ion temperatures only by Coulomb collisions. It would therefore appear that some of the enery of the shock wave has been diverted elsewhere, with particle acceleration being a persuasive explanation.

X-RAY DYNAMICAL STUDIES OF REMNANTS

A frontier that will be explored productively, especially by Chandra, is the study of the dynamics of the X-ray emitting gas in remnants. The first X-ray proper motion studies of the youngest and brightest supernova remnants were carried out using the high resolution imaging experiments on the Einstein and ROSAT Observatories, generally for broad radial or azimuthal sectors (Kepler's and Tycho's SNRs by Hughes [34,35], Cas A by Vink et al. [36] and Koralesky et al. [37]). The general trends in these results are two-fold. First, in comparison to similar radio expansion studies [38–41], the X-ray expansion results have yielded velocities that are systematically higher by roughly a factor of two. Second, in spatially resolved expansion studies carried out in both the radio and in X-rays, the expansion is observed to be significantly asymmetric, with expansion rates varying with azimuthal angle by factors of two or even significantly more. Both these points are illustrated in Figure 3 of Rudnick's review of Cas A elsewhere in these proceedings.

The disparity between the radio and X-ray expansion results is disconcerting. In most instances, the extent of the radio and X-ray emission from the remnants is similar, and yet the rate of radio expansion is only about half the rate of the X-ray expansion. Further observations with Chandra over the next several years should help to shed light on this puzzle. In particular, dynamical studies for the brightest remnants could be carried out for individual X-ray knots, and moreover the analysis could be done for each element separately, as has now been done extensively (at least for Cas A and a few other remnants) at optical and radio wavelengths.

The dynamics in the direction along the line of sight can be studied by measuring this velocity component through Doppler shifts in the energies of known emission lines. For Cas A, the presence of asymmetries in the line of sight velocities was known from early observations with the Focal Plane Crystal Spectrometer (FPCS) on the Einstein Observatory, a dispersive spectrometer that was used in this case with a $3' \times 30'$ aperture. Markert et al. [42] show that the FPCS spectral scans of He-like Si and S resonance and H-like Ly α lines from the northwest half of Cas A have lower line energies than those from the southeast, implying bulk motion of the gas in a roughly ring-shaped geometry. A Doppler map by Holt et al. [8] based on the centroid of the Si He α blend using more recent ASCA data (therefore with lower spectral resolution than the FPCS) confirms and extends this result with an angular resolution of about $1'$.

The Chandra version of the Doppler velocity map of Cas A obtained from the centroids of the Si He α blend is shown in Figure 3. The energy shifts measured correspond to velocities as high as 2500 km/s, with blue-shifted material generally located in the southeast and red-shifted material in the northwest, consistent with previous measurements. The Chandra map gives significantly more detail than previous maps: in the northwest, a concave arc of moderately blueshifted material is seen just south of a convex arc of redshifted material. These structures qualitatively correspond very well with the velocity structure of the dense ejecta knots measured with optical echelle spectra by Lawrence et al. [43], though there is more X-ray emission in the southeast than in the optical. A range of ionization ages is seen in the Si-emitting material in Cas A, and in principle this could affect the centroid of the Si blend, but fortunately the Si energy centroid remains dominated by the resonance line for the relevant range of ionization age and temperature in Cas A. The energy shifts therefore are dominated by Doppler shifts due to the bulk motion of the gas. Even with moderate energy resolution CCD spectrometers, Chandra provides more details of the line-of-sight velocity structure in the X-ray emitting gas than was previously available.

CONCLUSIONS

Chandra and XMM Newton have opened the way for true spatially resolved X-ray spectroscopy of supernova remnants. Though the CCD spectrometers are limited in their spectral resolution, the ability to disentangle spatial components that change on very small angular scales provides a new window in understanding these complex sources. The results presented in this review are just the beginning!

Acknowledgements: It is a pleasure to acknowledge Steve Holt, Rob Petre, and Andy Szymkowiak as my collaborators on the Chandra observation of Cas A, and to thank Anne Decourchelle for generously providing the results of her XMM study of Tycho's supernova remnant prior to publication. We are grateful to Kazik Borkowski for fruitful scientific discussions and for making his spectral models publicly available.

FIGURE 3. Doppler velocity map of Cas A based on the Si He α blend observed with ACIS on Chandra. Black corresponds to blueshifted velocities, white to redshifted velocities. The range of velocity shifts measured is roughly ±2500 km/s.

REFERENCES

1. Becker, R. H., et al. 1979, ApJL, 234, L73
2. Winkler, P. F., Canizares, C. R., Clark, G. W., Markert, T. H., & Petre, R. 1981, ApJ, 245, 574
3. Seward, F. D. 1990, ApJS, 73, 781
4. Aschenbach, B., Egger, R., & Trümper, J. 1995, Nature, 373, 587
5. Levenson, N. A., et al. 1997, ApJ, 484, 304
6. Williams, R. M., et al. 1999, ApJS, 123, 467
7. Tanaka, Y., Inoue, H., & Holt, S. S. 1994, PASJ, 46, L37
8. Holt, S. S., Gotthelf, E. V., Tsunemi, H., & Negoro, H. 1994, PASJ, 46, L151
9. Fujimoto, R., et al. 1995, PASJ, 47, L31
10. Bleeker, J., et al. 2001, A&A, in press
11. Canizares, C. R. 2000, this proceedings
12. Hwang, U., & Gotthelf, E. V. 1997, ApJ, 475, 665
13. Decourchelle, A., et al. 2001, A&A, in press
14. Wang, C.-Y., & Chevalier, R. A. 2000, ApJ, submitted, astro-ph/0005105
15. Vancura, O., Gorenstein, P., & Hughes, J. P. 1995, ApJ, 441, 680
16. Fesen, R. A., Becker, R. H., & Goodrich, R. W. 1988, ApJ, 329, L89
17. Fesen, R. A., & Gunderson, K. S. 1996, ApJ, 470, 967
18. Tananbaum, H., et al. 1999, IAU Circ 7246
19. Pavlov, G., et al. 2000, ApJL, 531, L53

20. Chakrabarty, D., et al. 2000, astro-ph/0001026 preprint
21. Hughes, J. P., Rakowski, C. E., Burrows, D. N., & Slane, P. O. 2000, ApJL, 528, L109
22. Hwang, U., Holt, S. S., & Petre, R. 2000, ApJ, 537, L119
23. Borkowski, K. J., Lyerly, W. J., & Reynolds, S. P. 2000, astro-ph/0008066
24. Becker, R. H., et al. 1980 ApJ, 25, L5
25. Fabian, A. C., et al. 1980, MNRAS, 193, 175
26. Seward, F. D., Gorenstein, P., & Tucker, W. 1983, ApJ, 266, 287
27. Pravdo, S. H., & Smith, B. W. 1979, ApJL, 234, L195
28. Koyama, K., et al. 1995, Nature, 378, 255
29. Gotthelf, E. V., et al. 2000, ApJ, submitted
30. Rudnick, L. 2001, these proceedings
31. Hughes, J. P. 1988, in IAU Colloq. 101, Supernova Remnants and the Interstellar Medium, ed. R. S. Rogers & T. L. Landecker (Cambridge: Cambridge Univ. Press), 125
32. Gaetz, T., et al. 2000, ApJL, 534, 47
33. Hughes, J. P., Rakowski, C. E., & Decourchelle, A. 2000, ApJL, 543, L61
34. Hughes, J. P. 1999, ApJ, 527, 298
35. Hughes, J. P. 2000, ApJ, 545, L53
36. Vink, J., Bloemen, H., Kaastra, J. S., & Bleeker, J. A. M. 1998, A&A, 339, 201
37. Koralesky, B., Rudnick, L., Gotthelf, E. V., & Keohane, J. W. 1998, 505, L27
38. Dickel, J. R., Sault, R., Arendt, R. G., Korista, K. T., & Matsui, Y. 1988, ApJ, 330, 254
39. Reynoso, E. M., Moffett, D. A., Goss, W. M., Dubner, G. M., Dickel, J. R., Reynolds, S. P., & Giacani, E. B. 1997, ApJ, 491, 816
40. Anderson, M. C., & Rudnick, L. 1995, ApJ, 441, 307
41. Agüeros, M. A., & Green, D. A. 1999, MNRAS, 305, 957
42. Markert, T. H., Canizares, C. R., Clark, G. W., & Winkler, P. F. 1983, ApJ, 268, 134
43. Lawrence, S. S., et al. 1995, AJ, 109, 2635

UV/Optical Observations of Young Supernova Remnants

William P. Blair

Department of Physics and Astronomy, Johns Hopkins University, Baltimore, MD, 21218, USA

Abstract.
Supernova remnants emit over much of the electromagnetic spectrum, and by taking advantage of that fact, we can strive to understand their global characteristics in ways that are not always possible for other kinds of objects. Observations in the optical, ultraviolet, and especially over this combined range (when possible) can provide a wealth of information on chemical abundances, kinematics, and physical conditions and processes in the shock fronts and associated emission of young supernova remnants. I briefly review some of the historical and recent work in this spectral range as it pertains to young remnants of core collapse and Type Ia supernovae.

OPENING REMARKS

In preparing for this talk, the first question I pondered was simply, "What constitutes a 'young' supernova remnant?" Interestingly, I found that 'age' was only one of several criteria I thought were of interest in defining the class. Kinematics, the presence of a central pulsar or X-ray point source, the presence of apparently enhanced abundances or ejecta, and sometimes the characteristics of the circumstellar/circumnebular region all contribute to a perception of 'youth'. Also, as one goes back beyond an age of 1000 years, the ages for many objects are not well known because they were not observed directly, and many factors besides 'time since the supernova explosion' can govern the physical sizes of the supernova remnants (SNRs) we observe today.

For instance, consider three of the primary young SNRs we discuss in the oxygen-rich class: Cas A in our Galaxy, 1E0102-7219 in the Small Magellanic Cloud (SMC; hereafter E0102), and N132D in the Large Magellanic Cloud (LMC). All three of these are at relatively well determined distances and have reasonably well-determined physical sizes. All three show rapidly expanding ejecta from which kinematic ages of 320, 1100, and 3150 years have been determined, respectively. The outer shell of N132D is some 25 pc in diameter, attributed to a pre-existing cavity wall. This is quite similar to the situation for the Cygnus Loop in our Galaxy, the prototypical 'middle-aged' SNR, which at a newly revised distance of 440 pc [1]

CP565, *Young Supernova Remnants: Eleventh Astrophysics Conf.,* edited by S. S. Holt and U. Hwang
© 2001 American Institute of Physics 0-7354-0001-6/01/$18.00

is some 22 pc by 27 pc in extent and is also thought to represent a cavity explosion [2]. The only thing missing in the Cygnus Loop is some O-rich ejecta! While I will stop short of classifying the Cygnus Loop as a 'young SNR', it may not be that much older than N132D, which we do classify as young.

Other objects that show some signs of youth but are somewhat older include Puppis A and the Vela SNR. Vela has recently moved closer as well (d = 250 pc; [3]) but still has a sizable diameter of 35 pc or so. Yet projected near its center is an active pulsar. The spin-down age of the pulsar, 11400 years [4], may or may not provide a relevant reference as pulsar spin-down ages have been called into question recently [5]. If pulsars remain close to the centers of their SNRs for this long, one might think there should be many more instances of pulsar-SNR associations than there are. Puppis A contains an X-ray point source [6], and although the outer shell shows nitrogen enhancement and low velocities, this SNR also contains rapidly moving O-rich debris, from which a kinematic age of 3700 years is determined [7].

The O-rich class of objects are, of course, attributed to core collapse SN progenitors, and hence very massive stars. The other extreme are the young SNRs like Tycho and SN1006, which are thought to arise from Type Ia SNe, and hence exploding White Dwarf stars. Perhaps between these extremes are the intermediate mass SN progenitors that result in 'plerionic' or Crab Nebula-like SNRs. The Crab Nebula is thought to have arisen from a progenitor in the 8 - 10 solar mass range ([8],[9]). However, it is unclear how far this simple picture extends. The SNR 0540-69.3 in the LMC has been called the 'Crab twin' because of its fast pulsar and optical synchrotron nebula, and yet its ejecta are O-rich and it has other abundance patterns similar to Cas A, thought to have arisen from a WN star [10].

So are there patterns in the existing set of young SNRs that we can interpret in a meaningful way, or is this just a small set of interesting (and possibly extreme) individual objects? A piece of relevant information may come from studies of nearby galaxies. Multiwavelength surveys of M33 and M31 have not turned up anything like a Cas A or Crab Nebula [11]. Surveys of NGC 6946 ([12], [13]) and M83 [14], each of which have had six SNe in the last century, have not turned up any bona fide examples of young O-rich SNRs with extreme optical, radio, or X-ray luminosities. It could be that we are stymied in our attempts to understand the patterns and systematics of the class of young SNRs by a combination of small number statistics and a set of 'extreme' objects that we try to treat as 'prototypical.' In any event, the way to address this issue is to learn as much as we possibly can about the individual objects and see whether the pieces of the puzzle come together or not. UV/optical observations contribute some key pieces of the puzzle.

IMPORTANCE OF UV/OPTICAL AND LIMITATIONS

One characteristic of SNRs is their penchant for producing emission over the entire electromagnetic spectrum. UV and optical emission tends to arise in a fairly dense component of the shocked gas. In young SNRs, this component is either

154

shocked ejecta, shocked ISM (or CSM), or a combination of the two. Optical studies, including longslit spectra and especially Fabry-Perot imaging, produce the best kinematic models of the expanding ejecta ([15], [16]). The intermediate mass elements have numerous forbidden, intercombination, and resonance lines and multiple ionization stages across the UV-optical spectral region. This allows one to determine abundances of these elements and reasonable physical conditions for comparison with results from other wavelength regions. When observations in the far-UV are possible, one obtains access to lines of elements such as carbon and silicon, which have no strong lines in the optical, and to lines from generally hotter material such as N V λ1240 and O VI λ1035. Combined UV/optical data have proven much more effective at determining shock velocities, physical conditions, and abundances in SNRs than optical data alone.

Unfortunately, access to the UV spectral region has been limited by a number of factors. The worst culprit is interstellar extinction, which is higher in the UV and rises higher still into the far-UV. Since most SNRs are in or near the galactic plane, only a handful of galactic SNRs have even been detected in the UV, let alone studied thoroughly. The Crab Nebula has been detected with the International Ultraviolet Explorer (IUE, [17]) and the Hopkins Ultraviolet Telescope (HUT; [18]), but just barely through an extinction of E(B-V)=0.5. With its sensitivity to diffuse emission, HUT was also used to detect SN1006 [19] and Puppis A [20], but at just a single position in each object and at fairly low signal-to-noise. Even in the Cygnus Loop and Vela, comparisons of UV and optical data have been difficult due to differing aperture sizes or shapes (in combination with the spatially variable emission). An object such as Cas A, lying behind A_v=4.3, will probably never be detected in the UV at a reasonable level (and even if it was detected, meaningful interpretation would be difficult).

The Magellanic Clouds are much farther away than galactic SNRs, but suffer relatively little foreground extinction. A few of these SNRs were studied with IUE, but the poor spatial scale made detailed studies difficult ([21], [22]). The advent of HST has provided reasonable spatial resolution for studying these objects, but relatively little of HST's resources have been directed toward studies of SNRs. The Far Ultraviolet Spectroscopic Explorer satellite (FUSE; [23], [24]) was launched over one year ago, bringing high spectral resolution and good sensitivity to far ultraviolet studies ([25]). At the same time, however, limited spectral coverage and poor spatial resolution make it applicable primarily to specific situations.

Below I provide an overview of some of the UV/optical work done to date, and suggest some avenues for future work. I will skip certain objects such as SN 1987A, Kepler, and the Crab Nebula, for which separate papers have been given at this conference, concentrating on recent results for some of the lesser known objects.

REMNANTS OF CORE COLLAPSE SUPERNOVAE

UV and optical observations of O-rich SNRs have two primary goals. One is to determine the primary physical processes responsible for the emission we see, and the other is to determine the relative abundances of chemical elements, especially in the ejecta. This in turn should allow us to compare to nucleosynthesis models of massive stars and either test these models or use them to determine the mass of the precursor stars (depending on your outlook).

The two primary objects of this sub-class available for UV/optical study are the two Magellanic Cloud objects mentioned earlier, N132D and E0102. Early IUE and optical work on these two objects ([26], [27]) raised some questions as to the dominant ionization mechanism. That is, while the dynamics of these objects point naturally to shock heating, the extreme luminosities in soft X-rays (and by implication, in the EUV) can in principle compete effectively in ionizing or at least pre-ionizing the material we see. Coupled with the difficulty of calculating shock models for extremely metal-rich material, this question has remained largely unsettled.

One of the real problems with the early work in the UV was the lack of spatial resolution. In N132D, for instance, where both O-rich and shocked interstellar material is present, the IUE large aperture could not successfully separate the various components [26]. Also, lying in the stellar bar of the LMC, N132D suffers from extreme stellar crowding. In E0102, no shocked interstellar component is evident, but blending O-rich material from many separate knots and filaments still caused potential confusion [27]. Likewise, comparing optical longslit spectra with the emission from the 10×20 arcsec IUE aperture when spatial variability is known to be present compromised the comparison of these objects with model calculations.

The power of HST has made possible a significant step forward for these objects. HST/WFPC2 imaging of N132D has permitted the stellar contamination to be removed effectively and has cleanly separated the emission from ejecta and ISM [28]. Furthermore, the spatial ionization structure in the outer shell has been modeled, pointing toward X-ray and EUV photoionization of this material by shocks in the ejecta and in the outer wall, strengthening the cavity wall picture for this component. More recently, E0102 has also been imaged, and the Faint Object Spectrograph (FOS) on HST has been used to observe several ejecta knots (two in N132D, and one in E0102) plus one emission knot in the outer shell of N132D across the entire UV/optical region available to HST [29]. This has largely eliminated the confusing effects of differing slit sizes and placements.

The knot in the outer shell of N132D shows a very rich emission line spectrum (see Figure 1, top panel). Modeling indicates, however, that even at the spatial scale available to HST, there must be multiple shock velocities mixing in the observed region. Also, the abundances determined are consistent with mean LMC interstellar abundances. In particular, there is no apparent enrichment in nitrogen abundance, as might be seen from a Wolf-Rayet or other massive progenitor.

The spectra of the two ejecta knots in N132D show minor differences, but ba-

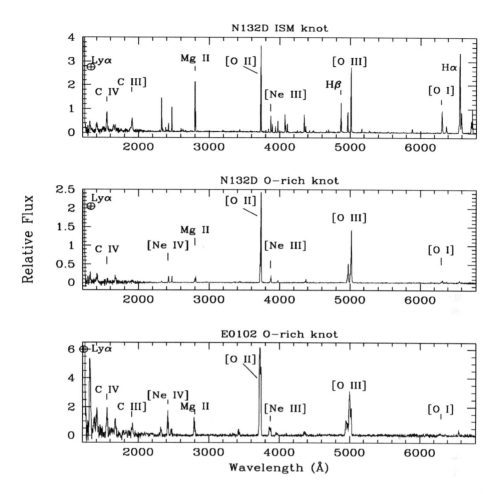

FIGURE 1. HST FOS UV/optical spectra of filaments in two Magellanic Cloud supernova remnants (from [29]). The top panel is the spectrum of a shocked interstellar cloud in the outer shell of N132D in the LMC. The middle panel shows a spectrum of an O-rich knot in the same remnant. For comparison, the bottom panel shows an O-rich knot from E0102 in the SMC.

sically support the idea that the O-rich ejecta we see emitting today are shock heated and of relatively uniform ionization and abundances. Systematic differences are more apparent, however, between N132D and E0102, as shown in the middle and bottom panels of Figure 1. Modeling these spectra has presented a number of challenges [29]. As with the ISM knot in N132D, it was necessary to include a range of shock velocities to approximate the observed relative line intensities. Despite the uncertainties in the models, the range of ionization stages available with UV/optical coverage permit a reasonable assessment of the relative abundances of key intermediate mass elements such as, O, Ne, Mg, and especially C.

Comparison with nucleosynthesis models, however, does not yet yield a definitive answer to the mass of the progenitors for these objects. The derived abundances can be matched reasonably by various models ([30], [31]), but these models also predict Si-group and Fe-group elements in the ejecta that simply are not seen. We are left with two tracks to pursue: fairly low mass progenitors that did not produce significant abundances of more processed material, or very massive progenitors which lost their outer layers before exploding, revealing the O-rich mantle material (i.e., Type Ib or Ic supernovae). Each of these cases has pros and cons, and it will take more work to sort out these possibilities.

REMNANTS OF TYPE IA SNE

UV/optical observations of young SNRs from Type Ia SNe have the primary purpose of understanding the physical processes involved in the fast, non-radiative shock waves associated with these objects. With one notable exception (see below), the ejecta from these remnants has not been visible in the UV/optical. Having said this, I should also mention that fast non-radiative shock waves are not restricted to remnants of Type Ia supernovae, as they are evident in objects such as Kepler's SNR [32] and even on the limb of the Cygnus Loop ([33], [1]). But Type Ia remnants such as Tycho and SN1006 in our Galaxy provide the cleanest examples for study.

Several recent observations are beginning to shed light on the elusive problem of how electrons and ions come into equilibrium in the post-shock gas. Raymond et al. [34] made the first successful UV observation of the shock on the northwestern limb of SN1006 using HUT, and measured O VI, N V, C IV, and He II lines from the $\sim$2000 km s^{-1} shock. The line widths were consistent with a single velocity width independent of the ion, indicating that the ions were not being thermalized very effectively. Furthermore, the line intensities were best matched by assuming a low level of equilibration between ions and electrons [35].

The optical Balmer lines are present in these fast, non-radiative shocks when the preshock material is at least partially neutral [36]. If observed with sufficient spectral resolution, the profiles at Hα show a two component structure with a narrow core and broad wings [37]. In a recent optical high resolution study, Ghavamian and collaborators were able to observe the broad and narrow components on both Hα and Hβ [38]. Using Monte Carlo techniques to model these data, they find a

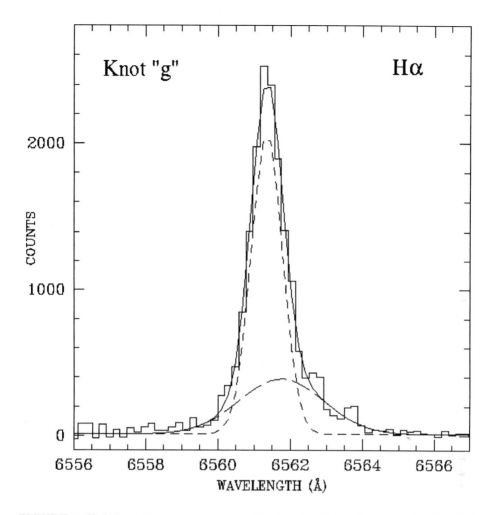

FIGURE 2. High dispersion optical spectrum of Hα from knot 'g' on the eastern side of Tycho's SNR (from [39]). The dashed lines show the components and the smooth solid line shows the total fit to the data (histogram). The width of the narrow component corresponds to 44 km/s, while the expected thermal width is only ∼15 km/s. This points to a non-thermal source of heating in the pre-shock material.

best match with a decreasing equilibration fraction with increasing shock velocity. Ghavamian et al. [39] also find the width of the narrow line component in Tycho that is systematically higher than expected for the preshock thermal conditions of the gas. This indicates that some non-thermal process, possibly cosmic ray heating or a magnetic precursor, is heating the preshock gas.

Nature has provided one example where we can probe the cool interior of a young Type Ia remnant using UV observation. The so-called "Schweizer-Middleditch star" [40] is a sub-dwarf B star that is seen in projection behind SN1006 ([41], [42]). FUV spectra of this star ([42], [43], [44]) have been used to identify broad absorptions of Fe II and several ionization stages of Si attributed to the intervening SNR. The favored model [45] has a cold, rapidly expanding center of Fe ejecta (interior to the reverse shock) that are being ionized up by the X-ray/EUV emission from the blast wave. However, only a small portion of the expected $\sim$0.5 solar mass of Fe is in the form of Fe^+. Searches to date for the strongest Fe^{++} resonance line at 1123 Å have been unsuccessful [44]. A FUSE observation of this star is in the works to perform a more detailed search for the missing iron.

The only other similar situation identified to date is the case of SN1885 in M31. This $\sim$115 year old supernova remnant has been detected in absorption against the bulge of M31 using both HST imaging and spectroscopy ([46], [47]). In this case, modeling of the absorptions is consistent with the expected mass of Fe from a Type Ia supernova, but the error bars are very large.

FUTURE DIRECTIONS

With the relatively small sample of objects with which we have to work, it is important for us to derive the best basic physical data for these objects and to use this information appropriately across various wavebands for interpreting the available observations. This may seem obvious, but with distances uncertain by factors of two or more (and different people using different values), comparing basic information such as luminosities or derived properties such as ages becomes very difficult to do systematically. This is very important not only for understanding the individual objects, but for investigating the relative properties of these objects and studying the big picture.

Another point which may be obvious but deserves mention is that we need to use the capability to make good UV/optical combined measurements while it is available. HST/STIS, with its longslit capability and excellent spatial resolution, can provide much better data more quickly and for more positions than was possible with HST/FOS, and yet very little time has been allocated to SNRs. When HST is no longer available, it may be a long time before the capability is available again. FUSE is making selected high resolution far-UV (sub-Lyα) measurements of interest for objects in our Galaxy, but has very limited spatial resolution for objects in the Magellanic Clouds and beyond. One interesting possibility for FUSE is to make additional absorption measurements of hot stars behind young SNRs if additional examples can be found.

Finally, it may be fruitful to obtain deeper, high spatial resolution emission line surveys of nearby galaxies to search for more young SNRs. This could tell us whether the objects we all know and love are really typical young SNRs or instead represent just the extreme, high luminosity tail of a larger distribution.

My attendance at this conference has been supported by STScI grant GO-08222-0297A to the Johns Hopkins University.

REFERENCES

1. Blair, W. P., Sankrit, R., Raymond, J. C., and Long, K. S. 1999, AJ, 118, 942
2. Levenson, N., et al. 1997, ApJ, 484, 304
3. Cha, A. N., and Sembach, K. R. 1999, ApJS, 126, 399
4. Reichley, P. E., Downes, G. S., and Morris, G. A. 1970, ApJ, 159, L35
5. Gaensler, B. M., and Frail, D. A. 2000, Nature, 406, 138
6. Petre, R., Becker, C. M., and Winkler, P. F. 1996, ApJ, 465, L43
7. Winkler, P. F., Tuttle, J. H., Kirshner, R. P., and Irwin, M. J. 1988, in IAU Coll. 101, ed. R. S. Roger and T. L. Landecker, (Cambridge: Cambridge Univ. Press), p. 65
8. Blair, W. P., et al. 1992, ApJ, 399, 611
9. Nomoto, K., et al. 1982, Nature, 299, 803
10. Kirshner, R. P., Morse, J. A., Winkler, P. F., and Blair, W. P. 1989, ApJ, 342, 260
11. Gordon, S. M., et al. 1998, ApJS, 117, 89
12. Matonick, D. M., and Fesen, R. A. 1997, ApJS, 112, 49
13. Lacey, C., Duric, N., and Goss, W. M. 1997, ApJS, 109, 417
14. Blair, W. P. 2000, unpublished.
15. Morse, J. A., Winkler, P. F., and Kirshner, R. P. 1995, AJ, 109, 2104
16. Lawrence, S. S., et al. 1995, AJ, 109, 2635
17. Davidson, K., et al. 1982, ApJ, 253, 696
18. Blair, W. P., et al. 1992, ApJ, 399, 611
19. Raymond, J. C., Blair, W. P., and Long, K. S. 1995, ApJ, 454, L39
20. Blair, W. P., Raymond, J. C., Long, K. S., and Kriss, G. A. 1995, ApJ, 454, L35
21. Benvenuti, P., Dopita, M., and D'Odorico, S. 1980, ApJ, 238, 601
22. Vancura, O. Blair, W. P., Long, K. S., and Raymond, J. C., 1992, ApJ, 394, 158
23. Moos, H. W., et al. 2000, ApJ, 538, L1
24. Sahnow, D. J., et al. 2000, ApJ, 538, L7
25. Blair, W. P., et al. 2000, ApJ, 538, L61
26. Blair, W. P., Raymond, J. C., Danziger, J., and Matteucci, F. 1989, ApJ, 338, 812
27. Blair, W. P., Raymond, J. C., and Long, K. S. 1994, ApJ, 423, 334
28. Morse, J. A., Blair, W. P., Dopita, M. A., Hughes, J. P., Kirshner, R. P., Long, K. S., Raymond, J. C., Sutherland, R.,S., and Winkler, P. F. 1996, AJ, 112, 509
29. Blair, W. P., Morse, J. A., Raymond, J. C., Kirshner, R. P., Hughes, J. P., Dopita, M. A., Sutherland, R.,S., Long, K. S., and Winkler, P. F. 2000, ApJ, 536, 667
30. Woosley, S. E., Langer, N., and Weaver, T. A. 1995, ApJ, 448, 315

31. Nomoto, K., Hashimoto, M., Tsujimoto, T., Thielemann, F.-K., Kishimoto, N., Kubo, Y., and Nakasoto, N. 1997, Nuc. Physics, A616, 79c
32. Blair, W. P., Long, K. S., and Vancura, O. 1991, ApJ, 366, 484
33. Levenson, N. A., Graham, James R., Keller, Luke D., Richter, and Matthew J. 1998, ApJS, 118, 541
34. Raymond, J. C., Blair, W. P., and Long, K. S. 1995, ApJ, 454, L31
35. Laming, J. M., Raymond, J. C., McLaughlin, B. M., and Blair, W. P. 1996, ApJ, 472, 267
36. Chevalier, R. A., and Kirshner, R. P. 1979, ApJ, 233, 154
37. Smith, R. C., Kirshner, R. P., Blair, W. P., and Winkler, P. F. 1991, ApJ, 375, 652
38. Ghavamian, P., Raymond, J., Smith, R. C., and Hartigan, P. 2001, ApJ, in press
39. Ghavamian, P., Raymond, J. C., Hartigan, P., and Blair, W. P. 2000, ApJ, 535, 266
40. Schweizer, F., and Middleditch, J. 1980, ApJ, 241, 1039
41. Wu, C. C., Leventhal, M., Sarazin, C. L., and Gull, T. R. 1983, ApJ, 269, L5
42. Fesen, R. A., Wu, C. C., Leventhal, M., and Hamilton, A. J. S. 1988, ApJ, 327, 178
43. Wu, C. C., Crenshaw, D. M., Fesen, R. A., Hamilton, A. J. S., and Sarazin, C. L. 1993, ApJ, 416, 247
44. Blair, W. P., Long, K. S., and Raymond, J. C. 1996, ApJ, 468, 871
45. Hamilton, A. J. S., and Fesen, R. A. 1988, ApJ, 327, 178
46. Fesen, R. A., Gerardy, C. L., McLin, K. M., and Hamilton, A. J. S. 1999, ApJ, 514, 195
47. Hamilton, A. J. S., and Fesen, R. A. 2000, ApJ, 542, 779

Young SNRs: IR Observations

Richard G. Arendt[1,2]

[1] *Raytheon ITSS*
[2] *NASA/GSFC, Greenbelt, MD 20771, USA*

Abstract.
Infrared observations of SNRs began with the *IRAS* mission in 1983. Now as more detailed *ISO* and 2MASS observations of SNRs are being analyzed, we are beginning to see more clearly what the IR has to reveal. For young SNRs this has most notably meant a detailed look at SN ejecta before it is mixed into the ISM. The IR can reveal the presence of elements and gas phases that are undetectable at other wavelengths. The IR emission is also essential for identifying the particular composition and mineral structure of dust grains that form in the SN ejecta.

INTRODUCTION

Infrared (IR) observations ($\sim 1 - 1000~\mu$m) of young supernova remnants (SNRs) provide views of these dynamic objects that can be difficult or impossible to attain at other wavelengths. A basic observational asset of the IR is that extinction is much lower than at optical and UV wavelengths. In addition, line emission in the IR provides diagnostic information on the temperature, density, and composition of various gas phases in SNRs. These diagnostics can probe different physical states compared to similar measures at other wavelengths. Unique to the IR however, is the fact that the IR is the only spectral region where thermal emission from dust is observed. Examination of the emission from dust in and around young SNRs can: provide direct evidence of dust formation and destruction; reveal the composition and mineralology of the dust; indicate the amount of mixing that takes place in the SN ejecta on both macroscopic and microscopic scales; and yield indirect diagnostics of the local ambient energy in the gas or radiation. On a cosmological scale, knowledge of the dust production (and destruction) in SNRs is important in determining the opacity of the ISM in the early epochs of star formation, which in turn is of critical importance in determining the relationship between the Cosmic IR Background and the history of star formation through the life of the Universe [1].

CP565, *Young Supernova Remnants: Eleventh Astrophysics Conf.*, edited by S. S. Holt and U. Hwang
© 2001 American Institute of Physics 0-7354-0001-6/01/$18.00

IR EMISSION MECHANISMS

IR emission in SNRs can arise through a wide variety of mechanisms, involving particles ranging from cold dust grains, to warm atoms and ions, to relativistic electrons. Thermal continuum emission from dust, ground-state fine structure lines of atoms and ions, free-free emission from hot ionized gas, and synchrotron emission from accelerated electrons all produce emission at IR wavelengths.

Following the broadband *IRAS* observations of SNRs, early modeling of dust emission typically involved blackbody emission modified by power-law grain emissivities [2]. The temperature of a dust grain immersed in a hot plasma will be determined primarily by the rate and energy of collisions with hot electrons, i.e. the density and temperature of the gas. In some cases (e.g. the Crab Nebula) radiative heating of the dust may also be important. In many cases it was clear that a range of dust temperatures was indicated by the observations. With more modern instruments (e.g. *ISO, SIRTF*) it has been and will be possible to detect very distinguishing emission features of the dust grains in SNRs. Such features include the traditional 10 and 20 μm astronomical silicate features, as well as more detailed and specific silicate emission bands. For carbonaceous grains, the nearly ubiquitous $3 - 12$ μm PAH features provide evidence of their carbon composition. Other IR emission bands may additionally reveal the presence of more obscure grain compositions (e.g. metal oxides, sulfides, nanodiamonds, and fullerenes.)

IR line emission from a variety of atoms, ions, and molecules have been observed in SNRs. Many of the strongest lines are ground-state fine structure lines of elements from C to Ni. In many cases several low to moderately ionized species of the same element can be observed in the IR (e.g. O I, O II, O IV, Ne II, Ne III, Ne V, S I, S III, S IV, Ar II, Ar III, Ar IV). For some species, IR line pairs provide density diagnostics. Excitation temperatures for these lines are typically several hundred degrees K. For molecular emission, IR wavelengths less than 30 μm can be used to directly trace H_2 through observations of the pure rotational and ro-vibrational transitions of the H_2 molecule.

While IR free-free emission has yet to be reported in SNRs, observations of IR synchrotron emission in remnants are of interest since extrapolations of observed radio and X-ray synchrotron emission often point to a spectral break occurring in the IR. Thus, IR observations of this break can help constrain detailed models of the electron acceleration and energy loss mechanisms.

IRAS RESULTS

The full sky survey carried out by *IRAS* [3] provided the first IR observations of SNRs. While low extinction is generally considered an observational asset, it also led to a high degree of confusion between IR emission of SNRs and that of the rest of the ISM on the line of sight. Largely limited by this confusion, surveys of SNRs in the *IRAS* data detected only roughly 1 in 3 SNRs at best [4,5]. Large and bright

SNRs, such as IC 443 [6,7], the Cygnus Loop [8,9], and Puppis A [10,11], stood out clearly, especially if they lie at relatively high galactic latitude. Many of the young SNRs were also seen, though many are too small to be well resolved with the $\sim 1'$ resolution provided by *IRAS*. Cas A [12,13], Kepler [13], Tycho [13], and the Crab [14,15] are all seen as bright compact sources, though other young SNRs are not (3C 58 [16], SN 1006 [4,5], and RCW 103 [4,5]).

The *IRAS* observations showed older SNRs to be slightly warmer than the typical ISM, with IR colors similar to those of H II regions. Younger SNRs are typically much warmer than the ISM, with colors that resemble those of planetary nebulae [4]. In most cases, the IR emission of SNRs was modeled as thermal emission from dust, though in some cases it was recognized or suspected that strong line emission may provide substantial emission even in the broadband *IRAS* observations [7,11], as has been confirmed for IC 443 by *ISO* spectral observations [17].

MORE RECENT RESULTS

Since *IRAS*, the most substantial advances in IR observations of SNRs have come from the *ISO* mission and the 2MASS project. *ISO* operated from 1995 to 1998, providing few-arcsecond imaging and spectral capabilites in the 2.5 – 200 μm wavelength range [18]. Sadly, *ISO* observations are limited to a relatively small number of SNRs. As the 2MASS project is an all-sky survey [19], it will eventually observe all SNRs in the J, H, and K$_s$ bands at $\sim 2''$ resolution. The remainder of this article summarizes some of the recent results from these two projects.

Cas A

Cas A is the ~ 320 year old remnant of a Type II (or Ib) supernova. *ISO* imaging of Cas A in the mid-IR (~ 10 μm) has now clearly revealed the large scale (few arcsecond) structure of the IR emission in Cas A. The clumpy ring of emission most strongly correlates with the optical structure of the SNR, though there are some resemblances to radio and X-ray structures as well [20,21]. Low resolution CVF spectra revealed continuum emission as well as strong lines of Ar II, Ar III, S IV, Ne II, and Ne III [20,22]. Neon in particular had been notoriously difficult to detect in Cas A at optical or UV wavelengths due to high extinction, but in the IR the neon is unmistakable. At slightly longer wavelengths, spectra of Cas A revealed a relatively sharp dust emission feature at ~ 22 μm (Figure 1) [23]. This feature was in good agreement with that indicated by earlier KAO observations. The position and shape of the feature is not quite right for most typical silicate minerals, however a certain class of relatively amorphous "protosilicate" minerals was found to fit the feature well [23], although there may yet be evidence that the protosilicates fail to fit the spectrum at ~ 10 μm [24]. In any case, the spectra are clearly indicative of silicate, not carbonaceous, dust in Cas A. Correlation between the continuum emission and the line emission indicates that the dust is associated

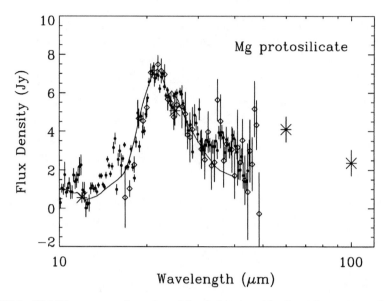

FIGURE 1. Mid–IR spectrum of a region of Cas A dominated by Fast Moving Knots [23]. Dots are *ISO* SWS data, diamonds are KAO data, asterisks are *IRAS* data. The line is the modeled emission from Mg protosilicate dust at a temperature of 169 K.

with the fast moving knots (FMKs) in Cas A, and thus is true ejecta of the SN explosion [23]. However, in greater detail, the dust and Ne emission is moderately anticorrelated, suggesting that the mixing of the Ne ejecta and the silicate-forming ejecta was far from complete [25]. Finally, *ISO* has provided evidence of a new component of relatively cold and slow-moving ejecta in Cas A. This component is suggested by low velocity Si II emission (Figure 2) and O III emission [26] in spectra towards the center of the SNR, and by a more filled-center morphology at far–IR wavelengths [21].

The 2MASS images of Cas A [27] have yet to be analyzed in detail, though the morphology of the emission seems to most strongly resemble that of the radio emission.

The Crab Nebula

Like Cas A, the Crab Nebula is the remnant of the supernova of a massive star. Unlike Cas A, however, the Crab Nebula is primarily powered by an energetic neutron star at its center. The J, H, and K_s 2MASS images of the Crab Nebula [27] are dominated by synchrotron emission, with a similar appearance to the optical continuum. Some of the brightest optical filaments of ejecta are faintly visible in the J and H bands (or more clearly, in the differences between the J, H, and K_s bands). Again, these data have yet to be analyzed in detail.

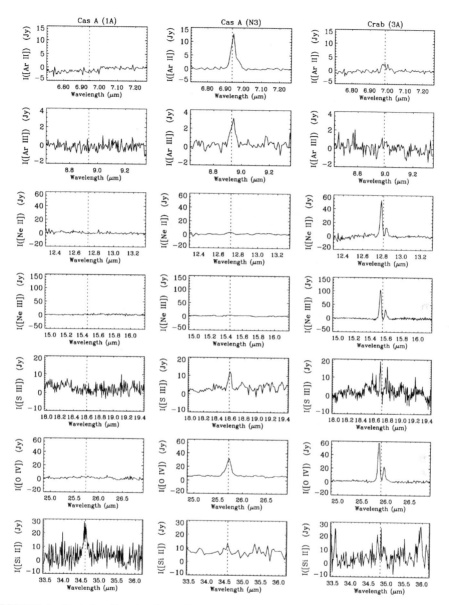

FIGURE 2. Emission line details in Cas A and the Crab Nebula. The first column shows spectra at a position near the center of Cas A, where apparently only a relatively low velocity ($v = $ -1850 km s^{-1}) [Si II] line is seen. The middle column shows spectra at an FMK position in the northern part of Cas A, where a variety of Ne, Ar, S, and O lines are observed, though the Cas A neon lines are hard to see on the same intensity scale as the Crab Nebula lines. The last column shows spectra from one of the brightest Crab Nebula filaments. The vertical dashed lines are drawn at a constant velocity, matching the [O IV] line, for each region.

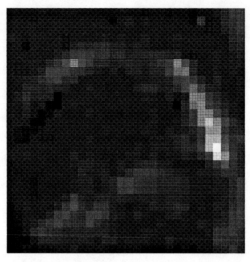

ISOCAM 14.9 μm

FIGURE 3. This ISOCAM image of Kepler's SNR (from the *ISO* data archive with no further processing).

ISO observations of the Crab nebula reveal synchrotron and line emission [28]. The line emission in the Crab filaments observed with the *ISO* SWS show much stronger Ne emission and much weaker Ar emission than the FMKs in Cas A (Figure 2). No dust emission is reported at wavelengths less than $\sim 12\ \mu$m. Other plerions (G 21.5–0.9 and 0540–69.3) are also reported to be dominated by synchrotron rather than dust emission [29]. However, the 60 and 100 μm observations do detect dust emission in the Crab, apparently associated with the optical filaments [30]. This relation is consistent with the previously detected extinction and H_2 emission of the optical filaments [31,32].

Kepler's SNR

Recent *ISO* and ground-based near-IR observations of Kepler's SNR show [Ar II] and [Ne II] lines, and emission in the J and H but not K bands [28]. Good correlation between the IR emission and the optical emission, along with relatively low inferred dust temperatures, is taken as evidence that the dust is swept–up circumstellar or interstellar material rather than ejecta, thus contrasting with the cases of Cas A and the Crab Nebula. The progenitor of Kepler's SNR may have been a Type II SN, but this is not certain.

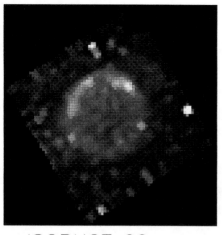
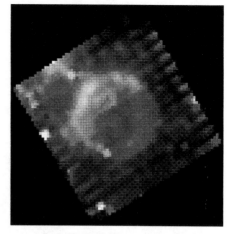

ISOPHOT 60 μm   ISOPHOT 90 μm

FIGURE 4. ISOPHOT images of Tycho's SNR (from the *ISO* data archive with no further processing).

Tycho's SNR

Tycho's SNR is usually assumed to the the result of a Type Ia SN. *ISO* observations of Tycho's SNR show that the 10.7 – 12 μm emission is correlated with the optical emission, and thus, like Kepler's SNR, the associated dust is expected to be circumstellar or interstellar in origin [28]. Longer wavelength 60 and 90 μm ISOPHOT images (Figure 4) also show their brightest emission in the north and east where the optical filaments are found. The correlation with the optical appears somewhat better at 60 μm (perhaps due to 63 μm [O I] emission) than at 90 μm, while both bands exhibit a nearly complete and circular shell.

RCW 103

RCW 103 is probably the oldest of the SNRs discussed here, but still relatively young at an age of ~ 1000 yr. An unusual X-ray point source at its center [33] suggests that it is the remnant of a Type II SN. Despite being too confused to be detected in the *IRAS* images, RCW 103 has yielded most impressive results from *ISO*. Spectroscopy reveals detection of nearly all the emission lines that might be expected within the wavelength range accessible with the SWS [34]. Observed lines include the "usual" O, Ne, Si, S, Ar, and Fe lines, lines of Cl and P, and dozens of H_2 lines. Abundances are estimated to be roughly solar, indicating that the observed IR–emitting material is swept–up, not ejecta, and that grain destruction has been efficient at returning refractory elements to the gas phase [34].

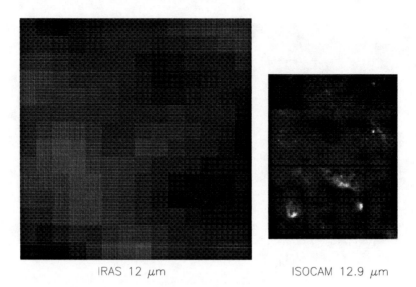

IRAS 12 μm   ISOCAM 12.9 μm

FIGURE 5. *IRAS* and ISOCAM images of RCW 103, demonstrating the great value of high resolution in distinguishing SNR emission apart from the confusion of the ISM. The *IRAS* image covers a larger area, but is reproduced at the same angular scale as the ISOCAM image.

Some Fe and H_2 lines had been previously observed in the near–IR [35,36]. The 2–arc barrel-shaped morphology of the 12.9 μm emission (probably dominated by [Ne II] 12.8 μm; Figure 5), resembles that of the near–IR. The 2MASS observations of RCW 103 provide a wide–field view with good resolution. The J and H band emission are dominated by [Fe II] lines while the K_s band traces the H_2 emission [37]. A clear separation of the molecular and Fe emission reveal the precursor structure of the fast shocks in RCW 103 [34,37]. The older SNR IC 443 shows similarly distinct (and much stronger) regions of Fe and H_2 emission, although in this case the shocks are slower than in RCW 103, and part of the SNR is very clearly impacting very dense molecular material [38].

CONCLUSION

For young remnants of Type II SNe, i.e. Cas A and the Crab Nebula, IR observations are beginning to reveal the nature of both the gas and dust ejecta, and the processes occurring as freshly synthesized elements make their way into the ISM. However, it may be that neither Cas A nor the Crab are typical SNRs. For older remnants, and for those of Type Ia SNe, the IR observations seem to reveal more about the shocked ambient medium than the ejecta directly. Further analysis of IR observations by *ISO*, 2MASS, and future projects (e.g. *SIRTF* and SOFIA) should continue to clarify and perhaps reform our understanding of the ejecta in young SNRs.

REFERENCES

1. Dwek, E., Fioc, M., and Városi, F., 2000, ISO Surveys of a Dusty Universe, eds. D. Lemke, M. Stickel, K. Wilke, (Berlin: Springer), 157
2. Dwek, E., and Arendt, R. G., 1992, ARAA, 30, 11
3. Neugebauer, G., et al., 1984, ApJ, 278, L1
4. Arendt, R. G., 1989, ApJS, 70, 181
5. Saken, J. A., Fesen, R. A., and Shull, J. M., 1992, ApJS, 81, 715
6. Braun, R., and Strom, R. G., 1986, A&A, 164, 193
7. Mufson, S. L., McCollough, M. L., Dickel, J. R., Petre, R., White, R., and Chevalier, R., 1986, AJ, 92, 1349
8. Braun, R., and Strom, R. G., 1986, A&A, 164, 208
9. Arendt, R. G., Dwek, E., and Leisawitz, D., 1992, ApJ, 400, 562
10. Arendt, R. G., et al., 1990, ApJ, 350, 266
11. Arendt, R. G., Dwek, E., and Petre, R., 1991, ApJ, 368, 474
12. Dwek, E., Hauser, M. G., Dinerstein, H. L., Gillett, F. C., and Rice, W. L., 1987, ApJ, 315, 571
13. Braun, R., 1987, 171, 233
14. Marsden, P. L., Gillett, F. C., Jennings, R. E., Emerson, J. P., de Jong, T., and Olnon, F. M., 1984, ApJ, 278, L29
15. Strom, R. G., and Greidanus, H., 1992, Nature, 358, 654
16. Green, D. A., and Scheuer, P. A. G., 1992, MNRAS, 258, 833
17. Olive, E., Lutz, D., Drapatz, S., and Moorwood, A. F. M., 1999, A&A, 341, L75
18. Kessler M. F., et al., 1996, A&A, 315, L27
19. Skrutskie, M., et al., 1997, The Impact of Large–Scale Near–IR Sky Survey, ed. F. Garzón, et al. (Dordrecht: Kluwer), 25
20. Lagage, P. O., et al., 1996, A&A, 315, L273
21. Tuffs, R. J., et al., 1999, The Universe as Seen by ISO, eds. P. Cox, M. F. Kessler, (ESA SP–427), 241
22. Lagage, P. O., et al., 1999, Solid Interstellar Matter: The ISO Revolution, eds. L. d'Hendecourt, C. Jobelin, A. Jones, (Berlin: Springer), 285
23. Arendt, R. G., Dwek, E., and Moseley, S. H., 1999, ApJ, 521, 234
24. Douvion, T., Lagage, P. O., and Pantin, E., 2000, A&A, submitted
25. Douvion, T., Lagage, P. O., and Cesarsky, C. J., 1999, A&A, 352, L111
26. Unger, S. J., et al. 1997, Proceedings of the First ISO Workshop on Analytical Spectroscopy, eds. A. M. Heras, K. Leech, N. R. Trams, M. Perry, (ESA SP–419), 305
27. http://www.ipac.caltech.edu/2mass/gallery/images_snrs.html
28. Douvion, T., Lagage, P. O., Cesarsky, C. J., and Dwek, E. 2001, in preparation.
29. Gallant, Y. A., and Tuffs, R. J., 1998, Mem. Soc. Astron. Ital., 69, 693
30. Tuffs et al. 1998, AGM, 14, 57
31. Hester, J. J., Graham, J. G., Beichman, C. A., and Gautier, T. N. 1990, ApJ, 357, 539
32. Graham, J. R., Wright, G. S., and Longmore, A. J., 1990, ApJ, 352, 172
33. Gotthelf, E. V., Petre, R., and Vasisht, G., 1999, ApJ, 514, L107

34. Oliva, E., Moorwood, A. F. M., Drapatz, S., Lutz, D., and Sturm, E., 1999, A&A, 343, 943
35. Oliva, E., Moorwood, A. F. M., and Danziger, I. J., 1989, A&A, 214, 307
36. Oliva, E., Moorwood, A. F. M., and Danziger, I. J., 1990, A&A, 240, 453
37. Rho, J., and Reach, W. R., 2001, this volume
38. Rho, J., Jarrett, T. H., Cutri, R. M., and Reach, W. R., 2001, ApJ, in press

Gamma-Ray Line Emission from Supernova Remnants

Peter A. Milne

NRC/NRL Research Associate, Code 7650, Washington, D.C. 20375, USA

Abstract. Gamma-ray line emission from the decays of radio-isotopes produced in supernovae should be detectable by three future gamma-ray telescopes. The 511 keV, 1157 keV, 1173 keV and 1809 keV lines produced in young supernova remnants are a powerful probe of the mass yields and ejecta kinematics of these objects. Nineteen supernova remnants have been studied to determine which gamma-ray lines will be detected by these three instruments. The results of this study and the scientific potential of these observations are discussed.

INTRODUCTION

A number of different radioactive isotopes are produced in both thermonuclear and core-collapse supernovae (SNe). The decays of many of these nuclei produce gamma-ray line emission, with lifetimes that range from less than a day to over a million years. Short-lived, or prompt emission occurs within the first few years after the SN event, during which time the escape of gamma-ray line emission is critically dependent upon the opacity of the ejecta. Observations of prompt gamma-ray lines probe the distribution of the radio-isotope as well as the kinematics of the ejecta's expansion for extra-galactic SNe [1,2]. Once the ejecta has thinned enough to be effectively thin to gamma-ray lines, I term the emission "SNR emission". Observations of SNR emission studies the overall production of the medium-lived radio-isotopes, and if the remnant is spatially-resolved, the distribution of these radio-isotopes as well as the kinematics of the ejecta's expansion. This probe of the ejecta is more straight-forward than studies at other wavelengths, where other factors, such as the level of ionization, are important. At late epochs, the ejecta has become too extended to be observed as a remnant. Very long-lived radio-isotopes that survive until this epoch and decay on million year lifetimes will produce a diffuse emission. This emission is observed as a Galaxy-wide emission, and is composed of the combined emission from thousands of SNe. Studies of diffuse emission study the galactic production of these long-lived radio-isotopes over million-year timescales.[1] Although prompt and diffuse emissions are scientifically interesting,

---

[1] Knodlseder et al. 1999 details CGRO/COMPTEL mapping of diffuse 1809 keV emission. [3]

CP565, *Young Supernova Remnants: Eleventh Astrophysics Conf.*, edited by S. S. Holt and U. Hwang
© 2001 American Institute of Physics 0-7354-0001-6/01/$18.00

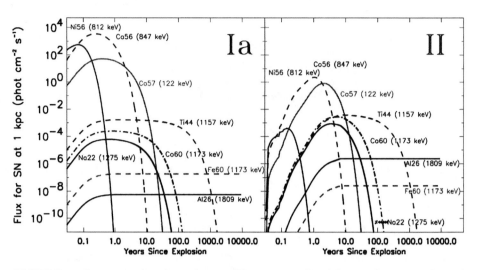

FIGURE 1. Gamma-ray line fluxes for two SNe at 1 kpc. The left panel shows line fluxes for the SN Ia model, W7. The right panel shows line fluxes for the SN II model, W10HMM.

only SNR emission will be discussed in this work.

Shown in Figure 1 are gamma-ray line fluxes for a Type Ia SN model (W7) [4] and a Type II SN model (W10HMM) [5] at a distance of 1 kpc. The prompt emissions from the short-lived decays of ^{56}Ni, its daughter nucleus, ^{56}Co, and from ^{57}Co are apparent. At a lower flux level are the gamma-ray lines from the medium-lived radio-isotopes ^{44}Ti, ^{22}Na and ^{60}Co. The COMPTEL instrument on-board CGRO has observed 1157 keV gamma-ray line emission from the SNR Cas A, and potentially from a second SNR, RX J0852.0-4622 (a.k.a. Vela Jr.) [6,7]. The large production of ^{44}Ti from both SNRs has challenged calculations of nuclear burning in SNe II. The distance and age of Cas A appears well constrained, so investigations have concentrated upon the lifetime of ^{44}Ti both in the laboratory [8] and in astrophysical environments [9]. The distance and age are far less certain for Vela Jr., but the observed flux levels (if confirmed) suggest a recent, nearby SN II. Though not yet observed galactically, ^{22}Na and ^{60}Co (which produce gamma-ray lines at 1275 keV and the 1173 keV), are two more medium-lived radio-isotopes produced in SNe. In addition, long-lived radio-isotopes such as ^{26}Al and ^{60}Fe emit at a sufficient level to be observable in nearby SNRs. Type II SNe produce more ^{26}Al than ^{60}Fe, Type Ia SNe more ^{60}Fe than ^{26}Al. Collectively, the mass yield estimates that result from gamma-ray line studies of SNRs determine the SN type.

In addition to nuclear de-excitation lines, the positron-electron annihilation line (at 511 keV) contributes gamma-ray line emission. The decays of ^{56}Ni $\rightarrow$ ^{56}Co $\rightarrow$ ^{56}Fe, ^{44}Ti $\rightarrow$ ^{44}Sc $\rightarrow$ ^{44}Ca, and ^{26}Al $\rightarrow$ ^{26}Mg all produce positrons. Assuming a favorable magnetic field geometry, positrons can escape in quantity from the SN ejecta [10]. Estimating that 5%Ia(0%-II), 100%(Ia/II) and 100%(Ia/II) of the decay positrons escape the SN ejecta, thermalize and annihilate with a mean lifetime of 10^5 years, the 511 keV line from positron-electron annihilation is also a SNR and diffuse gamma-ray line emission.[2] Per SN, Type Ia SNe produce almost 40 times more escaping positrons than Type II SNe [12].

SNR EMISSION

Three future instruments will be capable of studying the gamma-ray line emission from nearby SNRs. ESA's INTEGRAL satellite, due to be launched April 2002, will feature two instruments sensitive to gamma-rays, the spectrometer (SPI) and the imager (IBIS). Assuming one month observations (2 x 10^6 s on-source time), these are anticipated to achieve 2.9 x 10^{-5} and 4.1 x 10^{-4} phot cm^2 s^{-1} (5σ) sensitivity to the 1157 keV line (assumed to have a 30 keV FWHM), and 5.0 x 10^{-5} and 1.9 x 10^{-4} phot cm^2 s^{-1} (5σ) sensitivity to the 511 keV line (assumed to have a 5 keV FWHM).[3] SPI will have an angular resolution of $\sim$2.5°, IBIS will have an angular resolution of 12′. For both SPI and IBIS, these SNRs will only be observed if they are chosen as a target. An Advanced Compton Telescope being studied by NRL, will accumulate $\sim$2 x 10^7 seconds over a two-year (minimum) lifetime. Current simulations suggest that this instrument could achieve 7.4 x 10^{-8} phot cm^2 s^{-1} (5σ) sensitivity to the 1157 keV line and 1.1 x 10^{-7} phot cm^2 s^{-1} (5σ) sensitivity to the 511 keV line for two year observations. The ACT design simulated in this work features a wide ($\sim$120°) FoV and would operate in an equatorial orbit, always pointed at the celestial equator. The sensitivity would thus be optimal for remnants with low celestial declinations.

These instrument specifications were combined with distance, age, SN type and angular size estimates for 19 SNRs to estimate which gamma-ray lines would be detectable by the three instruments.[4] These estimates, shown in Table 1, suggest that many SNRs will have detectable emission.

REFERENCES

1. Matz, S.M. et al., *Nature* **331**, 416 (1988).
2. Kurfess, J.D. et al., *ApJ* **399**, L137 (1992).
3. Knodlseder, J., *AJ* **510**, 915 (1999).
4. Nomoto, K., Thielemann, F.-K., Yokoi, K., *ApJ* **286**, 644 (1984).
5. Pinto, P.A., Woosley, S.E. 1988, *ApJ* **329**, 820 (1988).
6. Iyudin, A.F. et al., *A&A* **284**, L1 (1994).
7. Iyudin, A.F. et al., *Nature* **396**, 142 (1998).

[2] Milne et al. (1998) details CGRO/OSSE mapping of 511 keV line emission.
[3] These values were derived from the "observation time estimator" at http://astro.estec.esa.nl/Integral/isoc/operations/html/OTE.html.
[4] For SNRs with angular sizes larger than the instrument's angular resolution, the sensitivity was scaled as the square-root of the size-to-resolution ratio. Clumpiness of the ejecta might lead to better sensitivities to those clumps than result from this assumption.

TABLE 1. Detectability of Gamma-Ray Lines (5σ)

SNR	D [kpc]	Age [yrs]	Size [']	SN Type	Ref.	Detectable Lines[a] [b] SPI	IBIS	ACT
Cas A	2.9	320	1.8	II	[13–15]	T		T,A,e+
Tycho	2.3	428	8	Ia	[16]			A,e+
SN 1006	1.8	994	26	Ia	[17,13,18]	e+		e+
Vela	0.4	11000	270	II	[13,19]	A		A,F,e+
Vela Jr.	0.2	680	120	II?	[7]	T,A,e+	T,e+	T,A,F,e+
Kepler	5	396	3.3	II?	[20,21]			T,A,e+
Cyg Lp	0.6	35000	180	II	[13,19]			A,F,e+
Monoc	0.6	46000	270	II	[13,19]			A,F,e+
LupusLp	0.5	49000	180	Ia	[18,22,23]	e+	e+	F,e+
Crab	2	946	3.6	II	[13]			T,A,e+
IC443[c]	1.5	2500	40	II	[13,24,25]			A,e+
G6.5-12	0.5	16000	480	Ia	[26]	e+	e+	F,e+
CTB 13	0.6	32000	210	II	[13,19]			A,F,e+
SN1987A	48	13	$\leq$12	II	[27]			T,C

[a] T=Ti44 (1157 keV), A=Al26 (1809 keV), e+ = 511 keV, F=Fe60 (1173 keV), C=Co60 (1173 keV)

[b] The Ti44,Na22 & Co60 lines were assumed to have 30 keV FWHM widths, the 511, Al26, & Fe60 lines were assumed to have 5 keV FWHM widths.

[c] The SNRs: Pup A, W44, RCW 103, RCW 86 & HB21 are similarly detectable.

8. Ahmad, I. et al., *Phys Rev Lett* **80**, 2550 (1998).

9. Mochizuki, Y. et al., *A&A* **346**, 831 (1999).

10. Chan, K.-W., Lingenfelter, R.E., *ApJ* **405**, 614 (1993).

11. Milne, P.A. et al., *Astro. Lett. & Comm.* **38**, 441 (1998).

12. Milne, P.A., and The, L.-S., and Leising, M.D., *ApJS* **124**, 503 (1999).

13. Lang, K.R., *Astrophysical Formulae*, Berlin: Springer-Verlag, 1980, 4, 478-481.

14. Ashworth, W.B., *J. Hist. Astron.* **11**, 1 (1980).

15. Dickel, J.R., Murray, S.S., Morris, J., Wells. D.C., *ApJ* **257**, 145 (1982).

16. Hughes, J.P., *ApJ* **545**, L53 (2000).

17. Laming, J.M., Raymond, J.C., McLaughlin, B.M., Blair, W.P., *ApJ* **472**, 267 (1996).

18. Leahy, D.A., Nousek, J., Hamilton, A.J.S., *ApJ* **374**, 218 (1991).

19. Dorman, L., *in Proc. of 4th Compton Symposium*, ed. C.D. Dermer, M.S. Strickman, J.D. Kurfess, New York: AIP, 1997, pp. 1172-1175.

20. Reynoso, E.M., Goss, W.M., *AJ* **118**, 926 (1999).

21. Hughes, J.P., *ApJ* **527**, 298 (1999).

22. Milne, D., *Australian J. Phys.* **24**, 757 (1971).

23. Clark, D.H., Caswell, J.L., *MNRAS* **174**, 267 (1976).

24. Fesen, R.A., *ApJ* **281**, 658 (1984).

25. Bhattacharya, et al., *in Proc. of 4th Compton Symposium*, ed. C.D. Dermer, M.S. Strickman, J.D. Kurfess, New York: AIP, 1997, pp. 1137-1140.

26. Combi, J.A., et al., *in Proc. of 5th Compton Symposium*, ed. M.L. McConnell, J.M. Ryan, New York: AIP, 1999, pp. 69-72.

27. Hughes, J.P., *in Proc. of ROSAT Science Symposium*, CfA pre-print 3863, (1993).

Chandra Discovery of Ejecta-Dominated X-ray Emission from the Old SNR Candidate Sgr A East

Y. Maeda[1], F. K. Baganoff[2], E. D. Feigelson[1], M. W. Bautz[2], W. N. Brandt[1], D. N. Burrows[1], J. P. Doty[2], G. P. Garmire[1], M. Morris[3], S. H. Pravdo[4], G. R. Ricker[2], L. K. Townsley[1]

[1] *Department of Astronomy and Astrophysics, 525 Davey Laboratory, The Pennsylvania State University, University Park, PA 16802-6305, U.S.A.*
[2] *Massachusetts Institute of Technology, Center for Space Research, 77 Massachusetts Avenue, Cambridge, MA 02139-4307, U.S.A.*
[3] *Division of Astronomy, Box 951562, UCLA, Los Angeles, CA 90095-1562, U.S.A.*
[4] *Jet Propulsion Laboratory, MS 306-438, 4800 Oak Grove Drive, Pasadena, CA 91109, U.S.A.*

Abstract.
We report on the X-ray emission from the shell-like, non-thermal radio source Sgr A East (SNR 000.0+00.0). The X-ray emitting region is concentrated within the central $\simeq 2$ pc of the larger radio shell. The spectrum shows strong $K\alpha$ lines from highly ionized ions of S, Ar, Ca, and Fe. The plasma appears to be rich in heavy elements, over-abundant by roughly a factor of four with respect to solar abundances, and shows a spatial gradient of elemental abundance: the spatial distribution of iron is more compact than that of the lighter elements. The X-ray properties support the long-standing hypothesis that Sgr A East is an old supernova remnant (SNR).

INTRODUCTION

The radio emission from the central few parsecs of the Galaxy has several components, including a compact non-thermal source (Sgr A*) thought to be associated with the central massive black hole, a spiral-shaped group of thermal gas streams (Sgr A West), and a $3'.5 \times 2'.5$ shell-like non-thermal structure (Sgr A East) (for a recent review [1]). A *Chandra* observation of the central few parsecs of the Galaxy was carried out on 1999 September 21 using the ACIS-I array of four abutted, frontside-illuminated CCDs. The effective exposure time was 40.3 ks. Baganoff et al. (2001) [3] discovered complex extended structures, one of which was found to be associated with the SNR candidate Sgr A East. This paper focuses on a detailed analysis of the X-ray emission from Sgr A East. Throughout this paper

CP565, *Young Supernova Remnants: Eleventh Astrophysics Conf.*, edited by S. S. Holt and U. Hwang
© 2001 American Institute of Physics 0-7354-0001-6/01/$18.00

we adopt a distance of 8.0 kpc to Sgr A East [4]. Details of the analysis procedures are presented in Maeda et al. (2000) [5].

X-RAY MORPHOLOGY AND SPECTRUM

In Fig 1, we show the smoothed broad-band (1.5–7 keV) X-ray image overlaid with radio contours from a 20 cm VLA image of Sgr A provided to us by F. Yusef-Zadeh. Dozens of point sources and complex extended structures are clearly seen, as reported in [3]. The outer oval-shaped contours are due to synchrotron emission from the shell-like non-thermal radio source Sgr A East. The thermal radio source Sgr A West, an HII region surrounding Sgr A*, appears on the western side of the Sgr A complex. Several bright compact X-ray sources can be seen in the vicinity of Sgr A West. One of these sources is coincident to within $0''.35$ with the radio position of the compact non-thermal radio source Sgr A* [3]. In addition to the compact sources, the Sgr A region shows bright diffuse X-ray emission superposed on a broader region of extended emission which peaks $\sim 1'$ east of Sgr A*, and which appears to fill the central ~ 2 pc of the Sgr A East radio shell (Fig. 1). This broader feature is especially conspicuous in the 6–7 keV band (Fig. 2), where the flux is dominated by iron-K line emission (see below). Based on its spectral and spatial properties, we associate the source of this diffuse X-ray emission with a hot optically thin thermal plasma located within the Sgr A East radio shell. From the smoothed images in different energy bands, we found that the structure of the Sgr A East emission is spectrally dependent. The half-power radius of the emission is $\sim 20''$ in the 6–7 keV band compared to $\sim 30''$ in the lower energy bands. This result indicates a spatial gradient of elemental abundance: the spatial distribution of iron is more compact than that of the lighter elements.

The spectrum of Sgr A East, shown in Fig 3, exhibits a continuum plus emission lines. The emission lines can be attributed to the $K\alpha$ transitions of the helium-like ions of sulfur, argon, calcium and iron. The line width for each line is consistent with being unresolved. The existence of the highly ionized ions confirms the presence of an optically thin thermal plasma. The line equivalent width is very large for iron ($EW \simeq 3.1$ keV). Since the high ionization degree of the iron-K line can be reproduced for the plasma in collisional ionization equilibrium [6], we fit the spectrum to models of an isothermal plasma with variable elemental abundances (MEKA; [7]) modified by the interstellar absorption. The model with solar elemental abundances [8] was rejected with $\chi^2/\text{d.o.f} = 309/185$ because it did not reproduce the large equivalent width of the iron line. The best-fit ($\chi^2/\text{d.o.f} = 217/184$) was obtained with $\simeq 4$ times solar abundances of the heavy elements (Fig 3). Assuming energy equipartition between electrons and ions, the best-fit model gives the following physical properties for the plasma, assuming a spherical volume with radius 1.6 pc and an unknown filling factor η: electron density $n_e \simeq 6\eta^{-\frac{1}{2}}$ cm^{-3}, gas mass $M_g \simeq 2\eta^{\frac{1}{2}}$ M$_\odot$, absorption-corrected X-ray luminosity $L_x \simeq 8 \times 10^{34}$ erg s^{-1} in the 2–10 keV band, and total thermal energy $2 \times 10^{49}\eta^{\frac{1}{2}}$ ergs.

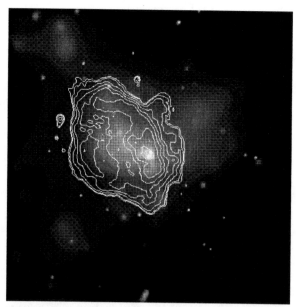

FIGURE 1. Smoothed X-ray image (1.5–7.0 keV) with 20 cm radio contours (white: Yusef-Zadeh private communication). Both exposure and vignetting corrections were applied. The field center is at Sgr A*, and the panel size is $8'.4 \times 8'.4$.

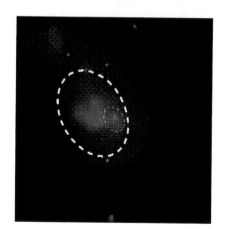

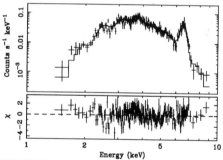

FIGURE 2. Same as Figure 2 but in the 6.0–7.0 keV band. The flux in this band is dominated by iron-K lines. The large and small white dashed ellipses approximately represent the Sgr A East non-thermal shell and an outer boundary of the Sgr A West region, respectively.

FIGURE 3. The ACIS-I spectrum of Sgr A East. Error bars are 1 σ. The solid line corresponds to the best-fit value with the MEKA model. Fit residuals are shown in the bottom panel.

179

DISCUSSION

The X-ray spectrum enriched by heavy elements suggests that the X-ray emitting plasma is dominated by SN ejecta. The small gas mass of $2\eta^{\frac{1}{2}}$ $M_\odot$ and thermal energy $\sim 10^{49}$ ergs are consistent with the ejecta from a single SN explosion. These results straightforwardly support the long-standing hypothesis that Sgr A East is a single SNR [2]. With the centrally concentrated X-ray emission we find here and the well-established non-thermal radio shell, Sgr A East is a new member of the class of "mixed morphology" (MM) SNRs [9]. No MM SNRs are seen in the LMC or SMC, in which interstellar matter is known to be less dense than in our Galaxy. We also found that Sgr A East is the most compact of the MM SNRs, probably indicating that Sgr A East is evolving into ambient materials denser than those in typical MM SNRs.

Since the electron density of the X-ray plasma is $\sim 6\eta^{-\frac{1}{2}}$ cm^{-3}, the ionization parameter $n_e t$ becomes $\simeq 2 \times 10^{12}$ $(t/10,000$ yr$)$ $\eta^{\frac{1}{2}}$ cm^{-3} s. Since the characteristic timescale for realizing collisional equilibrium is 10^{12} cm^{-3} s, the age of the plasma is expected to be $\geq 10,000$ yrs [6]. This age is consistent with the estimated age of a few 10^4 yrs based on the shear effect due to the differential Galactic rotation [10]. The sound crossing length, an effective length for heat conduction, is ~ 8 pc at 10^4 yrs, which is longer than the radius of the plasma (~ 1.6 pc). The radiative cooling time of the plasma is 10^6 yr. Our phenomenological success in fitting an isothermal MEKA model (a single temperature plasma in collisional ionization equilibrium) to the spectrum is completely consistent with those plasma conditions. Therefore, Sgr A East is probably an old SNR ($\sim 10^4$ yr) and its X-ray emission is dominated by the SN ejecta.

We express our thanks to the entire *Chandra* team for their many efforts in fabricating, launching, and operating the satellite, and for their work in developing software for calibrating and analyzing the data. Rashid Sunyaev shared valuable thoughts with us on Galactic Center issues. Farhad Yusef-Zadeh and his collaborators kindly provided us with an unpublished radio image. This research was supported by NASA contract NAS 8-38252 and in part (SHP) by JPL, under contract with NASA.

REFERENCES

1. Yusef-Zadeh, F., Melia, F. & Wardle, M. 2000, Science, 287, 85
2. Ekers, R. D., et al. 1983, A&A, 122, 143
3. Baganoff, F., et al. 2001, ApJ, in preparation
4. Reid, M. J. 1993, ARA&A, 31, 345
5. Maeda, Y., et al. 2000, ApJ, submitted
6. Masai, K. 1994, ApJ, 437, 770
7. Mewe, R., Gronenschild, E.H.B.M., & van den Oord, G.H.J. 1985, A&AS, 62, 197
8. Anders, E., & Grevesse, N. 1989, Geochimica et Cosmochimica Acta, 53, 197
9. Rho, J. & Petre, R. 1998, ApJL, 503, L167
10. Uchida, K. I., et al. 1998, IAU Symp. 184: p317

Exploring the Small Scale Structure of N103B

Karen T. Lewis[1], David N. Burrows[1], John A. Nousek[1],
Gordon P. Garmire[1], Patrick Slane[2], John P. Hughes[3]

[1] *The Pennsylvania State University, 525 Davey Laboratory, University Park, PA 16801, USA*
[2] *Harvard-Smithsonian Center for Astrophysics, 60 Garden Street, Cambridge, MA 02138, USA*
[3] *Department of Physics and Astronomy, Rutgers University,*
136 Frelinghuysen Road, Piscataway, NJ 08854-8019, USA

Abstract. We present the preliminary results of a 40.8 ks *Chandra* ACIS observation of the young supernova remnant (SNR) N103B located in the Large Magellanic Cloud. The image reveals structure at the sub-arcsecond level, including several bright knots and filamentary structures. The remnant has the characteristic spectrum of a Type Ia SNR, containing strong lines of Fe, He-and H-like Si and S, Ar, and Ca. Narrow band images reveal non-uniformities in the remnant.

INTRODUCTION

N103B, one of the brightest X-ray and radio sources in the Large Magellanic Cloud (LMC), is a young, compact supernova remnant (SNR). The remnant, located on the north-eastern edge of the HII region N103, is ∼ 40 kpc from the young star cluster NGC 1850. [1] The radio and X-ray morphologies are strikingly similar; the western edge (near the HII region) shows considerable structure and is ∼ 3 times brighter than the eastern edge. [1,2] In both bands, the emission arises from a region 7 pc in diameter (d=50kpc to LMC assumed). Due to its proximity to a star forming region, it was originally suspected that N103B was the result of the core collapse of a massive object. However, the ASCA spectrum shows no evidence for K-shell emission from O, Ne or Mg while Si, S, Ar, Ca, and Fe features are strong, indicating that the remnant is more likely the result of a Type Ia SN explosion. [3]

OBSERVATION AND DATA CLEANING PROCEDURE

N103B was observed for 40.8 ks with ACIS S-3, one of the back-illuminated CCDs in the ACIS-S array. We found that ∼35% of the frames had a high background count rate and that during these times, the spectrum of the background changed

CP565, *Young Supernova Remnants: Eleventh Astrophysics Conf.,* edited by S. S. Holt and U. Hwang
© 2001 American Institute of Physics 0-7354-0001-6/01/$18.00

significantly. These frames were removed from the data set. The standard ASCA grade filter (g 02346) was used.

ANALYSIS AND RESULTS

Our observation of N103B has confirmed many earlier results. The *Chandra* X-ray and the Australian Telescope Compact Array (ATCA) radio images [1] are quite similar (Figure 1). Since the radio and X-ray flux is depressed in the eastern half of the remnant we conclude that this intensity difference is due to a real density variation, as opposed to an absorption effect. While the bright radio knots are in the general vicinity of the X-ray knots, they do not overlap. Further investigation is required to determine whether there is a relative rotation between the two images, which may explain the offset. The spectrum of N103B as a whole (Figure 2) confirms the earlier ASCA results. We see no evidence for O or Ne in the spectrum, although a Mg line is present.

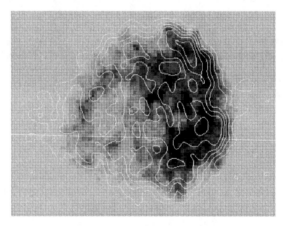

FIGURE 1. Plot of X-ray image with 2.5 cm ATCA contours [1]. Contours are 1.7, 1.3, 0.9, 0.6, 0.3, and 0.1 mJy/beam. The spatial resolution of the radio data is 1.2", which is slightly larger than the Chandra's.

Continuum subtracted (CS) and equivalent width (EW) images show significant variation throughout the remnant. The images for He-like Si are shown in Figure 3. The interval 1700-1865 eV was used to create the He-like Si line image. To estimate the continuum, we performed a linear interpolation between images in the 1515-1680 eV and 2050-2215 eV bands, which are free of any significant lines. All images were smoothed with a 3 pixel (1.5") boxcar to reduce the number of pixels with no counts. We were concerned that the CS and EW images were unreliable, as each pixel contained only a few counts at most, so the CS and EW images were collapsed into radial and azimuthal dimensions by binning the line and continuum images into rings and sectors. The CS and EW flux were calculated for each ring and sector

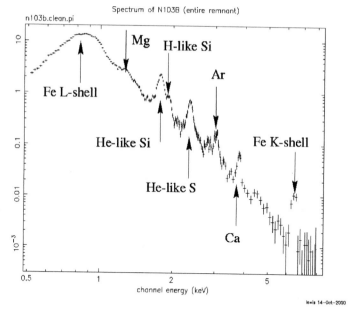

FIGURE 2. Spectrum of N103B as a whole. The data are plotted so that each data point has a 6 sigma significance with a maximum bin size of ∼ 65eV

as shown in bottom of Figure 3. The radial and azimuthal plots confirm what is seen in the images. The emission from He-like Si is greatest along the western edge, where the entire remnant is bright. However, as shown by the EW images and plots, it is only in the outer rim of the remnant that the He-like Si emission drastically dominates over the continuum emission. Throughout the remainder of the remnant, the ratio of line to continuum is relatively flat. We have not yet determined whether the change in emissivity is due to a difference in abundance, temperature, ionization or a combination of these.

CONCLUSIONS

Preliminary analysis of this *Chandra* observation reveal that N103B shows significant structure at the arcsecond level. The variations in X-ray flux are consistent with those in radio and are due to real density variation within the remnant. The spectrum of the remnant is that of a typical Type Ia SNR, however future spatially resolved spectroscopy will allow for a better classification. Finally, there are dramatic spatial variations in the emissivity of He-like Si in particular. The radial variation in the equivalent width of the remnant is particularly intriguing. Detailed spectral analysis of concentric rings will enable us to determine the source of these and other similar variations.

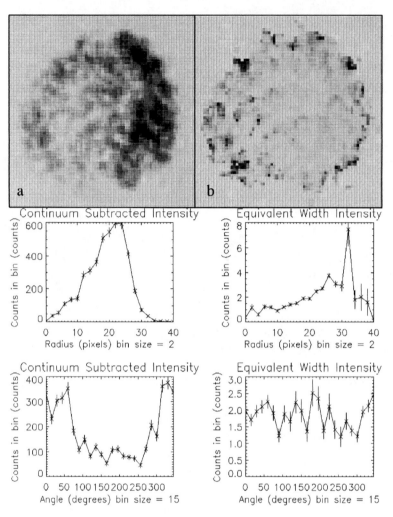

FIGURE 3. Plot of He-like Si continuum subtracted (a) and equivalent width images (b). Below are radial and azimuthal plots. 2 pixels = 1" ~ 0.2 pc

REFERENCES

1. Dickel, J. and Milne, D. 1995, AJ, 109, 1
2. Mathewson, D.S., Ford, V.L., Dopita, M.A., Tuohy, I.R., Long, K.S., & Helfand, D.J. 1983, ApJS, 51,345
3. Hughes, J.P, Hayashi, I., Helfand, D., Hwang, U., Itoh, M., Kirshner, R., Koyama, K., Markert, T., Tsunemi, H., & Woo, J., 1995 ApJL, 444, L81

A Look at SNR 0519-69.0 With *Chandra*'s ACIS

Rosa Williams[1], Robert Petre[1], and Stephen S. Holt[2]

[1] *NASA's GSFC, Code 662, Greenbelt, MD 20771, USA*
[2] *F. W. Olin College of Engineering, Needham, MA 02492, USA*

Abstract.
The young, Balmer-dominated SNR 0519–69.0 in the Large Magellanic Cloud is of the class of objects typified by the Tycho supernova remnant. Recent observations with *Chandra*'s Advanced CCD Imaging Spectrometer have shown the X-ray structure of this object in unprecedented detail. ACIS's combination of spatial and spectral resolution allow us to compare the X-ray spectra of the outermost limb emission to that from the rest of the remnant, revealing intriguing differences in spectral features.

INTRODUCTION

The supernova remnant (SNR) 0519–69.0, in the Large Magellanic Cloud (LMC), has an optical spectrum dominated by the presence of Balmer lines, and has therefore long been thought to be the result of a Type Ia supernova explosion [1]. This suggestion was strengthened by a study of the remnant in X-rays by Hughes et al. [2], who compared *ASCA* spectra of the SNR to model calculations of the nucleosynthetic yields from various types of supernovae and concluded that the spectra were most consistent with the products of a Type Ia explosion.

SNR 0519–69.0 is also one of the younger remnants in the LMC, with an estimated expansion velocity of 1000 – 1900 km s^{-1} corresponding to an age of perhaps 500 – 1500 yr [3]. As such, it is not surprising that the optical and X-ray spectra should show abundances consistent with those expected from SN ejecta; the SNR has not had time to sweep up sufficient circumstellar matter for the emission from this swept-up mass to predominate.

OBSERVATIONS AND ANALYSIS

SNR 0519–69.0 was observed with the *Chandra* Advanced CCD Imaging Spectrometer (ACIS; sequence number 500005), centered on the back-illuminated S3 chip. The ACIS (using a back-illuminated chip) has an angular resolution of 1″,

CP565, *Young Supernova Remnants: Eleventh Astrophysics Conf.*, edited by S. S. Holt and U. Hwang

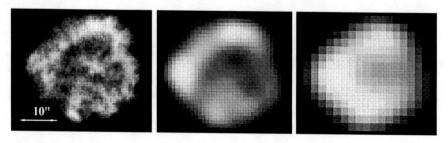

FIGURE 1. *Chandra* ACIS and ATCA 13 and 20 cm images of SNR 0519–69.0 (same scale)

an energy resolution of 100 eV at 1.0 keV (E/ΔE =9), an energy range of 0.2 – 10 keV, and an effective area of 600 cm^{-2} at 1.0 keV. Data reduction and analysis were performed using the CIAO, FTOOLS, and XSPEC data-processing routines.

Morphological Features

The ACIS images show a reasonably circular outline centered at the J2000 coordinates 05:19:34.9, -69:02:07, with a diameter of 32″ (7.9 pc at the LMC distance of 50 kpc). The emission is clearly shell-like, with small extensions to the south, east, and northeast of the shell; the northeastern extension appears as faint diffuse emission (Fig. 1, left). The emission from the SNR can be divided into four radial regions: a central bright patch of extended emission; the fainter intermediate interior with "spoke" structures; a bright X-ray ring with northern and southern "hotspots;" and a faint ridge of outermost limb emission beyond this ring.

We compare *Chandra* images to Australia Telescope Compact Array (ATCA) radio maps [4] shown in Fig 1. The central X-ray emission has no radio counterpart, indicating that this emission probably does not originate from a pulsar-powered nebula. Radio peaks appear to correspond to the X-ray ring and hotspots, which suggests the radio emission arises primarily from reverse-shocked ejecta.

A Change in Spectral Character

While the inner regions show comparable spectra, the outer limb shows markedly different spectral characteristics. The most notable difference is that at 0.5-0.7 keV, where the limb spectrum shows a peak (probably from O lines) that the inner regions do not. The outer limb also has a narrower Fe peak than the rest of the SNR, with no Fe K line. The strong He α lines from Mg, Si, and S so prominent in the inner spectrum are still visible at the limb, though (possibly due to the decreased signal-to-noise ratio of the limb spectrum) the Ca and Ar lines are not nearly so pronounced (Fig. 2).

The variation between the inner spectra and limb spectra imply that the bright emission in the X-ray ring is from reverse-shocked ejecta. The outer emission

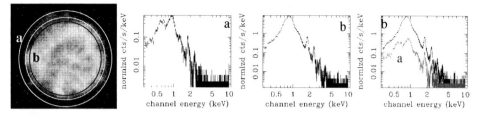

FIGURE 2. *Chandra* ACIS image and radial spectra of SNR 0519–69.0

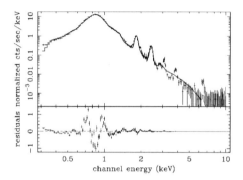

FIGURE 3. Spatially integrated ACIS spectrum and residuals for SNR 0519–69.0

is thought to be from forward-shocked, O-rich interstellar medium (ISM), though there are still still enhanced metal abundances at the limb. The internal similarities imply that there is no evidence for marked stratification of the SN ejecta within the interior. Likewise, we find no spectral evidence for nonthermal emission from a central source.

The Spatially Integrated Spectrum

We fit several models to the emission from SNR 0519–69.0. The prominent line emission suggests a significant thermal component. We attempted fits of various thermal plasma models, both collisional equilibrium (CIE) and nonequilibrium ionization (NEI) variants. NEI models consistently gave more reliable fits; considering the youth of the remnant, this is far from surprising. All of these thermal models deviate from the data above 1.5 keV, suggesting the presence of more than one spectral model component.

Our best fit[1] was provided by a model including NEI and synchrotron compo-

---

[1] Subsequent analysis has shown that models with two thermal NEI components can provide as good or better fits to the data as do NEI + synchrotron models. Table 1 values should thus be treated as parameters of only one of several acceptable solutions.

TABLE 1. Model Parameters

Component	param.	units	value	90% error
wabs	n_H	cm^{-2}	2.2×10^{21}	$\pm\ 1.1 \times 10^{19}$
vnpshock	kT_a	keV	1.1	$\pm\ 0.027$
vnpshock	kT_b	keV	0.21	$\pm\ 0.48$
vnpshock	O	% solar	0.51	$\pm\ 0.038$
vnpshock	Ne	% solar	0.72	$\pm\ 0.080$
vnpshock	Mg	% solar	0.57	$\pm\ 0.069$
vnpshock	Si	% solar	3.8	$\pm\ 0.28$
vnpshock	S	% solar	12	$\pm\ 0.97$
vnpshock	Fe	% solar	3.1	$\pm\ 0.21$
vnpshock	Ni	% solar	6.1	$\pm\ 0.80$
vnpshock	Tau_u	cm^{-3} s	2.0×10^{11}	$\pm\ 1.7 \times 10^{10}$
vnpshock	norm		3.0×10^{-3}	$\pm\ 1.1 \times 10^{-5}$
srcut	alpha		0.44	fixed
srcut	break	Hz	2.3×10^{16}	$\pm\ 2.8 \times 10^{14}$
srcut	norm	Jy (1 GHz)	0.11	fixed

nents, modified by the effects of photoelectric absorption. The radio spectral index and 1 GHz flux were set from ATCA data [4]. Some abundances were fixed at LMC values, with O, Ne, Mg, Si, S, Fe and Ni allowed to vary. The fit had $\chi^2 = 1887$ for 662 degrees of freedom ($\chi^2_{red} = 2.8$); fitted parameters are shown in Table 1 below. The results for the thermal fit are consistent with previous findings [2]. The emission measure for this model implies plasma density ~ 3.4 cm^{-3}. From the fitted ionization timescale, we estimate an ionization age of 1800 yr.

CONCLUSION

The new *Chandra* ACIS observations of SNR 0519–69.0 offer many intriguing features. The difference between the outermost emission and the inner emission shows a clear separation between the forward-shocked, swept-up ISM and the reversed-shocked ejecta. From the spectrum of 0519–69.0 we find an ionization age of about 1800 yr, consistent with the remnant's size and the inferred ejecta-dominated emission. While synchrotron emission may make up a small fraction of the SNR's hard X-rays, it is difficult to determine whether that emission originates at the shock front(s) or from a central source, or both. Clearly, *Chandra* has opened a new chapter in the detailed examination of extragalactic supernova remnants.

REFERENCES

1. Tuohy, I. R. et al. 1982, ApJ, 261, 473
2. Hughes, J. P., et al. 1995, ApJL, 444, 81
3. Smith, R. C. et al. 1993, ApJ, 407, 564
4. Dickel, J. R. et al. 1995, AJ, 109, 200

Far Ultraviolet Spectroscopy of a Nonradiative Shock in the Cygnus Loop: Implications for the Postshock Electron-Ion Equilibration

Parviz Ghavamian[1], John C. Raymond[2] and William P. Blair[3]

[1] *Department of Physics and Astronomy, Rutgers University, Piscataway NJ, 08854*
[2] *Harvard-Smithsonian Center for Astrophysics, Cambridge, MA 02138*
[3] *Department of Physics and Astronomy, Johns Hopkins University, Baltimore, MD 21218*

Abstract. We present far ultraviolet spectra of a fast (~ 300 km s^{-1}) nonradiative shock in the Cygnus Loop supernova remnant. Our observations were performed with the Far Ultraviolet Spectroscopic Explorer (FUSE) satellite, covering the wavelength range 905–1187 Å. The purpose of these observations was to estimate the degree of electron-ion equilibration by (1) examining the variation of O VI $\lambda\lambda$1032, 1038 intensity with position behind the shock, and (2) measuring the widths of the OVI lines. We find significant absorption near the center of the O VI λ1032 line, with less absorption in O VI λ1038. The absorption equivalent widths imply an O VI column density greatly exceeding that of interstellar O VI along the Cygnus Loop line of sight. We suggest that the absorption may be due to resonant scattering by O VI ions within the shock itself. The widths of the O VI emission lines imply efficient ion-ion equilibration, in agreement with predictions from Balmer-dominated spectra of this shock.

INTRODUCTION

A shock wave is termed nonradiative if the cooling time of the postshock gas exceeds the dynamical time scale. Nonradiative shocks lack cooling zones, therefore the optical and ultraviolet line emission is excited close to the shock front. The emission line fluxes and line widths are sensitive to the shock velocity and the degree of postshock electron-ion/ion-ion equilibration. This property makes the optical and ultraviolet lines valuable tools for investigating collisionless heating in high Mach number, low density shock waves.

For an ion of mass m_i, the jump conditions for a strong shock in an ideal gas medium ($\gamma = \frac{5}{3}$) give a postshock temperature $T_i = \frac{3}{16} \frac{m_i V_s^2}{k}$. In an unequilibrated shock, the electron and proton temperatures are in the ratio 1/2000:1, while the temperatures of O ions and protons are in the ratio of 16:1. In contrast, the temperatures of all three species are equal in an equilibrated shock. Collisionless heating

CP565, *Young Supernova Remnants: Eleventh Astrophysics Conf.*, edited by S. S. Holt and U. Hwang
© 2001 American Institute of Physics 0-7354-0001-6/01/$18.00

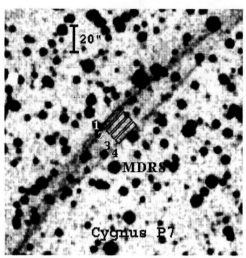

FIGURE 1. POSS image of the Cygnus P7 filament. The field of view is approximately $2.5' \times 2.5'$ square. The FUSE slit positions are marked (PA $= 315°$).

can produce equilibrations anywhere between these two extremes. In principle, the amount of collisionless heating from plasma waves, MHD turbulence and other processes is sensitive to such parameters as the magnetosonic Mach number and angle between the magnetic field and shock front. Understanding this relationship is of critical importance in the interpretation of X-ray, ultraviolet and optical data of high Mach number collisionless shocks.

A nonradiative shock in partially neutral gas emits a pure Balmer line spectrum generated by collisional excitation. Each Balmer line consists of a narrow component produced by excitation of cold H I and a broad component produced by proton-H I charge exchange. The shock velocity and electron-ion equilibration can be gauged from the broad component line width and ratio of broad component to narrow component flux [1] [2]. The ion-ion equilibration can be gauged in turn by measuring the width of the O VI $\lambda\lambda 1032$, 1038 lines and the variation of O VI flux with position behind the shock.

OBSERVATIONS

The target of our FUSE observations was Cygnus P7, a nonradiative shock in the NE corner of the Cygnus Loop supernova remnant [1]. We acquired ultraviolet spectra of Cygnus P7 in June 2000 for a total of 64 ks (all during orbital night) and four slit positions behind the shock (Fig. 1). The spectra presented here were acquired through the $4'' \times 20''$ medium resolution (MDRS) slit. To observe the intensity variation of O VI, we centered slits 1, 2, 3 and 4 at $5''$, $10''$ and $15''$ behind the shock, respectively (Fig. 1). To minimize the background contribution, the 1-D

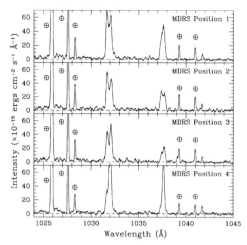

FIGURE 2. FUSE spectra of Cygnus P7 at slit positions 1, 2, 3 and 4. Spectra have been smoothed over 0.1 Å, and are presented without reddening correction.

spectra (Fig. 2) were extracted from the LiF 1A channel (one of four available with FUSE), the detector yielding the highest effective area at $\lambda > 1020$ Å. The spectral resolution is 0.08 Å(25 km s^{-1}).

ANALYSIS

The 1-D spectra of Cygnus P7 (Fig. 2) reveal three important features: (1) The O VI lines are considerably broader than the airglow lines, confirming the detection of O VI emission in the Cygnus P7 shock. The velocity width of the O VI $\lambda 1032$ line is $\sim 120 \text{ km s}^{-1}$. Combining this result with the shock velocity of 300–365 km s^{-1} predicted from Balmer line spectroscopy of Cygnus P7, we obtain $T_e/T_O \sim 0.7$. This is roughly consistent with the electron-proton equilibration $T_e/T_p \sim 0.7$–1.0 predicted by the Balmer line analysis; in an earlier analysis [1] of Balmer line emission from Cygnus P7, we measured an Hα broad component width of 262 km s^{-1}. This corresponds to a shock velocity of 265–365 km s^{-1} for cases of no equilibration and full equilibration, respectively. (2) the O VI emission lines are strongly affected by absorption. The absorption features tend to shift blueward at slit positions farther and farther behind the shock. The variation of O VI flux with position is masked by the O VI absorption features; (3) the shape of the $\lambda 1038$ absorption profile differs from that of the $\lambda 1032$ profile, mainly due to galactic H$_2$ absorption at the position of the O VI $\lambda 1038$ line.

INTERPRETATION

There are two possible explanations for the oberved O VI absorption: either the bulk of the absorption is produced by interstellar matter (O VI, H_2, C II*, etc.) or the absorption is produced locally by O VI ions within the shock. There are three lines of evidence in favor of the latter interpretation:

(1) The velocity width of the O VI $\lambda 1032$ absorption feature is ~ 90 km s^{-1}, comparable to the width of the $\lambda 1032$ emission line. The O VI column density and temperature required to produce the observed absorption are $N_{O\ VI} \gtrsim 2 \times 10^{14}$ cm^{-2} and $T_{O\ VI} \sim 3 \times 10^6$K. These values are $\gtrsim 10$ times larger than the O VI column density and temperature expected for the ISM along the Cygnus Loop line of sight [3]. (2) The progressively larger blueshift of the O VI absorption lines with distance behind the shock cannot be explained by interstellar O VI absorption. (3) The line center ratio $I_{1038}(v=0)/I_{1032}(v=0)$ varies significantly with slit position, from around 0.6 (Position 2) to 1.1 (Position 4). These localized variations are consistent with O VI resonant scattering within the shock.

The appearance of the O VI spectrum may be due to a combination of internal resonant scattering and a curved shock geometry. On the one hand, the O VI column density and temperature from the FUSE spectra are consistent with the range of shock velocities (300–365 km s^{-1} from the Balmer lines) and preshock densities (n $\sim$ 0.1–0.3 cm^{-3} from ROSAT X-ray observations) estimated for Cygnus P7. On the other hand, a curved shock geometry could explain both the increasing value of $I_{1038}(v=0)/I_{1032}(v=0)$ and the increasing absorption blue shift with position behind the shock. In a curved shock, O VI photons produced farther downstream could be absorbed by O VI ions on the near (blue shifted) side of the shock.

CONCLUSIONS

In FUSE observations of a nonradiative shock in the NE Cygnus Loop, we find O VI emission line widths consistent with moderate to high electron-ion equilibration. We also find strong O VI absorption near the centers of the O VI emission lines. A quantitative explanation of the interplay between the O VI emission and resonant scattering requires (1) a careful modeling of the shock geometry, and (2) a Monte Carlo simulation to adequately treat resonant scattering at moderate optical depths ($0 \le \tau \le 1$). We will address these issues in future work.

This research project is supported by NASA grant NAG5-9019.

REFERENCES

1. Ghavamian, P. Raymond, J., Smith, R. C. & Hartigan, P. 2001, ApJ 547, in press
2. Shelton, R. L. & Cox, D. P. 1994, ApJ 434, 599
3. Blair, W. P., Sankrit, R., Raymond, J. & Long, K. S. 1999, AJ 118, 942

Fabry-Perot [O III] $\lambda 5007$ Å Observations of the SMC Oxygen-rich SNR 1E 0102-72.9

K.A. Eriksen[1], J.A. Morse[2], R.P. Kirshner[1], and P.F. Winkler[3]

[1] *Harvard-Smithsonian Center for Astrophysics, Cambridge, MA 02138*
[2] *Center for Astrophysics and Space Astronomy, University of Colorado, Boulder, CO 80309*
[3] *Department of Physics, Middlebury College, Middlebury, VT 05753*

Abstract.
We present Rutgers/CTIO Fabry-Perot Imaging Spectrometer observations of the [O III] 5007 Å line of the oxygen-rich supernova remnant 1E 0102-72.9 in the Small Magellanic Cloud. Our data cube covers velocities from -2100 km/s to $+2300$ km/s with 100 km/s resolution, and for the first time reveals the complex three-dimensional structure of this remnant. The most prominent features are a number of rings of emission that are continuous in (X, Y, V) space, and bear striking resemblance to the Fe-Ni bubbles studied in recent numerical models. Using a simple kinematic model, we derive a free expansion age of 2100 years for E0102.

INTRODUCTION

The oxygen-rich supernova remnants (OSNRs) form a rare class of young SNRs believed to be the products of core-collapse supernovae. Only seven are known, with Cas A as the prototype. The optical emission from the OSNRs typically consists of two components: narrow-line knots with approximately interstellar abundances (the "quasi-stationary flocculi" or QSFs), thought to be shocked circumstellar material, and broad-line knots with strong lines of oxygen and with little or no evidence of hydrogen. These "fast-moving knots" (FMKs) are thought to be pure nuclear-processed material from the interior of the progenitor heated by the reverse shock. These filaments, lasting at most a few thousand years before being diluted by circumstellar material, allow us an unparalleled view of the newly synthesized elements from the cores of massive stars and structure of supernova ejecta.

Among the oxygen-rich remnants, 1E 0102-72.9 (E0102) is unique in that, while is several times older than Cas A, it has no QSFs, indicating that there has been little interaction with circumstellar material. Blair *et al.*[1] compared *HST* and *ASCA* spectra of E0102 with nucleosynthesis models and deduced that its progenitor was likely a Wolf-Rayet star. The FMKs in E0102 lie interior to a diffuse high-excitation

CP565, Young Supernova Remnants: Eleventh Astrophysics Conf., edited by S. S. Holt and U. Hwang

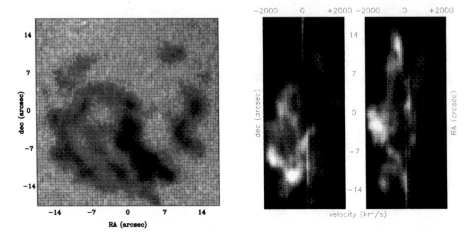

FIGURE 1. (left) A sum of 44 F-P interferograms covering -2100 km/s to +2300 km/s, displayed with square root intensity scaling. (right) The data cube summed along the X (left) and Y (right) dimensions, displayed with square root intensity scaling. These figures are the equivalent of viewing the the (X, Y, V) cube from the side and top, respectively. The surrounding H II region has been imperfectly subtracted from these plots, resulting in low level artifacts near zero velocity.

H II "halo" that may be the ejected WR wind photoionized by either the UV-flash of the supernova, or (more likely) intense EUV radiation produced in the shocked knots of ejecta. Recent *Chandra* observations (e.g. Gaetz *et al.*[2]) show that the main blast wave has not yet reached this material.

OBSERVATIONS AND KINEMATICS

We observed E0102 with the Rutgers/CTIO Fabry-Perot Imaging Spectrometer on the CTIO 4 meter in a manner similar to that of Morse, Winkler, and Kirshner [3]. The F-P etalon was stepped in wavelength increments corresponding to 100 km/s at 5007Å. We assembled the 44 individual reduced interferograms into a single data cube covering -2100 km/s to +2300 km/s. The sum of these images is presented in Figure 1, as well as renderings of the cube summed along the X and Y axes. These give views of the cube from the "side" and "top", showing the velocity structure versus declination and RA. From these plots it is clear that the brightest emission is in well structured blueshifted filaments, while the redshifted emission is fainter and less ordered.

The most prominent feature in the data cube is a bright ring of blueshifted emission that covers the southeast quadrant of the remnant. This continuous structure is ~4 pc in diameter and runs from at least -2100 km/s to nearly 0 km/s in velocity. A number of smaller, less luminous rings also exist in the cube, including a feature near zero-velocity seen in projection against the interior of the remnant. Blondin, Borkowski, and Reynolds [4] constructed numerical simulations of young supernova remnants with ejecta containing iron-nickel bubbles formed by radioactive heating. As these bubbles pass through the reverse shock, their dense walls contain most of the emission measure and result in ring-like filaments. The velocity structure of the bright E0102 ring in particular suggests a bubble passing through the reverse shock at an oblique angle to the line of sight. While this mechanism is attractive in its elegance, there are a number of potential complications in its application to our data. First, the Blondin *et al.* simulations focus on X-ray emitting material, and it is not clear that their mechanism can produce knots with the much higher densities required to produce optically emitting filaments (nor is it clear that it can not). Moreover, as the observed optical filaments emit in oxygen and neon lines, the nickel and iron ejecta would need to be well-mixed with these lighter elements to produce the bubbles. However, Hughes *et al.* [5] observe a clear separation of the iron-rich ejecta from the lighter elements in Cas A. Finally, and perhaps most troubling, there is little evidence for iron in the *ASCA* or *Chandra* X-ray spectra of E0102 [6,2]. Nevertheless, *some* mechanism appears to be at work in E0102 that gives its ejecta a "bubbly" structure.

Another feature that is apparent from Figure 1 is the complex non-uniform expansion of E0102. Though some circumstellar interaction is necessary to produce a reverse shock, given the purity of the FMKs and the lack of QSFs it is clear that little interaction has taken place, and it is almost certain that E0102 is still in the free expansion phase of evolution. In this case, one might expect filaments seen (in projection) farthest from the center to be moving nearly transverse to the line of sight, and thus to have little Doppler shift, while the most extreme shifts would be observed near the projected center of the remnant. The most striking departure from this picture is the zero-velocity ring that sits near the center of the remnant. One must invoke a particularly complex structure for the reverse shock in order for it to have reached back to the center of the remnant in some places while it is still illuminating redshifted filaments seen at much larger radii. This is indeed puzzling, as approximately uniform expansion has been found to be a reasonably good description of the velocity fields of Cas A [7] and N132D [3], two OSNRs that exhibit substantially more circumstellar interaction. To explore this further, we plot velocity versus radius in Figure 2. In this figure, a freely expanding shell would have an elliptical appearance, with the characteristic expansion velocity and size as axes. While much of the emission (particularly the blueshifted filaments) is fairly well approximated by an ellipse, emission from the central ring is clearly visible at nearly zero-velocity. We have also plotted our best fit velocity ellipse. Our parameters of $v = 1800$ km/s and $r = 4$ pc give a free expansion age of ~2100 years. This value is twice as large as that calculated by Tuohy and Dopita [8] from long-slit spectra, and Hughes, Rakowski, and Decourchelle [9] from the X-ray proper motion.

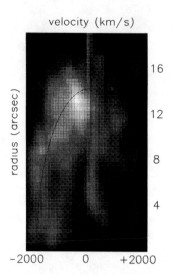

FIGURE 2. A plot of radius versus velocity with our best velocity ellipse. Note the strong shell like appearance of the blueshifted material, the central bright filament at nearly zero-velocity, and fainter redshifted material.

CONCLUSION

Our Fabry-Perot observations of E0102 reveal the complex velocity field of this young remnant for the first time with seeing-limited resolution. We find a surprising array of rings and bubbles, though the mechanism producing these structures is uncertain. Using a simple kinematic model, we derive a free expansion age of 2100 years for this SNR, roughly twice that determined from other observational methods.

P.F.W acknowledges the support the National Science Foundation (NSF) through Grant No. AST-9618465.

REFERENCES

1. Blair, W.P. *et al.* 2000, ApJ, 537, 667
2. Gaetz, T.J. *et al.* 2000, ApJ, 534, L47
3. Morse, J.A., Winkler, P.F., Kirshner, R.P. 1995, AJ, 109, 2104
4. Blondin, J.M., Borkowski, K.J., Reynolds, S.P. in press
5. Hughes, J.P. *et al.* 2000, ApJ, 528, L109
6. Hayashi, I., *et al.* 1995, PASJ, 46, L121
7. Reed, J.E., *et al.* 1995, ApJ, 440, 706
8. Tuohy, I.R., Dopita, M.A., 1983, ApJ, 268, L11
9. Hughes, J.P., Rakowski, C.A., Decourchelle, A. 2000, ApJ, 543, L61

2MASS Near-infrared images of RCW 103: A young SNR in a dense medium

J. Rho*, W. T. Reach*, B.-C. Koo†, and L. Cambresy*

*Infrared Processing and Analysis Center/California Institute of Technology
†Seoul National University, Korea

Abstract. We present near-infrared images of the supernova remnant RCW 103 (G332.4-0.4) from the Two Micron All Sky Survey (2MASS). In all three 2MASS bands—J, H, K_s—emission is detected, revealing a barrel-shaped, shell-like morphology. However, the J and H band shells are displaced with K_s band shell. The J and H band emission are mostly from [Fe II] lines while the K_s band emission is mostly from molecular hydrogen, as inferred from previous near-infrared spectroscopy. The 2MASS images show that the displacement between the [Fe II] and H_2 is not just due to a projection effect; instead, it appears that the layers emitting [Fe II] and H_2 lines are intrinsically double shells.

We compared the 2MASS images with radio and X-ray images at comparable resolutions. The X-ray and radio shells coincide with the J and H band shells, while the K_s band shell is located outside of the shock front. We discuss the origin of the molecular hydrogen shell in terms of a magnetic precursor or excitation by a radiative precursor. We also discuss the ISM environment of RCW 103 using physical parameters inferred from near-infrared lines.

Introduction

RCW 103 (G332.4-0.4) is a young supernova remnant (SNR) with an age $\sim 10^3$ yr (Nugent et al. 1984), showing a barrel-shaped, shell-like morphology. The remnant is bright in all wavelengths of the radio, optical, near and far infrared and X-rays (Dickel et al. 1996, Meaburn & Allen 1986, Oliva et al. 1999; Nugent et al. 1984). RCW 103 is one of a few SNR whose infrared emission is well studied. Near-infrared observations detected [Fe II] and H_2 lines and mid- and far-infrared observations show a rich spectrum of atomic lines (Oliva et al. 1999). Multiple OH masers were detected from RCW 103, suggesting it is a candidate for molecular cloud interaction. A compact X-ray source, 1E 161348-5055, was previously detected at the center of RCW 103 (Tuohy et al. 1983). The spectra of the source shows blackbody emission, and the source is known to have variable flux without pulsation (Gotthelf, Petre & Vasisht 1999).

CP565, *Young Supernova Remnants: Eleventh Astrophysics Conf.*, edited by S. S. Holt and U. Hwang
© 2001 American Institute of Physics 0-7354-0001-6/01/$18.00

Observation

The Two Micron All Sky Survey (2MASS) is being carried out by a pair of identical, fully dedicated, 1.3 meter Cassegrain equatorial telescopes that are located on Mt. Hopkins, near Tucson, AZ, and Cerro Tololo, near La Serena (CTIO), Chile. The imaging is taken simultaneously in the J (1.25μm), H (1.65μm) and K_s (2.17μm) bands with a three-channel camera. The telescope scans smoothly in declination at a rate of 57″ per second, and the telescope's secondary mirror tilts opposite the scan direction to momentarily freeze the focal plane image to reduce the dead-time (<0.1 sec) between the frames. The telescope scans are designed to cover "tiles" 6° long in the declination direction and one camera frame (8.5′) wide in right ascension. The images have 3.5″ spatial resolution with a pixel size of 1″, and the camera images each point on the sky six times for a total integration time of 7.8 sec (Cutri et al. 2000). The observation of RCW 103 took place on June 13, 1999, at CTIO and four 2MASS Atlas images were combined to produce the mosaiced image of RCW 103 in Fig. 1a.

Results

Emission from RCW 103 was detected in all three bands, showing a barrel shape, shell-like morphology. The K_s (dashed line in Fig. 1b) band emission is outside the J and H band (solid line in Fig. 1b) emission, and the displacement ranges from 10″ to 60″; it is smaller in the south and larger in the southeast. All three bands are bright along the southern shell and fainter along the northwestern shell. The surface brightnesses of the J and H band emission is consistent with the sum of ∼10 [Fe II] lines observed using spectroscopy (Oliva et al. 1991); the primary lines at 1.257μm and 1.64μm which make more than 70 % of the total emission. The surface brightnesses of K_s band emission is consistent with the sum of ∼15 molecular hydrogen lines observed by spectroscopy (Oliva et al. 1991). Thus the near-infrared emission of J, H and K_s emission is mostly composed of [Fe II] and H_2 line emission. After the extinction correction using A_v=4.5 mag, the surface brightness ranges from (24-100), (15-34), and (6-21) $\times$ 10^{-4} erg s^{-1} cm^{-2} sr^{-1}, for J, H and K_s band, respectively. The J and H band shell emission coincides with optical, radio and X-ray shell, indicating that the [Fe II] emission is close to the shock front. The 2MASS three color images superposed on the radio contours confirm the displacement; the J and H emission peaks coincide better with radio and optical peaks than with X-ray peaks, and the K_s emission is outside of the radio shell.

The surface brightness across the southern filaments for J and K_s band images is shown in Fig. 1b. The profiles confirm that the J and K_s band emission filaments ([Fe II] and H_2, respectively) are indeed displaced. The displacement is 20″ in the south (Fig. 3) while it was as large as 60″ in the southeast. Fig. 1b also shows that both [Fe II] and H_2 emission is split into a few thin filaments respectively, which can be thinner than the 2MASS spatial resolution of 3.5″. We have generated a color-color diagram using the stars in the field of RCW 103. The extinction is up to A_v=30 mag, so stars with the line of sight 6×10^{22} cm^{-2} are observed in 2MASS image, which are orders of magnitude deeper than optical observations. To 3-6 kpc distance of RCW 103 (Dickel et al. 1996; Burton et al. 1993), the extinction is A_v= 5 mag. Using the star counts and H-K_s color maps we derived extinction maps which shows a high extinction area (a few arcmin) to the immediate south of the remnant, where bright H_2 emission appears. There are four 2MASS sources near the pulsar position 1E 161348-5055 (Tuohy et al. 1983; Gotthelf et al. 1997) that are within the X-ray error box. In particular a red 2MASS star is present at that position, but the association with the pulsar is not clear.

Discussion

The J and H emission is mostly [Fe II] lines, and the [Fe II] spectral observations infer a temperature of 7000 K and a density range of 800 - 4000 cm^{-3} depending on the level transition of [Fe II] (Oliva et al. 1999). The [Fe II] image morphology in Fig. 1a is most similar to that of optical emission (Meaburn & Allen 1984), showing it is radiative cooling lines. An unusual fact on RCW 103 is that there is

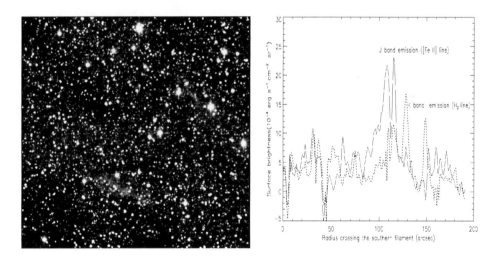

FIGURE 1. (a: left) 2MASS mosaic images of RCW 103. J (blue), H (green), and K_s (red) images show double shells with displacement between J & H and K_s emission. The brightnesses range (7.6-36), (7.5-16), and (4-13) × 10^{-4} erg s^{-1} cm^{-2} sr^{-1}, for J, H and K_s, respectively. (b: right) The surface brightness profiles of J and K_s band images crossing the southern filaments, showing the displacement of the two shells, and the resolved thinner filaments of [Fe II] and H_2 emission.

no evidence of enrichment of metal abundances, which is expected from a young SNR of 10^3 yr due to ejecta material. The infrared, X-ray, and optical spectra show no evidence of metal enrichment (Oliva et al. 1999; Nugent et al. 1984 using *Einstein* SSS data; Meaburn & Allen 1986).

We have explored hypotheses to explain the displament between the [Fe II] and H_2 shells as listed in the followings: (1) projection effect, (2) shocked H_2 lines due to dense gas as observed in IC 443, (3) radiative precursor, (4) reverse shock and (5) magnetic precursor. The 2MASS images covering the entire remnant show that the displacement between the [Fe II] and H_2 is not just due to a projection effect; instead, it appears that the layers emitting [Fe II] and H_2 lines are intrinsically double shells. If H_2 is a cooling gas as observed in IC 443, it should be at the shock front coinciding with the radio, optical and X-ray shells, suggesting that it is unlikely due to cooling lines, or reverse shock. The H_2 luminosity of a few 10^{35}erg s^{-1} is too high to be radiative precursor or irradiation by X-rays. The extinction map derived from star catalog and star counts suggests that there is a large cloud south of the remnant likely interacting with the remnant. The detection of OH masers (Frail et al. 1996) is also consistent with the idea that the remnant is interating with a cloud in the south, where the H_2 lines are bright.

The magnetic precursor is the most likely possible explanation for the displacement of the H_2 emission from the shock. The magnetic precursor occurs when a magnetic field is moderately strong with a J-type shock. The precursor heats and compress the medium ahead of the shock front where neutral gas undergoes a discontinuous change of state. The condition for a magnetic precursor is that the ion magnetosonic speed is greater than the shock speed ($B > n_H^{1/2} v_7 (\chi/3 \times 10^{-4})^{-1/2}$ μG, where n_H is a pre-shock density, v_7 is a shock velocity in units of 100 km s^{-1} and χ is an ionization fraction). The magnetic precursor is observed at shocks in dense molecular gas such as Herbig-Haro objects and star-forming cores (Hartigan et al. 1987) as well as in a low density medium as observed in the Cygnus Loop (Graham et a. 1991). H_2 emission of RCW 103 is likely enhanced due to the magnetic precursor in dense gas where the SNR is hitting a dense molecular cloud. However, the hypothesis of reverse shock may not be ruled out due to lack of detail studies of abundances, considering RCW 103 is a young age SNR. Further calculations are needed to investigate the mechanism and physical conditions of the mysterious H_2 emission outside the shock front.

REFERENCES

1. Hartigan, P.,Raymond, J., & Hartmann, L., 1987, ApJ, 316, 323
2. Oliva, E., Moorwood, A. F. M., Drapatz, S., Lutz, D. &Sturm, E., 1999b, A&A, 343, 943
3. Graham, J. R.; Wright, G. S.; Hester, J. J.;Longmore, A. J., 1991, AJ, 101, 175

Young SNR II

Emission Processes in Young SNRs

J.C. Raymond

Center for Astrophysics, 60 Garden St., Cambridge, MA 02138, USA

Abstract. This paper summarizes the emission processes that produce the radiation we use to study young supernova remnants. We discuss the relative importance of various processes and the accuracy with which the emissivities or rate coefficients are known, especially in cases where uncertainties in the theoretical or laboratory rates may limit the interpretation of observations. Common assumptions that may break down, such as thermal and ionization equilibrium, are discussed.

INTRODUCTION

Everything known about young supernova remants is derived from a broad range of observations from radio to gamma ray wavelengths. Some processes are important in only one wavelength band, while others make significant contributions over many orders of magnitude in wavelength. Most of the relevant physics is described in books such as Jackson [23], Tucker [43] or Rybicki and Lightman [34]. Here we summarize the processes that account for most of the observable emission, then discuss the circumstances under which common assumptions such as Maxwellian velocity distributions or equal ion and electron temperatures are valid.

ELECTRON-ION COLLISIONS

Bremsstrahlung

When an electron collides with an ion, a photon can be emitted as the electron is deflected from its path. The emission spectrum has constant intensity up to a cutoff that depends of the electron velocity. For most purposes the emissivity is integrated over a Maxwellian distribution at temperature T to give

$$j(\nu) = 6.8 \times 10^{-38} Z^2 N_e N_Z T^{-1/2} \bar{g} e^{-h\nu/kT} \quad ergs/(cm^3 \ s \ Hz) \qquad (1)$$

where ν is the photon frequency, Z is the ion charge, N_e and N_Z are the ion and electron densities and all the complications are buried in the Gaunt factor $\bar{g}$. The

CP565, *Young Supernova Remnants: Eleventh Astrophysics Conf.*, edited by S. S. Holt and U. Hwang
2001 American Institute of Physics 0-7354-0001-6

same process occurs when the electron distribution is not Maxwellian, of course, and it is called non-thermal bremsstrahlung.

Bremsstrahlung usually dominates the soft X-ray continua of young SNRs. It may contribute significantly at hard X-ray, radio and gamma ray energies as well. Free-free radio emission is detected from shocked gas in Herbig-Haro objects [14], but it is usually buried under synchrotron emission in young SNRs. In principle is should be straightforward to differentiate between thermal and non-thermal bremsstrahlung in SNRs because the non-thermal spectrum should curve upward rather than downward [3]. In practice this requires well-calibrated spectra and accurate Gaunt factors. Until recently, the non-relativistic Gaunt factors used in most analyses of X-ray spectra introduced errors of order 10% at temperatures and photon energies of order 10 keV. New calculations by Itoh et. al. [21] remedy this.

H and He generally dominate the bremsstrahlung emission except in highly enriched SN ejecta. Electron-electron bremsstrahlung begins to contribute as the electrons become relativisitic. Inverse bremsstrahlung might potentially contribute, but Baring et al. [4] show that it seems unimportant in practice.

Radiative Recombination

Radiative recombination produces continuum emission at optical through X-ray wavelengths, and recombination to excited levels can result in emission lines. The rate coefficients decline roughly as $T^{-1/2}$. In collisionally ionized gas near ionization equilibrium, the recombination continua make a modest contribution to the X-ray spectrum and recombination to excited levels makes significant, and diagnostically important, contribution to lines such as the forbidden $1s^2 - 1s2s$ ^{3}S lines of He-like ions. The direct excitation rate from the ground state is relatively small because the line is forbidden, while the high statistical weight of the triplets and the cascade paths that channel most of the recombinations to triplet levels through the $1s2s^3S$ state make recombination competetive.

Recombination is much more important in overionized plasmas, such as photoionized plasmas and plasmas that cool rapidly from a high temperature. In Young SNRs, the gas for the most part is underionized because the time needed to reach equilibrium ionization states is comparable to the age of the SNR (e.g. Itoh [19]), so radiative recombination is less important than usual. Exceptions are the Quasi-Stationary Flocculi, in Cas A (shocked gas cooling quickly and producing H and He recombination lines), photoionized filaments in the Crab Nebula, and the O I recombination lines at 7774 and 1350 Å [6] in E0102.2-7219, where the enormous radiative cooling rate in the ejecta lead to ionized oxygen at temperatures below 100 K. The O I recombination lines are rather weak compared with the predictions of shock models [20].

Radiative recombination rates to the ground levels of astrophysically important ions are computed from the photoionization cross sections and the Milne relation, so the accuracy is a good as that of the photoionization cross sections. Verner's

204

[45] recombination rates and spectra based on the cross sections ought to be good to about 20%. For higher levels, a hydrogenic approximation is often used. More sophisticated rates based on Los Alamos cross sections [11] differed by at most 15% [2], but further study for important cases such as He-like emission lines would be valuable.

Collisional Excitation

The optical, UV and X-ray spectra of most young SNRs are dominated by emission lines of atoms and ions. Fine structure lines of the neutrals and moderately ionized species, along with molecular lines, are seen in the IR. Electrons generally dominate the collision rates, and the coefficients are given by

$$q_{ex} = 8.63 \times 10^{-6} \, \Omega \omega_l^{-1} T^{-1/2} \, e^{(-E/kT)} \quad cm^3/s \tag{2}$$

Here the collision strength, Ω, is a slowly varying quantity related to the excitation cross section, ω_l is the statistical weight of the initial state, and E is the excitation threshold energy. For a Maxwellian distribution, this formula is as accurate as the value of Ω. The computation of Ω is made difficult by the presence of sharp resonances in the cross sections. For strong lines allowed in LS coupling these resonances may only affect the integrated excitation rate by 5-10%, but for forbidden or intercombination lines they can dominate the total excitation rate. In Distorted Wave computations the resonances are generally ignored, while Close Coupling calculations include them. Resonances are most important near threshold, so they dominate when E > kT. Shock heated plasma in young SNRs tends to be underionized, so resonances are somewhat less important than in equilibrium plasmas.

Uncertainties are typically estimated at 10-30% for theoretical calculations of Ω, but it is difficult to know. The uncertainties are larger for complicated ions (e.g. N-like, P-like or ions more complex than Ar) than the simple ions of the He isosequence whose bright lines have been studied in detail. Comparison with laboratory measurements shows a heartening level of agreement in many cases (e.g. the crossed beams experiment of Reisenfeld [33] compared with Close-Coupling calculations by Griffin et al. [18] for Si^{2+} excitation or the EBIT measurements of Brown et al. [9] compared with theoretical excitation rates for the Ne-like ion Fe XVII). Attempts to match the relative intensities of large numbers of emission lines in solar spectra (e.g. Brickhouse et al. [8]) tend to show a distressing number of outliers in the observed to predicted ratios in the complex spectra of Fe IX-XVI ions. Rates for high quantum number levels tend to be poorly known, but HULLAC calculations are filling in the gaps (Liedahl et al. [26]).

The state of the excitation rates used in general model codes is rapidly improving. The CHIANTI database includes up to date Close-Coupling calculations, especially for the optical through EUV bands [25]. The APEC code, which initially concentrates on X-ray emission [42], uses CHIANTI rates and rates computed by

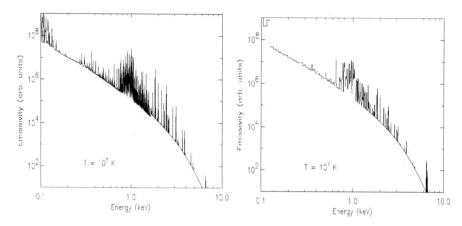

FIGURE 1. Comparison of X-ray spectra predicted by APEC (left) and by the aging Raymond and Smith [29] code (right). Plots courtesy of Randall Smith.

HULLAC. At the moment, the spectral range between 25 and 150 Å is probably the most in need of improvement, partly because relatively few observations have been available in that region. Molecular excitation rates, including H_2 are also improving [28].

Figure 1 compares the theoretical X-ray emission spectra from a 10^7 K plasma computed with the 1990 version of the Raymond and Smith [29] code and the current version of APEC. The older codes had the basic shape of the spectrum correct, but they lacked the spectral resolution needed to interpret spectra from CHANDRA or XMM, and they predicted some line ratios far from modern values.

An unusual feature of some young SNRs is the importance of excitation by protons. In SN1006, the intensities of the UV lines of C IV, N V and O VI are mostly due to proton excitation [30,24].

Dielectronic Recombination

Dielectronic recombination is primarily important for determining the ionization state of the plasma, but it also produces satellites to the resonance lines of some ions (He-like ions in particular) and it can also populate excited levels in much the same way as radiative recombination. DR populates doubly excited autoionizing levels which can Auger ionize or emit a photon. These doubly excited levels lie at lower energies than the associated singly ionized levels, so that they are favored by the difference in Boltzmann factor at low temperatures. The rate varies as

$$q_{DR} = q_0 T^{-3/2} e^{-E_s/kT} \tag{3}$$

where E_s, the energy of the doubly excited level, is lower than the energy E of the associated singly excited level. The most important cases tend to be satellites to

206

the resonance lines of He-like ions, and the ratio of the

As with radiative recombination, the underionization of plasma in a young SNR makes DR less important than in a plasma in ionization equilibrium. This is fortunate, as the accuracy of DR rates is relatively poor. While a rule-of-thumb value of 30% is often quoted, discrepancies among sophisticated calculations are often larger [35]. Laboratory measurements will help to remedy this situation, but the densities and electromagnetic fields in laboratory situations are much higher than in interstellar space, so the laboratory rates are used primarily to benchmark theoretical calculations.

Neutral-Ion collisions

Charge transfer into excited levels is significant in producing Balmer line emission from very fast shocks in partially neutral gas [15]. It contributes to the broad component of Hα and is therefore important to the interpretation of the broad to narrow intensity ratio [17].

Charge transfer between neutrals and highly charged ions has been suggested as a means to populate excited levels and produce X-rays, for instance

$$O^{7+} + H \rightarrow O^{6+}(nl) + p \tag{4}$$

which will usually produce one of the O VII 22 Å lines as the recombined ion decays to the ground state. Cravens [12] proposed this mechanism to explain the unexpected X-ray emission from comets, and it may contribute to the soft X-ray background [13]. Wise and Sarazin [47] computed the contribution to the X-ray emission line spectra in young SNRs, and they found it to be small. The process is only likely to contribute significantly if neutrals are mixed rapidly into very highly ionized material. Turbulent mixing layers [41] generated by encounters between dense clouds and the blast wave or reverse shock might show this effect.

Ionization State

It is safe to assume that the plasma in a young SNR is far from collisional ionization equilibrium. It is easy to compute the time-dependent ionization state if the temperature and density are known from the coupled equations

$$\frac{dn_z}{dt} = n_e n_{z-1} q_i(Z, z - 1) - n_e n_z(q_i(Z, z) + \alpha_r(Z, z)) + n_e n_{z+1}\alpha_r(Z, z + 1) \tag{5}$$

for all the ions z of element Z. The q_i are collisional ionization rates and α_r are recombination rates. The inner shell ionization contribution to q_i tends to be relatively important in young SNRs, adding a term proportional to n_{z-2}.

One approach to dealing with time-dependent ionization is to assume a constant temperature and simply add the product of density times time, $n_e t$, to the fit. The

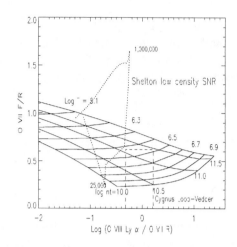

FIGURE 2. Diagnostic for time-dependent ionization after Vedder et al. [44].

advantage is that only one free parameter is added and it is straightforward to compute a model grid. Figure 2 shows a diagnostic grid patterned after Vedder et al. [44] based on oxygen lines. The ratio of forbidden to resonance lines depends on a combination of temperature and ionization state because of the recombination contribution. It is plotted against the ratio of O VIII to O VII resonance lines. The downward sloping solid lines in the figure connect models of different temperatures, and the upward sloping lines follow an increase in $n_e t$ at single temperature. The dotted line, based on a model of an SNR in a very low density gas by Shelton [38] follows the average state from very underionized to very overionized. The dashed line shows the Einstein FPCS observation of the Cygnus Loop [44].

The other approach is to model the SNR as a whole, using a Sedov solution or a numerical hydro model and following the individual elements of gas as they are heated and cooled. Such models are now available within XSPEC [7]. They have the advantage of a much more physically realistic basic model, but they do tie the result to a specific evolutionary scheme.

SYNCHROTRON EMISSION

Synchrotron emission has long been known to dominate the radio emission of SNRS and the X-ray and gamma ray emission of the Crab and similar remnants. It has recently been shown to be important in SN1006 X-ray emission [32]. The emission from electrons spiralling in a magnetic field can be computed directly, and the main difficulty lies in sorting out the magnetic field strength from the number of emitting electrons. It is frequently assumed that the electrons are randomly distributed in pitch angle, and the electron distribution is generally taken to be a power law.

ELECTRON-PHOTON COLLISIONS

Inverse Compton scattering off the cosmic microwave background probably dominates the gamma ray emission of SN1006 and other shell-type SNRs [4]. It is thoroughly treated in the reference books.

NUCLEAR INTERACTIONS

Two types of nuclear processes are of interest. Cosmic ray protons generated by the shock can collide with nuclei in the ambient ISM to produce pions which can decay to gamma rays. Collisions can also excite the nuclei. These processes must occur, and they should be substantial in cases where a young SNR occurs in or near a molecular cloud, but there is no definite detection as yet.

The second process is decay of nuclei formed in the explosion. Iyudin et al. [22] have reported the detection of ^{44}Sc gamma rays from Cas A resulting from decay of ^{44}Ti. The process has in interesting twist: ^{44}Ti decays by capture of an orbital electron. Slane et al. [40] have searched for the scandium $K\alpha$ X-rays that should be produced as a result of the resulting hole in the K shell. Mochizuki et al. [27] point out that if the titanium is fully ionized, the half life of ^{44}Ti can be extended beyond its normal 60 years.

WARM DUST EMISSION

Dust grains are heated by collisions with ions and electrons in the shock-heated gas of young SNRs. If they behaved as blackbodies, the analysis would be simple and dull. Their emission spectra are determined by emissivities which show strong peaks due to such dust components as silicates. Emissivities are generally determined by laboratory measurement of samples designed to mimic interstellar dust [16], and they often do quite well in reproducing observed spectra [1]. While the strong dust emission features seem to provide unambigous identification of the overall composition of the dust, it is difficult to know how well the surfaces of laboratory samples match the surfaces of interstellar grains, and it is therefore difficult to assess the accuracy of the absolute emissivities.

ION-ELECTRON THERMAL EQUILIBRATION

It is usually safe to assume that in young SNRs neither ions nor electrons have Maxwellian distributions and the the Maxwellian cores of the ion electron distributions have different temperatures. Unfortunately, the opposite assumptions are usually made owing to lack of adequate diagnostic information.

The measured particle distributions in the solar wind almost always show non-thermal velocity distribtions, typically a Maxwellian core with a power law tail, and typically anisotropic. Radio, and more recently X-ray and gamma ray, observations clearly show a non-thermal tail in young SNRs. Models of particle acceleration in shocks indicate that in shell-type remnants the non-thermal population of ions ought to be at least as strong as that of electrons. The Maxwellian assumption is usually built in to computation of ionization, recombination and excitation rates, so it is well to ask when this causes serious error.

The electrons that dominate the collisional ionization rate are generally those at 1-2 times the ionization potential, χ. In collisional equilibrium, those are typically electrons at 5-10 kT, a very small fraction of the total electron population, so departure from the Maxwellian at 10 kT has a large effect on the ionization state. In the rapidly ionizing conditions of recently shocked plasma, most of the electrons have energies above the ionization potentials of the ions, so departure from a Maxwellian will have a small effect. An exception might occur if inner shell ionization is important, for instance in the ionization of low and intermediate ions of iron by K-shell excitation and ionization if kT is around 1 keV. Similarly, X-ray emission lines will be affected by departures from Maxwellian distributions near their excitation potentials, around 1 keV for the oxygen and iron L-shell lines, 7 keV for iron Kα.

A possible exception might be the presence of nonthermal particles in really cold gas, as might occur in the shock precursor where cosmic rays are accelerated. These particles would have to have a rather large energy density to compete with photoionization heating. K shell photoionization by X-rays in neutral gas also produces a non-Maxwellian distribution [39,46], but even a modest ionization fraction provides enough electrons for Coulomb collisions to quickly establish a Maxwellian.

After the question of the existence of a thermal distribution comes the question of whether ions and electrons share the same temperature, as they would in thermal equilibrium. To the extent that a collisionless shock simply thermalizes some of the kinetic energy of the incoming particles, the temperature of a species is proportional to its mass. Coulomb collisions will eventually equalize the temperatures of the different species, but the time required increases as v^3, and it may be longer than the age of the SNR. Theoretical models predict heating of the electrons by wave processes [10] or by drift in the electric field of the shock [37]. Either model seems compatible with measurements of shocks in the solar wind showing less efficient electron heating in faster shocks [38]. Fewer measurements are available for SNR shocks, but they indicate that $T_e \simeq T_p$ in shocks slower than 300 km/s (Cygnus Loop), while $T_e \leq T_p$ in shocks around 2000 km/s (Tycho, SN1006) [17, 24].

An even more extreme departure between T_e and T_i occurs in the optically bright filaments of O-rich SNRs. Even if electrons are heated efficiently in the shock, as long as oxygen and other abundant elements retain L-shell electrons, the radiative cooling rate is enormous. Much of the optical and UV emission arises from a broad region in which $T_e \simeq 10^5$ K, with radiative cooling of the electrons balanced by Coulomb collisional heating with million K ions [20].

It appears that shocks do not even produce equilbrium among different ions. The kinetic temperatures of He and O ions are typically 25% higher than the proton temperatures behind solar wind shocks of modest Mach number [5]. Fast shocks, both in the solar wind and young SNRs, seem to randomize the velocities of the different particles independently, producing $T_i = m_i T_p$ [30, 31]

Other Random Processes

For completeness, we mention conversion of plasma waves into radio emission, as this has been extensively studied in shocks in the solar corona. Suprathermal electrons produced by the shock generate Langmuir turbulence through the bump-on-tail instability. Interaction of plasma waves with ion acoustic waves can produce radio waves close to the plasma frequency, while interaction of two plasma waves can produce the harmonic at twice the plasma frquency (Type II radio burst). This process must occur in young SNRs, but at extremely low frequencies, and emission by this process has not been reported.

CONCLUSION

The atomic and electromagnetic processes that produce the radiation from young SNRs are basically understood quite well. In cases such as bremsstrahlung, synchrotron emission and inverse Compton scattering, the rate coefficients are known at the percent level and can be used with confidence to infer physical parameters of the SNR and physical processes in the shocked plasma. A rough estimate of the accuracy of atomic emission line predictions would be 20%, but there is an enormous range.

REFERENCES

1. Arendt, R.G., Dwek, E., and Moseley, S.H. 1999, ApJ, 521, 234.
2. Arnaud, M., and Raymond, J.C. 1992, ApJ, 398, 384.
3. Baring, M.G., Ellison, D.G., Reynolds, S.P., Grenier, I.A., and Goret, P. 1999, ApJ, 513, 311.
4. Baring, M.G., Jones, F.C., and Ellison, D.C. 2000, ApJ, 528, 776.
5. Berdichevsky, D., Geiss, J., Gloeckler, G., and Mall, U. 1997, JGR, 102, 2623.
6. Blair, W.P., et al. 2000, ApJ, 537, 667.
7. Borkowski, K. 2000, private communication.
8. Brickhouse, N.S, Raymond, J.C., and Smith, B.W. 1994, ApJS, 97, 551.
9. Brown, G.V., Beiersdorfer, P., Liedahl, D.A., Widmann, K., and Kahn, S.M. 1998, ApJ, 502, 1015.
10. Cargill, P.J., and Papadopoulos, K. 1988, ApJL, 329, L29.
11. Clark, R.E.H., Cowan, R.D., and Bobrowicz, F.W. 1986, ADNDT, 34, 415.
12. Cravens, T. 1997, Geophys. Res. Lett., 24, 105.

13. Cravens, T. 2000, ApJL, 532, 153.
14. Curiel,S., Cantó, J., amd Rodriquez, L.F. 1987, Rev. Mex. A&A, 14, 595.
15. Chevalier, R.A., Kirshner,R.P., and Raymond, J.C. 1980, ApJ, 235, 186.
16. Dorschner, J., Friedmann, C., Gürtler, J., and Duley, W.W. 1980, Ap&SS, 68, 159.
17. Ghavamian, P., Raymond, J., Smith, R.C., and Hartigan, P. 2000, ApJ, in press.
18. Griffin, D.C., Pindzola, M.S., Robicheaux, F., Gorzycka, T.W., and Badnell, N.R. 1994, Phys. Rev. Lett., 72, 3491.
19. Itoh, H. 1978, PASJ, 30, 489.
20. Itoh, H. 1988, PASJ, 40, 673.
21. Itoh, N., Sakamoto, T., Kusano, S., Nozawa, S., and Kohyama, Y. 2000, ApJS, 128, 125.
22. Iyudin, A.F., et al. 1994, A&A, 284, L1.
23. Jackson, J.D., 1962, *Classical Electrodynamics* (Wiley: New York)
24. Laming, J.M., Raymond, J.C., McLaughlin, B.C., and Blair, W.P. 1996, ApJ, 472, 267.
25. Landi, E., Landini, M., Dere, K.P., Young, P.R., and Mason, H.E. 1999, A&AS, 135, 339.
26. Liedahl, D.A., Osterheld, A.L., and Goldstein, W.H. 1995, ApJL, 438, 115.
27. Mochizuki, Y., Takahashi, K., Janka, H.-Th., Hillebrandt, W., and Diehl, R. 1999, A&A 346, 831.
28. Liu, X, Shemansky, D.E., Ahmed, S.M., James, G.k., and Ajello, J. 1998, JGR, 103, 26739.
29. Raymond, J.C., and Smith, B.W. 1977, ApJS, 35, 419.
30. Raymond, J.C., Blair, W.P., and Long, K.S. 1995, ApJL, 454, L51.
31. Raymond, J.C., et al., 2000, Geophys. Res. Lett., 27, 1439.
32. Reynolds, S.P. 1996, ApJ, 459, L13
33. Reisenfeld, D.B. 1998, PhD Thesis, Harvard University.
34. Rybicki, G.B., and Lightman, A.P. 1979, *Radiative Processes in Astrophysics* (New York: Wiley)
35. Savin, D.W. 2000, Rev. Mex. A&A Ser. Conf., 9, 115.
36. Schwartz, S.J., Thomsen, M.F., Bame, S.J., & Stansberry, J. 1988, JGR, 93, 12923
37. Scudder, J.D. 1995, Adv. Sp. Res., 15, No. 8/9, 181.
38. Shelton, R.L. 1999, ApJ, 521, 217.
39. Shull, J.M., and van Steenberg, M.E. 1985, ApJ, 298, 268.
40. Slane, P. et al., this conference
41. Slavin, J.D., Shull, J.M., and Begelman, M.C. 1993, ApJ, 407, 83.
42. Smith, R.H., and Brickhouse, N.S. 2000, Rev. Mex. A&A Ser. Conf., 9, 134.
43. Tucker, W.H. 1975, *Radiation Processes in Astrophysics*, (MIT Press: Cambridge)
44. Vedder, P.W., Canizares, C.R., Markert, T.H., and Pradhan, A.K. 1986, ApJ, 307, 269.
45. Verner, D., Ferland, G.J., Korista, K.T., and Yakovlev, D.G. 1996, ApJ, 465, 487.
46. Victor, G.A, Raymond, J.C., and Fox, J.L. 1994, ApJ, 435, 864.
47. Wise, M.W., and Sarazin, C.L. 1989, ApJ, 345, 384.

High Resolution Spectroscopy of Two Oxygen-Rich SNRs with the Chandra HETG

C.R. Canizares, K.A. Flanagan, D.S. Davis, D. Dewey, J.C. Houck, M.L. Schattenburg

Center for Space Research, Massachusetts Institute of Technology
Cambridge, MA 02139

Abstract. The HETG can be used to obtain spatially resolved spectra of moderately extended sources. We present preliminary results for two well studied, oxygen rich supernova remnants in the Magellanic clouds, E0102-72 and N132D. The dispersed spectrum of E0102-72 shows images of the remnant in the light of individual emission lines from H-like and He-like ions of O, Mg, Ne and He-like Si with no evidence of Fe. The diameters of the images for various ions, measured in the cross-dispersion direction, increase monotonically with the ionization age for the given ion. This shows in detail the progression of the reverse shock through the expanding stellar ejecta. We see clear evidence for asymmetric Doppler shifts across E0102-72 of $\sim \pm2000\ km\ s^{-1}$. These can be modelled approximately by a partially-filled, expanding shell inclined to the line of sight. The dispersed spectrum of N132D is more affected by spatial/spectral overlap but also shows monochromatic images in several strong lines. Preliminary spectra have been extracted for several bright knots. Some regions of oxygen-rich material, presumably stellar ejecta, are clearly identified. Additional details on E0102-72 are presented by Flanagan *et al.* and Davis *et al.* in these proceedings, and further analysis is in progress.

INTRODUCTION

The High Energy Transmission Grating (HETG) consists of an array of periodic nanostructures that can be inserted behind the *Chandra* mirrors in order to disperse the focused X-ray beam into a spectrum at the focal plane. The HETG array includes two grating types, MEG and HEG, which together cover the energy band 0.4-8 keV with resolving powers of up to 1000. The dispersed spectrum is read out by the Advanced CCD Imaging Spectrometer (ACIS-S). A complete description is given in [1] and details can be found in the *Chandra* Proposers guide and at http://space.mit.edu/HETG.

As a "slitless" dispersive spectrometer, the HETG is most straightforward when used to observe point sources: a zeroth order image is formed at the normal focal

CP565, *Young Supernova Remnants: Eleventh Astrophysics Conf.*, edited by S. S. Holt and U. Hwang
© 2001 American Institute of Physics 0-7354-0001-6/01/$18.00

point, with dispersed spectra on either side (the HEG and MEG spectra are offset from one another by 10 degrees to form a shallow "X" centered on zeroth order). For point sources, the resolution of the complete spectrometer is 0.022 Å and 0.012 Å for MEG and HEG, respectively. For moderately extended sources, the spatial and spectral information are mixed in the dispersion direction, but not in the cross dispersion direction. Although this mixing complicates the analysis, moderately extended sources like those presented here are very amenable to simultaneous spectral/spatial analysis, particularly if their emission is dominated by distinct spectral lines. In that case the dispersed spectrum is analogous to a spectro-heliogram, showing a series of monochromatic images of the source in the light of individual spectral lines.

This paper gives a brief report of HETG observations of two relatively compact supernova remnants (SNRs) in the Magellanic clouds, E0102-72 and N132D. Both are members of the oxygen-rich class of SNRs, of which Cas A is often taken as the prototype and which are thought to be products of Type Ib or II supernovae in massive stars [2]. More detailed reports of our results on E0102-72 are included in the proceedings of this meeting (see [3] [4]).

E0102-72

The SNR 1E0102.2-7219 is a well studied member of the oxygen rich class of supernova remnants located in the SMC. It has a radius of $\sim$20 arc sec (6.4 pc), which is a perfect size for spectral imaging with the HETG. Recently, moderate resolution X-ray spectra integrated over the remnant were obtained with ASCA [5]. Gaetz et al. [6] reported spectrally resolved imaging from *Chandra's* ACIS detector, which shows an almost classic, text-book SNR with a hotter outer ring identified with the forward shock surrounding a cooler, denser inner ring which is presumably the reverse-shocked stellar ejecta. Hughes et al. [7] combined the *Chandra* image with earlier *Einstein* and *ROSAT* images to measure X-ray proper motions, which give an expansion age of 1000 yr, consistent with earlier estimates (they also deduce that a significant fraction of the shock energy has gone into cosmic rays). Blair et al. [2] analyzed extensive observations with HST of the optical/UV filaments. At this meeting, Eriksen et al. [8] report the analysis of new Fabry-Perot observations that yield an age of $\sim$2100 yr for the remnant.

We observed E0102 for a total of 140 ksec on two occasions with the *Chandra* HETG. A portion of the dispersed spectrum is shown in Fig. 1. It shows multiple images of the SNR in the light of individual spectral lines from various ions. Most prominent are the lines of H-like and He-like O, Ne, Mg. We see no clear evidence for any Fe emission (although weak Fe lines are reported by Rasmussen et al. [9] at this meeting from an observation with the *XMM-Newton* Reflection Grating Spectrometer). The O VIII image alone suggests that this remnant contains at least several solar masses of oxygen in the reverse-shocked ejecta.

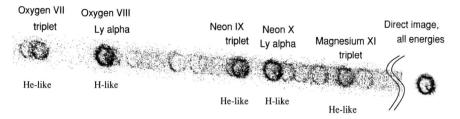

FIGURE 1. Dispersed high resolution X-ray spectrum of E0102-72. Shown here is a portion of the -1 order formed by the medium energy gratings (MEG). The zeroth order, which combines all energies in an undispersed image is at the right. Images formed in the light of several strong X-ray emission lines are labeled.

Imaging the Reverse Shock

The time scale for the ions in E0102 to reach ionization equilibrium is comparable to or longer than its age, so one would expect the plasma not to have achieved ionization equilbrium [10]. Hayashi *et al.* [5] found evidence for departures from ionization equilibrium for the integrated spectrum of the SNR (which ASCA could not resolve spatially) based on the relative intensities of He-like and H-like lines. They could not find a consistent model to explain the line ratios for different ions, even allowing for multiple components, and concluded that abundance inhomogeneities must be present. If we integrate the flux over the remnant for several of the strong, relatively isolated lines, we also find line ratios that indicate departures from ionization equilibrium [4].

Gaetz *et al.* [6] find direct evidence for progressive ionization of oxygen in the spectrally resolved *Chandra* image. They compared monochromatic images (at CCD spectral resolution) at the energies of the O VII and O VIII lines, and found that the O VII emitting region lies inside that of O VIII. This is what one would expect if the ejecta are being subject to a reverse shock propagating backwards (in a Lagrangian sense) towards the center of the remnant (ionization at smaller radii lags behind the ionization at larger radii).

We observe this progressive ionization quantitatively by comparing the images from the He-like resonance lines and H-like Lyα lines of O, Ne and Mg, and of He-like Si. Details are given elsewhere in these proceedings [3], but the qualitative evidence can be seen directly from Fig 1. For both O and Ne, the images of the He-like lines are noticeably smaller than the H-like images. Cuts across the spectrum in the cross-dispersion direction (to avoid any spectral/spatial confusion) show that, for each ion, the diameter of the He-like image is smaller than that of the H-like image. Furthermore, for a given temperature, one can find a characteristic ionization time scale or ionization age (ionization age is given by $\tau = n_e t$ for electron density n_e and time t) at which a given ion reaches its peak population fraction, and therefore its peak emissivity in an ionizing plasma. Doing this for the ions in question (assuming $\log T = 7.05$) shows a monotonic increase in the diameter of the

215

emitting region with increasing characteristic ionization timescale This suggests a simple interpretation, in which the differences in image diameters for the different ions are caused entirely by the progressive passage of the reverse shock through relatively homogeneous stellar ejecta. A more comprehensive analysis, which will take into account the likely density gradient in the ejecta, for example, is now underway.

Differential Doppler Shifts

We have found clear evidence for Doppler shifts in several of the emission lines, and evidence that these vary systematically across the remnant. This can be seen from a comparison of the two (plus and minus order) dispersed images from a strong line, such as Ne X Ly α, with each other and with the zeroth order image (constructed by selecting a narrow range of ACIS pulse heights at the corresponding line energy). For example, even in Fig. 1 it is clear that the Ne X image appears slightly elongated in the dispersion direction compared to the zeroth order image.

The fact that the HETG has both plus and minus order images breaks the spectral/spatial degeneracy that would otherwise confuse the signature of a Doppler shift. A dispersed image in the light of a single line that is distorted due to intrinsic spatial variations will look identical on either side of zeroth order, whereas a distortion due to a wavelength (Doppler) shift will appear with opposite offsets in the plus and minus orders, (i.e. a shift to longer wavelength moves to the right in the plus order but to the left in the minus order image, showing reflectional symmetry about zeroth order). Furthermore, comparison of plus and minus orders also allows one to identify and avoid confusion from the overlapping images of nearby lines.

We have examined the relatively clean dispersed $\pm 1^{st}$ order images in the Ne X Lyα line with the corresponding zeroth order image and find clear evidence for distortions with the mirror symmetry expected for Doppler shifts in wavelength (some positive, others negative). More than a dozen bright regions around the remnant were analyzed to determine the sense and magnitude of the shifts.

The result of the Doppler analysis for Ne X Ly α is shown in Fig. 2, where the arrows pointing left indicate blue-shifts and those pointing right indicate red-shifts. The magnitudes of the shifts range from -1600 to +2300 $km\ s^{-1}$. The velocity structure in the X-ray is systematic and clearly asymmetric, showing only redshifts on the eastern side but both red and blue shifts on the western side. The velocities measured for some of the optical knots, which generally lie interior to the brightest portions of the X-ray bright ejecta, are comparable to those found here and also show complex, asymmetric structure [11]

We have made a preliminary effort at modeling the asymmetric velocity structure shown in Fig 2. Rough correspondence is achieved if the emission comes from a partially filled spherical shell with azimuthal symmetry whose axis is inclined to the line-of-sight. The inner radius of the shell is 4.3 pc, its thickness is 1.1 pc, and the velocity increases with radius. The emissivity is concentrated toward the

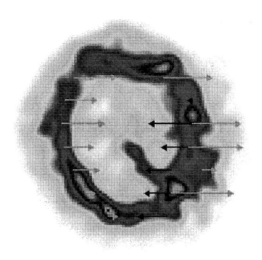

1000 km/s

FIGURE 2. Doppler velocities of E0102-72 Ne x Ly α. The lengths and locations of the arrows indicate the relative velocities and approximate locations of red/blue shifted ejecta. Arrows pointing left represent blue shifts; arrows pointing right represent red shifts.

equatorial plane (we assume a Gaussian distribution in the cosine of the latitude on the spherical shell with $\sigma=0.35$). Interestingly, Hughes [13] deduced that the surface brightness distribution measured with the *ROSAT* HRI also indicated that the emission was concentrated in a thick ring rather than a spherical shell.

N132D

N132D is located in the LMC at a distance of 50 kpc, with a radius of ~50 arc sec (~12 pc) and an estimated age of 3000 yr. Several recent X-ray studies have been presented from Beppo-SAX [14] and ASCA [15]. Previously, Hwang *et al.* [16] performed a non-equilibrium ionization analysis using moderate resolution spectra from the Solid State Spectrometer combined with high resolution spectroscopy of selected lines obtained with the Focal Plane Crystal spectrometer (both instruments were on the *Einstein* observatory). All these studies suggest the overall elemental abundances for N132D are roughly similar to those for the LMC as a whole. This indicates that N132D is dominated by swept-up material rather than ejecta, and is thus at a later stage of evolution than E0102-72. The most recent optical/UV data [2] support this conclusion.

We observed N132D for 100 ksec with the *Chandra* HETG with a roll angle that places the dispersion direction parallel to the narrow axis of this U-shaped SNR. A portion of the MEG spectrum is shown in Fig 3. The combination of larger size and more numerous emission lines in N132D result in greater spectral/spatial overlap than in E0102-72. Nevertheless, we can distinguish clear monochromatic images in the light of individual emission lines of oxygen, neon and iron, for example (iron is largely responsible for the enhanced number of lines). Even a preliminary examination indicates strong spatial variations in the concentrations of these species, with some features richer in iron and others in oxygen. The fact that we see oxygen rich material is particularly significant, since, contrary to the situation in E0102, no direct X-ray evidence for oxygen rich ejecta had previously been seen [15] (Behar *et al.* [17] also see regions of enhanced oxygen with *XMM-Newton*)

We have begun to measure the spectra of selected knots in N312D by extracting slices through the dispersed spectral image centered on the specific features of interest. Typically, a given feature will have some "tilt" with respect to the dispersion direction, so spectra in narrow slices perpedicular to the dispersion direction are shifted to a common origin (e.g. a stack of slices through the dispersed images of a curved feature whose zeroth order is shaped like '(' is shifted slice-by-slice and aligned so that its zeroth order becomes a straight line perpendicular to the dispersion direction). The stacked, shifted slices of dispersed spectra are then projected

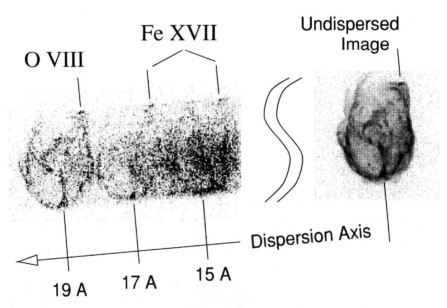

FIGURE 3. Dispersed high resolution X-ray spectrum of N132D. A portion of the MEG spectrum is shown at left, with zeroth order image at right.

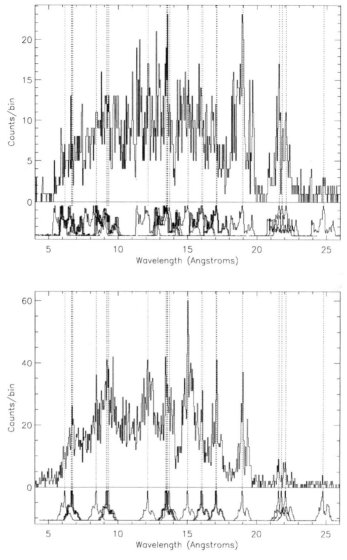

FIGURE 4. "Convolved" spectra (see text for explanation) of selected X-ray bright features in N132D Top: a knot in the uppermost part of the complex of filaments in the interior (it is the topmost filament that appears prominently above the lower vertical marker at ∼19 Å in Fig 3. Bottom: a knot in the bottom outer edge (marked with the lower vertical line in the zeroth order image and in several dispersed images) that appears enhanced at ∼17 Å in Fig 3. Spectra are uncorrected for instrumental efficiency. Under each spectrum the effective line shape function is plotted centered at the wavelengths of selected emission lines (marked with vertical dotted lines) which are likely to be strong.

onto the dispersion axis (so the zeroth order of our example would be projected to a point). Spatial extent in the dispersion direction within each narrow slice (e.g., if the '(' varies in thickness) is not removed, so the projection of the shifted blob onto the dispersion axis becomes the equivalent line shape function. The resulting spectrum is thus the true spectrum convolved with this equivalent line shape function (if there are strong gradients of abundance or temperature across the knot, then of course the line shape function could be different for different spectral lines). The next step, not yet undertaken, is to deconvolve this shape function from the spectrum.

We display in Fig. 4 such "convolved" MEG spectra extracted for two different X-ray bright features in N132D one from the interior (top panel) and the other from the rim (bottom panel) of the SNR image (see figure caption). Below each spectrum, the effective line shape function is plotted repeatedly centered at the wavelengths of selected emission lines that are likely to be strong in N132D. The differences in composition of the two blobs are evident. The interior knot shows strong O VIII Ly α emission at $\sim$19 Å and a prominent O VII triplet at $\sim$22 Å, whereas the rim feature shows relatively weak O emission and a strong Fe XVII 2p-3d lines at 15 and 17 Å. We anticipate being able to do a very detailed analysis of the physical conditions and composition of many such features.

CONCLUSIONS

The remnants Cas A, E0102-72 and N132D, with respective ages of 300 yr, 1000-2000 yr, and 3000 yr, provide an interesting sequence of SNRs at progressive stages in their evolution. While they have clear differences, such as the presence of O-burning products in the ejecta of Cas A but not of the other two (see the discussion in [2]), they also share significant similarities. All show oxygen rich optical filaments and, with the evidence described here for N132D, all show regions of enhanced oxygen X-ray emission. E0102-72 provides the first clear evidence for a progressive reverse shock that is also implied in the other two SNRs from non-equilibrium analyses.

The two younger members, Cas A and E0102-72, show asymmetric Doppler velocities in the X-ray. Markert *et al.* [18] discovered the Cas A Doppler velocities and asymmetry using the *Einstein* Focal Plane Crystal spectrometer, and these were subsequently confirmed and mapped with ASCA [12], with moderate angular resolution. The full velocity range for Cas A is roughly half that seen in E0102, which could well be the result of differences in projection. Thus, the only two young, oxygen-rich SNRs for which such measurements could have been made show the same kind of kinematic behavior. Our approximate model for E0102, like that of [18] for Cas A, suggests that the bright ejecta occupy a partially filled spherical shell in a ring-like geometry (as also suggested by Hughes [13] for E0102). While it is possible that the ejecta themselves were expelled in a ring, it seems more plausible to attribute the geometry to a ring-like distribution of circumstellar material which is decelerating and shocking a nearly spherical shell of ejecta, causing it to appear non-spherical. In that case, a significant fraction of the ejecta would still be relatively faint in X-rays. This would fit with the suggestion of Blair *et al.* [2], who argue that E0102 and N132D may be the remnants of very massive O stars that underwent substantial mass loss of their outer layers prior to exploding as Type Ib supernovae.

Further study study of these three bright objects is likely to give us new insights into the detailed structure and evolution of the remnants of supernovae in massive stars.

ACKNOWLEDGEMENTS

We thank the other members of our HETG/CXC group at MIT for their many contributions. This work was supported by NASA contract NAS8-38249 and SAO SV1-61010.

REFERENCES

1. Canizares, C.R., *et al. in preparation.*
2. Blair, W.P., Morse, J.A., Raymond, J.C., Kirshner, R.P., Hughes, J.P., Dopita, M.A., Sutherland, R.S., Long, K.S. & Winkler, P.F., *ApJ*, **537**, 667 (2000).
3. Flanagan, K.A., Canizares, C.R., Davis, D.S., Dewey, D. Houck, J.C., Schattenburg, M.L., *these proceedings.*
4. Davis, D.S., Flanagan, K.A., Houck, J.C., Allen, G.E., Schulz, N.S., Dewey, D., Schattenburg, & M.L., *these proceedings.*
5. Hayashi, I., Koyama, K., Masanobu, O., Miyata, E., Tsunemi, H., Hughes, J.P. & Petre, R., *PASJ*, **46**, L121 (1994).
6. Gaetz, T.J., Butt, Y.M., Edgar, R.J., Eriksen, K.A., Plucinsky, P.P., Schlegel, E.M. & Smith, R.K., *ApJ*, **534**, L47 (2000).
7. Hughes, J.P., Rakowski, C.E. & Decourchelle, A. *ApJ*, (2000) in press.
8. Eriksen, K.A. *et al., these proceedings.*
9. Rasmussen, A. *et al., these proceedings.*
10. Hughes, J.P. & Helfand, D.J., *ApJ*, **291**, 544 (1985).
11. Tuohy, I.R. & Dopita, M.A.,*ApJ*,**268**, L11 (1983)
12. Holt, S.S., Gotthelf, E.V., Tsunemi, H. & Negoro, H., *PASJ*, **46**, L151 (1994).
13. Hughes, J.P. 1988, in *Supernova Remnants and the Interstellar Medium*, ed. R.S. Roger & T.L. Landecker Cambridge: Cambridge Univ. Press, p. 125.
14. Favata, F., vink, J., Parmar, A.N., Kaastra, J.S., & Mineo, T., *A&A*, **324**, L45, (1997).
15. Hughes, J.P., Helfand, D.J.Hayashi, I., & Koyama, K., *ApJ*, **505**, 732 (1998).
16. Hwang, U., Hughes, J.P., Canizares, C.R. & Markert, T.H., *ApJ*, **414**, 219, (1993).
17. Behar, E. *these proceedings.*
18. Markert, T.H., Canizares, C.R., Clark, G.W., Winkler, P.F., *ApJ*, **268**, 134 (1983).

An Approximation to the Cooling Coefficient of Trace Elements

Robert A. Benjamin[†], Bradford A. Benson[*], and Donald P. Cox[†]

[†]*Department of Physics, University of Wisconsin, Madison, WI 53706, USA*
[*] *Department of Physics, Stanford University, Stanford, CA 94305, USA*

Abstract.
   An important ingredient in hydrodynamical models of SNR evolution is the incorporation of radiative cooling. Unfortunately, the cooling of hot ($T = 10^4 - 10^8$ K) plasmas depends sensitively upon the ionization state of the individual ions. This frequently departs from collisional ionization equilibrium. As a result, calculating an accurate cooling generally requires the solution of over a hundred differential equations. This computational requirement is generally prohibitive for three dimensional simulations. We discuss here progress on the development of a simple 2-dimensional solution for the cooling coefficient due to elements heavier than H and He. The cooling depends only on electron temperature and the mean charge of the ion population. In the test cases we consider, it provides a good approximation to the non-equilibrium cooling coefficent in cases for which the cooling due to these trace elements exceeds 10% of that due to H and He.

The goal of our project is to produce a reliable but compact approximation to the high temperature cooling coefficient of heavy elements in diffuse plasmas under non-equilibrium conditions. It is essential that such an approximation be developed for use with multi-dimensional hydrocodes where the additional burden of following the many differential equations of the ionization evolution severely restricts the available spatial resolution.

With only a few exceptions, hydrodynamic models that incorporate radiative cooling characterize the cooling with a single temperature: $L(T_e)$. These cooling curves assume either that the ionization state at a given temperature is characterized by collisional equilibrium, or that all gas follows a particular pre-calculated ionization history [1]. Since gas can have different types of ionization history, we were motivated to ask whether the cooling could be accurately characterized by two variables, mean charge and temperature:$L(T_e, \bar{z})$. The time evolution of $\bar{z}$ can then be obtained by integrating, $\frac{d\bar{z}}{dt} = n_e D(T_e, \bar{z})$, where electron density is n_e, the abundance weighted ionization state for elements heavier than helium is $\bar{z} = \Sigma A_Z f_{Z,z} / \Sigma f_{Z,z}$, and the ionization evolution function is $D(T_e, \bar{z})$. Therefore, we must generate two grids of coefficients, one for the cooling rate and one for the

CP565, *Young Supernova Remnants: Eleventh Astrophysics Conf.*, edited by S. S. Holt and U. Hwang
© 2001 American Institute of Physics 0-7354-0001-6/01/$18.00

ionization evolution. We have calculated a manifold of non-equilibrium isothermal evolutions to form the basis for our approximation. We have used the atomic data of [2], updated by [3].

A parcel of gas is initialized with equilibrium ionization appropriate to some intial temperature T_o. Its temperature is then suddenly changed to T_e and held fixed as the ionization scrambles through non-equilibrium states between T_o and T_e, finally approaching equilibrium at T_e. By choosing T_o to be low (e.g., $10^{4.0}$ K), and performing the calculation for a dense set of temperatures T_e between $10^{4.0}$ and $10^{8.0}$ K, the runs sample the full range of conditions possible in underionized gases. By repeating the whole set once again with very high T_o (e.g. $10^{8.0}$ K), conditions representing overionized gases are explored. Together these cases sample the full range of $\bar{z}$, above and below the equilibrium value of $\bar{z}_{eq}(T_e)$.

A particular moment in the evolution can be characterized by T_o, T_e, and the fluence ($n_e t$, or f). At such a moment the ionization vector is known, along with the mean charge $\bar{z}$, the cooling function $L(T_e, T_o, f)$. The indexing (T_o, f) can be

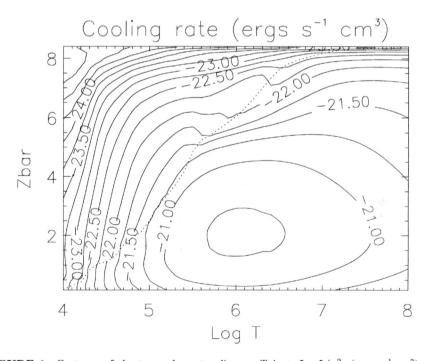

FIGURE 1. Contours of the trace elementcooling coefficient, $\mathrm{Log} L/n_H^2$ (ergs s^{-1} cm^3) as a function of gas temperature and mean charge $\bar{z}$ for a grid of isothermal evolutions. The dashed line shows the mean charge as a function of gas temperature, $z_{eq}(T_e)$. Gas was either started with ionization fractions corresponding to $T_o = 10^4$ K or $T_o = 10^8$ K. The gas then ionized up or recombined down to $\bar{z}_{eq}(T_e)$ at a range of fixed temperatures T_e. The resultant cooling coefficients were tabulated.

aliased with the mean charge, so that all quantities can be examined on the $\bar{z}, T_e$ plane. These runs allow us to develop our desired functions, $L(T_e, \bar{z})$ and $D(T_e, \bar{z})$. The former is shown in Figure 1, and summarizes the results of 80 calculations of isothermal ionization/recombination for 40 temperatures.

Are two variables truly enough to describe the cooling evolution? In examining these isothermal evolutions, we find that the distribution of ionization levels for the different elements in an ionizing plasma is well described by a single mean charge. For underionized plasmas, therefore we expect this method to be quite accurate. For a recombining plasma however, individual elements will develop a bimodal ionization distribution if it recombines through an ionization state with a closed electron shell. This is because of the very low rates of dielectron recombination for these ions. In this case, if the true jump in temperature for a plasma is less than our assumed initial temperature, inaccuracy in the cooling will result.

Nevertheless, a comparison of the ionization states of cases of isothermal evolution show significant difference from the equilibrium states and from each other. As a result, we expect our approximation to be more accurate than those currently in use. We now discuss a test of our approximation.

COOLING BLAST WAVE APPROXIMATION

We considered the evolution of a single parcel of gas overtaken and shock heated by a blast wave. We solve the energy equation

$$\frac{dE}{dt} = -(Ln_e n_H)V - p\frac{dV}{dt} \tag{1}$$

while assuming that the initial conditions are given by a Sedov blast wave in a medium with ambient density n_o and postshock temperature, T_2. The pressure is constrained to evolve according to $p(t) = p_2(t_2/t)^\alpha$, where we take $\alpha = 1.9$. The initial conditions set the value of p_2 and t_2. The resulting equation for the temperature evolution is

$$\frac{dT}{dt} = -\frac{2}{5}\left(\frac{Ln^2}{p} + \frac{\alpha T}{t}\right) \tag{2}$$

In order to test the widest possible range of conditions, we choose three values of T_2, with three different values of n_o for each. The values of density were spaced to alter the ratio of radiative to expansion cooling from radiation dominated (effectively steady state shock), to approximately equal expansion and radiative cooling, and then to very rapid expansion with the often freezing in of the ion population before low temperatures are reached. Table 1 shows our set of intial conditions, with $E_o = 10^{51}$ ergs for all nine cases.

During the course of a run, we calculate the evolving ionization structure and cooling coefficient, and record the particle density and temperature at each time

TABLE 1. Shock conditions for test cases

T_2	n_1	n_2	n_3
$3\times10^{5.0}$	$10^{-7.0}$	$2\times10^{-3.0}$	0.1
$3\times10^{6.0}$	$3\times10^{-4.0}$	10	$3\times10^{2.0}$
$3\times10^{7.0}$	1	$2\times10^{4.0}$	$10^{6.0}$

step. In parallel, we use our approximation to calculate the evolving $\bar{z}_{approx}$ and apply it to find the approximate cooling function $L_{approx}(\bar{z}_{approx}, T_e)$.

Whenever the gas is underionized the results are in excellent agreement with reality. In this regime, where the cooling function is quite large, it is also very rapidly evolving in time through the lower stages of ionization. In seperate investigations we found that even the use of equilibrium ionization vectors and $\bar{z}(T_{ioniz})$ produced good results.

Results for overionized gas are largely within the errors of the correctly calculated cooling function, and are largest in cases of rapid free expansion. In this regime, with the ionization states frozen, the approximation tends to freeze them at an incorrect value. The effect on the cooling rate is not important because the rate itself is negligible compared to that of H and He. The excess ionization has no dynamical significance: if the gas were subsequently reheated to a higher temperature, the rapid adjustment of ionization would erase the discrepancy.

By introduction of a table of $D(T_e, \bar{z})$ and $L(T_e, \bar{z})$, the ionization evolution of the trace elements can be successfully followed and the contribution to radiative cooling accurately approximated for temperatures in excess of 10^4 K. In most cases, the resulting approximate cooling function will be within the uncertainties of the true cooling function, limited by the accuracy of atomic transition rates. In rapid expansion cooling, where the approximation is less accurate at late times, radiative cooling is insignificant anyway, and dominated by the contributions of hydrogen and helium.

This work was supported by NASA Astrophysical Theory Grant NAG5-8417.

REFERENCES

1. Edgar, R.J. and Chevlier, R.A. 1986, ApJL, 310, 27
2. Raymond, J.C. and Smith, B.W. 1977, ApJS, 35, 419
3. Benjamin, R.A. and Shapiro, P.R. 2000, ApJS, submitted

Ionization Structure and the Reverse Shock in E0102-72

K.A. Flanagan, C.R. Canizares, D.S. Davis, D. Dewey, J.C. Houck, M.L. Schattenburg

Center for Space Research, Massachusetts Institute of Technology
Cambridge, MA 02139, USA

Abstract. The young oxygen-rich supernova remnant E0102-72 in the Small Magellanic Cloud has been observed with the High Energy Transmission Grating Spectrometer of Chandra. The high resolution X-ray spectrum reveals images of the remnant in the light of individual emission lines of oxygen, neon, magnesium and silicon. The peak emission region for hydrogen-like ions lies at larger radial distance from the SNR center than the corresponding helium-like ions, suggesting passage of the ejecta through the "reverse shock". We examine models which test this interpretation, and we discuss the implications.

BACKGROUND

1E0102.2-7219 (hereinafter referred to as E0102-72) is a young ($\sim$1000 years) oxygen-rich supernova remnant (SNR) in the Small Magellanic Cloud. E0102-72 has been observed at all wavelengths: in the optical [1] [2], UV [3] [4] , radio [5], and in the X-ray [6] [7] [8] [9] [10] [11] [12] [13]. Gaetz *et al.* have looked at direct Chandra images of E0102-72 and noted the radial variation with energy bands centered on OVII and OVIII, suggesting an ionizing shock propagating inward. This paper addresses this issue through analysis of the high resolution dispersed spectrum, and reinforces this interpretation.

THE HIGH RESOLUTION X-RAY SPECTRUM

Shown in Figure 1 is the dispersed high resolution X-ray spectrum of E0102-72. The Chandra observation was made using the High Energy Transmission Gratings (HETG) in conjunction with the Advanced CCD Imaging Spectrometer (ACIS-S). The HETG separates the X-rays into their distinct emission lines, forming separate images of the remnant with the X-ray light of each line. The dispersed spectrum contains lines of highly ionized oxygen, neon, magnesium and silicon (not shown

CP565, *Young Supernova Remnants: Eleventh Astrophysics Conf.,* edited by S. S. Holt and U. Hwang
© 2001 American Institute of Physics 0-7354-0001-6/01/$18.00

in the figure). Detailed study of the line images shows they are distorted along the dispersion axis due to Doppler shifts associated with high velocity material. Details and an overview of results are given in Canizares *et al.* elsewhere in these proceedings. Here we examine the apparent differences in ring radii seen in Figure 1. As discussed in detail below, these are likely caused by the changing ionization state, suggesting the passage of the supernova ejecta through the reverse shock.

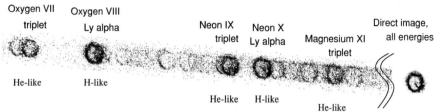

FIGURE 1. A portion of the high resolution spectrum formed by the medium energy gratings. At right in the figure is the zeroth order, which combines all energies in an undispersed image.

THE IONIZATION/SHOCK STRUCTURE OF E0102-72

Figure 2 compares helium-like and hydrogen-like lines of oxygen. The dispersed OVII Resonance line from the helium-like triplet (left) is displayed next to the hydrogen-like OVIII Ly α line (right) on the same spatial scale. The ring diameter of the hydrogen-like line is obviously larger than that of the helium-like line. The ring diameters of these lines were measured by tracing the distribution in the direction orthogonal to the dispersion axis, as shown in Figure 3. (By taking the distribution in the cross-dispersion direction, the result is largely independent of Doppler shift.) The He-like OVII distribution is nestled cleanly within the OVIII, and is measurably narrower in Figure 3. The ring diameters for all the bright lines were measured

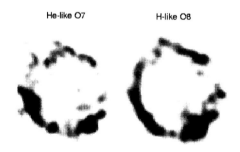

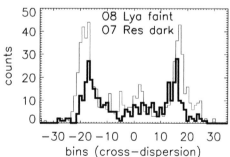

FIGURE 2. The OVII Resonance line is emitted from a region of smaller radius than that of OVIII Lyman α.

FIGURE 3. The OVII Resonance distribution is narrower than that of OVIII Lyman α.

similarly, and the results are shown on Figure 4. *All of the hydrogen-like lines lie outside the corresponding helium-like lines.* After passage of a shock, an ionizing

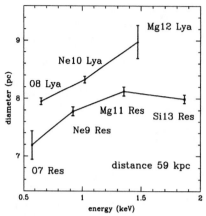

FIGURE 4. Ring diameter vs energy for the bright lines. Note that all the hydrogen-like lines (connected by the top curve) lie outside their helium-like counterparts.

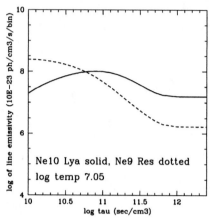

FIGURE 5. Helium-like NeIX Resonance line reaches its peak emissivity at $\log \tau = 10.0$, earlier than hydrogen-like NeX Lyman α, which peaks at about $\log \tau = 11.0$ sec/cm^3

plasma at a fixed electron temperature T_e will achieve a He-like state before it reaches a H-like state. This is shown in Figure 5, where the appropriate timescale is the ionization parameter $\tau = n_e t$, where t is the time since passage of the shock and n_e is the electron density [14]. Since the H-like state lies outward of the He-like state in E0102-72, this suggests the action of a shock moving *inward* relative to the ejecta - the 'reverse shock' which is the standard model for the mechanism which heats SNR ejecta to X-ray temperatures.

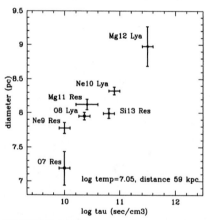

FIGURE 6. Measured diameter vs τ at peak emissivity. O7 is an upper limit for τ.

We applied a simple model in which a homogenous mix of ejecta is shocked to a fixed T_e of about 1 keV or log $T_e = 10^{7.05}$ (as suggested by our global NEI analysis [11]). We estimated τ by finding the peak emissivity (as in Figure 5). A plot of measured diameters as a function of τ is shown in Figure 6. Although any workable model must consider many parameters, the monotonic behavior suggests that these differences in ring diameter are attributable to the ionization structure resulting from the reverse shock.

228

WORK IN PROGRESS AND FUTURE WORK

Our current model calls for further work. For example, by taking the ratio of the histograms for O8 Lyman α and O7 Resonance from Figure 3, we can examine the allowed region of paramater space defined by T_e and τ *as a function of radial position*. (This approach as it pertains to the global spectrum is described by Davis *et al* [11] elsewhere in these proceedings.) By considering a likely density gradient in the ejecta and obtaining an appropriate estimate for electron density, we might obtain a corresponding estimate for the reverse shock velocity. We plan to apply this shock velocity information to see how it fits with the spatial extent of the bright ejecta and the presumed age of the remnant.

Acknowledgements We thank Glenn Allen, Norbert Schulz and Sara-Anne Taylor for helpful discussions. We are grateful to the CXC group at MIT for their advice and assistance. This work was prepared under NASA contract NAS8-38249 and SAO SV1-61010.

REFERENCES

1. Dopita, M.A., Tuohy, I.R. & Mathewson, D.S., 1981, *ApJ*, **248**, L105 (1981)
2. Tuohy, I.R. & Dopita, M.A.,*ApJ*,**268**, L11 (1983)
3. Blair, W.P., Raymond, J.C., Danziger, J. & Matteucci, F., *ApJ*, **338**, 812 (1989)
4. Blair, W.P., Morse, J.A., Raymond, J.C., Kirshner, R.P., Hughes, J.P., Dopita, M.A., Sutherland, R.S., Long, K.S. & Winkler, P.F., *ApJ*, **537**, 667 (2000)
5. Amy, S.W. & Ball, L., *ApJ*, **411**, 761 (1993)
6. Seward, F.D. & Mitchel, M., *ApJ*, **243**, 736 (1981)
7. Hayashi, I., Koyama, K., Masanobu, O., Miyata, E., Tsunemi, H., Hughes, J.P. & Petre, R., *PASJ*, **46**, L121 (1994)
8. Gaetz, T.J., Butt, Y.M., Edgar, R.J., Eriksen, K.A., Plucinsky, P.P., Schlegel, E.M. & Smith, R.K., *ApJ*, **534**, L47 (2000)
9. Hughes, J.P. 1988, in *Supernova Remnants and the Interstellar Medium*, ed. R.S. Roger & T.L. Landecker Cambridge: Cambridge Univ. Press, p. 125
10. Hughes, J.P., Rakowski, C.E. & Decourchelle, A. *ApJ*, (2000) in press
11. Davis, D.S., Flanagan, K.A., Houck, J.C., Canizares, C.R, Allen, G.E., Schulz, N.S., Dewey, D. & Schattenburg, M.L., *these proceedings*.
12. Canizares, C.R, Flanagan, K.A., Davis, D.S., Dewey, D., & Houck, J.C., *these proceedings*.
13. Rasmussen, A. *et al.*, *these proceedings*.
14. Hughes, J.P. & Helfand, D.J., *ApJ*, **291**, 544 (1985)

Spectral Line Imaging Observations of 1E0102.2-7219

D. S. Davis, K. A. Flanagan, J. C. Houck, C. R. Canizares, G. E. Allen, N. S. Schulz, D. Dewey and M. L. Schattenburg

Massachusetts Institute of Technology
Center for Space Research
77 Massachusetts Ave
Cambridge, MA 02139, USA

Abstract. E0102-72 is the second brightest X-ray source in the Small Magellanic Cloud and the brightest supernova remnant in the SMC. We observed this SNR for ~140 ksec with the High Energy Transmission Gratings (HETG) aboard the *Chandra* X-ray Observatory. The small angular size and high surface brightness make this an excellent target for HETG and we resolve the remnant into individual lines. We observe fluxes from several lines which include O VIII Lyα, Lyβ, and O VII along with several lines from Ne X, Ne IX and Mg XII. These line ratios provide powerful constraints on the electron temperature and the ionization age of the remnant.

1E0102.2-7219 (hereinafter E0102-72) is the brightest X-ray SNR in the Small Magellanic Cloud (SMC) and was discovered with the Einstein Observatory's IPC [1]. Higher resolution observations showed that this SNR exhibits a shell-like structure [2]. Optical emission from E0102-72 was detected revealing that this is an oxygen-rich SNR [3] and thus the result of the explosion of a massive progenitor.

A moderate resolution CCD spectrum of this remnant was obtained with ASCA which observed this object for ~35 ksec. The X-ray spectrum shows clear lines of oxygen, neon, and magnesium [4]. Hayashi et al. use the NEI modeling code of Hughes & Helfand [5] and find that the data require at least two NEI plasmas. The assumption that each observed element has an independent nt and temperature lead them to the conclusion that the elements are not well mixed. Hayashi et al. find the NEI parameters for oxygen are $\log(nt) = 10.31$ s cm^{-3} and $\log(T) = 7.03$ °K (0.93 keV) and those for Ne are $\log(nt) = 11.37$ s cm^{-3} and $\log(T) = 6.78$ °K (0.52 keV). The fact that the neon has a lower temperature and higher nt relative to oxygen leads them to the conclusion that the oxygen emission is related to the forward shock and the neon to the reverse shock. Our analysis below assumes that the emission is from a single τ plasma model.

CP565, *Young Supernova Remnants: Eleventh Astrophysics Conf.*, edited by S. S. Holt and U. Hwang
© 2001 American Institute of Physics 0-7354-0001-6/01/$18.00

CHANDRA HETG SPECTRA OF E0102-72

The Chandra X-ray Observatory (CXO) observed E0102-72 for a total of ~140 ksec with the High Energy Transmission Grating Spectrometer (HETGS) in the optical path of the telescope. The HETGS contains two types of gratings, the High and Medium Energy Gratings (HEG & MEG). These gratings provide an $E/\Delta E$ of 100-1000 for point sources over the energy range of 0.4 - 10 keV. The HEG and MEG are rotated with respect to each other so that the dispersed spectra form an "X" pattern on the detector. The problem of overlapping orders is resolved using the moderate energy resolution of the ACIS detector. Additional details of the analysis and results can be found in these proceedings [6] [7]. Here we present results from the MEG spectrum of E0102-72.

We extracted the line fluxes using an annular aperture that encloses the observed ring. The background was taken from an annular region outside the dispersed image of interest but with the same center as the region for the line flux.

X-RAY MORPHOLOGY

Figure 1 shows the dispersed image of the SNR in the O VIII Lyα line (18.97 Å). While the overall shape of the remnant can be described as a ring it is clear that this is not true on small scales. The ring has gaps on the eastern side and a bright knot can be seen in the southwest quadrant. The image of the SNR in the Ne X Lyα line (12.13 Å) is shown in Figure 2. The Ne X image shows fewer gaps in the rim than the O VIII image, but like the oxygen image shows deviations from a ring structure. We find no evidence that the different elements are stratified. However, we do find ionization structure in the remnant [7]. Despite these complexities a global analysis is useful as it allows us to determine integrated line fluxes for comparison with *ASCA* and *XMM* results, and to compare our models with earlier results.

LINE DIAGNOSTICS

Line ratios were calculated for pairs of hydrogen-like and helium-like lines and these are compared to the NEI model in XSPEC 11. Here we present results (Table 1) for the flux ratio of ions from the same element so that the results are independent of the abundances. We use the column density of 8.0×10^{20} atoms cm^{-2} [8] to correct the flux for interstellar absorption.

We derive tight constrains of the physical state of the plasma from measurements of O VIII Lyβ/O VIII Lyα and from the ratio of O VII Forbidden + IC to O VII Resonance. Using Figure 3 we find that the temperature must be above ~0.68 keV and $\log(\tau)$ is between 9.85 and 10.05 s cm^{-3}. The ratios of Ne X Lyα to Ne IX Res along with Ne X Lyβ/Ne X Lyα also provide tight constraints on the physical state of the plasma. They do not overlap the allowed region for oxygen. The allowed region of parameter space is shown in Figure 4; the width of the regions

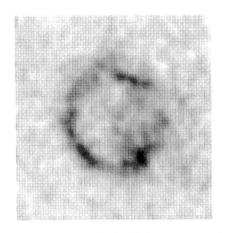

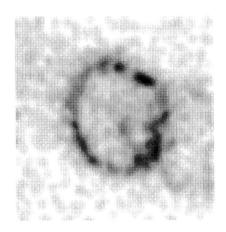

FIGURE 1. The OVIII Lyα image of E0102-72 from the MEG -1 order. Note the bright knot in the southwest.

FIGURE 2. The Ne X Lyα image of E0102-72. This is the data from the MEG -1 order.

is determined by the three σ limits on the line ratios. The lower limit on the temperature of the plasma is 0.54 keV from the oxygen ratio and 0.97 keV from the neon ratio and is consistent with the *ASCA* analysis of this remnant [4]. The lower limits for the temperature are robust even for a multi-τ plasma.

Line Ratio	Flux Ratio	3σ range
O VIII Ly β/O VIII Lyα	0.138	0.122 – 0.154
O VII (For + IC)/ OVII Res	0.593	0.422 – 0.764
O VII (4 -> 1)/O VIII Lyα	0.338	0.089 – 0.587
Ne X Lyβ/Ne X Lyα	0.141	0.124 – 0.158
Ne X Lyα/Ne IX Res	0.723	0.603 – 0.843

Table 1: The flux ratios for oxygen and neon.

CONCLUSIONS

We have measured the strong emission lines of O and Ne in the SNR E0102-72. The detected flux ratios from these lines are consistent with emission from a young SNR that has not yet reached ionization equilibrium. Our global analysis yields an electron temperature that is above ~ 0.54 keV and an ionization age between $9.85 < \log(\tau) < 10.6$, consistent with this being a young remnant.

We find that the line emission from each element, measured globally, can be described by an NEI plasma using a simple plane parallel shock model. Our analysis is consistent with the oxygen and neon emission both being from the reverse shock and we find no compelling evidence that oxygen and neon are not well mixed. Future work will include analysis of other lines present in this spectrum. We also

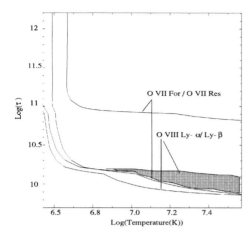

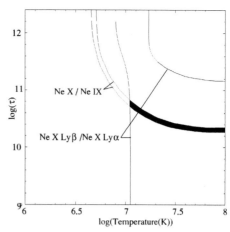

FIGURE 3. The allowed parameter space in τ Temperature (K) from the oxygen lines is shown as the shaded region.

FIGURE 4. The allowed parameter space in τ Temperature (K) from the neon lines. Note that the ranges for the x and y axis are not the same as those in Fig 3.

will use the imaging capability of the HETGS to explore how the plasma diagnostics vary around E0102-72 and map the temperature and ionization age of this remnant.

We would like to thank the HETG/CXC group at MIT for their assistance and useful discussions which contributed to this work. This research is funded under NASA contract NAS8-38249 and SAO SV1-61010

REFERENCES

1. Seward, F. D. & Mitchell, M. 1981, ApJ 243, 736
2. Hughes, J. P. 1988 IAU Colloq. 101, Supernova Remnants and the Interstellar Medium, ed R.S. Rodger, T.L. Landecker (Cambridge University Press), p 125
3. Dopita, M. A., Tuohy, I. R., Mathewson, D. S. 1981, ApJL, 248, L105
4. Hayashi, I., Koyama, K., Ozaka, M., Miyata, E., Tsunemi, H., Hughes, J. P. & Petre, R. 1994, PASJ, 46, L121
5. Hughes, J. P. & Helfand, D. J. 1985, ApJ, 291, 544
6. Canizares, C. R., Flanagan, K. A., Davis, D. S., Dewey, D. & Houck, J. C. 2000 (these proceedings)
7. Flanagan, K. A., Canizares, C. R., Davis, D. S., Dewey, D., Houck, J. C. and Schattenburg, M. L. 2000 (these proceedings)
8. Blair, W. P., Raymond, J. C., Danzinger, J., Matteucci, F. 1989, ApJ, 338, 812

Young SNR III

Radio Emission from SNe and Young SNRs

Kurt W. Weiler[1], Nino Panagia[2], Marcos J. Montes[1], Schuyler D. Van Dyk[3], Richard A. Sramek[4], and Christina K. Lacey[5]

[1] NRL, Code 7213, Washington, DC 20375-5320, USA
[2] STScI, 3700 San Martin Drive, Baltimore, MD 21218, USA & Ap. Div., Space Sci. Dept. of ESA
[3] IPAC/Caltech, Mail Code 100-22, Pasadena, CA 91125, USA
[4] P.O. Box 0, NRAO, Socorro, NM 87801, USA
[5] Univ. of South Carolina, Dept. of Phys. and Astron., Columbia, SC 29208, USA

Abstract. Study of radio supernovae (RSNe), the earliest stages of supernova remnant (SNR) formation, over the past 20 years includes two dozen detected objects and more than 100 upper limits. From this work we are able to identify classes of radio properties, demonstrate conformance to and deviations from existing models, estimate the density and structure of the circumstellar material and, by inference, the evolution of the presupernova stellar wind, and reveal the last stages of stellar evolution before explosion. It is also possible to detect ionized hydrogen along the line of sight, to demonstrate binary properties of the stellar system, and to show clumpiness of the circumstellar material. More speculatively, it may be possible to provide distance estimates to RSNe.

INTRODUCTION

A series of papers published over the past 20 years on radio supernovae (RSNe), the earliest stages of supernova remnant (SNR) formation, has established the radio detection and, in a number of cases, radio evolution for approximately two dozen objects: 3 Type Ib supernovae (SNe), 5 Type Ic SNe, and the rest Type II SNe. A much larger number of more than 100 additional SNe have low radio upper limits.

In this extensive study of the radio emission from SNe and young SNRs, several effects have been noted: 1) Type Ia SNe are not radio emitters to the detection limit of the Very Large Array[1] (VLA); 2) Type Ib and Ic SNe are radio luminous

---

[1] The VLA telescope of the National Radio Astronomy Observatory is operated by Associated Universities, Inc. under a cooperative agreement with the National Science Foundation.

CP565, *Young Supernova Remnants: Eleventh Astrophysics Conf.*, edited by S. S. Holt and U. Hwang
© 2001 American Institute of Physics 0-7354-0001-6/01/$18.00

with steep spectral indices (generally $\alpha < -1$; $S \propto \nu^{+\alpha}$) and a fast turn-on/turn-off, usually peaking at 6 cm near or before optical maximum; and 3) Type II SNe show a range of radio luminosities with flatter spectral indices (generally $\alpha > -1$) and a relatively slow turn-on/turn-off, usually peaking at 6 cm significantly after optical maximum. Type Ib and Ic may be fairly homogeneous in some of their radio properties while Type II, as in the optical, are quite diverse.

There are a large number of physical properties of SNe which we can determine from radio observations. VLBI imaging shows the symmetry of the explosion and the local circumstellar medium (CSM), estimates the speed and deceleration of the SN blastwave propagating outward from the explosion and, with assumptions of symmetry and optical line/radio-sphere velocities, allows independent distance estimates to be made (see, e.g., [1,2]). Measurements of the multi-frequency radio light curves and their evolution with time show the density and structure of the CSM, evidence for possible binary companions, clumpiness or filamentation in the presupernova wind, mass-loss rates and changes therein for the presupernova stellar system and, through stellar evolution models, estimates of the ZAMS presupernova stellar mass and the stages through which the star passed on its way to explosion. It has also been proposed by Weiler, *et al.* [3] that the time from explosion to 6 cm radio maximum may be an indicator of the radio luminosity and thus an independent distance indicator for Type II SNe and that Type Ib and Ic SNe may be approximate radio standard candles at 6 cm radio peak flux density.

MODELS

All known RSNe appear to share common properties of: 1) nonthermal synchrotron emission with high brightness temperature; 2) a decrease in absorption with time, resulting in a smooth, rapid turn-on first at shorter wavelengths and later at longer wavelengths; 3) a power-law decline of the flux density with time at each wavelength after maximum flux density (optical depth ~ 1) is reached at that wavelength; and 4) a final, asymptotic approach of spectral index α ($S \propto \nu^{+\alpha}$) to an optically thin, nonthermal, constant negative value [4,5]. Chevalier [6,7] has proposed that the relativistic electrons and enhanced magnetic field necessary for synchrotron emission arise from the SN blastwave interacting with a relatively high density CSM which has been ionized and heated by the initial UV/X-ray flash. This CSM is presumed to have been established by a constant mass-loss ($\dot{M}$) rate, constant velocity (w) wind (*i.e.*, $\rho \propto r^{-2}$) from a massive stellar progenitor or companion. This ionized CSM is the source of some or all of the initial absorption although Chevalier [8] has proposed that synchrotron self-absorption (SSA) may play a role in some objects.

A rapid rise in the observed radio flux density results from a decrease in these absorption processes as the radio emitting region expands and the absorption processes consequently decrease. Weiler, *et al.* [5] have suggested that this CSM can be "clumpy" or "filamentary," leading to a slower radio turn-on, and Montes, *et al.* [9] have proposed the possible presence of a distant ionized medium along the line-of-sight which is time independent and can cause a spectral turn-over at low radio frequencies. In addition to clumps or filaments, the CSM may be structured with significant density irregularities such as rings, disks, shells, or gradients.

Parameterized Radio Light Curves

Following [4,5,9], we adopt a parameterized model (N.B.: The notation is extended and rationalized here from previous publications. However, the "old" notation of τ, τ', and τ'', which has been used previously, is noted, where appropriate, for continuity.):

$$S(\text{mJy}) = K_1 \left(\frac{\nu}{5 \text{ GHz}}\right)^\alpha \left(\frac{t - t_0}{1 \text{ day}}\right)^\beta e^{-\tau_{\text{external}}} \left(\frac{1 - e^{-\tau_{\text{CSM}_{\text{clumps}}}}}{\tau_{\text{CSM}_{\text{clumps}}}}\right) \left(\frac{1 - e^{-\tau_{\text{internal}}}}{\tau_{\text{internal}}}\right)$$

$$\text{(1)}$$

$$\tau_{\text{external}} = \tau_{\text{CSM}_{\text{uniform}}} + \tau_{\text{distant}} = \tau + \tau'' \tag{2}$$

$$\tau_{\text{CSM}_{\text{uniform}}} = \tau = K_2 \left(\frac{\nu}{5 \text{ GHz}}\right)^{-2.1} \left(\frac{t - t_0}{1 \text{ day}}\right)^\delta \tag{3}$$

$$\tau_{\text{distant}} = \tau'' = K_4 \left(\frac{\nu}{5 \text{ GHz}}\right)^{-2.1} \tag{4}$$

$$\tau_{\text{CSM}_{\text{clumps}}} = \tau' = K_3 \left(\frac{\nu}{5 \text{ GHz}}\right)^{-2.1} \left(\frac{t - t_0}{1 \text{ day}}\right)^{\delta'} \tag{5}$$

with K_1, K_2, K_3, and K_4 corresponding, formally, to the flux density (K_1), uniform (K_2, K_4) and clumpy or filamentary (K_3) absorption, at 5 GHz one day after the explosion date t_0. The terms $\tau_{\text{CSM}_{\text{uniform}}}$ and $\tau_{\text{CSM}_{\text{clumps}}}$ describe the attenuation of local, uniform CSM and clumpy CSM that are near enough to the SN progenitor that they are altered by the rapidly expanding SN blastwave. $\tau_{\text{CSM}_{\text{uniform}}}$ is produced by an ionized medium that uniformly covers the emitting source ("uniform external absorption"), and the $(1 - e^{-\tau_{\text{CSM}_{\text{clumps}}}})\tau_{\text{CSM}_{\text{clumps}}}^{-1}$ term describes the attenuation produced by an inhomogeneous medium with optical depths distributed, with equal probability, between 0 and $\tau_{\text{CSM}_{\text{clumps}}}$ ("clumpy absorption"; see Natta & Panagia [10] for a more detailed discussion of attenuation in inhomogeneous media). The τ_{distant} term describes the attenuation produced by a homogeneous medium which uniformly covers the source but is so far from the SN progenitor that it is not affected by the expanding SN blastwave and is constant in time. All absorbing media are assumed to be purely thermal, singly ionized gas which absorbs via f-f transitions with frequency dependence $\nu^{-2.1}$. The parameters δ and δ' describe the time dependence of the optical depths for the local uniform and clumpy or filamentary media, respectively.

Since it is physically realistic and may be needed in some RSNe where radio observations have been obtained at early times and high frequencies, we have also included in Eq. 1 the possibility for an internal absorption term. This internal absorption (τ_{internal}) term may consist of two parts – SSA ($\tau_{\text{internal}_{\text{SSA}}}$), and mixed, thermal f-f absorption/non-thermal emission ($\tau_{\text{internal}_{\text{ff}}}$):

$$\tau_{\text{internal}} = \tau_{\text{internal}_{\text{ssa}}} + \tau_{\text{internal}_{\text{ff}}} \tag{6}$$

$$\tau_{\text{internal}_{\text{ssa}}} = K_5 \left(\frac{\nu}{5 \text{ GHz}}\right)^{\alpha-2.5} \left(\frac{t - t_0}{1 \text{ day}}\right)^{\delta''} \tag{7}$$

$$\tau_{\text{internal}_{\text{ff}}} = K_6 \left(\frac{\nu}{5 \text{ GHz}}\right)^{-2.1} \left(\frac{t - t_0}{1 \text{ day}}\right)^{\delta'''} \tag{8}$$

with K_5 and K_6 corresponding, formally, to the internal, non-thermal ($\nu^{\alpha-2.5}$) SSA (K_5) and the internal, mixed with nonthermal emission, thermal ($\nu^{-2.1}$) free-free absorption (K_6), respectively, at 5 GHz one day after the explosion date t_0. The parameters δ'' and δ''' describe the time dependence of the optical depths for the SSA and f-f internal absorption components, respectively.

A cartoon of the expected structure of an SN and its surrounding media is presented in Fig. 1 (see also, Lozinskaya [11]). The radio emission is expected to arise near the blastwave [12].

RESULTS

The success of the basic parameterization and model description can be seen in the relatively good correspondence between the model fits and the data for all three subtypes of RSNe, *e.g.*, Type Ib SN1983N (Fig. 2a, [13]), Type Ic SN1990B (Fig. 2b, [14]), Type II SN1979C (Fig. 3a, [15]) and Type II SN1980K (Fig. 3b, [16]). (Note that after day ~ 4000, the evolution of the radio emission from both SN1979C and SN1980K deviates from the expected model evolution and that SN1979C shows a sinusoidal modulation in its flux density [17–19].)

Mass-Loss Rate

From the Chevalier [6,7] model, the turn-on of the radio emission for RSNe provides a measure of the presupernova mass-loss rate to wind velocity ratio ($\dot{M}/w$).

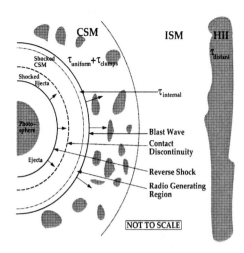

FIGURE 1. Cartoon, not to scale, of the SN and its shocks along with the stellar wind established CSM and more distant ionized material. The radio emission is thought to arise near the blastwave. The expected locations of the several absorbing terms in Eqs. 1 – 8 are illustrated.

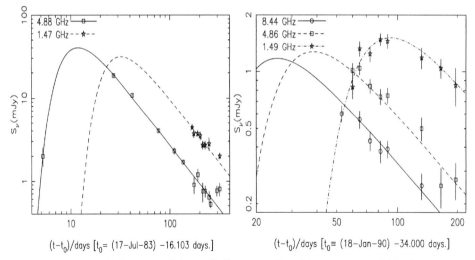

FIGURE 2. Type Ib SN1983N (left, 2a, [13]) at 6 cm (4.9 GHz; *squares, solid line*) and 20 cm (1.5 GHz; *stars, dashed line*) and Type Ic SN1990B (right, 2b, [14]) at 3.4 cm (8.4 GHz; *circles, solid line*), 6 cm (4.9 GHz; *squares, dashed line*), and 20 cm (1.5 GHz; *stars, dash-dot line*).

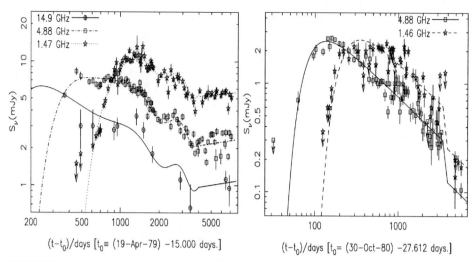

FIGURE 3. Type II SN1979C (left, 3a, [15]) at 2 cm (14.9 GHz; *crossed circles, solid line*), 6 cm (4.9 GHz; *squares, dash-dot line*), and 20 cm (1.5 GHz; *stars, dotted line*) (For discussion of the increase in the radio flux density after day ~4000 and the sinusoidal modulation of the radio emission, see [15,17–19]) and Type II SN1980K (right, 3b, [16]) at 6 cm (4.9 GHz; *squares, solid line*), and 20 cm (1.5 GHz; *stars, dashed line*) (For discussion of the sharp drop in flux density after day ~ 4000, see [16].).

For the case in which the external absorption is entirely due to a uniform medium, using the formulation of Weiler, *et al.* [4] Eq. 16, we can write

$$\frac{\dot{M}(\mathrm{M_\odot \ yr^{-1}})}{(w/10 \mathrm{\ km \ s^{-1}})} = 3 \times 10^{-6} \ \tau_{CSM_{\mathrm{uniform}}}^{0.5} \ m^{-1.5} \left(\frac{v_i}{10^4 \mathrm{\ km \ s^{-1}}}\right)^{1.5} \times$$

$$\left(\frac{t_i}{45 \mathrm{\ days}}\right)^{1.5} \left(\frac{t}{t_i}\right)^{1.5m} \left(\frac{T}{10^4 \mathrm{\ K}}\right)^{0.68} \tag{9}$$

Since the appearance of optical lines for measuring SN ejecta velocities is often delayed, we arbitrarily take $t_i = 45$ days. Because our observations have shown that, generally, $0.8 \le m \le 1.0$ ($m = -\delta/3$, [4]) and, from Eq. 9, $\dot{M} \propto t_i^{1.5(1-m)}$, the dependence of the calculated mass-loss rate on the date t_i of the initial ejecta velocity measurement is weak, $\dot{M} \propto t_i^{<0.3}$. Thus, we can take the best optical or VLBI velocity measurements available without worrying about the deviation of their exact measurement epoch from the assumed 45 days after explosion. For convenience, and because many SN measurements indicate velocities of ~ 10,000 km s^{-1}, we take $v_i = 10,000$ km s^{-1}. We also adopt values of $T = 20,000$ K, $w = 10$ km s^{-1} (which is appropriate for a RSG wind) and, generally from our best fits to SN radio data, $t = (t_{6\mathrm{cm \ peak}} - t_0)$ days, $\tau_{CSM_{\mathrm{uniform}}}$ from Eq. 3, and m from $m = -\delta/3$. With these assumptions, mass-loss rates, $\dot{M}$, can be calculated.

From Eq. 9 and its modification for inclusion of a clumpy CSM (see the following Section) the mass-loss rates from SN progenitors are generally estimated to be $\sim 10^{-6}$ $M_\odot$ yr^{-1} for Type Ib and Ic SNe and $\sim 10^{-4}$ $M_\odot$ yr^{-1} for Type II SNe. For the specific case of SN1993J, where detailed radio observations are available starting only a few days after explosion, Van Dyk, et al. [20] find evidence for a changing mass-loss rate for the presupernova star which was as high as $\sim 10^{-4}$ $M_\odot$ yr^{-1} approximately 1000 years before explosion and decreased to $\sim 10^{-5}$ $M_\odot$ yr^{-1} just before explosion.

Clumpiness of the Presupernova Wind

In their study of the radio emission from SN1986J, Weiler, Panagia, & Sramek [5] found that the simple Chevalier [6,7] model could not describe the relatively slow turn-on. They therefore added terms described mathematically by $\tau_{CSM_{clumps}}$ in Eqs. 1 and 5. This extension greatly improved the quality of the fit and was interpreted by [5] to represent the possible presence of filamentation or clumpiness in the CSM. Such a clumpiness in the wind material was again required for modeling the radio data from SN1988Z [21] and SN1993J [20]. Since that time, evidence for filamentation in the envelopes of SNe has also been found from optical and UV observations (see, e.g., [22,23]).

Binary Systems

In the process of analyzing a full decade of radio measurements from SN1979C, [17,18] found evidence for a significant, quasi-periodic, variation in the amplitude of the radio emission at all wavelengths of ~ 15 % with a period of 1575 days or ~ 4.3 years (see Fig. 3a at age < 4000 days). They interpreted the variation as due to a minor (~ 8 %) density modulation, with a period of ~ 4000 years, on the larger, relatively constant presupernova stellar mass-loss rate. Since such a long period is inconsistent with most models for stellar pulsations, they concluded that the modulation may be produced by interaction of a binary companion in an eccentric orbit with the stellar wind from the presupernova star.

This concept was strengthened by more detailed calculations for a binary model from Schwarz & Pringle [19]. Since that time, the presence of binary companions has been suggested for the progenitors of SN1993J [24] and SN1994I [25], indicating that binaries may be common in presupernova systems.

HII Along the Line-of-Sight

A reanalysis of the radio data for SN1978K from Ryder, et al. [26] clearly shows flux density evolution characteristic of normal Type II SNe. Additionally, the data indicate the need for a time-independent, free-free absorption component. Montes, *et al.* [9] interpret this constant absorption term as indicative of the presence of HII along the line-of-sight to SN1978K, perhaps as part of an HII region or a distant circumstellar shell associated with the SN progenitor. SN1978K had already been noted by [26] for its lack of optical emission lines broader than a few thousand km s^{-1} since its discovery in 1990, indeed suggesting the presence of slowly moving circumstellar material.

To determine the nature of this absorbing region, a high-dispersion spectrum of SN1978K at the wavelength range 6530 – 6610 Å was obtained by Chu, *et al.* [27]. The spectrum shows not only the moderately broad Hα emission of the SN ejecta, but also narrow nebular Hα and [N II] emission. The high [N II] 6583/Hα ratio of 0.8 – 1.3 suggests that this radio absorbing region is a stellar ejecta nebula.

Rapid Presupernova Stellar Evolution

SN radio emission that preserves its spectral index while deviating from the standard model is taken to be evidence for a change of the average circumstellar density behavior from the canonical r^{-2} law expected for a presupernova wind with a constant mass-loss rate, $\dot{M}$, and a constant wind velocity, w. Since the radio luminosity of a SN is proportional to $(\dot{M}/w)^{(\gamma-7+12m)/4}$ [6] or, equivalently, to the same power of the circumstellar density (since $\rho_{CSM} \propto M/w$), a measure of the deviation from the standard model provides an indication of deviation of the circumstellar density from the r^{-2} law. Monitoring the radio light curves of RSNe also provides a rough estimate of the time scale of deviations in the presupernova stellar wind density. Since the SN blastwave travels through the CSM roughly 1,000 times faster than the stellar wind velocity which established the CSM ($v_{blastwave} \sim$ 10,000 km s^{-1} *vs.* $w_{wind} \sim$ 10 km s^{-1}), one year of radio light curve monitoring samples roughly 1000 years of stellar wind evolution.

Sphericity of an SN Explosion

It has often been suggested that SN explosions are non-spherical, and there is evidence in a number of stellar systems for jets, lobes, and other directed mass-loss phenomena. Also, the presence of polarization in the optical light from SNe (including SN1993J) has been interpreted for non-sphericity (see, *e.g.,* [28]) and probably the most obvious evidence for non-spherical structure in an SN system is the very prominent inner ring around SN1987A. However, our most direct evidence for the structure of at least the blastwave from an SN explosion and the CSM with which it is interacting is from VLBI measurements on SN1993J. A series of images taken by Marcaide, *et al.* [1] over a period of two years from 1994 September through 1996 October show only a very regular ring shape indicative of a relatively spherical blastwave expanding into a relatively uniform CSM. The cause of such apparently conflicting results is still to be resolved.

Deceleration of Blastwave Expansion

Radio studies also offer the only possibility for measuring the deceleration of the blastwave from the SN explosion. So far, this has been directly possible for two objects, SN1987A and SN1993J, although in some cases the deceleration can be estimated from model fitting to the radio light curves. Manchester, *et al.* [29] have shown that the blastwave from the explosion of SN1987A traveled through the tenuous medium of the bubble created by the high speed wind of its BSG progenitor at an average speed of ~10% of the speed of light (~ 35,000 km s^{-1}), but has decelerated dramatically to only ~ 3,000 km s^{-1} since it has reached the inner edge of the prominent optical ring.

Marcaide, *et al.* [1], through the use of VLBI techniques, have been able to follow the expansion of the radio shell of SN1993J in detail, and also find that it is starting to decelerate, although the deceleration is not nearly as dramatic as that seen for SN1987A. While the expansion speed of SN1993J is quite high at ~ 15,000 km s^{-1} [1], the deceleration is much more gradual than that of SN1987A.

Peak Radio Luminosities and Distances

Long-term monitoring of the radio emission from SNe shows that the radio light curves evolve in a systematic fashion with a distinct peak flux density (and thus, in combination with a distance, a peak spectral luminosity) at each frequency and a well-defined time from explosion to that peak. Studying these two quantities at 6 cm wavelength, peak spectral luminosity ($L_{6cm\ peak}$) and time after explosion date (t_0) to reach that peak ($t_{6cm\ peak} - t_0$), Weiler, *et al.* [3] find that they appear related. In particular, based on five objects, Type Ib and Ic SNe may be approximate radio standard candles with a peak 6 cm spectral luminosity of

$$L_{6cm\ peak} \approx 1.3 \times 10^{27} \text{ erg s}^{-1} \text{ Hz}^{-1} \tag{10}$$

and, based on 17 objects, Type II SNe appear to obey a relation

$$L_{6\ cm\ peak} \simeq 6.0 \times 10^{23} \ (t_{6\ cm\ peak} - t_0)^{1.5} \text{ erg s}^{-1}\text{Hz}^{-1} \tag{11}$$

with time measured in days. (Note that the constants in Eqs. 10 and 11 are slightly different from those determined by [3] since there are now more objects available for inclusion.) If these relations are supported by further observations, they provide a means for determining distances to SNe, and thus to their parent galaxies, from purely radio continuum observations.

KWW, & MJM wish to thank the Office of Naval Research (ONR) for the 6.1 funding supporting this research. Additional information and data on radio emission from SNe can be found on *http://rsd-www.nrl.navy.mil/7214/weiler/* and linked pages.

245

REFERENCES

1. Marcaide, J. M. *et al.*, *Astrophys. J. Lett.* **486**, L31 (1997)
2. Bartel, N. *et al.*, *Nature* **318**, 25 (1985)
3. Weiler, K. W., Van Dyk, S. D., Montes, M. J., Panagia, N., & Sramek, R. A., *Astrophys. J.* **500**, 51 (1998)
4. Weiler, K., Sramek, R., Panagia, N., van der Hulst, J., & Salvati, M., *Astrophys. J.* **301**, 790 (1986)
5. Weiler, K., Panagia, N., & Sramek, R., *Astrophys. J.* **364**, 611 (1990)
6. Chevalier, R. A., *Astrophys. J.* **259**, 302 (1982)
7. Chevalier, R. A., *Astrophys. J. Lett.* **259**, L85 (1982)
8. Chevalier, R. A., *Astrophys. J.* **499**, 810 (1998)
9. Montes, M. J., Weiler, K. W., & Panagia, N., *Astrophys. J.* **488**, 792 (1997)
10. Natta, A. & Panagia, N., *Astrophys. J.* **287**, 228 (1984)
11. Lozinskaya, T. A., *Supernovae and Stellar Wind in the Interstellar Medium*, New York: AIP, 1992, p. 190
12. Chevalier, R. A. & Fransson, C., *Astrophys. J.* **420**, 268 (1994)
13. Sramek, R. A., Panagia, N. & Weiler, K. W., *Astrophys. J. Lett.* **285**, L59 (1984)
14. van Dyk, S. D., Sramek, R. A., Weiler, K. W. & Panagia, N., *Astrophys. J.* **409**, 162 (1993)
15. Montes, M. J., Weiler, K. W., Van Dyk, S. D., Panagia, N., Lacey, C. K., Sramek, R. A., & Park, R., *Astrophys. J.* **532**, 1124 (2000)
16. Montes, M. J., Van Dyk, S. D., Weiler, K. W., Sramek, R. A., & Panagia, N., *Astrophys. J.* **506**, 874 (1998)
17. Weiler, K., Van Dyk, S., Panagia, N., Sramek, R., & Discenna, J., *Astrophys. J.* **380**, 161 (1991)
18. Weiler, K., Van Dyk, S., Pringle, J., & Panagia, N., *Astrophys. J.* **399**, 672 (1992)
19. Schwarz, D. H. & Pringle, J. E., *Mon. Not. R. Astr. Soc.* **282**, 1018 (1996)
20. Van Dyk, S., Weiler, K., Sramek, R., Rupen, M., & Panagia, N., *Astrophys. J. Lett.* **432**, L115 (1994)
21. Van Dyk, S., Sramek, R. A., Weiler, K., & Panagia, N., *Astrophys. J. Lett.* **419**, L69 (1993)
22. Filippenko, A., Matheson, T., & Barth, A., *Astron. J.* **108**, 222 (1994)
23. Spyromilio, J., *Mon. Not. R. Astr. Soc.* **266**, 61 (1994)
24. Podsiadlowski, Ph., Hsu, J., Joss, P., & Ross, R., *Nature* **364**, 509 (1993)
25. Nomoto, K., Yamaoka, H., Pols, O. R., van den Heuvel, E., Iwamoto, K., Kumagai, S., & Shigeyama, T., *Nature* **371**, 227 (1994)
26. Ryder, S., Staveley-Smith, L., Dopita, M., Petre, R., Colbert, E., Malin, D., & Schlegel, E., *Astrophys. J.* **417**, 167 (1993)
27. Chu, Y.-H., Caulet, A., Montes, M. J., Panagia, N., Van Dyk, S. D., & Weiler, K. W., *Astrophys. J. Lett.* **512**, L51 (1999)
28. Hoeflich, P., Wheeler, J. C., Hines, D. C., & Tramaell, S. R., *Astrophys. J.* **459**, 307 (1996)
29. Manchester, R., Gaensler, B., Wheaton, V., Staveley-Smith, L., Tzioumis, A., Kesteven, M., Reynolds, J., & Bizunok, N., *Pub. Ast. Soc. Aust.*, in prep. (2001)

Cas A—A Y2K Status Report

L. Rudnick

University of Minnesota, Minneapolis, MN 55455, USA

Abstract.
Cas A has continued to be a rich source of information as new telescopes and detailed observations have become available. In this Y2K update, I focus on the observational findings over the last five years in the areas of ejecta mapping, hydrodynamic structure and kinematics, and high energy particles. Readers are unfortunately left to their own devices on the intriguing point source.

INTRODUCTION

1680 was a year of reckoning. It saw the birth of Edward Teach, more popularly known as the pirate Blackbeard, the launch of the Pueblo Revolt by southwestern US Native American tribes, the grant of New Caesarea (New Jersey) from the Duke of York, who believed he owned it, to a gentleman, a merchant, a citizen, a tanner and a maulster, as well as the report by Flamsteed [1] of the supernova that likely led to what we now know as Cas A.

In the intervening years, Cas A has proven to be an astronomical treasure trove, due to its very high brightness (in all but the obscured optical bands), its proximity (3.4 kpc, [2]), and its inconstant youth. Readers are recommended to a somewhat outdated, but nice overview of the remnant by Lozinskaya [3]. In this paper, I will highlight some of the notable observational findings primarily from the last five years. My focus is on three areas: Ejecta, Hydrodynamic Structure/Kinematics, and High Energy Particles, plus two important loose ends. Apologies are due, in advance, to the many developments unheralded here.

I THE EJECTA

Important progress has been made in the study of ejecta over a wide range of temperatures and shock conditions. The key finding from the emission-line images is that there has been a lot of mixing of the ejecta, either during the supernova explosion on in the assumed WR progenitor, on a variety of different scales. Most of the mass is in the X-ray emitting gas, with revised estimates of $4.3 M_\odot$ in ejecta and a swept-up mass of $\approx 9 M_\odot$ [4].

CP565, Young Supernova Remnants: Eleventh Astrophysics Conf., edited by S. S. Holt and U. Hwang
© 2001 American Institute of Physics 0-7354-0001-6/01/$18.00

Perhaps the most exciting information has come from the new Chandra observations of line emission showing the segregation of the various ejecta layers. The Si, S, Ar and Ca emission [5] show large scale distributions similar to those of the optical Fast Moving Knots (FMKs, [6]), likely deep core ejecta rich in heavy elements such as O, S and Ar, but weak in H, He. These X-ray and optical emitting regions therefore likely represent material heated by reverse shocks propogating back into the clumpy and diffuse ejecta. These shocked regions must be quite inhomogeneous in density, temperature and reverse-shock speeds to maintain such a variety of emission [5]. One major surprise is that the X-ray Fe emission, which should originate deeper in the core, is found outside much of the Si emission [5,7]. Large-scale bulk motions prior to or during the supernova explosion are likely responsible for this inversion, although individual small-scale clumps are not fully mixed, but retain clear signatures of their origin layer [7]. The origin of the small- and large-scale regions of line-free continuum [5,7] is not yet clear, and will be an important area of study over the next few years.

A new type of optical emission-line feature, "Mixed-Emission-Knots" (MEKs) have now been discovered [8]. They share the properties of both previously identified types of optical ejecta - the FMKs and the Fast Moving Flocculi, with strong [NII] and a range of $H\alpha$ strengths, thought to originate from the progenitor's outer layers [9]. The existence of MEKs may testify to the penetration of narrow fingers of high speed core ejecta into the surface layers of the progenitor [8]. If this picture is confirmed, it will provide a wonderful diagnostic into the explosion structure and dynamics.

Turning to the cooler material, mid-infrared observations using ISOCAM show a good correlation between the dust and gas and the FMKs, with most emission again coming from the ejecta [10]. The knots may thus be the site of dust formation [10,11]. Looking at the distribution of various elements, an anti-correlation is seen between Ne and Si, while S and Ar show extensive mixing [12]. The mixing during the explosion is therefore mostly on macroscopic scales, and quite heterogeneous. The anti-correlation between Ne (with Mg) and Si may limit how much core-collapse SNe can contribute to the galactic silicate population [12,13]. However, an unusual silicate mineral does appear to form in conjunction with the FMKs [11]. The 2MASS [14] mosaic by van Dyk [15] will likely also yield important insights when it is analysed.

II HYDRODYNAMIC STRUCTURE AND KINEMATICS

All of the material discussed above are ejecta which have been locally reverse-shocked as they are decelerated, either by encountering a swept-up mass shell or by the remnant's overall reverse shock (as in the bright ring), or by impacting the circumstellar material (e.g., as in the "jet" and other protuberances). But these multiwavelength images, as well as the radio, have left open the question for years

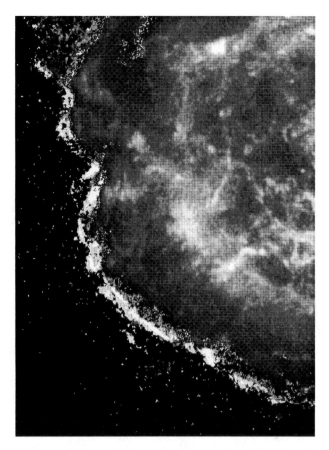

FIGURE 1. Radio greyscale image of the southeastern portion of Cas A overlaid with a burned out image of the outer rim from the 50ks GTO Chandra ACIS observation (Holt, P.I.).

about where the global outer blast wave and reverse shocks were located. The identification of these shocks is now clear from Chandra observations.

A thin, nearly circular rim of emission is seen most clearly in the line-free 4-6 keV continuum, forming an outer boundary to the diffuse radio and X-ray emission [16] (see Figure 1). One dimensional profiles of the radio and X-ray brightnesses are shown in Figure 2. The radio emission rises and its polarization angle changes at the outer shock, although there is no radio limb-brightening as in the X-ray continuum. When the radio and Si brightness distributions are decomposed into a series of underlying thin shells, the location of the reverse shock is found near the inner edge of the bright ring.

The X-ray spectra of the outer rim typically have much smaller equivalent widths than the bright, ejecta-dominated ring [16]. With the location of these shocks fixed, it will now be possible to go back to the emission at all wavelengths and study the connection between radiative and hydrodynamic properties.

The location of the shocks also puts strong constraints on the dynamical state of Cas A [17]. The ratio of outer blast wave to reverse shock radii is 3:2, indicating that the swept-up mass is of the same order as the ejected mass, consistent with the $\approx$ 2:1 ratio estimated from ASCA spectra [4]. As more mass is swept up, the reverse shock quickly moves back through the ejecta, separating from the blast wave. The centers of the rim (outer shock) and the bright ring (projected reverse shock) are offset by about 15% of their own radii [16]. This is likely due to an asymmetric ejection or a strong gradient in the circumstellar medium. The former explanation receives support from the kinematics and composition structure of the jet [8], while the latter is invoked to explain the offset velocity centroid [2].

The well-known discrepancy by a factor of three between the FMK and radio velocities has not been considered problematic. The FMKs are understood to be almost undecelerated when they are first shocked by encountering the dense material marked by the bright radio/X-ray ring [2,18]. This type of explanation is unlikely to work, however, for the recently discovered 2:1 ratio in the X-ray:radio bright ring expansion rate [19,20], since they consist of cospatial extended filamentary and diffuse emission. The paradox, illustrated in Figure 3, is currently unresolved.

Adding to this confusion is the suggestion that the spatially and kinematically distinct jet might be the only visible example of a dozen such plumes of ejecta [8]; there clearly are rings and other velocity substructures in the FMK population [21]. Also, recent low-frequency radio expansion measurements [22] appear to disagree with earlier results [23] on the large scale ring emission. Finally, there has been little attempt to follow up on the interesting model in which a circumstellar shell of material swept up by the pre-supernova wind regulates the structure and dynamics of the remnant [24]. Sorting this all out will become much easier with more detailed radio velocity analysis [25] and upcoming projects to make accurate, high resolution proper motion measurements of FMK and X-ray emitting material.

III   HIGH ENERGY PARTICLES

My original motivation for studying Cas A was to learn about the acceleration of relativistic particles. It is therefore exciting to report some important advances over the last several years in probing the relativistic electron and proton distributions, which may extend up to the 10^{14-15}eV "knee".

Figure 4 [26] summarizes the high energy spectrum of Cas A. The important radiation processes include electron synchrotron and bremsstrahlung, collisions of relativistic protons with circumstellar material leading to π^0 decay, and inverse Compton emission by synchrotron/bremsstrahlung electrons.

Perhaps the most exciting image of Cas A is shown in Figure 5 from the HEGRA collaboration [33,34]; although it lacks the dynamic range (with a signal:noise $\approx$ 5) and detail of the maps at other wavelengths, this "image" is only the first

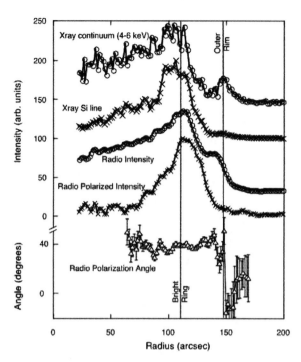

FIGURE 2. One dimensional plots of brightness as a function of distance from the center, in various bands and polarization angle, for a 55deg segment in the northwest [16].

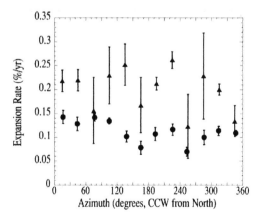

FIGURE 3. A comparison of the expansion velocities of the bright ring as seen in X-ray and radio maps [19].

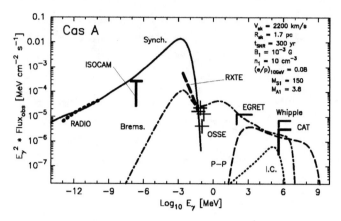

FIGURE 4. The emerging broadband spectrum of nonthermal emission in Cas A [26]. Observations include radio [27], mid-IR limits [28], keV-MeV [29], MeV-GeV [30], and TeV upper limits from Whipple and CAT (Cerenkov Array at Thémis) [31,32]

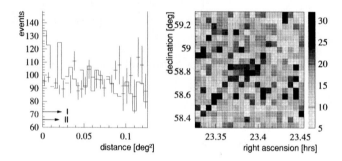

FIGURE 5. The next generation of Cas A studies, showing the HEGRA detection [33].

glimpse of a rich future resource. If the TeV emission comes from high energy p-p reactions, then we will learn about the strength and evolution of the blast wave; if it comes from inverse Compton emission, then it can be used to determine the SNR magnetic field strengths and relativistic electron densities, which are otherwise nearly impossible to disentangle.

Sliding down to $\approx 10 - 100 keV$ photons, Cas A is too bright for an extrapolation of the thermal emisssion that dominates at lower energies [37,38]. This high energy tail is consistent with a power law with a photon index of ≈ 3. Electrons with energies of order 10s of TeV could produce such a tail via synchrotron or nonthermal bremsstrahlung [39]. The emission from these electrons will also contribute significantly to the brightness at lower energies; mapping this non-thermal component [40] for comparison to the radio emission will be a critical step for particle acceleration studies. However, at the fine scales, there is only little correspondence between the radio and X-ray features [41], so the information will be subtle.

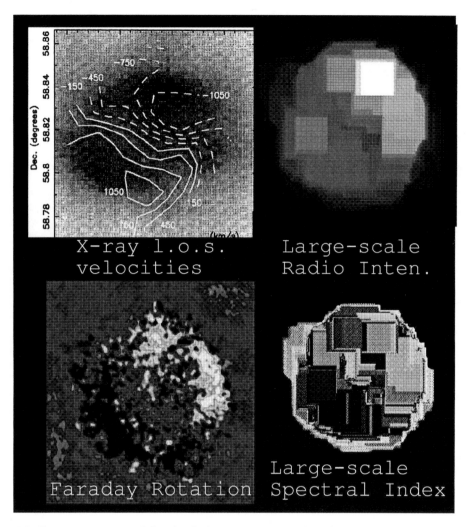

FIGURE 6. Four views of Cas A, all showing a clear NW-SE asymmetry. Upper left: X-ray line-of-sight velocity structure [35]. Bottom left: Polarization angle offset between radio bands due to Faraday Rotation. Upper right: Diffuse component of radio intensity [36]. Bottom right: Spectral index between $\lambda 6, 20$ cm for diffuse component.

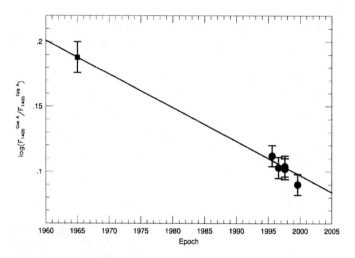

FIGURE 7. The fading of Cas A [44].

More hints about the acceleration of relativistic particles have come from spectral index variations across the remnant in the μeV part of the spectrum [42]. These were shown to arise from different power laws at different locations [43]. This likely indicates that particle acceleration conditions are responsible, with steeper values typically found in the plateau region, behind what we now know is the outer shock. With new Chandra observations, we will be able to characterize the local temperature and other conditions that may be responsible for this behavior. Another important area for further work will be separating the emission properties for diffuse and fine scale components [41]. Using a new multiresolution technique [36], previously hidden relationships between the spectral index varations and other properties become visible (see Figure 6). The diffuse component of the radio emission is steeper in the NW of Cas A, where the X-ray Doppler velocities [35] and Faraday rotation show the likely effects of the front-back asymmetry in the circumstellar density [2] on Cas A's partially illuminated shell.

IV   IMPORTANT LOOSE ENDS

The Point Source.   Finally, it was discovered [45]. It has no radio or optical counterpart. It might be a black hole, or a slowly accreting neutron star [46], one with hot polar caps [47] or a fallback accretion disk [48]. It's important; stay tuned.

Ti-44.   The 1.157 MeV line from ^{44}Ti was detected at $\approx 4\sigma$ using Comptel [49], but not at the corresponding level then expected at 68 and 78 keV using BeppoSAX [50]. These results could be consistent depending on the shape of the underlying X-ray continuum, but the matter is still unresolved. If the MeV detection holds, then the ^{44}Ti abundance (compared to ^{56}Ni) is high, and could be resolved with an asymmetric explosion model [51] (certainly appropriate given much of the above discussion) or high density Fe/Ti clumps with an advantageous size and density distribution [52].

Y3K? Cas A is fading, at least in the high frequency radio [44], by about $0.65\%/yr$ (see Figure 7), so the long term future looks dim. But if this short Y2K update is any guide, the future for the rich information and understanding from this remnant looks exceedingly bright.

REFERENCES

1. Ashworth, W. B., Jr., 1980, JHA, 11, 1
2. Reed, J.E., Hester, J. J., Fabian, A.C. & Winkler, P.F., 1995, ApJ 440, 706
3. Losinskaya, T.A., 1992, *Supernovae and Stellar Wind in the Interstellar Medium*, (AIP: New York)
4. Vink, J., Kaastra, J.S. & Bleeker, J.A.M 1996, AAp, 307, L41
5. Hwang, U., Holt, S. & Petre, R., 2000, ApJ, 537, 119
6. Kamper, K. & van den Bergh, S. 1976, ApJS, 32, 351
7. Hughes, J. P., Rakowski, C.E., Burrows, D. N. & Slane, P. 2000, ApJ, 528, L109
8. Fesen, R.A.,& Gunderson, K.S., 1996, ApJ, 470, 967
9. Fesen, R. A., Becker, R.H. & Blair, W.P. 1987, ApJ, 313, 378
10. Lagage, P.O., Claret, A., Ballet, J., Boulanger, F., Cesarsky, C. J., Fransson, C. & Pollock, A., 1996, AAp, 315, 273
11. Arendt, R.G., Dwek, E. & Moseley, S. H., 1999, AJ, 521, 234
12. Douvion, T., Lagage, P.O., Cesarsky, C. J. 1999, AAp, 352, L111
13. Dwek, E. 1998, ApJ 501, 643
14. Skrutskie, M.F., Schneider, S.E., Stiening, R., Strom, S.E., Weinberg, M.D., Beichman, C., Chester, T., Cutri, R., Lonsdale, C., Elias, J., Elston, R., Capps, R., Carpenter, J., Huchra, J., Liebert, J., Monet, D., Price, S. & Seitzer, P., 1997, in The Impact of Large Scale Near-IR Sky Surveys, eds. F. Garzon et al. (Kluwer: Netherlands), 25
15. http://www.ipac.caltech.edu/2mass/gallery/images_snrs.html
16. Gotthelf, E., Koralesky, B., Jones, T.W., Rudnick, L., Hwang, U., Petre, R. & Holt, S., 2000, submitted to ApJ
17. Truelove, J.K. & McKee, C.F., 1999, ApJS, 120, 299
18. Anderson, M. C., Jones, T. W., Rudnick, L., Tregillis, I. L., & Kang, H., 1994, ApJ, 421, 31
19. Koralesky, B., Rudnick, L., Gotthelf, E.V., Keohane, J. W., 1998, ApJ, 505, 27
20. Vink, J., Bloemen, H., Kaastra, J.S., Bleeker, J.A.M., 1998, AAp, 339, 201
21. Lawrence, S. S., MacAlpine, G. M., Uomoto, A., Woodgate, B. E., Brown, L. W., Oliversen, R. J., Lowenthal, J. D., Liu, C. 1995, AJ, 109, 2635
22. Agüeros, M.A. & Green, D. A., 1999, MNRAS, 305, 957
23. Anderson, M.C. & Rudnick, L. 1995, ApJ, 441, 307
24. Borkowski, K.J., Szymkowiak, A.E., Blondin, J.M., & Sarazin, C. L., 1996, ApJ, 466, 866

25. Koralesky, B. & Rudnick, L., 2001, these proceedings
26. Ellison, D.C., Goret, P., Baring, M., Grenier, I.A., & Lagage, P-O., 1999, in *26th ICRC Proceedings, Salt Lake City*, ed. D. Kieda, M. Salamon & B. Dingus, Vol. 3, 468
27. Baars, J.W.M., Genzel, R., Pauliny-Toth, I.I.K, & Witzel, A., 1977, AAp, 61, 99
28. Lagage, P.-O., 1999, in preparation
29. The, L.-S., et al., 1996, AApS, 120, 357
30. Esposito, J.A., Hunter, S.D., Kanbach, G., & Sreekumar, P., 1996, ApJ, 461, 820
31. Lessard, R., & Whipple Collaboration, 1999, in *Proc. 19th Texas Symposium, Paris, 1998*, eds. J. Paul, T. Montmerle, and E. Aubourg (CEA Saclay)
32. Goret, P, Gouiffes, C., Nuss E. (for the CAT collaboration), & Ellison, D.C., 1999, in *26th ICRC Proceedings, Salt Lake City*, ed. D. Kieda, M. Salamon & B. Dingus, Vol. 3, 496
33. Völk, H.J. and HEGRA Collaboration, 2000, astro-ph/0001301
34. Pühlhofer, G., Völk, H.J., Widner, C.-A. et al., 1999, in *26th ICRC Proceedings, Salt Lake City*, ed. D. Kieda, M. Salamon & B. Dingus, Vol. 3, 492
35. Holt, S.S., Gotthelf, E.V., Tsunemi, H., & Negoro, H., 1994, PASJ, 46, 151
36. Rudnick, L., 2001, preprint
37. Allen, G.E., Keohane, J.W. Gotthelf, E.V., Petre, R., Jahoda, K., Rothschild, R.E., Lingenfelter, R.E., Heindl, W.A., Marsden, D., Gruber, D.E., Pelling, M.R. & Blanco, P.R., 1997, ApJ 487, L97
38. Favata, F., Vink, J., Dal Fiume, D., Parmar, A.N., Santangelo, A., Mineo, T., Preite-Martinez, A., Kaastra, J.S., & Bleeker, J.A. M. 1997, AAp 324, L49
39. Laming, J. M., 2000 astro-ph/0008426
40. Vink, J., Maccarone, C., Kaastra, J.S., Mineo, T., Bleeker, J.A.M, Mineo, T., Bleeker, J.A.M., Preite-Martinex, A. & Bloemen, H., 1999, AAp, 344, 289
41. Koralesky, B., Rudnick, L., Gotthelf, E., Petre, R., Hwang, U. & Holt, 2001, preprint
42. Anderson, M.C. & Rudnick, L. 1996, ApJ, 456, 234
43. Wright, M., Dickel, J., Koralesky, B. & Rudnick, L. 1999, ApJ, 518, 284
44. Reichart, D. E. & Stephens, A.W. 2000, ApJ, 537, 904
45. Tananbaum, H., et al. 1999, IAU Circ. 7246
46. Umeda, H. & Nomoto, K., 2000, ApJ, 534, L193
47. Pavlov, G. G., Zavlin, V. E., Aschenbach, B., Trümper, J., & Sanwal, D., 2000, ApJ, 531, 53
48. Chakrabarty, D., Pivovaroff, M.J., Hernquist, L. E., Heyl, J.S. & Narayan, R., 2000, astro-ph/0001026
49. Iyudin, A. F. et al., 1994, AAp, 284, L1
50. Vink, J., Kaastra, J. S., Bleeker, J. A. M., & Bloemen, H., 2000, *Proc. of the E1.1 Symposium of COSPAR Scientific Commission E, 12-19 July, 1998, Nagoya, Japan*, ed. K. Makishima, L. Piro, and T. Takahashi, (Pergamon Press: New York), 689
51. Nagataki, S., Hashimoto, M., Katsuhiko, S., Yamada, S., & Mochizuki, Y.S., 1998, ApJ, 492, L45
52. Mochizuki, Y., Takahashi, K., Janka, H.-Th., Hillebrandt, W., & Diehl, R., 1999, AAp, 346, 831

Kepler's Supernova Remnant

Anne Decourchelle

Service d'Astrophysique, DAPNIA/DSM/CEA Saclay, L'Orme des Merisiers, F-91191
Gif-sur-Yvette, France

Abstract. In this paper, I intend to review our current knowledge on Kepler's supernova remnant. The observations are discussed with a special emphasis on the inputs
they provide to model this remnant. In particular, the X-ray band provides crucial
informations and constraints on the origin of Kepler's SNR, on its evolutionary stage
as well as on particle acceleration at the forward and reverse shocks.

INTRODUCTION

The supernova event, which gave birth to Kepler's supernova remnant, was widely
observed by Chinese and European astronomers. The supernova was first seen
on October 9, 1604 and last seen by Johannes Kepler in October 1605 [1]. The
different names of the supernova remnant are : Supernova Ophiuchi, SN 1604,
Kepler, G4.5+6.8, 3C 358.

With a Galactic latitude of 6.8 degrees and a distance of 4.8 ± 1.4 kpc [2],
the remnant locates at a large height above the Galactic plane ($\simeq$ 600 pc),
corresponding to a low density Interstellar Medium ($0.1 - 0.02$ part cm^{-3}, [3,4]).
Due to its large distance, Kepler's remnant has the smallest angular size (3 arcmin)
of all young historical remnants. The hydrogen column density in the line of sight
is in the order of $3 - 6 \ 10^{21}$ cm^{-2}.

The supernova type of Kepler is not as well established as the other historical
supernovae (Cas A, Tycho, SN 1006, Crab). The first arguments were in favor of a
type I supernova. Based on the historical light-curve, Baade [1] concluded that it
was a supernova of type I. Its location above the Galactic plane is consistent with
an old population. However, Doggett and Branch [5] have shown that the light
curve was also compatible with a supernova of type II-L.

CP565, *Young Supernova Remnants: Eleventh Astrophysics Conf.*, edited by S. S. Holt and U. Hwang
© 2001 American Institute of Physics 0-7354-0001-6/01/$18.00

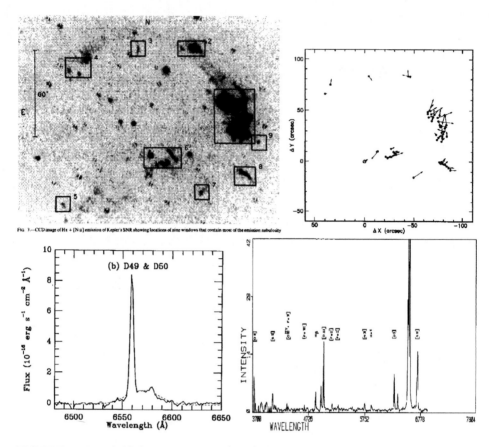

FIGURE 1. Top: (left) Optical image and (right) positions and 400 yr proper motions of optical knots from Bandiera and van den Bergh [16]. Bottom: (left) Hα line in a nonradiative filament from Blair *et al.* [15] and (right) spectrum of a bright optical knots from Dennefeld [10].

OPTICAL OBSERVATIONS

Radiative emission

The optical remnant of Kepler's SNR was discovered by Baade [1] from historical positional information. Spectroscopic and imaging observations were obtained by a number of authors [6–16]. Both the spectroscopy and the observed proper motions of the optical knots provide arguments in favor of a massive progenitor and a type II SN.

The optical emission is strongly asymmetric, with the brightest knots on the eastern and northern edges, as well as towards the projected center of the remnant as is shown in Figure 1. From the study of the proper motions of the optical knots in Kepler's SNR [7,8,16], a very low expansion velocity has been derived ($V_{exp} = 67 \pm 26$ km s^{-1}, [16]), as well as a high velocity global translation (278 ± 12 km s^{-1}, [16]). The 400 yr proper motions in Kepler's SNR (extrapolated from the period 1942-1989) are presented in Figure 1 [16]. Unlike the remnant of Cas A, no fast moving knots have been found in Kepler's SNR. The optical knots are attributed to circumstellar material. The high velocity of the center of the supernova shell is consistent with either a population II progenitor, or a massive runaway progenitor [8,16].

Spectrophotometric observations of the brightest optical knots [6,7,10,11,13,15] have revealed radiative shock emission lines of H$_\alpha$, [NII], [SII], [OIII], [OI]. High electronic densities of $\simeq 10^3 - 10^4$ cm^{-3} have been derived from the [S II] lines ratio ($I(6731) > I(6717)$) as well as a nitrogen overabundance by a factor of about 4 [10,11,15]. These densities are inconsistent with the expected interstellar density at Kepler's distance above the Galactic plane.

In summary, the brightest optical knots in Kepler's SNR have a low expansion velocity, a high density and nitrogen overabundance. This points to a circumstellar origin, which forced to reconsider Kepler's supernova being a type Ia.

Nonradiative shock emission

The presence of Balmer-dominated (or nonradiative) emission has been discovered by Fesen et al. [13] in the northern rim of Kepler's SNR. While the emission from the brightest knots is characteristic of cooling radiative shocks, the emission from faint knotty and diffuse components corresponds to Balmer-dominated nonradiative shocks [13,15]. This emission is characterized by the lack of other prominent emission lines and the presence of broad and narrow Hα emission. As for Tycho and SN 1006, it indicates the propagation of the shock in a partially neutral medium. Extensive imagery and spectroscopy of Kepler's SNR by Blair et al. [15] have revealed that most of the optical emission is dominated by Hα, except a few bright knots in the northeastern rim. This is clearly seen from Figure 3 (ratio of subtracted Hα and [SII] images) in Blair et al. [15].

Most of the arguments for a circumstellar origin of the optical knots were based on the spectrophotometry of the brightest knots, suggesting high densities and overabundances of nitrogen. This might have to be reconsidered in view of the much lower and uniform density medium required for most of the optical knots and diffuse emission.

Kepler's SNR exhibits bright quasi-stationary optical knots like Cas A, but no fast moving knots. Kepler's SNR has extended nonradiative emission, but instead of the filamentary structure observed in Tycho and SN 1006, its emission is diffuse and knotty.

INFRARED OBSERVATIONS

The recent data from ISOCAM onboard the ISO satellite have shown that the infrared thermal dust emission (in the band 10.7-12.0 micron) is in excellent agreement with Hα emission, as in Tycho [17]. For discussion on the infrared observations of Kepler's SNR, I refer the reader to papers by Braun [18], Arendt [19], Dwek et al. [20], Saken et al. [21] and Douvion [17].

RADIO OBSERVATIONS

The first radio observations of Kepler's SNR are from Hermann and Dickel [22], Strom and Sutton [23] and Gull [24]. Kepler's SNR has a limb-brightened morphology, especially prominent in its northern half (Matsui et al. [25], Dickel et al. 1988 [26]). The optical emission is slightly outside the radio and there is a good overall large scale correlation with X-rays. The magnetic field is radially aligned, which is a common feature in young SNRs. This alignement is interpreted as stretching due to Rayleigh-Taylor instabilities [27]. The magnetic field is amplified ($\simeq 20$ B_{ISM}). Mechanisms for this amplification are discussed in Gull [28], Reynolds and Chevalier [29] and Jun and Norman [27]. The expansion rate x, defined as $R \propto t^x$, has been measured in Kepler's SNR by Dickel et al. [26] and shows strong variations from the southern ($x = 0.65$) to the northern rim ($x = 0.35$).

X-RAY OBSERVATIONS

The X-ray emission from Kepler's SNR was discovered by HEAO1 in 1979 [30]. The presence of strong emission lines of Si, S, Ar, Ca and Fe have been revealed by the instruments with spectral capabilities: *Einstein/SSS* [31], *EXOSAT* [32], *GINGA* [33], *ASCA* [34], *XTE* [35]. The X-ray imaging instruments have shown the asymmetric X-ray morphology of Kepler's SNR, comparable to the radio emission: *Einstein HRI* [36], *EXOSAT CMA* [32], *ROSAT HRI* [37]. Using *ROSAT* and *Einstein* satellites, the X-ray expansion rate has been determined to be very close to free expansion ($x \simeq 0.93$) and nearly twice as fast as the mean radio expansion rate [37]. High energy observations with *RXTE* have revealed X-ray emission up to 25 keV in Kepler's SNR [35]. Note that an important lack in these X-ray observations of Kepler's SNR is the absence of spectro-imaging data, waiting for *Chandra* and *XMM-Newton* observations.

MODELING OF KEPLER'S SNR

Two approaches have been undertaken to model the remnant of Kepler, motivated by understanding either the (optical and X-ray) morphology, or the spectral X-ray properties of the emitting gas.

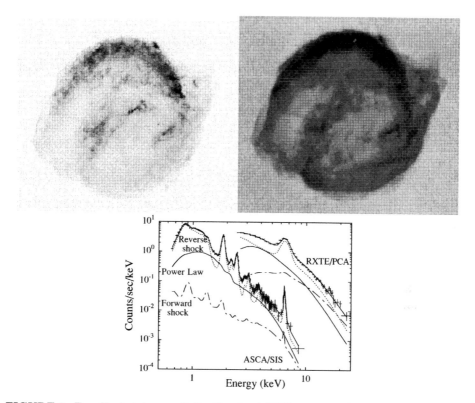

FIGURE 2. Top: Kepler's images. Left: *Chandra ACIS* X-ray image (courtesy of Una Hwang). Right: Radio 6 cm VLA image (courtesy of Tracey DeLaney). The scale of the two images is not exactly the same. Bottom: Broadband X-ray spectrum and model [35]

Morphological approach

The morphological approach has been initiated by Bandiera [38]. Based on the asymmetry of the remnant at all wavelengths, on the global velocity of $\simeq 300$ km s^{-1}, on the high density of the optical knots for the high latitude of Kepler, and on the overabundance of nitrogen, Bandiera proposed that the progenitor was a massive runaway star with strong mass loss, ejected from the Galactic plane $\simeq 2 \; 10^6$ yr ago [8,38]. The supernova exploded in a medium modified by the bowshock interaction of the stellar wind of the progenitor with the interstellar medium. The dynamics of this interaction has been studied in detail by Borkowski et al. [3,39] and leads to a complicated asymmetric structure with reflected and transmitted shocks due to the bowshock shell. The overall model is compatible with previous X-ray spectra from *Einstein SSS* and *EXOSAT/ME* assuming a partial equilibration between the electronic and ion temperatures [39]. This model needs a small Fe overabundance.

Spectral approach

The spectral study of the X-ray emission in young SNRs is a powerful approach to understand their origin and evolution stage. The temperature and ionization state distribution are strongly dependent on the initial density profile in the ambient matter and in the ejecta. In particular, the presence of a stellar wind around the progenitor can inverse the temperature behavior leading to low temperatures at the contact discontinuity (see Chevalier [40]). Also, the expected density profile out of the SN core is different for a SN Ia (exponential, see Dwarkadas and Chevalier [41]) and for SN II (power law, the index of the power law depends on the initial mass of the progenitor).

The Sedov model was the first model to be applied to Kepler's X-ray spectrum obtained from *Einstein* and *EXOSAT* [42,32]. The required overabundances of heavy elements (Si,S,Fe), as well as the derived ISM swept mass were inconsistent with the Sedov hypothesis and demonstrated the necessity of including the ejecta and a reverse shock in the model.

For reverse shock models, the simplest assumption on the density structure of the ejecta (flat profile) was investigated by Hughes and Helfand [42] and Rothenflug et al. [43]. These models are unable to fit together the Fe K line and the other lower energy lines (Fe L, Si K, S K,...), because the range of temperatures in the shocked ejecta is too tight. The derived temperature is too low to emit the iron K line in the ejecta and a large overabundance of Fe in the ISM is required to fit the broadband spectrum [43].

Power-law density profiles in the ejecta, expected from SN models (see Arnett [44] for SN 1987A), have been investigated using self-similar solutions [40] by Decourchelle and Ballet [4] and Decourchelle et al. [45]. A reasonable fit to the data is obtained over a large energy band (0.3-10 keV) and the emission is dominated by the ejecta. The improved data and modeling have considerably reduced the required ambient density from 7 to $\simeq 0.5$ cm^{-3}. Models with r^{-2} stellar wind are unable to reproduce the spectrum, because the temperature is too low in the ejecta for the Fe K line. Models with typical parameters of SN Ia require a very low energy of explosion ($\simeq 5 \; 10^{49}$ erg) and a lower continuum than observed. Only models of type II supernova without stellar wind explain well the whole spectrum with reasonable values of the energy of explosion and mass [4,45].

Higher energy observations with *RXTE* have provided crucial additional constraints for the temperature profile [35]. It confirms the presence of a high temperature component in the ejecta, both from the Fe K line and the high energy continuum. An additional power-law emission is required to fit correctly the continuum: its contribution is essentially important in the *ASCA* energy range and is not a dominant component of the high energy spectrum [35]. The power-law component is attributed to synchrotron emission and fits qualitatively with the expectation from Reynolds's synchrotron models [46] with a spectral index of $\simeq 2.5$ and a flux at 1 keV three times less than extrapolated from the radio synchrotron.

Apart from inhomogeneities in the ambient medium, the spherically symmetric assumption (implicitly used in the self-similar solutions) is disrupted by the development of hydrodynamic instabilities at the interface between the ejecta and the ambient medium [28,47,48]. Hily-Blant et al. [49] have shown the consequences of these instabilities on the X-ray morphology and spectrum of Kepler's SNR in the case of a uniform ISM. The iron K-line is emitted where the temperature is highest, close to the contact discontinuity, and thus traces well the instabilities. The oxygen K-line, corresponding to lower temperature, traces regions close to the reverse shock. While the X-ray morphology is modified compared to the spherically symmetric model, the overall X-ray spectrum is almost unchanged by the instabilities. Thus, the self-similar modeling of the global X-ray spectrum is still valid and consistent with models which include the development of hydrodynamic instabilities.

Nonlinear particle acceleration

Reynolds and Ellison [50] have shown that models of efficient first-order Fermi shock acceleration [51] are able to produce the mean radio spectral slope of -0.64 of Kepler's SNR, which differs substantially from the test-particle value of -0.5 expected for strong shocks. In addition, nonlinear shock acceleration predicts curvature of the electron spectrum which is consistent with Kepler's radio spectrum [50] (as well as Tycho) and can be taken as the first evidence of efficient acceleration of protons at the SNR shock.

The arbitrary power law component, required to fit the broadband X-ray spectrum, is interpreted as possible synchrotron emission from electrons accelerated at TeV energies [35,46]. In addition to the discovery of nonthermal X-ray emission in SN 1006 [52], Cas A [53] and G347.3-0.5 [54], this supports the case for efficient proton acceleration.

Decourchelle, Ellison and Ballet [55] have investigated the impact of efficient particle acceleration on the hydrodynamics and non-equilibrium ionization X-ray emission using analytical bi-fluid solutions for the hydrodynamics [56], coupled to nonlinear shock jump conditions [57]. With increased acceleration efficiency, the interaction region (between the forward and reverse shocks) gets thinner with much larger density and pressure gradients than in the usual test-particle case [55]. This model, applied to the modeling of the broadband X-ray spectrum of Kepler's SNR, imposes a low efficiency of the acceleration at the reverse shock (in order to have sufficiently high temperature to produce the iron K line), which might be related to the low magnetic field value in the ejecta, expected from the expansion. At the forward shock, we predict efficient acceleration for standard values of the ambient magnetic field. Spectro-imaging data are required to verify these predictions. A case for efficient acceleration at the forward shock has been made for the remnant of 1E 0102.2-7219 in the Small Magellanic Cloud by Hughes, Rakowski and Decourchelle [58,59] using *Chandra* observations.

NEW PERSPECTIVES AND CONCLUSION

Because of its south declination, Kepler's SNR was not as much observed in the optical and radio range as the other young Supernova Remnants. In the infrared and X-ray regimes, the observational limitation was its small angular size in comparison to the spatial resolution of the instruments onboard satellites. New X-ray and radio observations are revealing complex properties and morphology of Kepler's SNR.

Recent radio VLA observations of Kepler's supernova remnant have been carried out to derive accurate measurements of the proper motions (Koralesky *et al.* [60]), spectral indices (DeLaney *et al.* [61]) and polarization [62]. As in Cas A [63] and Tycho [64,65], the radio expansion is about half that of the X-rays [66]. The spectral index shows spatial variations within Kepler's remnant from -0.85 to to -0.6 [61,62].

Kepler's SNR has recently been observed with *Chandra* Observatory (Hwang *et al.* [67]). The northern asymmetry previously discussed arises mainly from the silicon emission, while the high energy continuum (4-6 keV) image is symmetric, which might indicate that the medium in the vicinity of Kepler's SNR is rather homogeneous. The iron K emission is at smaller radii than the silicon emission as in Tycho SNR [68,69]. The ears structure on the southeastern and northwestern edges are now clearly visible in X-rays and might be an indication of jets as in the case of G309.2-0.6 [70].

With the coming spectro-imaging X-ray observations with *Chandra* and *XMM-Newton*, we will be able to put strong new constraints on the origin of Kepler's SNR as well as to investigate open issues like the efficiency of particle acceleration at the shocks and the relation between thermal and nonthermal components in this and other young supernova remnants.

I would like to thank Tracey DeLaney, Larry Rudnick, Una Hwang and Steve Holt for providing me their recent results before publication, respectively on radio VLA and *Chandra* X-ray observations, and Jean Ballet for a careful reading of the manuscript. I also benefit from fruitful discussions during the conference. I thank the Ministère des Affaires Etrangères, the Comité National Français des Astronomes and Steve Holt for partial support to attend this conference.

REFERENCES

1. Baade, W., 1943, ApJ, 97, 119
2. Reynoso, E. M., Goss, W. M.,1999, AJ, 118, 926
3. Borkowski, K. J., Blondin, J. M., Sarazin, C. L., 1992, ApJ, 400, 222
4. Decourchelle, A., Ballet, J., 1994, A&A, 287, 206
5. Doggett, J. B., Branch, D., 1985, AJ, 90, 230
6. Minkowski, R., 1943, ApJ, 97, 128

7. van den Bergh, S., Marscher, A. P., Terzian, Y., 1973, ApJS, 26, 19

8. van den Bergh, S., Kamper, K. W., 1977, ApJ, 218, 617

9. van den Bergh, S., 1980, A&A, 86, 155

10. Dennefeld, M., 1982, A&A, 112, 215

11. Leibowitz, E. M., Danziger, I. J., 1983, MNRAS, 204, 273

12. D'Odorico, S., Bandiera, R., Danziger, J., Focardi, P., 1986, AJ, 91, 1382

13. Fesen, R. A., Becker, R. H., Blair, W. P., Long, K. S., 1989, ApJ, 388, L13

14. van den Bergh, S., Pritchet, C. J., 1991, PASP, 103, 194

15. Blair, W. P., Long, K. S., Vancura, O., 1991, ApJ, 366, 484

16. Bandiera, R., van den Bergh, S., 1991, ApJ, 374, 186

17. Douvion, T., Lagage, P. O., Dwek, E., Cesarsky, C. J., 2000, A&A, submitted

18. Braun, R., 1987, A&A, 171, 233

19. Arendt, R. G., 1989, ApJS, 70, 181

20. Dwek, E., Petre, R., Szymkowiak, A., Rice, W. L., 1987, ApJ, 320, L27

21. Saken, J. M., Fesen, R.A., Shull, J. M., 1992, ApJS, 81, 715

22. Hermann, B. R., Dickel, J. R., 1973, AJ, 78, 879

23. Strom, R. G., Sutton, J., 1975, A&A, 42, 299

24. Gull, S. F., 1975, MNRAS, 171, 237

25. Matsui, Y., Long, K. S., Dickel, J. R., Greisen, E. W., 1984, ApJ, 287, 295

26. Dickel, J. R., Sault, R., Arendt, R. G., Matsui, Y., Korista, K. T., 1988, ApJ, 330, 254

27. Jun, B.-I., Norman, M. L., 1996, ApJ, 465, 800

28. Gull, S. F., 1973, MNRAS, 161, 47

29. Reynolds, S. P., Chevalier, R. A., 1981, ApJ, 245, 912

30. Tuohy, I. R., Nugent, J. J., Garmire, G. P., Clark, D. H, 1979, Nature, 279, 139

31. Becker, R. H., White, N. E., Boldt, E. A., Holt, S. S., Serlemitsos, P. J., 1980, ApJ, 237, 77

32. Smith, A., Peacock, A., Arnaud, M., Ballet, J., Rothenflug, R., Rocchia, R., 1989, ApJ, 347, 925

33. Hatsukade, I., Tsunemi, H., Yamashita, K., Koyama, K., Asaoka, Y., Asaoka, I., 1990, PASJ, 42, 279

34. Kinugasa, K., Tsunemi, H., 1999, PASJ, 51, 239

35. Decourchelle, A., Petre, R., 1999, AN, 320, 203

36. White, R. L., Long, K. S., 1983, ApJ, 264, 196

37. Hughes, J. P., 1999, ApJ, 527, 298

38. Bandiera, R., 1987, 1987, 319,885

39. Borkowski, K. J., Sarazin, C. L., Blondin, J. M., 1994, ApJ, 429, 710

40. Chevalier, R. A., 1982, ApJ, 258, 790

41. Dwarkadas, V. V., Chevalier, R. A., 1998, ApJ, 497, 807

42. Hughes, J. P., Helfand, D. J., 1985, ApJ, 291, 544

43. Rothenflug, R., Magne, B., Chieze, J. P., Ballet, J.,1994, A&A, 291, 271

44. Arnett, W. D., 1988, ApJ, 331, 377

45. Decourchelle, A., Kinugasa, K., Tsunemi, H., Ballet, J.,1997,in "X-Ray Imaging and Spectroscopy of Cosmic Hot Plasmas", Proceedings of an International Symposium on X-ray Astronomy ASCA Third Anniversary, Edited by F. Makino and K. Mitsuda

(1997)., p.367

46. Reynolds, S. P., 1998, ApJ, 493, 375
47. Chevalier, R. A., Blondin, J. M., Emmering, R. T.,1992, ApJ, 392, 118
48. Chevalier, R. A., Blondin, J. M., 1995, ApJ, 444, 312
49. Hily-Blant, P., Decourchelle, A., Chièze, J. P., 1999, DEA report, University of Paris XI
50. Reynolds, S. P., Ellison, D. C., 1992, ApJ, 399, L75
51. Ellison, D. C., Reynolds, S. P., 1991, ApJ, 382, 242
52. Koyama, K., Petre, R., Gotthelf, E. V., Hwang, U., Matsura, M., Ozaki, M., Holt, S. S., 1995, Nature, 378, 255
53. Allen, G. E., Keohane, J. W., Gotthelf, E. V., Petre, R., Jahoda, K., Rothschild, R. E., Lingenfelter, R. E., Heindl, W. A., Marsden, D., Gruber, D. E., Pelling, M. R., Blanco, P. R., 1997, ApJ, 487, 97
54. Slane, P., Gaensler, B. M., Dame, T. M., Hughes, J. P., Plucinsky, P. P., Green, A., 1999, ApJ, 525, 357
55. Decourchelle, A., Ellison, D. C., Ballet, J., 2000, ApJ, 543, L57
56. Chevalier, R. A., 1983, ApJ, 272, 765
57. Berezhko, E. G., Ellison, D. C., 1999, ApJ, 526, 385
58. Hughes, J. P., Rakowski, C. E., Decourchelle, A., 2000, ApJ, 543, L61
59. Hughes, J. P., 2001, these proceedings
60. Koralesky, B., 2001, these proceedings
61. DeLaney, T., Koraleski, B., Rudnick, L., Dickel, J. R., 2001, these proceedings
62. DeLaney, T., Koraleski, B., Rudnick, L., Dickel, J. R., 2001, in preparation
63. Koralesky, B., Rudnick, L., Gotthelf, E. V., Keohane, J. W., 1998, ApJ, 505, L27
64. Reynoso, E. M., Moffett, D. A., Goss, W. M., Dubner, G. M., Dickel, J. R., Reynolds, S. P., Giacani, E. B., 1997, ApJ, 491, 816
65. Hughes, J. P., 2000, ApJ, 545, L53
66. Koralesky, B., Rudnick, L., DeLaney, T., 2000, ApJ, submitted
67. Hwang, U., *et al.*, 2001, in preparation
68. Hwang, U., Gotthelf, E. V., 1997, ApJ, 475, 665
69. Decourchelle, A., Sauvageot, J. L., Audard, M., Aschenbach, B., Sembay, S., Rothenflug, R., Ballet, J., Stadlbauer, T., West, R. G., 2001, A&A, 365, L218
70. Gaensler, B. M., Green, A. J., Manchester, R. N., 1998, MNRAS, 299, 812

Vela Z – so Young and so Exotic

Miroslav D. Filipović[1,2,3], Paul A. Jones[1] and Bernd Aschenbach[2]

[1] *University of Western Sydney Nepean, P.O. Box 10, Kingswood, NSW 2747 Australia*
[2] *Max-Planck-Institut für extraterrestrische Physik, Giessenbachstraße, D-85740 Garching, Germany*
[3] *Australia Telescope National Facility, CSIRO, P.O. Box 76, Epping, NSW 2121 Australia*
indexRX J0852.0-4622

Abstract. We present a multi-frequency study of the recently discovered supernova remnant (SNR) RX J0852.0-4622 (Vela Z) superimposed on the well-known Vela SNR. The structure of SNR RX J0852.0-4622 is shell-like (barrel-shaped/bilateral) with correlation between radio-continuum, EUV and ROSAT PSPC X-ray emission. The radio-continuum coinciding with X-ray confirms that synchrotron radiation is responsible for the north and south brightened X-ray limbs. The area of RX J0852.0-4622 has filaments of optical line emission in [O III] and [S II], but some of these may be associated with the surrounding Vela SNR.

INTRODUCTION

Aschenbach [1] and Iyudin et al. [2] have reported the discovery of the previously unknown galactic SNR RX J0852.0-4622 (G266.3-01.2; GRO J0852-4642), identified by its X-ray and γ-ray emission. The X-ray image obtained in the ROSAT all-sky survey (RASS) shows a disk-like, partially limb-brightened emission region, which is the typical appearance of a shell-like SNR (see Figure 1). The X-ray spectrum indicates that RX J0852.0-4622 is a young object and must be close-by because of the large angular diameter of $\sim 2°$. Comparison with historical SNRs limits the age to about $\sim 1\,500$ yrs and the distance to < 1 kpc. The case for RX J0852.0-4622 being a SNR was supported by the detection of γ-ray line emission from ^{44}Ti, which is produced exclusively in supernovae. Using the mean lifetime of ^{44}Ti (90.4 yrs), the angular diameter, and adopting a mean expansion velocity of 5000 km s^{-1} as well as a ^{44}Ti yield of 5×10^{-5} M$_\odot$, Iyudin et al. [2] derived an age of ~ 680 yrs and a distance of ~ 200 pc, which makes RX J0852.0-4622 the closest supernova to Earth to have occurred during recent human history. Rood et al. [3], in an analysis of NO$_3^-$ in an Antarctic ice core, found that the precipitation dates agree with the occurrence of historical supernovae, with the exception of one spike dated to 1320$\pm$20 AD. Recently, Burgess & Zuber [4] have suggested that this spike may be associated with the explosion of RX J0852.0-4622.

CP565, *Young Supernova Remnants: Eleventh Astrophysics Conf.*, edited by S. S. Holt and U. Hwang
© 2001 American Institute of Physics 0-7354-0001-6/01/$18.00

The analysis of the X-ray spectra shows that the emission from the northern limb of RX J0852.0-4622 may be thermal but it can also be represented by a power law spectrum created by synchrotron radiation with a photon power law index of $\nu = -2.6^{+0.3}_{-0.4}$. Meanwhile the power law has been confirmed by ASCA measurements [5], [6]. Both the morphology and the X-ray spectral characteristics are very similar to those of SN 1006. Chen & Gehrels [7] and Aschenbach et al. [8] pointed out that RX J0852.0-4622 was most likely to be created by a core-collapse SN. Two X-ray sources and one radio object close to the SNR centre have been found, either one of which could be the neutron star left by the SN [1] or background galaxy.

All confirmed SNRs within our Galaxy have been detected at radio wavelengths [9], i.e. all optical/X-ray SNR detections are associated with radio-continuum emission. The north-eastern limb of RX J0852.0-4622 appears to coincide with the 408 MHz emission peak of Vela Z [10], which is a non-thermal radio source. Combi et al. [11] presented a radio-continuum detection of RX J0852.0-4622 at 1400 MHz with low angular resolution (30′) and a revised analysis at 2420 MHz from the Duncan et al. [12] data. They concluded that the overall spectral index is $\alpha \sim -0.3$. Recently, Duncan & Green [13] observed RX J0852.0-4622 with the Parkes radio telescope at 1400 MHz and in combination with 2420 MHz observations [12] estimated the overall radio spectral index to be $\alpha = -0.4 \pm 0.5$.

RESULTS

The EUVE image of the Vela SNR at 50-180 Å wavelength [14] (image obtained through skyview http://skyview.gsfc.nasa.gov and smoothed to 10′ resolution) shows prominent emission from the area of RX J0852.0-4622 which appears to be a shell associated with this SNR rather then the Vela SNR. The EUV emission has an arc inside the southern X-ray arc of RX J0852.0-4622 (see Figure 1 – left). Because of the different X-ray spectra of RX J0852.0-4622 and surrounding Vela SNR the ROSAT data from the highest energies (E>1.3 keV) are used to distinguish the two. However, at all wavelengths there is a problem in distinguishing the two SNRs. It appears that the EUV emission within the shell of RX J0852.0-4622 is due to this SNR. There is EUV emission to the north-east of RX J0852.0-4622 corresponding to radio feature C in Combi et al. [11] which we attribute to the surrounding Vela SNR.

We also analyse optical line images in [O III], [S II] and Hα from M. Price (Mt. Stromlo) which show filamentary structure typical for SNRs, but again it is difficult to distinguish RX J0852.0-4622 from the Vela SNR. Figure 1 (right) shows the [O III] emission relative to the X-ray, but there is little correlation in structure between the two wavelengths. However, the prominent EUV arc inside the south side of RX J0852.0-4622 is associated with a [O III] arc.

Also, we compare the radio-continuum images with the ROSAT PSPC (E>1.3 keV) image [15] and find several features to coincide. We investigated several prominent features seen only at one frequency. All radio-continuum fea-

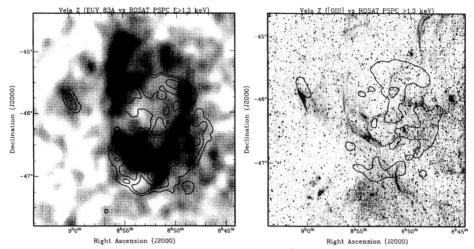

FIGURE 1. The EUV image of RX J0852.0-4622 at 83Å overlaid with contours of ROSAT PSPC (E>1.3 keV) data (left). X-ray contour levels (in black) are 0.8, 1.2, 2 and 3 in units of 10^{-4} ROSAT PSPC counts s^{-1} arcmin^{-2}. On the right is the [O III] image (gray scale) with one contour of the ROSAT PSPC X-ray image to show the position of RX J0852.0-4622.

tures of RX J0852.0-4622, except for the southernmost feature, have X-ray peak positions shifted relative to the radio in the outward radial direction by about 4′ (which is 0.23 pc or 48 000 AU at the distance of RX J0852.0-4622 – 200 pc). We also found in the MOST 843 MHz original (43″ resolution) image that this feature coincides with a point-like source which is probably a background object. However, the X-ray and radio emission at other frequencies are extended and therefore there is still emission in that area from RX J0852.0-4622.

The features A, B and C introduced by Combi et al. [11] can be seen in our radio images but less clearly distinguished from the emission of the surrounding Vela SNR. Duncan & Green [13] interpret features A and C as part of large Vela SNR filaments, and feature B as an isolated source not related to RX J0852.0-4622.

From the radio-continuum images and from the ROSAT PSPC X-ray image (E>1.3 keV) (see Figure 1) we conclude that the eastern limb of RX J0852.0-4622 is missing at both frequencies despite the two weak radio features. This makes RX J0852.0-4622 a so-called barrel-shaped (bilateral) remnant [16]. Similar morphological structure can be found in SN 1006 which has, however, a somewhat steeper radio spectrum of α=-0.6 [9]. In the direction of the missing eastern limb (some 25′ north-east from the expected position of the limb) we observe RCW 37 (feature D from [11]). RCW 37 is an Hα feature and classified as a H II region by Rogers et al. [17]. This feature is also seen in Fig. 1 as a sharp linear feature in [O III] and there is some evidence for the EUV emission as well. We do not have a radio spectral index for this object.

The radio diameter of RX J0852.0-4622 is measured from the 4850 MHz image

[15] to be $116'\pm 2'$ in Dec direction and $108'\pm 2'$ in RA direction. The brightest radio and X-ray component is at position angle $320°\pm 5°$. The long axis of the SNR ellipse is also at approximately $320°$; parallel to and $1.2°$ south of the Galactic plane. The X-ray diameter of RX J0852.0-4622 is larger than the radio diameter by about 10%. The difference of measured diameter, and the small displacement of radio to X-ray peaks, also makes RX J0852.0-4622 similar to SN 1006 [16].

The western limb of RX J0852.0-4622 is represented in the X-ray image as X-ray luminous feature. However, elongated and very weak emission at 2420 MHz is noticed in that region. This absence of stronger radio-continuum emission suggests that synchrotron radiation makes a negligible contribution to the dominant X-ray emission of this feature.

It is noted that for RX J0852.0-4622 the ratio of flux density at 2420 MHz (~ 16 mJy arcmin^{-2} for the strongest north feature) over that at 1 keV (1.2×10^{-8} Jy arcmin^{-2}) of 1.3×10^6 is similar to that of SN 1006 (7×10^5), thus supporting a synchrotron model (for the X-rays) but with significantly less non-thermal energy density.

REFERENCES

1. Aschenbach, B., 1998, *Nature* **396**, 141
2. Iyudin, A.F., Schönfelder, V., Bennett, K., et al., 1998, *Nature* **396**, 142
3. Rood, T.R., Sarazin, C.L., Zeller, E.J., Parker, B.C., 1979, *Nature* **282**, 701
4. Burgess, C.P., Zuber, K., 2000, *Astroparticle Physics* **14**, 1
5. Tsunemi, H., Miyata, E., Aschenbach, B., Hiraga, J., Akutsu, D., 2000, *PASJ* in press
6. Slane, P., Hughes, J.P., Edgar, R., et al., 2000, *ApJ* in press.
7. Chen, W., Gehrels, N., 1999, *ApJ* **514**, L103
8. Aschenbach, B., Iyudin, A.F., Schönfelder, V., 1999, *A&A* **350**, 997
9. Green, D.A., 1998, 'A Catalogue of Galactic Supernova Remnants (1998 September version)', Mullard Radio Astronomy Observatory, Cambridge, United Kingdom, http://www.mrao.cam.ac.uk/surveys/snrs/
10. Milne, D.K., 1968, *Aust. J. Phys.* **21**, 201
11. Combi, J.A., Romero, G.E., Benaglia, P., 1999, *ApJ* **519**, L177
12. Duncan, A.R., Stewart, R.T., Haynes, R.F., Jones, K.L., 1995, *MNRAS* **277**, 36
13. Duncan, A.R., Green D.A., 2000, *A&A* in press
14. Edelstein, J., Vedder, P.W., Sirk, M., 1993, *BAAS* **182**, 41.19
15. Filipović, M.D., Aschenbach, B., Iyudin, A.F., et al., 2001, *A&A* submitted
16. Winkler, P.F., Long, K.S., 1997, *ApJ* **491**, 829
17. Rodgers, A.M., Campbell, C.T., Whiteoak, J.B., 1960, *MNRAS* **121**, 103

High frequency radio observations of the composite historical SNR G11.2-0.3

Roland Kothes[1] and Wolfgang Reich[2]

[1] *National Research Council, HIA, DRAO, Penticton, Canada*
[2] *Max Planck Institut für Radioastronomie, Bonn, Germany*

Abstract.

The historical supernova remnant G11.2-0.3 shows evidence of an evolutionary gradient from ejecta dominance on one side to the dominance of the swept up material on the other. This change seems to be gradually implying an almost homogenous ambient medium structure with a density gradient. We found evidence for this behaviour in the radio continuum (total power and polarisation), the magnetic field structure, and the distribution of the soft X-ray emission. In the center we discovered a small flat spectrum source which is most likely a synchrotron nebula powered by the central pulsar. There are two jetlike extensions to the east and west best visible at lower frequencies indicating a collimated outflow of particles.

THE OBJECT

G11.2-0.3, the remnant of the historical supernova of 386 AD, is one of only a small number of young supernova remnants (SNRs) known in the Galaxy. The SNR shows the typical structure of a shell-type SNR with a spectral index $\alpha = -0.5$ ($S \sim \nu^{\alpha}$) in the radio continuum. The lack of a sharp outer boundary suggests an early evolutionary state [1]. Morsi & Reich [2] found evidence for a flat spectrum core within the remnant. HI absorption suggests a distance of 5 kpc [1] [3] which translates to a diameter of 6 pc, a typical size for a young SNR. X-ray observations reveal a soft shell [4] with a hard spectrum core [5]. Torii et al. [6] [7] discovered a young pulsar in the center with a period of 65 ms and a rotational energy loss rate of $\dot{E} = 8.8 \cdot 10^{36}$ erg/s.

For a better understanding of the properties of G11.2-0.3, in particular its central component and its polarisation structure, we conducted a number of new observations at high radio frequencies with the Effelsberg 100-m radiotelescope at 4.85 GHz, 10.45 GHz, 14.7 GHZ, and 32 GHz including linear polarisation measurements except at 14.7 GHz. Maps of total intensity and polarised intensity at 32 GHz are displayed in figure 1. The overall spectrum of G11.2-0.3 gives a spectral index of $\alpha = -0.50 \pm 0.02$ which is rather flat compared to the other young shell-

CP565, *Young Supernova Remnants: Eleventh Astrophysics Conf.*, edited by S. S. Holt and U. Hwang
© 2001 American Institute of Physics 0-7354-0001-6/01/$18.00

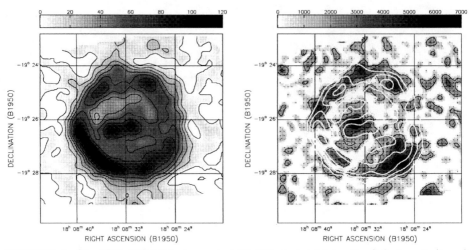

FIGURE 1. The SNR G11.2-0.3 observed at 32 GHz in total intensity (left, in mJy/beam) and polarised intensity (right, in μJy/beam). Vectors in E-field direction are indicated. White contours in the PI map represent total intensity.

type SNRs like Kepler, Tycho, Cas A, and SN1006, which show spectral indices between -0.6 and -0.7 [8].

THE STRUCTURE OF THE SHELL

G11.2-0.3 shows an almost circular shell, which is divided in 4 prominent sectors (fig. 1). These sectors reveal a clumpy structure in total intensity except the southeastern one (Sector I), which is narrower with a higher peak flux. This indicates that this part of the remnant is further evolved than the others, since the compression of the ambient medium increases with an evolving SNR.

A map of polarised intensity at 10.45 GHz with vectors in B-field direction corrected for Faraday rotation is shown in figure 2. G11.2-0.3 shows a radial magnetic field structure which indicates ejecta dominated evolution. However, Sector I reveals a more disturbed magnetic field with much lower polarisation. This supports the earlier assessment that this part of the remnant is further evolved. Percentage polarisation of an SNR will decrease up to the point where the contribution of the swept up material to the synchrotron emission equals that of the ejecta since the magnetic field structures of those expanding layers are perpendicular to each other.

The ROSAT HRI image in figure 2 [4] reveals that most of the soft X-ray emission is coming from Sector I. The intensity is decreasing away from it. This indicates increasing X-ray emitting mass towards Sector I. The temperature is decreasing since lower temperatures give a higher soft X-ray emissivity. The most likely reason for this is a density gradient around the SNR with increasing density away from the Galactic plane. X-ray observations also indicate swept up material between 3 and

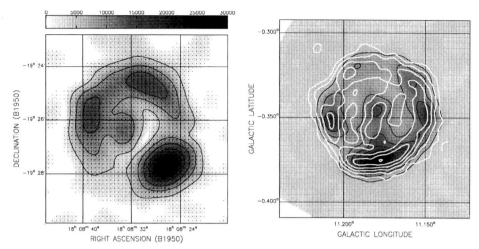

FIGURE 2. Left: Map of polarised intensity at 10.45 GHz with vectors in intrinsic B-field direction corrected for Faraday rotation at a resolution of 1.15' (μJy/beam). Right: ROSAT HRI image of G11.2-0.3. White contours represent radio continuum at 32 GHz.

4 $M_\odot$ [4] [5]. The mass of the ejecta should be significantly higher since it is still dominating the overall behaviour of the remnant. Together with the lack of any HI bubble around it and the recently discovered pulsar [6] this indicates a type II supernova with an early B-type progenitor star.

THE INTERIOR STRUCTURE

Within the SNR there are two slightly extended sources visible at 32 GHz (fig. 1). In order to get an idea of the spectral behaviour we made use of the NVSS data at 1.4 GHz [9]. Unresolved structures in the remnant are shown in figure 3 at 1.4 GHz and 32 GHz at a common resolution of 45". Diffuse structures were removed by using the background filtering method invented by Sofue & Reich [10]. Obviously the inner structure has a much flatter spectrum than the shell. The possibility of confusion with extragalactic background objects can be ruled out. The high resolution maps of Green et al. [1] resolve the compact sources into diffuse extended emission about 1' in diameter. Using the correlation between a pulsar's energy loss rate $\dot{E}$ and the radio surface brightness of its synchrotron nebula found by Kothes [11] we get a flux density of about 1 Jy at 1 GHz with $\dot{E} = 8.8 \cdot 10^{36}$ erg/s [7]. Assuming $\alpha = 0.0$ for the pulsar nebula we get $\alpha = -0.57$ for the shell, which is more typical for a young SNR than the -0.5 calculated before. The position of the pulsar has an uncertainty of 1', however, it is very likely that its position coincides with the X-ray source in the very center of the remnant and thus with the central point source in the 32 GHz map (figure 2). The second point source visible

273

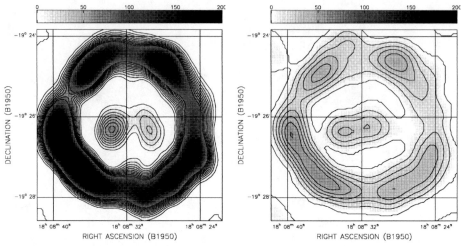

FIGURE 3. Unresolved structures in G11.2-0.3 at 1.4 GHz and at 32 GHz (right) in mJy/beam.

at 32 GHz is located at the same position as the stronger one in the NVSS map. However, the weaker NVSS source is located further west than the center source at 32 GHz. We note similarities with the plerionic SNR G21.5-0.9, where a compact X-ray source [12] is seen in the center of axisymmetric radio structures suggesting some collimated outflow of particles [13].

REFERENCES

1. Green D.A., Gull S.F., Tan S.M. Simon A.J.B., 1988, MNRAS 231, 735
2. Morsi H.W., Reich W., 1987, A&AS 71, 189
3. Becker R.H., Markert T., Donahue M., 1985, ApJ 296, 461
4. Reynolds S.P., Lyutikov M., Blandford R.D., Seward F.D., 1994, MNRAS 271, L1
5. Vasisht G., Aoki T., Dotani T., Kulkarni S.R., Nagase F., 1996, ApJ 456, L59
6. Torii K., Tsunemi H., Dotani T., Mitsuda K., 1997, ApJ 489, L145
7. Torii K., Tsunemi H., Dotani T., Mitsuda K., Kawai N., Kinugasa K., Saito Y., Shibata S., 1999, ApJ 523, L69
8. Green D.A., 2000, A Catalogue of Galactic Supernova Remnants. (http://www.mrao.cam.ac.uk/surveys/snrs)
9. Condon J.J., Cotton W.D., Greisen E.W., Yin Q.F., Perley R.A., Taylor G.B., Broderik J.J., 1998, AJ 115, 1693
10. Sofue Y., Reich W., 1979, A&AS 38, 251
11. Kothes R., 1998, MmSAI 69, 971
12. Slane P., Chen Y., Schulz N.S., Seward F.D., Hughes J.P., Gaensler B.M., 2000, ApJ 533, L29
13. Fürst E., Handa T., Morita K., Reich P., Reich W., Sofue Y., 1988, PASJ 40, 347

Radio Proper Motions: Cas A & Kepler

Barron Koralesky[1,2] and L. Rudnick[1]

[1] *University of Minnesota, Minneapolis, MN 55455, USA*
[2] *Macalester College, Saint Paul, MN 55105, USA*

Abstract.

We present a preliminary analysis of a new generation of proper motion measurements of the radio emission from the Cas A and Kepler supernova remnants. We utilized a cross-correlation type of measurement which is sensitive to features with a wide range of sizes and shapes. Initial results show a great deal of velocity structure in both remnants, superposed on their overall expansion. We find many inward-moving features in Cas A, and show the X-ray/radio velocity discrepancy in Kepler.

INTRODUCTION

Proper motion measurements of SNRs provide important information on their dynamics, MHD structure, evolutionary state, and interactions with their circumstellar and interstellar environments. Cas A's motions have been studied in the most detail, with the overall expansion first visible in the motion of the FMKs [1], with more recent work on these emission line features showing that they illuminate a 5300 km/s sphere with a center offset by 770 km/s [2]. Radio ring components show a much slower expansion (≈ 1500 km/s, on average) with superposed random motions [3,4]. Paradoxically, the coincident bright X-ray and radio rings in Cas A have average velocities which differ by a factor of two [5,6]. Kepler's radio ring has been significantly decelerated to an expansion parameter of $m \approx 0.5$ from the initial free expansion ($m = 1.0$) and shows significant azimuthal asymmetries [7] (both properties are also shared by Cas A). Like the Quasi-Stationary-Flocculi in Cas A, Kepler's optical emission is from shocked and heated circumstellar material, with very slow expansion velocities [8,9]. With the current investigation, we intend to give a more complete analysis of the dynamical state of the remnants, including both the radial dependencies of the expansion rate and a characterization of the various scales on which velocity structure occurs.

CP565, *Young Supernova Remnants: Eleventh Astrophysics Conf.,* edited by S. S. Holt and U. Hwang

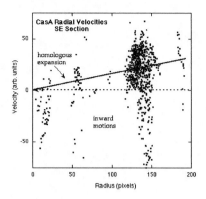

FIGURE 1. Radial velocity as a function of distance from center (both units arbitrary) for Southeastern section of Cas A. Negative velocities imply inward motions.

I MEASUREMENTS

We compared matched VLA [1] λ 6 cm images at two epochs, with time intervals of 11 and 13 years, for Cas A and Kepler, respectively. Proper motion measurements were made within sliding boxes of 7″ (at 1.3″ resolution) for Cas A and 9″ (at < 3″ resolution) for Kepler. These measurements are sensitive to motions of arbitrarily shaped structures in arbitrary directions, and complement other studies that have measured compact features only or the average velocities of the bright rings.

For each box, we subtracted a shifted version of the same box from the earlier epoch, and minimized the rms difference as a function of the two dimensional shift. This yields proper motions which are typically accurate to sub-pixel levels whenever there is adequate structure within the box. Typical motions are on the order of 0.5″ − 2″ over the time intervals studied. The velocity vectors shown in this paper are generally of sufficient signal:noise to be trustworthy, but the detailed statistical errors are still being determined.

A potential source of "error" in the velocities, especially for Cas A, comes from subtle changes in structure and/or relative brightness of features within a box, which can then masquerade as proper motions. However, elimination of measurements where there is a large residual after subtraction (evidence of structural change) does not change any of the velocity patterns seen.

II INITIAL CAS A RESULTS

The two-dimensional plot of velocities appears very similar to those previously published [3,4] with much better coverage. The highly turbulent nature of the flow mirrors the extensive fine structure seen in the remnant, especially interior to the

---

[1] The VLA is a facility of the U.S. National Science Foundation, operated by the National Radio Astronomy Observatory on behalf of Associated Universities, Inc.

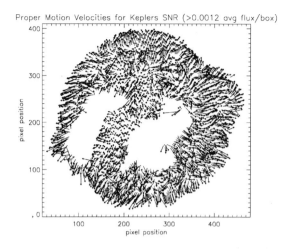

FIGURE 2. Measured proper motions for Kepler's SNR.

bright ring. There are also some large scale regions which show coherent non-radial flows. Using this expanded data set, we see the general linear rise in velocity with increasing radius that is expected from either a homologous expansion of a thin, uniform velocity shell (Figure 1, showing data from the "clean" Southeast region). Note, however, the many *inward* moving features (negative velocities); most of these are statistically significant. There is a large inward moving population at the outer edge of the bright ring. The origin of these inward moving features is not yet known, although they are probably related to ejecta decelerated and then swept up by the inward moving reverse shock. At radii less than about 50 pixels ($\approx 20''$), the expansion velocities become confused with transverse motions.

III   INITIAL KEPLER RESULTS

Since Kepler's motions have been less extensively studied, we show several of our new key results. Figure 2 shows the two dimensional velocity vectors, in which considerable structure, with typical deviations from radial of $\approx 40\,$deg, is seen on top of an overall expansion. In Figure 3, we take a more detailed look at the velocity as a function of radius for two sections of the remnant. In the East, we see the signature of homologous expansion. In the North however, where the emission is strongest in the optical, IR, and X-ray, as well as in the radio, there is a strong *decrease* in velocity at large radii. The outer edge of the radio ring has been significant decelerated, probably by the dense thermal plasma. (There may be some radial dependence to the X-ray velocities as well [10].)

A comparison of Kepler's velocity structures and the local magnetic field direction [7, 11] shows a number of regions that are tightly correlated. This indicates that local shear motions have probably been responsible for amplifying the field. Had

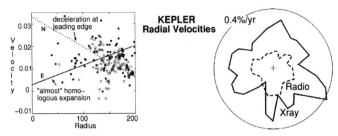

FIGURE 3. Left: Radial velocity as a function of radius for two segments of Kepler's SNR, showing very different behaviors. Right: Kepler expansion rates for X-ray (solid, [10]) and radio (dashed) as a function of azimuth. Zero velocity (the cross) is well-defined for radio velocities, but X-ray values could be shifted around the center position by misalignments between epochs.

the magnetic field simply been frozen into independent blobs of plasma, there would be no general preferred orientation. Alternatively, had shock amplification been the dominant process, then we would have seen stronger field transverse to the motion. Local shear motions are expected from the Rayleigh-Taylor instabilities that are thought to provide net radial fields in young SNRs. In Kepler, it is possible that we are actually mapping out the velocity structure of these unstable plumes, although deflections by an inhomogeneous CGM are as plausible at this point.

Figure 3 also shows the same paradoxical discrepancy between Kepler's X-ray [10] and radio velocities as mentioned above for Cas A. Both wavebands show major variations as a function of azimuth, but the radio expansion rate is, on average, half that seen in the X-rays. We hope that our detailed analysis of the velocity structure will provide some insight into this important, puzzling result.

ACKNOWLEDGEMENTS. This work is supported at Minnesota, in part, by NSF grant AST-9619438 and by a NASA GSRP Fellowship to BTK. We thank T. Delaney for assistance with figures and the magnetic field comparisons.

REFERENCES

1. Kamper, K. and van den Bergh, S., 1976, ApJS, 32, 35
2. Reed, J. E., Hester, J. J., Fabian, A.C. & Winkler, F., 1995, ApJ, 440, 706
3. Tuffs, R., 1986, MNRAS, 219, 13
4. Anderson, M.C. & Rudnick, L. 1995, ApJ, 441, 307
5. Koralesky, B., Rudnick, L., G,otthelf, E., & Keohane, J., 1998, ApJ, 505, 27
6. Vink, J., Bloemen, H., Kaastra, J., Bleeker, J., 1998, AAp, 339, 201
7. Dickel, J., Sault, R., Arendt, R., Korista, K., & Matsui, Y., 1988, ApJ, 330, 254
8. Fesen, R., Becker, R., Blair, W., Long, K., 1989, ApJ, 338, 13
9. Blair, W. P., Long, K. S., & Vancura, O. 1991, ApJ, 366, 484
10. Hughes, J. P., 1999, ApJ, 527, 298
11. Delaney, T., Koralesky, B., Rudnick, L. & Dickel, J., 2001, preprint

Spectral Index Variations in Kepler's Supernova Remnant

Tracey DeLaney[1], Barron Koralesky[1], Lawrence Rudnick[1], and John R. Dickel[2]

[1] *Department of Astronomy, University of Minnesota, Minneapolis, MN 55455, USA*
[2] *Astronomy Department, University of Illinois at Urbana-Champaign, Urbana, IL 61801, USA*

Abstract.
A new epoch of VLA measurements of Kepler's supernova remnant was obtained to make accurate measurements of the proper motions, spectral indices, and other properties. We have combined these new images with Hα, infrared, and X-ray data to decipher the three dimensional structure and dynamics of Kepler, and to better understand the physical relationships between the various nonthermal and thermal plasmas. Spatial variations in the radio spectral index are observed between λ6 and λ20 cm. The spectral indices range from −0.85 to −0.6. The flat spectrum radio emission in the northern and southern parts of the ring is either coincident with or at a larger radius than the steep spectrum emission. In places, the X-ray emission is correlated with the steep spectrum radio emission, in other places, with the flat spectrum radio emission. The Hα and infrared images are remarkably similar. Their leading edges are coincident and are either in front of or coincident with the leading edges of the X-ray and radio emission.

INTRODUCTION

In order to understand supernova remnants (SNRs), we must know how the various thermal and non-thermal plasmas are coupled. These plasmas provide clues as to where shock structures are located. The location of shock structures, when coupled with spectral index measurements, will help us to understand particle acceleration in SNRs.

Kepler's supernova remnant is the second youngest SNR known in the Galaxy. As a young, bright SNR, it has been studied at many wavelengths. In this paper we wish to provide an improved look at the radio structure and spectral variations. We also look in more detail at the relationship between the radio emission and those at optical, infrared (IR), and X-ray wavelengths.

CP565, *Young Supernova Remnants: Eleventh Astrophysics Conf.,* edited by S. S. Holt and U. Hwang
© 2001 American Institute of Physics 0-7354-0001-6/01/$18.00

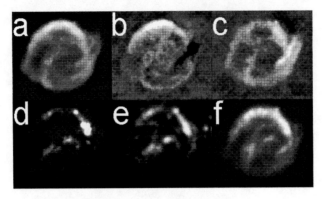

FIGURE 1. Gray-scale images of (a) λ6 cm radio continuum, (b) flat spectrum radio, (c) steep spectrum radio, (d) Hα, (e) IR, and (f) X-ray.

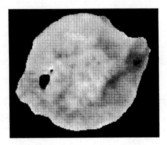

FIGURE 2. Radio spectral index. The gray-scale range is –0.85 (darkest) to –0.6 (lightest).

KEY FINDINGS

Gray-scale images of the λ6 cm radio, Hα, IR, and X-ray emission are shown in Figure 1 along with flat and steep spectrum radio emission. All of these images are at a resolution of $7\farcs2$.

The radio total intensity image was reconstructed from λ6 cm VLA data taken in 1997. We also show radio spectral images utilizing "spectral tomography," a technique designed to separate overlapping spectral features [1]. The "flat" tomography image represents radio emission flatter than $\alpha = -0.75$ while the "steep" image represents emission steeper than $\alpha = -0.65$.

To make the X-ray image, the 1997 ROSAT high resolution imager data were obtained from the ROSAT public archive. The Hα image was taken in 1987 [2] and was kindly provided by W. Blair. The IR image represents 12 μm data taken in 1996 with ISO [3] and kindly provided by P.-O. Lagage and T. Douvion.

Spectral index variations are observed between $\alpha = -0.85$ to –0.6. The protrusion ("bump") at a position angle of $-25°$ is steep in spectral index as is the "ear" on the western side of the remnant. Figure 2 is a gray-scale map of radio spectral index between $\lambda 6$ and $\lambda 20$ cm.

Plots of angle averaged radial profiles around selected regions of the remnant can be seen in Figure 3. The Hα and IR are at the leading edge and are $\approx 15''$ wide. In the north to northeast, the flat and steep spectrum radio and X-ray track each other, are $\approx 25''$ wide, and trail the Hα and IR by $\approx 5''$. In the northwest, the X-ray has a similar width to the Hα and IR but trails by several arcseconds. Also in the northwest, the flat spectrum radio is similar to the X-ray while the steep spectrum radio is $\approx 40''$ wide and trails the flat by $\approx 15''$. In the south, the steep radio tracks the X-ray and trails the flat radio by $\approx 15''$.

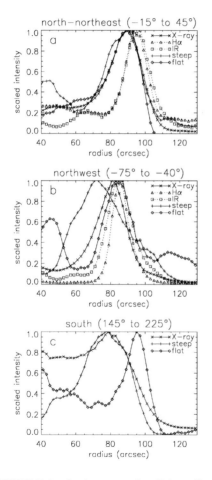

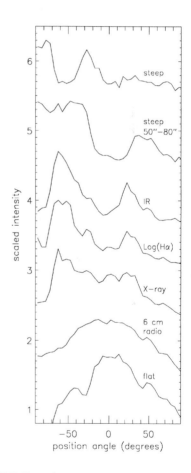

FIGURE 3. Angle averaged radial profiles at all wavebands for the specified regions. The center used is $\alpha_{50} = 17^{h}27^{m}41\overset{s}{.}42$ and $\delta_{50} = -21°27'15\overset{''}{.}35$.

FIGURE 4. Average intensity in an annulus from $70'' - 100''$ ($50'' - 80''$ for the steep) in radius around the northern ring. The center used is the same as that in Figure 3.

In the northern half of the ring, the Hα and IR emission is concentrated to the northwest and northeast. In contrast, the total radio emission is brightest directly north. The X-ray emission is a hybrid between these two. The flat spectrum radio is similar to the total radio emission in the north. The steep spectrum radio emission exhibits similar azimuthal dependence to the Hα and IR emission but at a smaller radius ($50'' - 80''$). The azimuthal plots are shown in Figure 4.

COMMENTS AND QUESTIONS

1. Why is there a segregation of flat and steep spectrum regions? The findings to date show no universal pattern of association between the X-ray plasma and the two different radio spectral components. It may be possible that the steep spectrum radio emission tracks the reverse shock while the flat radio tracks the primary shock. This would explain the correlation between the X-ray and steep radio in the south if the X-ray emission there results from the reverse shock.

2. In diffusive shock acceleration, spectral index depends only on compression. If the shock velocity is constant (little deceleration) around the remnant, then higher densities (lower temperatures) lead to higher Mach numbers and flatter expected spectral indices. Comparisons of spectral index with local velocity measurements (see paper by Koralesky *et al.*, these proceedings) may help to evaluate these effects.

3. If the flat spectrum radio emission originates from higher density regions, then a correlation with the X-ray, Hα and IR emission is expected. The X-ray has some bright emission in the north where the flat spectrum component is strong, but the Hα and IR do not, Similarly, the flat spectrum radio emission is not the brightest in the northwest where the X-ray, Hα, and IR are the brightest.

4. The spectral index variations may be due to other factors than just compression. Different seed particle distributions, different strength magnetic fields, and shock obliquity to the ambient magnetic field may also play a role.

This work has been conducted with support by NSF grant AST 96-19438 at the University of Minnesota. The VLA is operated by the National Radio Astronomy Observatory, which is a facility of the National Science Foundation, operated under cooperative agreement by Associated Universities, Inc.

REFERENCES

1. Katz-Stone, D. M. and Rudnick, L. 1997, ApJ, 488, 146
2. Blair, W. P. *et al.* 1991, ApJ, 366, 484
3. Douvion, T. *et al.* 1999, in *The Universe as seen by ISO: ESA proceedings SP-427*, ed. P. Cox & M. F. Kessler, (ESTEC, Noordwijk, The Netherlands: ESA Publications Division), 301

The Crab and Related Remnants

The Crab Nebula: The Gift that Keeps on Giving

J. Jeff Hester

Department of Physics & Astronomy
Arizona State University, Tempe AZ, 85287-1504, USA

Abstract. In many ways the Crab Nebula is the object that started off the study of supernova remnants, yet hundreds of years after its discovery it is still poorly understood in many respects. Recently, a number of observational and theoretical results have shed new light on longstanding questions and misconceptions about the Crab. We are taught in graduate school that the Crab is a freely expanding ejecta-dominated remnant, with the slight added complication that it contains a pulsar and synchrotron nebula. This conception of the Crab is incorrect. Instead, when we think of the Crab we should think *first* of the powerful and dynamic axisymmetric wind from an energetic pulsar as it powers a high pressure synchrotron nebula. That synchrotron nebula is sweeping up and concentrating thermal ejecta into dense, complex Rayleigh-Taylor filaments as it pushes its way out through a large, freely expanding remnant. This larger remnant is all but unseen, but probably carries the bulk of the mass and kinetic energy from the explosion. Every aspect of the visible Crab – from its overall size and shape, to the complex structure of its filaments, to the highly dynamical structure at its heart – is a direct result of the action of the wind from the Crab pulsar.

INTRODUCTION

The Crab Nebula is certainly one of the best known and arguably one of the most important objects in astrophysics. The Crab is the remnant of a supernova that was seen by Chinese astronomers in 1054 AD. After the impressively comprehensive review of the history of observations of the Crab given by Virginia Trimble (this volume), further review here would be redundant at best, and perhaps even presumptuous.

At the center of the Crab Nebula is the Crab pulsar. The spindown luminosity of this 33 ms pulsar ($\sim 5 \times 10^{38}$ ergs s^{-1}) is carried away by a relativistic wind. In the outer part of the Crab there are filaments of thermal material consisting of ejecta from the supernova. Between the two is the Crab synchrotron nebula. This cloud of magnetized ultrarelativistic electrons and positrons is powered by the wind from the pulsar and is in turn confined by the surrounding thermal ejecta. Despite the fact that the Crab was the first object unambiguously recognized as

CP565, *Young Supernova Remnants: Eleventh Astrophysics Conf.,* edited by S. S. Holt and U. Hwang

the remnant of a historical supernova, there are many longstanding problems with our understanding of the remnant. The first problem is that the thermal ejecta has a mass of $\sim 1 - 2M_\odot$, which is only around 20% of what would be expected. The velocity of the fastest moving ejecta in the Crab is around 1500 km/sec, which is much less than the 5-10,000 km/sec typical of young SNRs. Combining these, the kinetic energy of the visible ejecta is $\sim 3 \times 10^{49}$ ergs, which is much less than the canonical 10^{51} ergs of kinetic energy released by a supernova. An obvious question to ask when looking at these statistics on the Crab is whether we are seeing all that there is. Anticipating the conclusion of this presentation, the answer to that question is "no."

THINGS WE KNOW ABOUT THE CRAB

The Crab has been so well studied that it is impossible, especially in a short conference submission, to adequately review the literature on the object. Instead, we will approach the Crab by listing a number of interesting things that we know about the Crab.

1. The Crab is axisymmetric, and this axisymmetry is a consequence of the wind from the pulsar. The Crab is extended on the sky along an axis from the SE to the NW. X-ray observations of the Crab show a bright torus of synchrotron emission surrounding and bipolar jet-like features extending along this axis of symmetry. Recent *Chandra* observations of the Crab show that the torus breaks into many fine wisps, and also shows the presence of a bright ring of X-rays interior to the previously recognized torus.[1] The jets are quite prominent in these observations as well. Tracing this back, the observed structure must reflect the structure of the pulsar wind, which must include both a collimated polar jet and a strong equatorial wind. When thinking of the Crab it is this axisymmetric picture that should immediately come to mind.

The axisymmetric structure of the Crab is also seen in visible light images of the Crab taken with the *HST*[2]. Features along the spin axis include the jet and counterjet, as well as a number of bright knots. The inner knot is located approximately 0″.5 SE of the pulsar along the jet axis. A second knot is located along this same axis approximately 5″ from the pulsar, at the base of the optical and X-ray jet. There are also indications in the outer nebula that material flowing out along the jet has distorted the magnetic topology of the synchrotron nebula. The equatorial torus projects onto the sky as the region of the well known synchrotron wisps. There are also features associated with the high latitude wind from the pulsar. These include the "halo" in the *HST* observations, as well as the region to the SE referred to as the "anvil." The observed structure is *not* symmetrical between the two poles of the pulsar, and the axisymmetry of the system is also broken fairly close to the pulsar itself.

The filaments of ejecta share the overall axisymmetry imposed by the pulsar wind. This is perhaps best demonstrated by Lawrence et al., who obtained Fabry-Perot observations of the Crab[3]. Their data shows that when viewed along the symmetry axis, the Crab filaments array themselves in a thick "cigar band" around this axis of symmetry.

2. The Crab synchrotron nebula is highly dynamic.

The sound speed in the synchrotron plasma is $\sim c/\sqrt{3}$, which means that the sound crossing time for the wisps observed at *HST* resolution is around 4 days! Scargle documented the variability of the Crab Nebula, but his ground based observations had low resolution, and were taken with separations of months[4]. As a result, his findings were badly aliased. More recently, *HST* has been used to obtain observations of the Crab over much shorter time intervals[5]. These results show significant changes in the structure of the synchrotron nebula on time scales as short as one week. These data demonstrate the persistence of the inner knot and high latitude halo, the extremely dynamical nature of the knot $\sim 5''$ SE of the pulsar, and the very rapid motion of the equatorial wisps, which appear to move outward from the pulsar at speed of $\sim 0.5c$. The dynamical nature of the Crab is currently the target of an observational campaign using *HST*, *Chandra*, and the VLA.

The knot located $5''$ to the SE of the pulsar is perhaps the most dynamic structure in the nebula, earning it the nickname the "sprite." The sprite may look completely different in two images taken only a few days apart. The obvious interpretation of the sprite is to suggest that it is a shock in the pulsar's polar wind. The sprite is located at the base of the polar jet, and diffuse synchrotron features can be seen moving away from the back side of the sprite at $\sim 0.5c$ along the jet axis. One of the most fascinating structures along the spin axis of the pulsar is the inner knot, located $\sim 1.5 \times 10^{16}$ cm from the pulsar. This knot is highly polarized, with a magnetic field perpendicular to the spin axis. The emissivity of the knot is ~ 6000 times the nebula average. The polarization of the knot, along with its variability in position and brightness, identifies it with a stable synchrotron feature in the polar jet. It has been suggested that this might be a quasistationary shock[6], but if this is correct then somehow the polar wind from the pulsar must shock twice, both at the inner knot and at the outer sprite. If the inner knot is a shock, it is also surprising that there is no visible synchrotron emission extending from the knot further out into the nebula, as is seen on the back side of the sprite.

The equatorial wisps in the Crab are seen to move outward at $\sim 0.5c$. As they do so, the wisps sharpen and brighten, and then later fade and become more diffuse. The bright inner ring of X-ray emission lies immediately interior to the locus of the region where visible wisps form. It seems likely that the bright X-ray ring traces the location of the shock in the equatorial wind. (The sprite is also a bright X-ray feature.) Polarization observations with *HST* were able to isolate emission from individual wisps. These observations show that the magnetic field lies along the wisps, as expected if these are flux tubes.

The variability in the synchrotron nebula seen at visible wavelengths is also seen at other wavelengths. X-ray variations in the torus have been reported from

ROSAT observations[7], and similar activity in the torus is seen at radio wavelengths as well[8]. Bietenholz et al. conclude that previous claims of variations in the radio spectral index in the inner Crab are probably actually due to temporal evolution between the epochs during which observations at different frequencies were obtained. This should be taken as a warning when interpreting other claims of spectral index variations based on datasets taken at different times.

The structure and motion of the equatorial wisps can be modeled phenomenologically by assuming that they are circular features moving outward from the pulsar. The parameters for such a model are the wisp velocity, the radius at which a wisp becomes luminous, and the radius at which the wisps turn off. This, together with the tilt of the pulsar axis relative to the line of sight and the pulsar latitude into which a wisps moves, defines the model. It is interesting to note that the same light travel effects responsible for superluminal motion in radio jets also affect the observed structure of the wisps. Changes in arrival time delay as a wisps moves amplifies the apparent speed of wisps coming toward us, while reducing the apparent speed of wisps on the back side of the nebula. As a result, when traced all the way around the nebula, wisps form ellipses that are offset from the pulsar. Models with wisps moving outward into latitudes between ±5° above and below the equatorial plane of the pulsar, and the spin axis tilted by 65° with respect to the observer's line of sight, produce features that reasonably match both the geometry and speed of wisps seen in the *HST* observations.

Many models have been proposed for the Crab wisps over the years. Of recent note is the suggestion that wisps are Kelvin-Helmholtz instabilities [9]. Lou proposed that the wisps are compressible fluctuations in the wind[10], but this model predicts that the wisps should be quasistationary, contrary to observation. Gallant & Arons suggested that the wisps are internal structure in a shocked ion-dominated wind[11], but it has yet to be shown that this model is consistent with observation. Hester proposed that the wisps are due to synchrotron cooling instabilities in the relativistic material flowing outward near the equatorial plane[5]. This is similar to models proposed by Simon & Axford[12], and discussed by Bodo et al.[13]

The synchrotron emissivity of a plasma at a given pressure peaks when the plasma is near equipartition between the field and particle pressure. The material that emerges from the wind shock in the Crab, on the other hand, is strongly particle-dominated. If the synchrotron emissivity in a region within such a flow is slightly greater than its surroundings, then the region will lose particle pressure relative to those surroundings and be compressed. This compresses the particles in the region, and it also amplifies the magnetic field. The region (in which the field is now stronger) is pushed closer to equipartition, and so its synchrotron emissivity is even higher. What follows is a classical cooling instability as the original region grows both sharper and brighter. The instability continues until the region reaches the point that further losses cause the synchrotron emissivity to drop. The wisp fades and eventually blends into its background.

This scenario for wisp evolution is in good agreement with the observations. The time scale for wisp evolution should be set by the synchrotron lifetimes of the particles responsible for carrying the pressure of the plasma. In the case of the Crab, this would have to be the high energy particles responsible for the X-ray emission from the inner Crab. It is therefore significant that the volume in which the equatorial wisps are seen coincides with the volume of the equatorial X-ray torus, reflecting the fact that the time scale for synchrotron losses of the X-ray emitting particles and the time scale for dynamical evolution of the wisps are comparable. This supports the idea that radiative losses in the high energy electrons drive the wisps.

3. The Crab filaments result from Rayleigh-Taylor instabilities between the synchrotron nebula and thermal ejecta.
The synchrotron nebula in the Crab is confined by surrounding thermal material. *HST* observations of the thermal filaments show that they result from Rayleigh-Taylor instabilities at the surface of the synchrotron nebula[14]. Chandrasekhar originally worked out the theory for magnetic R-T instabilities by balancing the buoyant force driving the instability with magnetic tension, which acts to stabilize the interface[15]. This theory predicts a relationship between the magnetic field, the local gravity, the density, and the wavelength at which the instability grows most rapidly. These four quantities can each be determined in the Crab from observation, and turn out to satisfy the predicted relationship.

The R-T instability in the Crab arises as the pressure of the synchrotron nebula tries to push the heavy confining thermal material out of its way so that it can expand. If the synchrotron plasma had pushed all the way through the thermal ejecta, then filaments should be isolated radial fingers. Observations, however, show that they are not. Instead, they are connected by their neighbors by bridges of relatively faint, high ionization thermal material. In fact, around most of the nebula the synchrotron nebula is bounded on the outside by a thin skin seen in [O III] and other high ionization species. The physics of the R-T instability therefore demands that there must be a confining medium surrounding the visible Crab[14]. Consideration of the nature of this skin leads us to the fourth interesting statement about the Crab.

4. The visible Crab Nebula is surrounded by an extended, freely expanding ejecta-dominated remnant.
This is not a new idea. Chevalier originally suggested the existence of a larger, freely expanding remnant to account for the missing mass and kinetic energy in the Crab[16]. The theory for blowing a pulsar wind bubble inside of a freely expanding remnant was worked out by Reynolds & Chevalier[17]. Sankrit & Hester considered the properties of the [O III] skin that surrounds much of the Crab and connects the R-T fingers with their neighbors[18]. They realized that this skin must be the cooling region behind a radiative shock that is driven into the extended freely expanding remnant by the pressure of the confined synchrotron nebula. The properties of the shock around the Crab can be inferred directly from the known properties of the nebula. The pressure driving the shock is just the pressure of the synchrotron nebula. The velocity of the

shock is determined by comparing the observed expansion rate of the Crab with the expansion velocity predicted at that radius assuming free expansion beginning in 1054 AD. It has long been known that thermal ejecta in the Crab is expanding at a speed in excess of free expansion[19], and that the pressure of the synchrotron nebula is responsible for the required acceleration. The visible thermal ejecta is running into the larger, freely expanding remnant at a speed of approximately 150 to 200 km/sec. Balancing the driving pressure and the ram pressure at the shock then gives a mass density of the material around the visible Crab of around 10 amu cm^{-3}.

With the pressure, density, and shock velocity known, we have all we need to calculate a model of the shock at the boundary of the Crab. Sankrit & Hester constructed such a model and compared its predictions with predictions of a photoionization model for the skin[18]. The shock model correctly predicts the brightness of the [O III] skin around the Crab. While it is possible to construct a photoionization model that also matches the [O III] skin simply by adjusting the density, the strength of other lines predicted by the two models are very different. In particular, the shock model predicts much stronger emission from high ionization species than does the photoionization model. For example, emission in both C IV and [Ne V] is predicted to be about a factor of 10 brighter by the shock model than is expected by photoionization. The shock model prediction is in excellent agreement with *HUT* observations of strong, ubiquitous C IV from the Crab at both red- and blue-shifted velocities[20]. The shock model also correctly predicts the strength of [Ne V] in ground-based spectra of the boundary of the Crab[21]. Papers reporting both of these observations struggled unsuccessfully to find photoionization models that could explain the results.

More recently, van Tassell et al. have obtained [Ne V] images of the Crab that show the cooling region behind the shock to be traceable around the perimeter of roughly the southeastern 2/3 of the nebula[22]. This NW/SE asymmetry in the shock is a very interesting result. The NW end of the Crab, around which the shock cannot be traced in [Ne V], is the same region where Velusamy reported a "breakout" of the synchrotron emission beyond the confining thermal ejecta[23]. Rather than a true breakout, we suggest that the shock here as simply become nonradiative. As discussed by Sankrit & Hester, a lower preshock density means that the cooling time behind the shock is longer. When the postshock cooling time becomes greater than the dynamical age of the system, the shock becomes nonradiative, and the high ionization skin is no longer seen. Interestingly, this also means that filament formation has stopped in this direction. Filaments grow from R-T instabilities at the interface between the synchrotron nebula and cooled material on the back side of the shock. Without cooling and compression, the density of postshock material will be too low to drive the instability, and filament formation will cease. Hence in regions where the skin is absent, we should see the synchrotron nebula beginning to extend beyond the limit of the thermal filaments.

This is exactly what is happening in the NW portion of the nebula. We are catching the Crab at a very interesting time, just as the shock driven by the synchrotron nebula is undergoing a transition from radiative to nonradiative. The NW/SE asymmetry itself is likely the result of the motion of pulsar. The Crab pulsar is moving toward the NW[24]. As a result, the synchrotron nebula in the NW is pushing into ejecta that is located farther from the explosion center, and hence is lower density than the preshock material in the SE.

If a larger, ejecta-dominated remnant surrounds the visible Crab, it is reasonable to ask why evidence of this remnant is not seen directly. The answer in part is that the Crab sits within a large, low density cavity. Observations of 21 cm emission show that the Crab is located within a large cavity in the local H I density that may be as large as $3°.8$, or about 130 pc across[25]. The radius of such a bubble divided by the age of the remnant gives a speed of around 70,000 km/sec that would be required for ejecta to have hit the wall of the cavity by now. Given this fact, it is not surprising that no signs of such an interaction have yet to be seen. It is interesting to note, however, that the Crab-like remnant SN0540 in the LMC is seen to be surrounded by a large, faint shell remnant[26]. This is presumeably what the Crab would look like today if the explosion had occurred inside of a somewhat smaller bubble.

The evidence supporting the existence of an extended freely expanding remnant around the Crab is very strong. It is the opinion of the author that this extended remnant has been demonstrated to exist. Even so, searches for emission from an extended remnant around the Crab have consistently failed. These searches have been conducted in X-ray, radio, and especially visible light[27]. The difficulty is that the density of this material is low – less than about 7 cm^{-3} at the edge of the nebula and falling rapidly from there. The filling factor is high, owing to the fact that UV emission from the synchrotron nebula should have photoevaporated small structures not confined by the synchrotron pressure that confines the visible filaments. The temperature and ionization state of the material should be high. The abundance of hydrogen in the ejecta should also be low. All of these effects mean that recombination emission from hydrogen (the favored line for searches) should be very faint[18]. When combined with the fact that emission lines from the halo would be very broad, it is not surprising that searches have failed.

There *has* been a direct detection of the extended remnant around the Crab, however. Sollerman et al. present the results of *HST STIS* spectroscopy of the pulsar[28]. They detect the presence of C IV $\lambda1550$ in absorption. The C IV line is blue shifted, with the line appearing at around -1500 km s^{-1}, equal to the velocity of material at the outer edge of the synchrotron bubble. The line is broad, extending out to a blue shift of around -3000 km s^{-1}, and perhaps beyond. The relatively low signal to noise in the *STIS* spectrum makes it difficult to infer detailed properties of the extended remnant around the Crab. Sollerman et al. place a lower limit of 0.3 $M_\odot$ on the ejecta mass surrounding the Crab and a lower limit of 1.5×10^{49} ergs of kinetic energy, but find that their observations allow an ejecta mass of 8 $M_\odot$ or more, and a kinetic energy of 10^{51} ergs in the extended remnant. This is consistent with the model of Sankrit & Hester, who found that comparable amounts of material could be effectively hidden in the extended remnant without violating the observational constraints on Hα emission from the halo.

The implication is clear. What we see of the ejecta in the Crab Nebula is only the ejecta that has been swept up by the expanding synchrotron nebula and concentrated into filaments by Rayleigh-Taylor instabilities. Most of the remnant mass and kinetic energy is probably in a much more extended cloud of freely expanding ejecta.

5. Filaments in the Crab have a highly stratified ionization structure that greatly affects their composite spectrum.
Observations of the thermal filaments in the Crab obtained with *HST* have been discussed by several authors[14,18,29,30]. These observations show that the Crab filaments have a very complex and highly stratified ionization structure. Typically this structure consists of a low-ionization core, surrounded by a higher ionization envelope. The cores of the filaments in the Crab are remarkably cool and dense environments. For example, dust in the the dense filament cores can be seen in absorption against the background of the synchrotron nebula. Analysis of absorption profiles in *HST* data suggests that the ratio of dust mass to gas mass in the filament cores could easily be 0.1 or higher, as might be expected for material that is enriched in massive elements[30]. This is also the location of the molecular hydrogen emission emission that has been observed in the filaments[31].

Sankrit et al. calculated photoionization models that match the observed stratified ionization structure of the Crab filaments[30]. While it is clear that the basic picture of power-law photoionization of the filaments is correct, the presence of the observed small scale structure has significant implications for interpretation of spectra of Crab filaments. For example, the *HST* observations show that approximately 85% of the [O I] emission from the Crab comes from very sharp features with scales less than $\sim 0\overset{\prime\prime}{.}5$ (1.5×10^{16} cm). In comparison, less than 15% of the emission from [O III] comes from such small scale features. Different lines demonstrably originate in regions with very different physical properties. This is especially interesting when we consider that photoionization models of the Crab filaments show that the ratio of He I λ5876 to Hβ can vary by a factor of 5 between dense filaments and the surrounding low density regions in which diffuse emission arises. No extreme claims have been made, but a cautionary flag has been raised. The simple ratio of He I λ5876 to Hβ is not, by itself, a fool-proof indicator of helium abundance over the range of conditions that are known to exist across the emitting regions within the Crab. When looking at emission from Crab filaments, extreme stratification of physical conditions across very small scale structures is *not* a second order effect. It is the zeroth order effect.

THIS IS NOT YOUR ADVISOR'S CRAB

We started out to build a list of interesting things that we know about the Crab Nebula. The Crab is axisymmetric, reflecting the symmetry of the pulsar wind.

The Crab synchrotron nebula is far more dynamical than previously understood. Thermal filaments are formed by Rayleigh-Taylor instabilities at the surface of the synchrotron nebula. The Crab Nebula is surrounded by an extended, freely-expanding remnant that probably contains most of the mass and kinetic energy of the explosion. Crab filaments have a highly stratified ionization structure that complicates interpretation of their spectra. All of this adds up to a different picture of the Crab than the picture carried around by many workers in the field. We are taught to think of the Crab as a free expansion SNR with the slight added complication that it contains a pulsar and synchrotron nebula. This conception of the Crab is incorrect. Instead, recent work shows that when we think of the Crab we should think *first* of the powerful and dynamic axisymmetric wind from an energetic pulsar that is concentrating thermal material into dense, complex Rayleigh-Taylor filaments as it pushes its way out through a large, freely expanding remnant. This larger remnant is all but unseen, but probably carries the bulk of the mass and kinetic energy from the explosion. Every aspect of the visible Crab, from the dynamical synchrotron features at its center, to the structure of the filaments, to the shape and energetics of the nebula as a whole is a direct result of the action of the pulsar wind.

REFERENCES

1. Weisskopf, M. C., et al. 2000, ApJ, 536, L81
2. Hester, J. J. et al. 1995, ApJ, 448, 240
3. Lawrence, S. S. et al. 1995, ApJ, 109, 2635
4. Scargle, J. D. 1969, ApJ, 156, 401
5. Hester, J. J. 1998, in Proc. International Conf. Neutron Stars and Pulsars: Thirty Years after the Discovery, ed. N. Shibazaki et al. (Tokyo: Universal Academy Press), 431
6. Lou, Y.-Q. 1998, MNRAS, 294, 443
7. Greiveldinger, C., & Aschenbach, B. 1999, ApJ, 510, 305
8. Bietenholz, M. F., Frail, D., & Hester, J. J. 2001, in preparation.
9. Begelman, M. C. 1999, ApJ, 512, 755
10. Lou, Y.-Q. 1996, MNRAS, 279, 129
11. Gallant, Y. A., & Arons, J. 1994, ApJ, 435, 230
12. Simon, M., & Axford, W. I. 1967, ApJ, 150, 105
13. Bodo, G. et al. 1992, Astron. Astrophys. 256, 689
14. Hester, J. J. et al. 1996, ApJ, 456, 225
15. Chandrasekhar, S. 1961, Hydrodynamical and Hydromagnetic Stability (Oxford: Oxford Univ. Press)
16. Chevalier, R. A. 1977, in Supernovae, ed. D. N. Schramm (Dordrecht: Reidel), p 53
17. Reynolds, S. P., & Chevalier, R. A. 1984, ApJ, 278, 630
18. Sankrit, R., & Hester, J. J. 1997, ApJ, 491, 796
19. Trimble, V. 1968, AJ, 73, 535
20. Blair, W. P. et al. 1992, ApJ, 399, 611

21. Davidson, K. 1979, ApJ, 228, 179
22. van Tassell, H. et al., in preparation
23. Velusamy, T. 1984, Nature, 308, 251
24. Wyckoff, S., & Murray, C. A. 1977, MNRAS, 180, 717
25. Wallace, B. J., Landecker, T. L., Kalberla, P. M. W., & Taylor, A. R. 1999, ApJ(Suppl), 124, 181
26. Morse, J. A. 2000, private communication
27. Fesen, R. A., Shull, J. M., & Hurford, A. P. 1997, AJ, 113, 354
28. Sollerman, J. et al. 2000, ApJ, 537, 861
29. Blair, W. P. et al. 1997, ApJ(Suppl), 109, 473
30. Sankrit, R. et al. 1998, ApJ, 504, 344
31. Graham, J. R., Wright, G. S., & Longmore, A. J. 1990, ApJ, 352, 172

Plerions and Pulsar-Powered Nebulae

Bryan M. Gaensler[1]

Center for Space Research, Massachusetts Institute of Technology, Cambridge MA 02139, USA

Abstract.
In this brief review, I discuss recent developments in the study of pulsar-powered nebulae ("plerions"). The large volume of data which has been acquired in recent years reveals a diverse range of observational properties, demonstrating how differing environmental and pulsar properties manifest themselves in the resulting nebulae.

INTRODUCTION

All isolated pulsars have steadily increasing periods. However, only a very small fraction of the corresponding spin-down luminosity, $\dot{E} \equiv 4\pi^2 I \dot{P}/P^3$ (where P is the pulsar spin period and I is the pulsar's moment of inertia), can usually be accounted for by the pulsations themselves. It is rather assumed that most of this rotational energy is converted into a relativistic wind.

At some distance from the pulsar, the wind pressure and confining pressure balance, producing a shock at which relativistic particles in the wind are accelerated and have their pitch angles randomized. These particles consequently radiate synchrotron emission, generating an observable pulsar-powered nebula, or "plerion".[2] Plerions can be powerful probes of a pulsar's interaction with its surroundings — their properties can be used to infer the geometry, energetics and composition of the pulsar wind, the space velocity of the pulsar itself, and the properties of the ambient medium. Furthermore, the mere existence of a plerion indicates the presence of a pulsar within, even when the latter has not yet been directly detected.

In the following discussion, I briefly review our historical and theoretical understanding of plerions, and describe some of the highlights from recent work. There have been many recent developments in this field, and space precludes me from providing a more comprehensive review. I refer the reader to preceding reviews on plerions and their pulsars by Slane et al [1], Chevalier [2], Frail [3,4] and Gaensler [5].

---

[1] Hubble Fellow
[2] For simplicity, I will use the term "plerion" to refer generically to all forms of pulsar nebulae, and will make specific distinctions when I refer to particular sub-categories of object.

CP565, *Young Supernova Remnants: Eleventh Astrophysics Conf.,* edited by S. S. Holt and U. Hwang

HISTORICAL OVERVIEW

Any discussion of plerions and pulsar wind nebulae must of course begin with the Crab Nebula. While it has long been realized that the Crab Nebula is the product of a supernova explosion [6], its filled-center morphology, flat radio spectrum and high fraction of linear polarization all indicate that it is a very different source from "shell-type" supernova remnants (SNRs) like Cassiopeia A and the Cygnus Loop. A pulsar was discovered embedded in the Crab Nebula in 1968 [7], and it is the continual injection of energy into the nebula by this source which is believed to produce the Crab's unusual properties.

In the early 1970s, it was suggested that the filled-center SNR 3C 58 was similarly powered by an (as yet unseen) pulsar [8]. Before long, several other such sources were identified, and a whole class of "Crab-like" SNRs began to emerge [9–11]. Weiler & Panagia [10] proposed the name "plerion" for such sources, from the Greek "πληρης" meaning "full".[3]

Milne et al [13] introduced an additional complication, when they pointed out that the SNR MSH 15–56, while having a limb-brightened radio shell like most SNRs, also contains a central core which otherwise resembles a plerion (see Figure 1). They termed this source a "composite" SNR, and proposed that it combined the properties of shell-type SNRs and of plerions.

In the last 20 years, this simple classification of SNRs into shells, plerions and composites has largely remained unchanged. In Green's most recent Galactic SNR Catalog [14], 225 SNRs are listed, of which 23 are composites and nine are plerions. Of these 32 remnants which are presumed to be powered by a pulsar, the central pulsars have now been detected in ten, while a further five contain central X-ray point sources believed to be associated neutron stars. Adding to this list two pulsar nebulae in the LMC in which pulsars have been detected, we have thus now identified the central source in 50% of SNRs with plerionic components. The premise that plerions are powered by pulsars therefore remains essentially unchallenged.

OBSERVATIONAL PROPERTIES

At radio wavelengths, plerions (and plerionic components of composite SNRs) generally have an amorphous filled-center morphology, with a flat spectrum ($-0.3 \lesssim \alpha \lesssim 0$, $S_\nu \propto \nu^\alpha$) and a high degree of linear polarization. In cases where an associated pulsar has been identified, it is not necessarily near the center and/or brightest part of the plerion. Kothes [15] has proposed that the radio luminosity of a plerion, L_R, the diameter of the plerion, D, and $\dot{E}$, approximately follow the relation $L_R \propto \dot{E} D$.

---

[3] Shakeshaft [12] has pointed out that "πληριον" is not a Greek word, and that "plethoric supernova remnant" would be a more linguistically-correct term to describe these objects. However, this advice does not seem to have been heeded by the community!

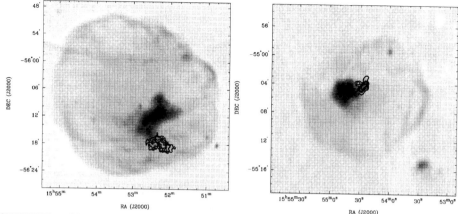

FIGURE 1. The composite SNRs MSH 15–5$6$ (left) and G327.1–1.1 (right). The greyscale shows the 843 MHz radio emission from each remnant [16], demonstrating their composite morphologies. The contours delineate X-ray emission as seen by *ASCA* SIS [17,18], and show the smaller extent and offset of the X-ray plerions with respect to their radio counterparts.

In the X-ray regime, plerions imaged at sufficient spatial resolution seem to contain axially-symmetric structures such as tori and jets. Plerions generally have smaller X-ray extents than they do in the radio (see Figure 1), presumed to be the result of the shorter lifetimes of synchrotron-emitting electrons at progressively higher energies. For the same reason, the spectrum of plerions in X-rays is steeper ($\alpha \sim -1$) than in the radio, and any associated pulsar is usually coincident with the brightest X-ray emission. Various efforts have been to look for correlations between $\dot{E}$ and the corresponding plerion's X-ray luminosity, L_X. These studies consistently suggest that these two quantities are correlated, showing that $L_X \propto \dot{E}^a$ with an exponent in the range $1 \lesssim a \lesssim 1.4$ [19,20].

A PLETHORA OF PLERIONS

While we assume that all plerions are "Crab-like" in that they are similarly powered by pulsars, in most respects other plerions have very different properties to the Crab Nebula. Indeed, on looking through the available sample, it becomes apparent that plerions form a very heterogeneous set of objects. Without wanting to suggest any definitive categories[4], some of the different types of object include:

- "Standard" composites, in which the plerionic component contains a detected pulsar and is surrounded by a shell. The SNR 0540–69.3 in the LMC [21] is a good example of such a source.

---

[4] See Chevalier [2] for a suggested evolutionary sequence for some of these categories.

- "Naked" plerions, typified by the Crab Nebula, in which no surrounding SNR shell can be seen. In the case of the Crab itself, a good argument can be made that that a SNR blast-wave is present, but that it simply can't be seen [22]. However, Wallace et al [23,24] have shown that some of these "naked" plerions are interacting directly with the ambient ISM, and have argued that such sources are produced by low-energy explosions in which no fast-moving ejecta are produced.

- "Bow-shock" plerions such as those associated with PSRs B1853+01 and B1757–24 [25,26]. These nebulae have a cometary morphology resulting from their pulsar's high space-velocity.

- "Radio-quiet" plerions such as MSH 15–5$2$ and CTA 1 [27,28]. In these cases, a pulsars powers a nebula which is prominent in X-rays but is not seen at radio wavelengths. It has been proposed that these plerions are produced by pulsars with high magnetic fields, $B \gtrsim 10^{13}$ G [29,30]. Such plerions suffer severe adiabatic losses early in their lives, and so are radio-bright only at the earliest stages of their evolution.

- "Hyper-plerions", such as G328.4+0.2 and N 157B, have very large diameters ($\lesssim 20$ pc) and radio luminosities higher than that of the Crab Nebula [31,32]. These plerions appear to be powered by a low field pulsar, $B \lesssim 10^{12}$ G, which generates a large and long-lived (~ 10 kyr) radio nebula [31].

- "Interacting composites", typified by CTB 80, in which a high-velocity pulsar catches up with and re-energizes its associated shell SNR [33].

- "Low frequency spectral break" plerions, such as 3C 58 and G21.5–0.9. These plerions show a sudden steepening of their spectra at radio or millimeter frequencies, thought to be caused by a "phase change" in the pulsar's energy output [34]. These sources will be discussed further below.

PLERION EVOLUTION

A "standard" model of evolution has emerged in which a plerion is modeled as a spherically symmetric expanding synchrotron bubble for which (in the simplest case) equipartition is assumed between particles and magnetic fields in the nebula [35,36]. The only source of energy input into the nebula is the spin-down of the pulsar, whose time-evolution is described by:

$$\dot{E} = \frac{L_0}{(1 + t/\tau_0)^p}, \quad p = \frac{n+1}{n-1} \tag{1}$$

where L_0 is the pulsar's initial spin-down luminosity, τ_0 is some characteristic time-scale (typically a few hundred years) and n is the pulsar's braking index. For times

$t < \tau_0$, the pulsar's rate of output is approximately constant, $\dot{E} \approx L_0$. At times $t > \tau_0$, the spin-down luminosity decays as a power-law, $\dot{E} \propto t^{-p}$.

Competing against this injection are two sources of energy loss: adiabatic losses due to expansion of the nebula, and synchrotron losses. Including all these terms, one can derive expressions for the evolution of the particle content, magnetic field, luminosity and spectrum in each phase of evolution. For $t < \tau_0$, a single break is seen in the plerion's spectrum, corresponding to the frequency at which synchrotron losses dominate at time t. At times $t > \tau_0$, this original spectral break moves to higher frequencies as the nebular magnetic field decays, while a second "fossil" break, resulting from the rapid decay of $\dot{E}$ beginning at $t = \tau_0$, appears at lower frequencies [35]. When modeling the evolution of a plerion, the presence of a surrounding shell-type SNR, and/or the actual detection of the associated pulsar, allows one to estimate the nebular magnetic field strength, the rate of energy input by the pulsar and the age of the system. The properties of the SNR, pulsar and plerion can then be used to jointly constrain the parameters of the system [1].

While this picture can explain the basic properties of the Crab Nebula and other plerions, there have been many subsequent refinements to take into account particular situations and details of the nebular physics. To conclude this section, I list below some of the recent work that has been carried out on plerion evolution, and discuss some of the more interesting developments in more detail.

- Amato et al [37] have taken into account the fact that the particle distribution within a plerion is not homogeneous, and develop a model in which synchrotron-emitting particles propagate away from the pulsar.

- van der Swaluw et al [38] have considered the interaction which occurs when a pulsar catches up with and penetrates its associated shell SNR.

- Luz & Berry [39] have modeled the interaction between a plerion and its surrounding shell SNR.

- Wilkin [40] has developed a detailed treatment of bow-shocks produced by anisotropic winds, as are likely to be produced by pulsars.

- Chevalier [2] has pointed out that the reverse shock produced by a SNR blastwave will reach the center of the SNR in $\sim 10^4$ yr. This can compress, brighten and distort a central plerion, and could account for the filamentary appearance and offset of the pulsar from the center of the plerion seen in Vela X.

- Chevalier [41] has also recently developed a simplified one-zone model for the X-ray emission from a plerion. He derives an analytic expression for the emission, so that the X-ray luminosity, L_X, depends only on the spin-down luminosity of the pulsar ($\dot{E}$), the photon index of the nebula (Γ), the wind magnetization parameter (σ), the Lorentz factor of the wind (γ_w) and the radius of the shock (r_s). This model successfully predicts the ratio $L_x/\dot{E}$ for most plerions in which pulsars have been detected.

- Various authors have considered plerions such as 3C 58, which have low-frequency spectral breaks [34,42,43]. These plerions are characterized by sharp spectral breaks ($\Delta\alpha \sim 0.8 - 1.0$) at frequencies $\nu_b \lesssim 50$ GHz, in sharp contrast to the $\Delta\alpha = 0.5$ break seen for the Crab Nebula in the infrared. The spectral breaks seen in these other plerions cannot be due to synchrotron losses, as the inferred magnetic field is so high that the energy in magnetic fields would then be larger than the kinetic energy of the plerion. Woltjer et al [34] show that a plerion powered by a pulsar with a low braking index ($n \ll 3$) can produce a low-frequency fossil break at times $t > \tau_0$, but that this break is not sharp enough to match observations. They instead consider a model in which there is a sudden "phase change" in the pulsar's energy output, when perhaps the pulsar's wind suddenly becomes magnetically dominated. In such a system a sharp low-frequency spectral break is indeed predicted. Such a model can also account for the low X-ray luminosity of 3C 58, and for the fact that its radio flux is increasing with time (rather than decreasing in the case of the Crab). While these arguments make a strong case that the central sources in these low-frequency break plerions are quite different from the Crab Pulsar[5], the only definitive resolution to this puzzle will be to actually detect their central sources.

NEW RESULTS

Gamma rays: It has long been thought that many of the unidentified γ-ray sources in the Galactic Plane correspond to young high-$\dot{E}$ pulsars and their associated plerions. Indeed, Halpern et al [44] have identified a new radio source, G106.6+3.0, which is coincident with the otherwise unidentified *EGRET* source 3EG J2227+6122 and also with the X-ray source AX J2229.0+6114. This radio source is polarized, has a flat spectral index, and has a possible bow-shock morphology. The properties of G106.3+3.0 all suggest that this source is a plerion powered by a pulsar with $\dot{E} \gtrsim 10^{36}$ erg s^{-1}. Other possible plerions associated with γ-ray sources have been reported by Roberts et al [45] and Oka et al [46].

Radio: Stappers & Gaensler have carried out an extensive search for radio nebulae associated with radio pulsars [47–49]. Of 31 pulsars observed, only one new pulsar nebula was found, indicating that pulsars reside predominantly in low-density environments. Meanwhile, the Australia Telescope Compact Array continues to image various plerions at high spatial resolution, highlighting the diversity of plerion properties and morphologies [31,32,50].

Optical: The morphology of an optical bow-shock around a pulsar contains a great deal of information about a pulsar's interaction with the ISM. Imaging and spectroscopy of such bow-shocks around PSRs B2224+65 ("the Guitar Nebula")

---

[5] For a different interpretation in terms of central pulsars with high magnetic fields, see the discussion by Frail [4].

and J0437–4715 have resulted in determinations of the 3D space velocities of these pulsars, and the densities and ionization fractions of their environments [51,52].

X-rays: *ASCA* observations of various pulsars and their plerions have recently been re-analyzed. Contrary to previous claims [53], there now seems to be no plerions apparent around PSRs B1610–50, B1055–52, B0656+14 or Geminga [49,54,55]. PSR B1046–58 has no extended plerion, but may be associated with a compact X-ray nebula [55].

With the recent launch of *XMM* and *Chandra*, it is unsurprising that there are many new results on pulsars and their nebulae. The high-resolution of *Chandra* has been brought to bear on the two prominent plerions in the Large Magellanic Cloud, SNRs 0540–693 and N 157B. X-ray data on the former suggest a possible Crab-like morphology with the hint of jets and a torus [56,57], while observations of the latter appear to confirm the cometary morphology for this nebula seen in *ROSAT* data [58].

The *Chandra* image of the Vela Pulsar is spectacular (Figure 2; [59]). The surrounding nebula is remarkably similar to the Crab, showing clear evidence for equatorial rings and axial jets, and appearing to rule out earlier interpretations of this system as a bow-shock. The orientation of the pulsar's proper motion, the pulsar's spin axis and the direction of outflow along the X-ray jets all seem to align. This alignment is similar to that seen for the Crab Pulsar, and provides important constraints on the origins of pulsar spin periods and space velocities [60,61]. Helfand et al [59] note that the X-ray properties of the Vela plerion suggest a magnetization parameter $\sigma \approx 1$, in sharp contrast to the Crab Nebula for which $\sigma \approx 3 \times 10^{-3}$.

Finally, *Chandra* and *XMM* observations of the plerion G21.5–0.9 have revealed both a compact central source and a surrounding halo (Figure 2). The central source is resolved by *Chandra*, and may be a wisp or termination shock which is hiding the pulsar itself [62]. Meanwhile, *XMM* observations clearly demonstrate that the surrounding halo has a power-law spectrum, and that this spectrum steadily steepens with radius [63]. It is not yet clear whether the plerion is much larger than previously thought, or if the halo is a surrounding SNR blast-wave so that G21.5–0.9 should be re-classified as a composite SNR. Deep radio observations of this source will be required to distinguish between these possibilities.

CONCLUSIONS

With the high-quality data now available across the spectrum, it has become abundantly apparent that the Crab Nebula is not typical of plerions. The large diversity in the observed and inferred properties of plerions appears to result from differences in their environments, ages and associated pulsars. Key to understanding this variety seems to be better modeling of the interaction of plerions and pulsars with their surrounding SNRs.

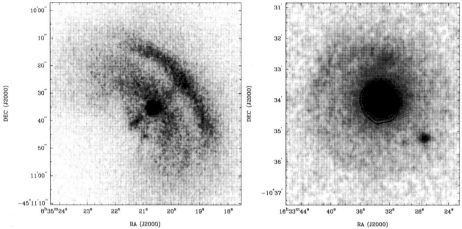

FIGURE 2. The Vela Pulsar and surrounds (left) and SNR G21.5–0.9 (right). The image of the Vela Pulsar was produced from archival *Chandra* HRC data, and has been convolved with a 0″.5 gaussian (see also [59]). G21.5–0.9 was observed with *XMM* EPIC MOS [63] — the contours show 1.4 GHz radio emission from the plerion indicating its previously-known extent, while the greyscale has been scaled logarithmically to show the faint surrounding X-ray halo.

It is also clear that many plerions are still waiting to be discovered. Approximately 20% of Galactic radio SNRs are yet to be imaged with a spatial resolution of better than ∼ 10 beams across their diameter, and imaging these SNRs at higher resolution could reveal central pulsar-powered components. Furthermore, many pulsar nebula are seen in hard X-rays but not at radio wavelengths, and so many apparently shell-type SNRs which have not yet been observed at higher energies may also harbor a central plerion.

A glance at the list of targets approved for *Chandra* and *XMM* shows that many plerions have been or are about to be observed with these new instruments. These data will obviously produce a great deal more information on some of the issues touched on here. Do all pulsar winds show a "torus + jets" morphology? Is the alignment between rotation axis and proper motion seen for the Crab and Vela pulsars a common characteristic? Do anomalous X-ray pulsars and soft γ-ray pulsars power associated nebula? Do "naked" plerions have faint surrounding shells? We can look forward to a whole new picture of plerions emerging in the near future.

I thank Lisa Townsley and Jasmina Lazendic for sharing with me their recent work on N 157B, Bob Warwick for providing an *XMM* image of G21.5–0.9, and all those with whom I work on plerions and pulsars for many interesting discussions and collaborations. This work was supported by NASA through Hubble Fellowship grant HST-HF-01107.01-A awarded by the Space Telescope Science Institute, and by NSF under Grant No. PHY99–07949.

REFERENCES

1. Slane, P., Bandiera, R., and Torii, K., 1998, Mem. Soc. Astron. It., 69, 945
2. Chevalier, R. A., 1998, Mem. Soc. Astron. It., 69, 977
3. Frail, D. A., in *Neutron Stars and Pulsars: Thirty Years after the Discovery*, edited by N. Shibazaki, N. Kawai, S. Shibata, and T. Kifune, Tokyo: Universal Academy Press, 1998 p. 423
4. Frail, D. A., in *The many faces of neutron stars*, edited by R. Buccheri, J. van Paradijs, and M. A. Alpar, vol. 515 of *NATO ASI Series*, Dordrecht: Kluwer, 1998 p. 179
5. Gaensler, B. M., in *Spin, Magnetism and Cooling of Young Neutron Stars*, Santa Barbara: Institute for Theoretical Physics, 2000 (http://online.itp.ucsb.edu/online/neustars_c00/gaensler)
6. Lundmark, K., 1921, Publ. Astr. Soc. Pacific, 33, 225
7. Staelin, D. H., and Reifenstein, III, E. C., 1968, Science, 162, 1481
8. Weiler, K. W., and Seielstad, G. A., 1971, ApJ, 163, 455
9. Lockhart, I. A., Goss, W. M., Caswell, J. L., and McAdam, W. B., 1977, MNRAS, 179, 147
10. Weiler, K. W., and Panagia, N., 1978, A&A, 70, 419
11. Weiler, K. W., and Shaver, P. A., 1978, A&A, 70, 389
12. Shakeshaft, J. R., 1979, A&A, 72, L9
13. Milne, D. K., et al, 1979, MNRAS, 188, 437
14. Green, D. A., *A Catalogue of Galactic Supernova Remnants (2000 August Version)*, Cambridge: Mullard Radio Astronomy Observatory, 2000. (http://www.mrao.cam.ac.uk/surveys/snrs/)
15. Kothes, R., 1998, Mem. Soc. Astron. It., 69, 971
16. Whiteoak, J. B. Z., and Green, A. J., 1996, A&AS, 118, 329, (http://www.physics.usyd.edu.au/astrop/wg96cat/)
17. Plucinsky, P. P., 1998, Mem. Soc. Astron. It., 69, 939
18. Sun, M., Wang, Z., and Chen, Y., 1999, ApJ, 511, 274
19. Seward, F. D., and Wang, Z.-U., 1988, ApJ, 332, 199
20. Becker, W., and Trümper, J., 1997, A&A, 326, 682
21. Manchester, R. N., Staveley-Smith, L., and Kesteven, M. J., 1993, ApJ, 411, 756
22. Sankrit, R., and Hester, J. J., 1997, ApJ, 491, 796
23. Wallace, B. J., Landecker, T. L., and Taylor, A. R., 1997, AJ, 114, 2068
24. Wallace, B. J., Landecker, T. L., Taylor, A. R., and Pineault, S., 1997, A&A, 317, 212
25. Frail, D. A., Giacani, E. B., Goss, W. M., and Dubner, G., 1996, ApJ, 464, L165
26. Frail, D. A., and Kulkarni, S. R., 1991, Nature, 352, 785
27. Gaensler, B. M., et al, 1999, MNRAS, 305, 724
28. Slane, P., et al, 1997, ApJ, 485, 221
29. Bhattacharya, D., 1990, J. Astrophys. Astr., 11, 125
30. Crawford, F., et al, 2001, ApJ, in press (astro-ph/0012287)
31. Gaensler, B. M., Dickel, J. R., and Green, A. J., 2000, ApJ, 542, 380
32. Lazendic, J. S., et al, 2000, ApJ, 540, 808

33. Shull, J. M., Fesen, R. A., and Saken, J. M., 1989, ApJ, 346, 860

34. Woltjer, L., Salvati, M., Pacini, F., and Bandiera, R., 1997, A&A, 325, 295

35. Pacini, F., and Salvati, M., 1973, ApJ, 186, 249

36. Reynolds, S. P., and Chevalier, R. A., 1984, ApJ, 278, 630

37. Amato, E., et al, 2000, A&A, 359, 1107

38. van der Swaluw, E., Achterberg, A., and Gallant, Y. A., 1998, Mem. Soc. Astron. It., 69, 1017

39. Luz, D. M. G. C., and Berry, D. L., 1999, MNRAS, 306, 191

40. Wilkin, F. P., 2000, ApJ, 532, 400

41. Chevalier, R. A., 2000, ApJ, 439, L45

42. Green, D. A., and Scheuer, P. A. G., 1992, MNRAS, 258, 833

43. Salvati, M., Bandiera, R., Pacini, F., and Woltjer, L., 1998, Mem. Soc. Astron. It., 69, 1023

44. Halpern, J. P., Gotthelf, E. V., Leighly, K. M., and Helfand, D. J., 2001, ApJ, in press (astro-ph/0007076)

45. Roberts, M. S. E., Romani, R. W., Johnston, S., and Green, A. J., 1999, ApJ, 515, 712

46. Oka, T., et al, 1999, ApJ, 526, 764

47. Gaensler, B. M., Stappers, B. W., Frail, D. A., and Johnston, S., 1998, ApJ, 499, L69

48. Gaensler, B. M., et al, 2000, MNRAS, 318, 58

49. Stappers, B. W., Gaensler, B. M., and Johnston, S., 1999, MNRAS, 308, 609

50. Dickel, J. R., Milne, D. K., and Strom, R. G., 2000, ApJ, 543, 840

51. Chatterjee, C., and Cordes, J. M., in *Pulsar Astronomy — 2000 and Beyond, IAU Colloquium 177*, edited by M. Kramer, N. Wex, and R. Wielebinski, San Francisco: Astronomical Society of the Pacific, 2000 p. 517

52. Mann, E. C., Romani, R. W., and Fruchter, A. S., 1999, BAAS, 195, 41.01

53. Kawai, N., and Tamura, K., in *Pulsars: Problems and Progress, IAU Colloquium 160*, edited by S. Johnston, M. A. Walker, and M. Bailes, San Francisco: Astronomical Society of the Pacific, 1996 p. 367

54. Becker, W., Kawai, N., Brinkmann, W., and Mignani, R., 1999, A&A, 352, 532

55. Pivovaroff, M., Kaspi, V. M., and Gotthelf, E. V., 2000, ApJ, 528, 436

56. Gotthelf, E. V., and Wang, Q. D., 2000, ApJ, 532, L117

57. Kaaret, P., et al, 2000, ApJ, in press (astro-ph/0008388)

58. Townsley, L., et al, 1999, BAAS, 195, 53.01

59. Helfand, D. J., Gotthelf, E. V., and Halpern, J. P., 2001, ApJ, submitted (astro-ph/0007310)

60. Lai, D., Chernoff, D. F., and Cordes, J. M., 2001, ApJ, 549, in press (astro-ph/0007272)

61. Spruit, H., and Phinney, E. S., 1998, Nature, 393, 139

62. Slane, P., et al, 2000, ApJ, 533, L29

63. Warwick, R. S., et al, 2001, A&A, in press (astro-ph/0011245)

Pulsars, AXPs and SGRs

R. N. Manchester

Australia Telescope National Facility, CSIRO
P.O. Box 76, Epping NSW 1710, Australia[1]

Abstract. Pulsars, including both those detected at radio wavelengths and those only detectable at high energies, Anomalous X-ray pulsars (AXPs) and Soft Gamma-ray Repeaters (SGRs) are all believed to be rotating neutron stars. Yet they have very different properties. In the past few years the number of known objects in all of these classes has increased dramatically. Many of these new discoveries are young neutron stars, leading to a significant increase in the number of believable associations with supernova remnants. In this review, I outline the main properties of each class of object and describe some recent results.

INTRODUCTION

The discovery of the Vela and Crab pulsars [1,2] and the observation that the Crab pulsar period is steadily increasing [3] quickly led to the identification of pulsars as rotating neutron stars [4]. It also confirmed the association of neutron stars with supernovae, a suggestion made much earlier by Baade & Zwicky [5], and established that the Crab Nebula is powered by electromagnetic spin-down of a central neutron star [6]. In the three decades since then, extensive searches have discovered more than 700 radio pulsars. These pulsars fall into two broad categories, the so-called 'normal' pulsars with pulse periods in the range 33 ms to 8.5 s and relatively rapid spin-down, and the 'millisecond' pulsars (MSPs), typically with pulse periods of between 1.5 and 20 ms and very low rates of spin-down. Most of the MSPs are in binary orbit about another star, either a white dwarf or another neutron star, leading to the suggestion that they acquire their rapid spin and weak magnetic fields as a result of accretion from a binary companion — the 'recycling' hypothesis [7,8].

Timing measurements made over an extended interval (typically months or years) allow accurate measurements of the pulse period, P, and its rate of increase, $\dot{P}$. Excepting glitches and other period irregularities, the intrinsic $\dot{P}$ is always positive, consistent with the notion that pulsars are powered by their rotational kinetic energy. For a dipole-dominated magnetic field, estimates of the pulsar age, known

---

[1] Email:rmanches@atnf.csiro.au

CP565, *Young Supernova Remnants: Eleventh Astrophysics Conf.*, edited by S. S. Holt and U. Hwang
© 2001 American Institute of Physics 0-7354-0001-6/01/$18.00

as the 'characteristic age', $\tau_c = P/(2\dot{P})$, and the value of magnetic field at the neutron star surface, $B_0 = 3.2 \times 10^{19}(P\dot{P})^{1/2}$, may be obtained. Normal pulsars have characteristic ages between about 10^3 and 10^7 years and surface magnetic field strengths B_0 in the range 10^{10} to 10^{13} G, whereas MSPs typically have ages of 10^8 to 10^{10} years and surface fields of between 10^8 and 10^9 G.

In the past couple of years, the number of known pulsars has nearly doubled, to more than 1400, largely as a result of two very successful searches using a multibeam receiver on the Parkes 64-m radio telescope.

Beginning with Geminga, an increasing number of pulsars which are detectable *only* at high energies (X-ray or γ-ray) have been discovered. No confirmed pulsed radio emission has been observed from these objects, despite sensitive searches.

Anomalous X-ray Pulsars (AXPs) also are detected at X-ray wavelengths, have no detectable radio emission and, in some cases, are associated with supernova remnants. However, they have very long pulse periods, in the range 6 – 12 seconds, and larger spin-down rates than even the youngest radio pulsars. Soft Gamma-ray Repeaters (SGRs) are an apparently related class of object. They are characterised by the emission of transient bursts of γ-rays and/or hard X-rays and by the fact that more than one burst is observed from a given source. For several of these sources, a periodic modulation has been observed in the burst emission and also in the persistent low-level X-ray emission observed from these objects.

In the following sections I review some recent results relating to each of these classes of object.

RADIO PULSARS

By mid-1997, there were a total of 731 known radio pulsars. Of these, 693 lie within the Galactic disk; the remainder are found in globular clusters (33) or in the Magellanic Clouds (5). At this time, a new multibeam receiver operating in the 20-cm band was installed on the Parkes radio telescope. This receiver has 13 beams, each with dual linear polarisation, arranged in a double hexagonal pattern around a central beam. The multiple beams, combined with a low system noise (~ 21 K) and wide bandwidth (~ 300 MHz), make this a very sensitive instrument for survey observations. Two major pulsar surveys been made using this instrument.

The first, known as the Parkes multibeam pulsar survey, is a large collaborative project between Jodrell Bank Observatory, MIT, University of Bologna and the ATNF, searching a region along the southern Galactic plane with $|b| < 5°$ between $l = 260°$ and $l = 50°$ [9,10]. With a sampling interval of 250 μs and an observation time per pointing of 35 min, the limiting flux density is about 0.2 mJy for pulsars which have a period ~ 1 s and lie off the Galactic plane. This sensitivity is at least five times better than that of any previous similar survey. The survey is currently about 80% complete and has been extremely successful, with 583 pulsars discovered so far. Of these, two have periods in the millisecond range and eight are members of binary systems.

The second survey, known as the Swinburne multibeam pulsar survey, complements the first in that it covers two adjacent 10°-wide strips with $5° < |b| < 10°$ [11]. With a sampling interval of 125 μs and covering a larger region at higher latitudes, this survey is more optimal for detection of MSPs. The observation time per pointing is 4 min, giving a limiting flux density of about 0.5 mJy. The survey is now complete and also was very successful, finding 58 previously unknown pulsars, of which seven have millisecond periods. Six of these are members of binary systems.

Figure 1 shows the period and DM distributions of previously known pulsars and those discovered in these two surveys. All but two of the Parkes multibeam pulsars have 'normal' periods. Their distribution is similar to that of previously known pulsars, except that there is a somewhat greater proportion of pulsars both with periods ~ 0.1 s and with very long periods. The relatively low proportion of MSPs can largely be attributed to the low Galactic latitude of the survey — the sensitivity to MSPs is reduced by dispersion smearing for distant pulsars lying close to the Galactic plane. In contrast, about 12% of the Swinburne pulsars have millisecond periods, a much higher proportion than $\sim 5\%$ for the previously known disk population. This reflects the higher sampling rate of the Swinburne survey compared to most previous surveys.

Pulsars from the Swinburne survey have modest DMs as expected from the higher Galactic latitude and somewhat lower sensitivity of this survey. The distribution for the Parkes multibeam pulsars is very different, peaking at a DM ~ 300 cm^{-3} pc. These large values are a consequence of the high sensitivity and low Galactic latitude of the survey and imply that most of the Parkes multibeam pulsars are relatively distant. The distances can be estimated using a model of the free electron

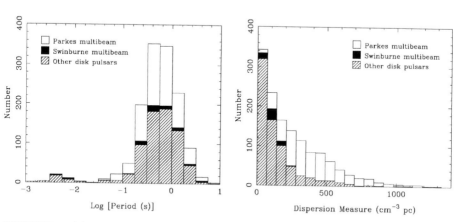

FIGURE 1. (a) Period distributions and (b) dispersion measure distributions of pulsars discovered in the Parkes multibeam survey, the Swinburne multibeam survey and of previously known Galactic disk pulsars.

density in the Galaxy, such as that by Taylor & Cordes [12]. This shows that many of the Parkes multibeam survey pulsars are at distances comparable to or beyond the Galactic Centre. However, the model is quite uncertain in these regions; the new discoveries will greatly help in its improvement.

After confirmation of a pulsar, timing observations are made over 12 – 18 months in order to determine an accurate position, period and period derivative. The timing observations also reveal any pulsars which are members of binary systems and allow determination of the binary parameters. Figure 2 is a plot of pulse period versus the rate of change of period, $\dot{P}$, for Galactic disk pulsars. Multibeam pulsars tend to lie higher in this diagram than previously known pulsars, indicating that, on average, they are younger and have stronger magnetic fields. Indeed, the four strongest known surface dipole magnetic fields are all for multibeam pulsars (cf. Camilo et al. [13]). One of these, PSR J1119–6127, has a period of 0.407 s, but a characteristic age of only 1600 years. Observations with the Australia Telescope Compact Array [14] have revealed an associated supernova remnant, which is also detected at X-ray wavelengths in archival ASCA data [15].

Figure 2 also shows that five of the multibeam pulsars with periods ~ 0.1 s are members of binary systems and have low values of $\dot{P} \sim 10^{-18}$. One of these, PSR J1811–1736, is most probably a double-neutron-star system [9], whereas the other four and the 9-ms binary PSR J1435–6100 have white-dwarf companions [16]. It is likely that all of these pulsars have undergone spin-up by accretion during the giant evolutionary phase of the companion star.

Another binary multibeam pulsar, PSR J1740–3052, lies above the spin-up line with a pulse period of 570 ms and $\dot{P} \sim 2.5 \times 10^{-14}$. This pulsar is in a highly eccentric orbit of period 231 days and has a very large companion mass, > 11 $M_\odot$ [17]. The companion must therefore be either a black hole or a massive non-degenerate star. One might expect to be able to detect a non-degenerate companion optically, but this pulsar lies very close to the Galactic Centre and at a comparable distance, so obscuration is likely to be very large. Observations at 2.2 μm do reveal a late-type star at the pulsar position. However, if this star were at the pulsar distance, it would be larger than the periastron radius of the orbit. Furthermore, the expected eclipses and advance of periastron are not observed. On the other hand, orbital variations of dispersion delay and rotation measure *are* observed, suggesting that the companion is non-degenerate. We believe that the most likely companion is a B main-sequence star hidden in the light of the late-type star.

ROTATION-POWERED X-RAY AND γ-RAY PULSARS

X-ray and γ-ray observations have revealed several pulsed sources which are clearly rotating neutron stars, similar in every respect to ordinary pulsars except that they have no detectable radio emission. Table 1 lists the six currently known sources of this type. Three of them are associated with supernova remnants and are clearly young pulsars — PSR J0537–6910 has the shortest period of any 'normal' pulsar and PSR J1846–0258 is the youngest known pulsar. The X-ray pulse profile for PSR J1846–0258 is wide, covering more than a third of the period. It is possible that, for this and other pulsars in this class, the radio beam is much narrower and not sweeping across the Earth. PSR J1846–0258 was first detected serendipitously in RXTE data, but the very large $\dot{P}$ value and the association with Kes 75 are based on archival ASCA data spanning about seven years [23]. This pulsar lies very close in the $P - \dot{P}$ plane to the multibeam radio pulsar PSR J1119–6127 (Figure 2).

For all but one of the pulsars in Table 1, the X-ray emission is magnetospheric in

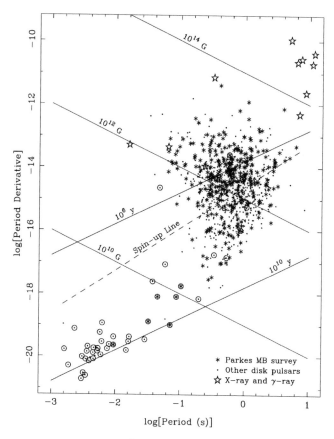

FIGURE 2. Plot of period derivative $\dot{P}$ versus P for Parkes multibeam pulsars, other known Galactic disk radio pulsars and objects that emit only at X-ray and γ-ray wavelengths and are believed to be rotating neutron stars. This latter group includes young pulsars such as the 16-ms pulsar associated with the supernova remnant N157B, and AXPs and SGRs which have periods of several seconds and very high values of $\dot{P}$. Binary pulsars are indicated by a circle around the point. Lines of constant characteristic age, τ_c, and dipole surface magnetic field, B_0, are shown on the plot, as is the line of minimum period a neutron star can attain by spin-up due to accretion.

TABLE 1. Rotation powered X-ray and γ-ray pulsars

PSR J	Period (ms)	Age (kyr)	Association	Reference
0537–6910	16.1	5.0	N157B in LMC	[18]
0633+1746	237	340	Geminga	[19]
0635+0533	33.8	–	Be-star binary	[20]
1808–3658	2.5	–	Low-mass binary	[21]
1811–1926	64.6	23	G11.2–0.3	[22]
1846–0258	323	0.72	Kes 75	[23]

origin, powered by the rotational kinetic energy of the neutron star. PSR J1808–3658 is different. It is a millisecond pulsar in orbit around a low-mass star with the X-ray emission powered by accretion and modulated by the rotation of the neutron star. This system is unique in that the neutron star almost certainly has been (and presumably is being) spun up by the accretion; it is the "missing link" between low-mass X-ray binary systems and radio MSPs.

ANOMALOUS X-RAY PULSARS AND SOFT GAMMA-RAY REPEATERS

AXPs are unusual X-ray sources in which the X-ray emission is modulated at periods within the relatively narrow range of 6 – 12 s, typically with a sinusoidal profile. Unlike high-mass X-ray binary systems, there is no evidence for a binary companion and the pulsational period has a very large spin-down rate. The X-ray emission has a very soft spectrum and is relatively steady with a luminosity in the range 10^{34} to 10^{36} erg s^{-1}, much less than the Eddington-limited value of 10^{38} erg s^{-1} characteristic of high-mass binary X-ray sources. The rapid spin-down suggests that these objects are young, and indeed, several are clearly associated with supernova remnants. Taken together, these properties suggest that AXPs are slowly rotating isolated neutron stars. If one assumes that the spin-down is due to magnetic-dipole braking as in normal pulsars, the implied magnetic fields are huge, $\gtrsim 10^{14}$ G. Because of these extremely high implied field strengths, these objects have been called 'magnetars' [24].

The six known objects which clearly belong in this class are listed in Table 2. Two of these, 1E 2259+586 and 1E 1841–045, are centrally located within supernova remnants and were clearly associated at the time of their discovery. For AX J1845–0258, the associated remnant, G29.6+0.1, was discovered in later VLA observations [25].

The spin-down luminosity $\dot{E} = 4\pi^2 I\dot{P}/P^3$ of AXPs is typically between 10^{32} and 10^{33} erg s^{-1} (where $I = 10^{45}$ g cm^2 has been assumed). This is much less than the X-ray luminosity, showing that, unlike radio pulsars and the young high-energy pulsars discussed above, the observed pulsed emission cannot be powered by rotational kinetic energy. Since these objects apparently have no binary companion,

TABLE 2. Anomalous X-ray Pulsars

Name	Period (s)	$\dot{P}$	B_0 (G)	Association	Ref.
1E 1048.1–5937	6.45	2.2×10^{-11}	3.8×10^{14}	–	[26]
AX J1845–0258	6.96	–	–	G29.6+0.1	[27]
1E 2259+586	6.98	7.3×10^{-13}	7.0×10^{13}	CTB 109	[28]
4U 0142+61	8.69	2.3×10^{-12}	1.4×10^{14}	–	[29]
1RSX 1708–4009	11.0	1.9×10^{-11}	4.6×10^{14}	–	[30]
1E 1841–045	11.8	4.1×10^{-11}	7.0×10^{14}	Kes 73	[31]

accretion cannot be the energy source. Thompson & Duncan [24] have suggested that they are powered by decay of the very strong magnetic fields; the total energy in the magnetic field is of order $B_0^2 R_{NS}^3 \sim 10^{46}$ erg, sufficient to power the X-ray source for several thousand years. Stronger fields inside the neutron star may provide an even larger reservoir of energy.

Coherent timing of AXPs has shown that the spin-down is relatively steady, lending support to the idea that AXPs are highly magnetised neutron stars [32]. Further support was given by the detection of a glitch in the period of 1RSX J1708–4009 [33]. Figure 3 shows timing residuals for this pulsar in which the glitch is clearly evident. Furthermore, there is a clear increase in $\dot{P}$ after the glitch with $\Delta\dot{P}/\dot{P} \sim 1\%$. These properties are characteristic of the glitches observed in radio pulsars.

It is interesting to note that 1E 2259+586 lies very close in the $P - \dot{P}$ plane to the radio pulsar PSR J1814–1744. Yet these two pulsars have very different properties.

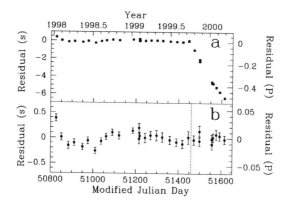

FIGURE 3. Residuals from a coherent timing fit to data for the AXP 1RSX J1708–4009 [33]. In the upper part of the figure, the timing model is from a fit to data up to MJD 51450; at about that time the period suddenly glitched by about six parts in 10^7. The lower part of the figure shows residuals from a model including the glitch.

TABLE 3. Soft Gamma-ray Repeaters

SGR 1900+14	5.16	$\sim 10^{-10}$	$\sim 8 \times 10^{14}$	G42.8+0.6 (??)	[35]
SGR 1806–20	7.48	$\sim 3 \times 10^{-10}$	$\sim 5 \times 10^{14}$	G10.0-0.3 (?)	[36]
SGR 0526–66	8.1	–	–	N49 in LMC (?)	[37]
SGR 1627–41	–	–	–	G337.0-0.1 (?)	[38]

The reasons for these differences are not obvious.

Soft Gamma-ray Repeaters (SGRs) are apparently closely related to AXPs, but their distinguishing characteristic is that they have frequent short-duration bursts of hard X-rays or low-energy γ-rays [34]. As Table 3 shows, there are four of these objects now known. Pulsations have been observed in three of them, with periods in the range 5 – 8 seconds. The pulsations are observed in intense longer-duration bursts (Figure 4) as well as in the persistent X-ray emission which is also detected from these objects. Timing observations show that the periods are rapidly increasing, corresponding to magnetic field strengths comparable to or even larger than those for AXPs. Coherent timing solutions have been obtained for SGRs 1806–20 [41] and 1900+14 [39] showing large changes in the spin-down rate, in some cases by more than a factor of two, over intervals of 50 – 100 days. This variability indicates that ages and magnetic fields deduced from spin-down rates may not bear a very close relationship to reality.

All of the SGR sources are located close to or within supernova remnants and

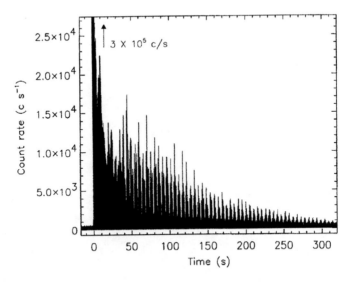

FIGURE 4. Gamma-ray burst from SGR 1900+14 observed by the Ulysses spacecraft in the 25 – 150 keV band showing the 5.16-s periodic modulation [40]. Figure by P.M. Woods (private communication)

physical associations were suggested [42,43]. However, SGR 0526–66 lies near the edge of N49, 1900+14 lies outside G42.8+0.6 [44] and recent work [45] has shown that 1806–20 lies outside the plerion component of G10.0–0.3. If the very young ages for these objects are to be believed, and it is assumed that the neutron star was born near the centre of the supernova remnant, these locations require implausibly large neutron-star velocities. Either the age estimates are grossly in error or the associations are not real.

SUMMARY

The last few years have seen a great increase in the number and diversity of objects associated with neutron stars. The number of known radio pulsars has nearly doubled, X-ray satellite observatories have discovered young neutron stars not detectable at radio wavelengths, and AXPs and SGRs have been identified with magnetars — highly magnetized and slowly rotating neutron stars. Associated with this growth, the number of believable associations with supernova remnants has climbed to about 20. Recent additions include the young X-ray pulsars PSR J0537–6910 in N157B and PSR J1846–0258 in Kes 75, and G292.2–0.5 associated with the radio pulsar PSR J1119–6127. The well-established associations with supernova remnants and the spin-down behaviour leave little doubt that AXPs are young neutron stars. Whether SGRs are also young neutron stars associated with supernova remnants and whether there is any evolutionary connection between them and AXPs remain an open questions.

I thank my colleagues for their many and varied contributions to the work described here. The Australia Telescope is funded by the Commonwealth of Australia for operation as a National Facility managed by CSIRO.

REFERENCES

1. Large M. I., Vaughan A. E., Mills B. Y., 1968, Nature, 220, 340
2. Staelin D. H., Reifenstein III E. C., 1968, Science, 162, 1481
3. Richards D. W., Comella J. M., 1969, Nature, 222, 551
4. Gold T., 1969, Nature, 221, 25
5. Baade W., Zwicky F., 1934, Proc. Nat. Acad. Sci., 20, 254
6. Pacini F., 1967, Nature, 216, 567
7. Alpar M. A., Cheng A. F., Ruderman M. A., Shaham J., 1982, Nature, 300, 728
8. Bhattacharya D., van den Heuvel E. P. J., 1991, Phys. Rep., 203, 1
9. Lyne A. G. et al., 2000, MNRAS, 312, 698
10. Manchester R. N. et al., 2001, MNRAS, submitted
11. Edwards R. T., 2000, in Kramer M., Wex N., Wielebinski R., eds, Pulsar Astronomy - 2000 and Beyond, IAU Colloquium 177. Astronomical Society of the Pacific, San Francisco, p. 33
12. Taylor J. H., Cordes J. M., 1993, ApJ, 411, 674

13. Camilo F. M., Kaspi V. M., Lyne A. G., Manchester R. N., Bell J. F., D'Amico N., McKay N. P. F., Crawford F., 2000, ApJ, 541, 367
14. Crawford F., Gaensler B. M., Kaspi V. M., Manchester R. N., Camilo F., Lyne A. G., Pivovaroff M. J., 2001, ApJ, in press
15. Pivovaroff M. J., Kaspi V. M., Camilo F., Gaensler B. M., Crawford F., 2001, ApJ, submitted
16. Camilo F. et al., 2001, ApJ, in press
17. Stairs I. H. et al., 2001, MNRAS, submitted
18. Marshall F. E., Gotthelf E. V., Zhang W., Middleditch J., Wang Q. D., 1998, ApJ, 499, L179
19. Halpern J. P., Holt S. S., 1992, Nature, 357, 222
20. Cusumano G., Maccarone M. C., Nicastro L., Sacco B., Kaaret P., 2000, ApJ, 528, L25
21. Wijnands R., van der Klis M., 1998, Nature, 394, 344
22. Torii K., Tsunemi H., Dotani T., Mitsuda K., 1997, ApJ, 489, L145
23. Gotthelf E. V., Vasisht G., Boylan-Kolchin M., Torii K., 2000, ApJ, 542, L37
24. Thompson C., Duncan R. C., 1996, ApJ, 473, 322
25. Gaensler B. M., Gotthelf E. V., Vasisht G., 1999, ApJ, 526, L37
26. Mereghetti S., Israel G. L., Stella L., 1998, MNRAS, 296, 689
27. Torii K., Kinugasa K., Katayama K., Tsunemi H., Yamauchi S., 1998, ApJ, 503, 843
28. Fahlman G. G., Gregory P. C., 1981, Nature, 293, 202
29. Israel G. L., Mereghetti S., Stella L., 1994, ApJ, 433, L25
30. Sugizaki M., Nagase F., Torii K. I., Kinugasa K., Asanuma T., Matsuzaki K., Koyama K., Yamauchi S., 1997, PASJ, 49, L25
31. Vasisht G., Gotthelf E. V., 1997, ApJ, 486, L129
32. Kaspi V. M., Chakrabarty D., Steinberger J., 1999, ApJ, 525, L33
33. Kaspi V. M., Lackey J. R., Chakrabarty D., 2000, ApJ, 537, L31
34. Gögüs E., Woods P. M., Kouveliotou C., van Paradijs J., Briggs M. S., Duncan R. C., Thompson C., 2000, ApJ, 532, L121
35. Kouveliotou C. et al., 1999, ApJ, 510, L115
36. Kouveliotou C. et al., 1998, Nature, 393, 235
37. Cline T. L. et al., 1980, ApJ, 237, L1
38. Hurley K., Kouveliotou C., Woods P., Mazets E., Golenetskii S., Frederiks D. D., Cline T., van Paradijs J., 1999, ApJ, 519, L143
39. Woods P. M. et al., 1999, ApJ, 524, L55
40. Hurley K. et al., 1999a, Nature, 397, 41
41. Woods P. M. et al., 2000, ApJ, 535, L55
42. Kulkarni S. R., Frail D. A., 1993, Nature, 365, 33
43. Rothschild R. E., Kulkarni S. R., Lingenfelter R. E., 1994, Nature, 368, 432
44. Hurley K. et al., 1996, ApJ, 463, L13
45. Hurley K., Kouveliotou C., Cline T., Mazets E., Golenetskii S., Frederiks D. D., van Paradijs J., 1999b, ApJ, 523, L37

NUV and FUV Spectroscopic Timing Observations of the Crab Pulsar with HST/STIS

Theodore R. Gull[1] , Jason Sollerman[2], Peter Lundqvist[2],
George Sonneborn[1], and Don Lindler[1]

[1] NASA's Goddard Space Flight Center, Code 681,Greenbelt, MD 20771, USA
[2] Stockholm Observatory, SE-133, 36 Saltsjobaden, Sweden

Spectroscopic ultraviolet observations of the Crab Nebula and its pulsar were accomplished using the Space Telescope Imaging Spectrograph (STIS) on the Hubble Space Telescope (HST). As the Multianode Microchannel Array (MAMA) detectors were used in time-tag mode, good time resolution (0.125 microseconds) and good spatial resolution (0.0244 "pixel^{-1} were obtained. The low dispersion grating modes G140L and G230L provided spectral resolving power of 600 from 1140 to 3200 Å [1,2]. Results are already published by Gull *et al.* (1998) for the NUV spectral region and by Sollerman *et al.* (2000) for the FUV spectral region and for combined spectroscopy for the NUV, FUV and visible spectrum recorded from the Nordic Optical Telescope.

The spectroscopy of the Pulsar is consistent with a flat spectrum dereddened with a standard extinction curve with the parameters $E(B-V) = 0.52$, $R = 3.1$. This dereddened spectrum of the Crab pulsar can be fitted by a power law with spectral index $\alpha_\nu = 0.11 \pm 0.04$. The main uncertainty in determining the spectral index is the amount and characteristics of the interstellar reddening, and we have investigated the dependence of α_ν on E(B-V) and R. The Ly α absorption indicates a column density of $(3.0 \pm 0.5) \times 10^{21}$ cm^{-2} of neutral hydrogen, which agrees well with our estimate of E(B-V)=0.52 mag. Other lines show no evidence of severe depletion of metals in atomic gas. Broad, blueshifted absorption is seen in C IV (1550Å), reaching a blueward velocity of 2500 km s^{-1}. While the high velocity absorption may be internal to the nebulosity, it is more likely from a fast outer shell surrounding the Crab Nebula, pre-existing from the progenitor wind phase.We adopt a spherically symmetric model for such a shell. From the line profile we find that the density appears to decrease outwardly. A likely lower limit to the shell mass is 0.3 M$_\odot$ with an accompanying kinetic energy of $\sim 1.5 \times 10^{49}$ ergs. A fast, massive shell with 10^{51} ergs cannot be excluded but is less likely if the density profile is much steeper than $\rho(R) \sim R^{-4}$ and the maximum velocity is < 6000 km s^{-1}.

With the time-tag mode of the MAMAs, we obtained the pulse profile in the

CP565, *Young Supernova Remnants: Eleventh Astrophysics Conf.,* edited by S. S. Holt and U. Hwang
© 2001 American Institute of Physics 0-7354-0001-6/01/$18.00

spectral region from 1140 to 3200 Å. The profile is marginally narrower in the FUV relative to the profile obtained in the NUV, and at visible wavelengths [3]. In the extended emission covered by our $25'' \times 0.5''$ slit in the far-UV, we detect C IV 1550 Å and He II 1640 Å emission lines from the Crab Nebula.

Based on observations with the NASA/ESA Hubble Space Telescope, obtained at the Space Telescope Science Institute, which is operated by the Association of Universities for Research in Astronomy, Inc. under NASA contract NAS5-26555.

REFERENCES

1. Gull T., *et al.*, Ap. J. **495**, L51 (1998).
2. Sollerman J., *et al.*, Ap. J. **537**, 861 (2000).
3. Percival J., *et al.*, Ap. J. **407**, 276 (1993).

A Brief Update on the First Balloon Flight of a Crystal Diffraction Lens Telescope, CLAIRE

R. Smither[1], P. V. Ballmoos[2], G. Skinner[3] and Balloon launch crew from Toulouse, France

[1] *Argonne National Laboratory, Argonne, IL 60439, USA*
[2] *CESR, Toulouse, France*
[3] *University of Birmingham, Birmingham, UK*

Abstract.

The first balloon flight of a crystal diffraction lens telescope, CLAIRE, took place on June 15, 2000 from Gap, France. The target for the telescope was the Crab Nebula. This is the first time that a Laue (transmission) type crystal diffraction lens[1-2] has been used to look an astrophysics source. It is the first of three balloon flights planned to test this concept for future astrophysics experiments and the possibility of a satellite experiment.

1. INTRODUCTION

The first balloon flight of a Laue type crystal diffraction lens telescope[1-2], CLAIRE, took place on June 15, 2000 from Gap, France. The target for the telescope was the Crab Nebula. The nebula is about 30 arc sec in diameter for visible light. The pulsar that illuminates the nebula is a very small object in the middle of the nebula. The shape of the Crab Nebula is quite different for low energy X-rays and is as yet undetermined for high energy gamma rays.

2. THE CRYSTAL DIFFRACTION LENS TELESCOPE

Figure 1a is a drawing of the telescope tilted at an angle of 45 degrees. The telescope is supported by a frame that is attached to the balloon. The telescope frame is constructed with carbon fiber tubes. The crystal diffraction lens is located at the top of the telescope. The 9-element Ge detector matrix is located at the bottom of the telescope. The distance from the lens to the detector is 279 cm. A photograph of the telescope, prior to launch, is shown in figure 1b.

CP565, *Young Supernova Remnants: Eleventh Astrophysics Conf.*, edited by S. S. Holt and U. Hwang
© 2001 American Institute of Physics 0-7354-0001-6/01/$18.00

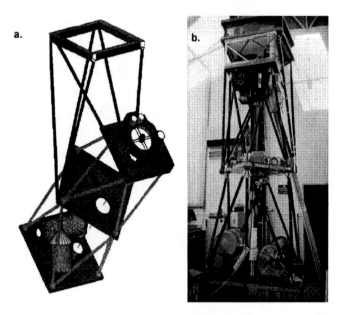

a.

b.

FIGURE 1. A drawing of the crystal diffraction telescope tilted at 45 degrees (a). A photo of the CLAIRE telescope just before launch (b).

The lens is shown in detail in figure 2a as it is mounted on the top of the telescope. It consists of 576 individual Ge-Si crystals, each of which is 1 cm x cm by 2-4 mm thick in the direction of the diffracted beam, and are mounted in 6 concentric rings. The orientation, and thus the Bragg diffraction angle of each crystal is adjustable, with an accuracy of a few arc seconds. The fine pointing of the lens is achieved by adjusting the alignment of the lens relative to the axis of the telescope. The lens is gimbaled so that it can follow the angular position of the target to a few arc seconds. This greatly reduces the needed accuracy of pointing the telescope as a whole. The crystal diffraction telescope acts just like a visible light telescope in that it collects radiation incident from a large area and concentrates it in a small focal spot on the detector; thus the background from a small volume in the well shielded detector does not increase as the collection area of the lens increases.

This approach results in a substantial improvement in the signal to background ratio. As the target moves to the left of the center of the telescope axis, the focal spot moves to the right on the detector surface.

3. CRYSTAL DIFFRACTION FORMULA

The crystal lens behaves like a simple convex lens for visible light. The relationship between the focal length of the lens (F) and the distance from the source to the lens (S) and the distance from the lens to the detector (D) is given by equation

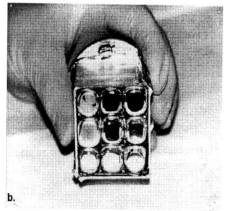

FIGURE 2. A photo of the crystal diffraction lens as it is mounted on the telescope (a). Nine element Ge gamma-ray detector matrix (b).

(1)

$$\frac{1}{F} = \frac{1}{S} + \frac{1}{D}$$

(1)

The focal length is a function of the energy (E) being focused and is given by equation (2) for this lens.

$$F(m) = constant \times E = 1.641 \times E(keV)$$

(2)

Thus for this lens when it is focused for a gamma-ray energy of 170 keV the focal length is 279 cm. The Bragg angle θ (the angle between the incident radiation and the surface of the crystalline planes of the Ge crystal, needed to diffract a gamma ray) is a function of the wavelength or the energy of the gamma ray. This angle is given by equation (3).

$$Sin(\theta) = \frac{wavelength}{2d}$$

(3)

where the wavelength and the spacing between the crystalline plans (d) are in the same units. The energy of the gamma ray in keV is given by equation (4).

$$E(keV) = \frac{12.39}{Wavelength(\mathring{A})}$$

(4)

For an energy of 170 keV the Bragg angles are quite small. The Bragg angle for the smallest ring [111], and for the largest ring [422] are 0.641 degrees and 1.813 degrees, respectively. Higher energies will have even smaller Bragg angles. When the telescope is tuned to an astrophysics source, (S) approaches infinity and the lens-detector distance (D) approaches (F). The lens is tuned to an energy of 170 keV and the corresponding distance from the lens to the detector is 279 cm.

4. THE CRYSTAL LENS AND GE GAMMA-RAY DETECTOR

The lens consists of 576 Ge-Si crystals, located in 6 concentric rings. The individual crystals are typically 1 cm x 1 cm and 3-4 mm thick. Figure 2a shows the lens as it is mounted in the telescope. The combination Ge-Si crystals are used to control the mosaic width of the crystals by controlling the amount of Si in the crystal as it is grown. One uses the control of mosaic widths to maximize the efficiency of the lens. The lens is held in a double gimbal telescope frame. Without this feature it would be very difficult to keep the telescope pointed at the Crab Nebula. The detector is a 9 element matrix of Ge gamma-ray detectors with a frontal area of 1.5 cm x 1.5 cm and a depth of 4 cm. This detector matrix is shown in figure 2b. The lens is tuned to focus 170 keV gamma rays into the detector with a focal spot of about 1.2 cm. The crystal diffraction lens acts just like a simple convex lens does for visible light, so as the target moves off the central axis, the focused beam moves off axis by the same angle. If the target moves off axis by 18 min. of arc, the focused beam will move from the center of the central detector to the center of one of the side detectors.

5. THE BALLOON FLIGHT

The balloon Launch occurred on June 15, at about 9:30 am and a float time of 5 hours at 100,000 feet. The launch used the French system of supporting the payload with 2 small balloons while the main balloon was being filled. Both singles spectra and coincidence spectra were telemetered back to the home base located on a high point in the Gap area. The pay load was dropped by parachute about 50 miles from Toulouse. The data analysis is still underway.

ACKNOWLEDGMENTS

The authors wish to thank the balloon crew and graduate students who helped with the launch and the graduate students that helped to finish the crystal diffraction lens and especially Phillipe Laporte who supervised much of this work.

REFERENCES

1. "New Method for Focusing X-Rays and Gamma-Rays", Robert K. Smither, Rev. Sci. Instr., 44, 131-141, (1982).
2. "Review of Crystal Diffraction and its Application to Focusing Energetic Gamma Rays", Robert K. Smither, et. al., Experimental Astrophysics, 6,47-56, (1995).

Young Collapsed Supernova Remnants: Similarities and Differences in Neutron Stars, Black Holes, and More Exotic Objects

James S. Graber

407 Seward Square SE
Washington, DC 20003, USA

Abstract. Type Ia supernovae are thought to explode completely, leaving no condensed remnant, only an expanding shell. Other types of supernovae are thought to involve core collapse and are expected to leave a condensed remnant, which could be either a neutron star or a black hole, or just possibly, something more exotic, such as a quark or strange star, a naked singularity, a frozen star, a wormhole or a red hole. It has proven surprisingly difficult to determine which type of condensed remnant has been formed in those cases where the diagnostic highly regular pulsar signature of a neutron star is absent. We consider possible observational differences between the two standard candidates, as well as the more speculative alternatives.

We classify condensed remnants according to whether they do or do not possess three major features: 1)a hard surface, 2)an event horizon, and 3)a singularity. Black holes and neutron stars differ on all three criteria. Some of the less frequently considered alternatives are "intermediate," in the sense that they possess some of the traits of a black hole and some of the traits of a neutron star. This possibility makes distinguishing the various possibilities even more difficult.

INTRODUCTION

Almost by definition, all supernovae leave an expanding remnant of their explosion, and these expanding remnants are the prime focus of this conference. In addition at least some supernovae leave a condensed remnant which is of interest in its own right, and which may also strongly influence the larger expanding remnant. We focus here on these condensed remnants. Typically, three alternatives are considered: a neutron star, a black hole, or no remnant at all. Type Ia Supernovae are thought to leave no remnant, but all other types are expected to leave either a black hole or a neutron star. However, other types of collapsed remnants have been considered by many authors. Among them are quark [1,2], strange [3–6] and boson stars [7,8], soliton stars [9], frozen stars [10–12], naked singularities [13,14], and wormholes [15–17]. Other, less well-known, possibilities include a type of object Mitra calls an eternally collapsing object(ECO) [18], and a somewhat similar object that I have discussed previously and called a red hole [19], but am now beginning to call a big red bag," because this latter term is more descriptive. In this brief conference poster summary, we will not go into the theoretical motivations for considering these more exotic objects, nor into their technical definitions. Instead, we will provide generic, phenomenological descriptions of these different types of object, and consider how they might be detected or rejected by observations of supernovae and especially of their condensed remnants. In most cases, these same techniques are used to detect the difference between neutron stars and black holes, and to constrain certain properties of the condensed objects, such as neutron star equations of state.

The primary purpose of this paper is to urge that alternate models of condensed remnants not be overlooked, and to point out that standard models are already experiencing some difficulties explaining the observations. Several significant observational constraints are already in hand and more are in the offing. The second purpose is to suggest, and partly demonstrate, that both the theoretical models and the observational constraints can be expressed in terms of largely theory-free phenomenological parameters, for standard as well as more exotic objects.

TRIPARTITE CLASSIFICATION OF REMNANTS

Standard interpretation of standard theory, i.e., General Relativity (GR), leads us to expect only black holes or neutron stars. We here consider in addition possibilities suggested by unusual forms of matter in standard GR, by nonstandard interpretations of GR, and by alternate theories of gravity. In the spirit of the PPN approach to gravity, we wish to consider in a categorical or parametric, theory-independent, way all possible condensed remnants, not just those that have already been considered in some detail. Therefore we here consider not specific models, but classes of models, based on three generic characteristics that strongly affect the behavior, appearance, and detectability of the class of objects.

These three characteristics are the presence or absence of singularity(s), event horizon(s), and hard surface(s). Black holes and neutron stars differ on all three of these criteria. Some of the other alternatives are "intermediate," in that they possess some of the traits of a black hole and some of the traits of a neutron star. This makes distinguishing the various possibilities even more difficult.

CLASSIFICATION OF CONDENSED OBJECTS

We begin with the standard possibilities. A black hole has an event horizon and a singularity, but no hard surface. Just the opposite, a neutron star has a hard surface, but no event horizon and no singularity. Among the less frequently considered alternatives, the strange, quark, boson and soliton stars all fall in the same category as the neutron stars; in fact they could be described as smaller, denser, and perhaps harder or stiffer neutron stars. The naked singularity is in a class by itself, with no hard surface and no event horizon, but obviously a singularity. Of course, the singularity can be pointlike or ringlike; spacelike, timelike or null; and of dimensions ranging from zero to three. Many different varieties of naked singularity have been discussed. Wormholes need to be subdivided to fit into our categories. The classical wormhole [15,16], which has been described as "two black holes glued together at the event horizon," would have an event horizon, but no singularity and no hard surface; the Einstein-Rosen bridge also falls in this category. The more modern "traversable" wormhole [17], (which requires exotic matter for its construction) has no event horizon, no hard surface, and no singularity. Also in this category are the red hole or the big red bag and the eternally collapsing object. The frozen star [11] concept, which was an earlier understanding of the Schwarzschild solution that has now been largely replaced by the black hole paradigm, did seemingly have a hard surface, as well as what we now call an event horizon, and the singularity was hidden by both the event horizon and the hard surface. In this version, the frozen star had all three characteristics: a hard surface, an event horizon and a singularity. In other interpretations (see e.g. Rosen [12]), the hard surface prevented the singularity from forming, and the frozen star had only an event horizon and a hard surface without a singularity.

OBSERVATIONAL EFFECTS OF SELECTED CHARACTERISTICS

A hard surface can absorb or reflect infalling energy in the form of matter or radiation. If it does absorb the infalling energy, it heats up and reradiates at least some of the energy, perhaps at a different wavelength and perhaps slowly over time, but all the infalling energy is at least potentially observable and, in principle, recoverable. Thus accretion energetics and cooling curves can the presence or absence of a hard surface.

An event horizon is a "one-way membrane" that absorbs and hides all infalling energy. The energy is lost from the view of the external observer and can not be seen or recovered (except quantum mechanically), although its mass can be detected gravitationally. Hence energy balance calculations and observations can be a critical indicator of an event horizon.

A point singularity can be treated as a hard surface of radius zero that immediately reflects or reradiates all infalling energy. Or it can be treated as an event horizon of radius zero that absorbs without (nongravitational) trace all infalling energy in whatever form. Or it can be treated as a source of total unpredictability, leading to totally random and unpredictable results, possibly even including nonconsevation of normally conserved quantities. Or it can be treated as an indication that the theory has broken down, and must be modified. Higher dimensional and more complex singularities can be treated analogously.

If the singularity is not hidden, it might be indicated by higher temperatures, faster re-radiation, and a smaller apparent size.

DISTINGUISHING COLLAPSED REMNANTS

It has been surprisingly difficult to detect the difference between black holes and neutron stars. So far the only ironclad technique has been the detection of highly regular pulsar radiation, which is conclusively diagnostic for the presence of a neutron star. Other observational surprises include the inability to detect any compact remnant in many non-type Ia supernovae remnants, and the difficulty of conclusively detecting pulsars whose beam is not directed at us. Detecting intermediate objects whose properties are less well understood can only be substantially more difficult. Nevertheless, there have been observations that are hard to interpret with standard neutron star and black hole models, which have led to the suggestion that perhaps one of these less familiar candidates is being observed.

Possible means of observing or constraining the condensed remnant include: direct and indirect observation of size and shape, collapse energetics, early and late cooling curves, and chemical composition of SNR ejecta.

REFERENCES

1. Drago, A. *Nucl. Phys.* **A616**, 659 (1999).
2. Schaab, C., et. al. *Ap. J.* **480**, L111 (1997).
3. Masden, J., *Phys. Rev. Lett.* **81**, 3311 (1998).
4. Cheng, K. S., et. al. *Science* **280**, 407 (1998).
5. Schwab, C., et. al. *J. Physics* **G23**, 2029 (1997).
6. Cheng, K. S., and Dai, Z. G., *Phys. Rev. Lett* **77**, 1210 (1996).
7. Torres, D. F., *Phys. Rev.* **D56**, 3478 (1997).
8. Hawley, S. H., and Choptuik, M. W., *Phys. Rev.* **D62**, 104024 (2000).
9. Seidel, E., and Suen, W., *Phys. Rev. Lett* **72**, 2516 (1994).
10. Oppenheimer, J. R., and Snyder, H., *Phys. Rev.* **56**, 455 (1939).
11. Thorne, K. S., *Black Holes and Time Warps*, New York: Norton, 1970, pp. 217-218
12. Rosen, N., in Carmeli,M. et. al., Eds. *Relativity*, New York: Plenum (1970)
13. Ishibashi, A., and Hosoya, A., *Phys. Rev.* **D60**, 104028 (1999).
14. Chakrabarti, S. K., and Joshi, P. J., *Int. J. Mod. Phys* **D3**, 647 (1994).
15. Misner, C., *Phys. Rev.* **118**, 1110 (1960).
16. Brill, D., and Lindquist, R. W., *Phys. Rev.* **131**, 471 (1963).
17. Bracelo, C., and Visser, M., *Class. Quant. Grav.* **17**, 3843 (2000).
18. Mitra, A., *astro-ph/9910408*.
19. Graber, J. S., *astro-ph/9908113*.

A Zoo of Composite SNRs

John R. Dickel[1,2]

[1] *Astronomy Department, University of Illinois at Urbana-Champaign, Urbana, IL 61801, USA*
[2] *ASTRON (Netherlands Foundation for Research in Astronomy), Dwingeloo, The Netherlands*

Abstract.
Composite supernova remnants show a great variety of interactions between the pulsar, whether seen or not, and the surrounding remnant. A large range of progenitor characteristics and circumstellar environments must be needed to produce the differences among these supernova remnants.

INTRODUCTION

It is thought that the explosion of a massive star as a supernova should produce both a gaseous supernova remnant (SNR) and a collapsed neutron star in the core. In turn, composite SNRs consist of two parts: a supersonically expanding shell preceded by a shock and a compact nebula powered by the pulsar, a rotating magnetized neutron star. The shell is visible in radio wavelengths as non-thermal synchrotron emission from relativistic electrons accelerated in strong magnetic fields at the interface of the expanding ejecta with the surrounding medium and at the shocks. Shock heating will also produce thermal X-rays. The compact nebula, or plerion, produces strong synchrotron emission at all wavelengths.

Most SNRs do not show all these characteristics, however. Sometimes we do not see a plerion; other times we see the shell around the plerion; and sometimes just a pure plerion without the shell remnant around it. In principle all plerions should eventually become composite SNRs when the ejecta have swept up enough surrounding material to provide the shock heating and particle acceleration. While we don't know the density of the local environment around all plerions, there are definite cases, such as G74.9+1.2 [1] and G63.7+1.1 [2], in which no shell is observed but a naked plerion appears to lie within and/or interacting with dense circumstellar and interstellar material. In these two cases, the authors argue that the pure plerions may arise from supernovae with less energy than those which produce composites or pure shell SNRs.

To evaluate these differences further it is important to understand the evolution and other properties of the composite SNRs. Even in this limited class of objects, there is a great variety among individual members. We shall discuss gen-

CP565, *Young Supernova Remnants: Eleventh Astrophysics Conf.,* edited by S. S. Holt and U. Hwang
© 2001 American Institute of Physics 0-7354-0001-6/01/$18.00

eral properties of the synchrotron emission from the plerion components, mention some interesting examples, and suggest possible explanations of some of the unusual features observed.

SYNCHROTRON RADIATION FROM THE PLERIONS

The polarized synchrotron radiation from plerions is presumably produced by injection of particles from the pulsar. Sometimes the pulsar is readily detected, e.g. the 16-msec pulsar [3] in N157B [4] in the Large Magellanic Cloud, but in others there is no indication of a central compact object, e.g. MSH 15-56 [5,6]. Intermediate cases also exist; for example, G21.5-0.9 has a small-diameter x-ray source in the center but no detectable pulsations [7].

Most plerions are rather elliptical with axial ratios of about 2/1. This asymmetry is probably related to the injection process which may not be isotropic. For example, the Chandra image of the region of the pulsar in the Crab Nebula shows a ring with apparent jets perpendicular to the plane of the ring [8].

The radio spectra of the continuum synchrotron radiation of the plerions are nearly flat with a break to steeper spectral indices in the x-ray. This characteristic is generally attributed to synchrotron losses at the higher energies, but it is difficult to fit with uniform energy losses throughout the entire source. Simple losses should result in a kink in the energy spectrum of the radiating particles with a change by a power of 1.0, or a change in the spectral index of the emission by 0.5 [9]. This break will generally occur in the infrared for remnants with ages between a few hundred and a few thousand years. Some plerions in composites may have larger changes [10], suggesting possible time variations in the injection, acceleration rates, and initial energy spectra of the particles. Although there are few accurate infrared flux densities, most spectra of the integrated emission from these SNRs also show that the spectral breaks are not sharp but have some curvature, e.g. [10]. A curved interface between two line segments would be found for a collection of sharp breaks over a range of energies from different locations.

SOME SPECIFIC EXAMPLES

The prototype of the composite SNRs is MSH 15-56, a large filamentary radio and optical shell with a radius of about 20 pc [6,11]. A bright radio plerion, with axes of about 11 x 5 pc, is offset about 8 pc from the center of the remnant in a direction approximately perpendicular to the long axis of the plerion. The plerion also has an unusual striated appearance with the magnetic field directed along the striations. There are soft x-rays across the entire shell SNR, but the most interesting x-ray feature, is a small, definitely extended, hard x-ray spot beyond one corner of the radio plerion [12]. No pulsar or point x-ray source has been detected anywhere in the remnant. Perhaps MSH 15-56 is an older version of N157B, discussed next.

N157B is an example of a SNR apparently just entering the composite stage. There is a hint of shell structure in the high-resolution radio images [13] and a significant component of thermal emission in the x-ray radiation from this source [14]. The brightest part of the radio plerion is offset northward from both the center of the outline of the entire remnant and from the (also non-central) position of the pulsar [3,15]. The x-ray plerion is certainly brightest right around the pulsar and then appears to trail off toward the brightest part of the radio emission. It looks as if the pulsar began its life at the centroid of the radio feature and has moved off about 5 pc to the southeast. This would infer a motion of about 1000 km sec^{-1} for the estimated age of 5000 years. The decrease of x-ray emission away from the pulsar could be attributed to synchrotron losses of this high energy component while the radio emission has not yet decayed. There should still be bright radio emission from the region just around the pulsar, however, which is not visible. Perhaps a high magnetic-field pulsar (a magnetar) might die quickly, so that the current emission from the near surroundings of the pulsar could still be visible in x-rays but be too faint to see in the radio.

A different appearing composite SNR is G21.5-0.9 which consists of a somewhat elliptical plerion with an x-ray half-width along the major axis of about 40" and a near point x-ray peak in the center. The radio half-width in the same direction is 70" [16]. Around this, bright plerion is a faint (about 2% of the plerion brightness) x-ray ring with some shell characteristics; it extends to about 4 plerion diameters [7]. A radio shell at this brightness level would be difficult to detect in the available data. Radio images of G21.5-0.9 appear essentially identical at 5, 22, and 94 GHz [16], suggesting that the break frequency in the synchrotron radiation spectrum may lie above the previously suggested value of 50 GHz [17].

Other notable examples of composite SNRs include: W50 [18], an extended shell SNR containing a binary system with helical jets, SS433 [19], [see 20,21,22,23]. CTB 80 [24] has a pulsar, a plerion, a bright emission plateau, and multiple unexplained jets emanating from the central region of the SNR [25,26]. The Vela SNR also has its pulsar offset from the plerion [27]. The plerion in W44 is barely distinguishable being the smallest and faintest known plerion relative to its shell [28]. The object 0540-693 in the LMC contains a pulsar, plerion, and a shell visible at all wavelengths [29,30]. Thus, with the possible exception of 0540-693, all composite SNRs seem to have some unusual asymmetries between the pulsar and the plerion and/or irregular shell components which have yet to be understood.

DISCUSSION

In summary, while the plerionic and composite SNRs are considered to be powered by pulsars, many show no direct evidence of a pulsar or its specific effects on the SNR. Those with observed interactions between the pulsar and the SNR have a great variety of properties. Explanations of the differences can include differences in the surroundings caused by the pre-supernova evolution of the progenitor and its neighbors, a range of magnetic field strengths in the pulsars, different explosion energies, and possible membership in a binary system. The latter condition may account for rapidly moving pulsars, precessing or other irregular jet morphology, and perhaps the hiding of the pulsar.

I thank Richard Strom and Eric van der Swaluw for valuable discussions and the Netherlands Organization for Scientific Research for support.

REFERENCES

1. Wallace, B., Landecker, T., Taylor, A. R., and Pineault, S. 1997, A&A, 317, 21
2. Wallace, B., Landecker, T, and Taylor, A. R. 1997, AJ, 114, 2068
3. Gotthelf, E., Zhang, W., Marshall, F., Middleditch, J, and Wang, Q. D. 1998, Memorie della societa Astronomia Italiana, 69, 825
4. Henize 1956, ApJS, 2, 315
5. Mills, B., Slee, O., and Hill, E 1961, AustJPhys, 14, 497
6. Dickel, J., Milne, D., and Strom, R. 2000, ApJ, 543, 840
7. Bandiera, R. 2000, presented to JENAM, Moscow
8. http://chandra.harvard.edu/photo/0052/index.html
9. Reynolds, S. and Chevalier, R. 1984, ApJ, 278, 630
10. Gallant, Y. and Tuffs, R. 1999, The Universe as Seen by ISO, (ESA SP-427) 313
11. van den Bergh, S. 1979, ApJ, 227, 497
12. Plucinsky, P. 1998, Memorie della societa Astronomia Italiana, 69, 939
13. Lazendic, J., Dickel, J., Haynes, R., Jones, P., and White, G. 2000, ApJ, 540, 808
14. Dennerl, K. et al. 2001, A&A, in press
15. Wang, Q. D. et al. 2000, in preparation
16. Bock, D., Wright, M., and Dickel, J. 2000, Imaging at Radio Through Submillimeter Wavelengths, ASP Conference Series 217, 362
17. Salter, C., Reynolds, S., Hogg, D., Payne, H., and Rhodes, P. 1989, ApJ, 338, 171
18. Westerhout, G. 1958, BAN, 14, 215
19. Stephenson, C. and Sanduleak, N. 1977, ApJS, 33, 459
20. Crampton, D., Cowley, A., and Hutchings, J. 1980, ApJL 235, 131
21. Hjellming, R. and Johnston, K. 1981, ApJL, 246, 141
22. Downes, A., Pauls, T., and Salter, C. 1981, A&A, 103, 277
23. Safi-Harb, S. and Ogelman, H. 1997, ApJ, 483, 868
24. Wilson, R. and Bolton, J. 1960, PASP. 72, 331
25. Angerhofer, P. Strom, R., Velusamy, T, and Kundu, M. 1981, A&A, 94, 313
26. Strom, R. 1988, Supernova Remnants and the Interstellar Medium, IAU C. 101, 343
27. Bock, D., Turtle, A., and Green, A. 1998, AJ, 116, 1886
28. Giacani, E. et al. 1997, AJ, 113, 1379
29. Manchester, R., Staveley-Smith, S., and Kesteven, M. 1993, ApJ, 411, 756
30. Gotthelf, E. and Wang, Q. D. 2001, ApJ, in press

Millimetric Observations
of Plerionic Supernova Remnants

Rino Bandiera[*], Roberto Neri[†], and Riccardo Cesaroni[*1]

*Oss. Astrofisico di Arcetri, Largo E.Fermi, 5, 50125 Firenze, Italy
†IRAM, 300 rue de la Piscine, 38406 St Martin d'Hères, France

Abstract. We present results of observations of the Crab Nebula and G21.5–0.9, performed at 1.3 mm with the MPIfR bolometers arrays at the IRAM 30 m telescope. In the Crab Nebula we measure spatial variations of the average spectral index between 20 cm and 1.3 mm. Since the electrons emitting at mm wavelengths are affected by negligible synchrotron losses, such variations imply the presence of at least two different populations of injected particles. By subtracting the emission extrapolated from the radio a residual component appears, similar in size and shape to the soft X-ray map as well as to the flatter-spectrum optical component (Véron-Cetty and Woltjer 1993). Moreover near the major synchrotron filaments we measure a spectral bending consistent with a break at a frequency lower than the average break frequency in the Nebula: this indicates that near the filaments the magnetic field is typically 6 times higher than the average. For G21.5–0.9 we derive a spectral break at ~540 GHz, in contrast with the previously accepted value of 40 GHz. Therefore this object does not strictly belong to the class of "plerions with a low-frequency break".

INTRODUCTION

The synchrotron emission from plerionic supernova remnants is typically observed in the radio and X-ray spectral windows, which provide complementary information on the source. Radio emission samples electrons with synchrotron lifetimes larger than the remnant age, and therefore their energy distribution well reproduces that of the injected particles, averaged over the remnant's age. X-ray emission comes instead from high-energy, fast-burning particles: it is then tightly related to the present rate of particles injection. The short lifetimes of these particles determine the size of the X-ray nebular emission, which is usually smaller than the actual size of the magnetic bubble filling the plerion's volume.

Although millimetric (mm) wavelengths may be expected to be just an extension of the "classical" radio domain, sampling particles in a similar regime, there are cases in which they provide a qualitatively different information from that in radio.

---

[1] This work was partly supported by the Italian Ministry for University and Research under Grant Cofin99-02-02, and by the National Science Foundation under Grant No. PHY99-07949

We report here about results of recent observations at 1.3 mm (i.e. 230 GHz), with the IRAM 30 m telescope, equipped with bolometer arrays developed by the Max-Planck-Institut für Radioastronomie. The spatial resolution is ~10 arcsec, and the sensitivity ~3 mJy/beam.

CRAB NEBULA: DISCOVERY OF A NEW SYNCHROTRON COMPONENT

In the classical radio range spectral maps of the Crab Nebula are extremely "boring", with an upper limit of 0.01 [1] to spatial variations of the spectral index α (except for the wisps area). The homogeneity of α is taken as the evidence of a single acceleration process for the relativistic electrons. In the optical and X-ray ranges spatial variations of α have been measured: for instance, in the optical α steepens outwards, from $\simeq -0.6$ near the center to over -1.0 at the edge [2,3]. This steepening may be explained in terms of synchrotron losses of the electrons.

In the integrated spectrum of the Crab Nebula the "evolutive break" (where adiabatic and synchrotron losses balance) is at $\sim 10^4$ GHz, and the mm emitting particles are not subject to substantial synchrotron losses: therefore the mm map was expected to simply scale with the radio one. However when our mm map is compared with a VLA map at 1.4 GHz (kindly provided us by M.F. Bietenholz) some morphological differences can be noticed. The spectral map between the two frequencies shows significant spectral variations (up to 0.07 in α), with flatter spectra in the inner region. In the inner region the spectral index increases from radio to mm wavelengths. Such behavior (just the opposite of what expected in the case of synchrotron downgrading) can be the sign of a further spectral component.

With this in mind we have subtracted from the 1.3 mm map an extrapolation of the radio map, using the radio spectral index $\alpha = -0.27$, and in the residuals map we found a clear excess in the inner region, together with some less prominent negative features in positions corresponding to those of radio filaments. The latter ones indicate that in the spectra of filaments a spectral bending occurs at a frequency (ν_b) a factor ~260 lower than the average break frequency in the Nebula. We interpret this as the evidence that the magnetic field in the main filaments is a factor ~6 higher than the average ($\nu_b \propto B^{-3}$, for particles trapped into the filaments).

THE NATURE OF THE INNER COMPONENT

The size (~1.5 arcmin) and shape of the inner component are similar to those of the optical flatter spectrum region discussed by Véron-Cetty and Woltjer [2], which in turn is similar in size and shape to the X-ray nebula. While in the optical and X rays this smaller size can be accounted for, at least qualitatively, in terms of synchrotron downgrading, the puzzle is why a similar component is observed also at 1.3 mm, where the particles are in the adiabatic regime. This suggests that the observed pattern is not uniquely determined by synchrotron losses.

What about the nature of this component? We can exclude free-free emission, because it requires a large mass of ionized gas (which would be opaque at low radio frequencies). We can also exclude dust emission, because a fit to the integrated spectrum of the Crab Nebula requires very cold dust ($T < 5$ K). Therefore this emission is likely to be nonthermal, and its spectrum must have a lower cutoff near the mm wavelengths, in order to be compatible with the absence of spatial variations of α in radio [1].

The total number of particles in this newly discovered component can be evaluated as about $\sim 2 \times 10^{48}(3 \times 10^{-4}\, G/B)$, a number close to what predicted by Kennel and Coroniti [4] on the basis of a model in which both particles and magnetic field are provided by a relativistic pulsar wind. That model has been quite successful, except for the fact that it cannot explain the presence of the radio emitting electrons. This new mm component matches some requirements of Kennel and Coroniti model (like on the total number of particles, and on the presence of a low energy cutoff) better than the main radio component does. However no model is available yet which reproduces in detail what observed: for instance there is no

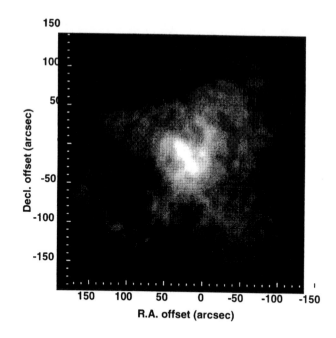

FIGURE 1. Crab Nebula: map of the residuals at 1.3 mm after subtracting an extrapolation from the radio range and after correcting for spectral bending in synchrotron filaments.

explanation why the mm emitting particles, having long lifetimes, are still confined in the inner region.

It has been argued (Hester, this meeting) that the new component we have found may be a mere artifact, originated from having compared observations at different epochs: the Crab Nebula is in fact known to contain highly dynamic structures. However variable structures look typically wave-like at small scales [5], while average to zero on scales larger than the spatial resolution of our observations. Studies on the secular changes of the radio flux [6] give an upper limit below 1% to the variability of the integrated flux; also our maps at 1.3 mm are consistent with no change over ≃1 year. For these reasons we believe that radio variability is small on large scales, and that the new synchrotron component we have discovered is real.

G21.5–0.9: A "NOT-TOO-LOW FREQUENCY BREAK"

Previous high-frequency flux measurements of this plerion, in the range 84–142 GHz [7], are below the extrapolation from lower radio frequencies, and imply a spectral break at ~ 40 GHz. For this reason it has been catalogued among the "low frequency break" plerions (like CTB87, 3C58, etc.), also known as "non Crab-like" plerions [8]. The problem with these plerions is that, if the measured break is the evolutive one, the energy stored in the nebular field is huge.

From our maps we derive a total flux of 3.9±0.4 Jy at 230 GHz (about twice that expected from the previous high-frequency radio data), which is also nicely fitted by the interpolation between radio and infrared data [9]. This leads to a revised estimate (~540 GHz) for the break frequency, and then to less extreme estimates for the nebular field, as derived from the break position. Finally, a comparison of the radial profile at 230 GHz with that from a map at 22 GHz (kindly provided by W. Reich) indicates that the effective size of the nebula starts shrinking in the mm range, an effect likely related to synchrotron losses.

REFERENCES

1. Bietenholz, M. F., Kassim, N., Frail, D. A., Perley, R. A., Erickson, W. C., and Hajian, A. R., *Ap. J.* **490**, 291 (1997).
2. Véron-Cetty, M. P., and Woltjer, L., *Astr. Ap.* **270**, 370 (1993)
3. Bandiera, R., Amato, E., and Woltjer, L., *Mem. SAIt* **69**, 901 (1998).
4. Kennel, C. F., and Coroniti, F. V., *Ap. J.* **283**, 710 (1984).
5. Bietenholz, M. F., and Kronberg, P. P., *Ap. J.* **393**, 206 (1992).
6. Aller, H. D., and Reynolds, S. P., *Ap. J.* **293**, L73 (1985).
7. Salter, C. J., Emerson, D. T., Steppe, H., and Thum, C., *Astr. Ap.* **225**, 167 (1989).
8. Woltjer, L., Salvati, M., Pacini, F., and Bandiera, R., *Astr. Ap.* **325**, 299 (1997).
9. Gallant, Y. A., and Tuffs, R. J., *Mem. SAIt* **69**, 963 (1998).

G291.0–0.1: Powered by a Pulsar?

David Moffett[1], Bryan Gaensler[2], and Anne Green[3]

[1] *Furman University Physics Department, Greenville, SC 29613, USA*
[2] *MIT Center for Space Research, Cambridge, MA 02139, USA*
[3] *University of Sydney School of Physics, Australia*

Abstract.

As part of a continuing study of plerionic supernova remnants, we have closely examined the SNR G291.0–0.1 (MSH 11–62) at radio wavelengths following ASCA X-ray observations of this source. The high angular resolution, full-polarization radio observations at 13 and 20 cm support its plerionic nature; the remnant has an interior dominated by a bright, linearly-shaped, polarized central feature, and has an exterior composed of filamentary structure. No compact source suggestive of a pulsar or associated wind nebula was found. In addition, we observed HI absorption toward the remnant, which places it at or beyond 3.5 kpc.

WHERE'S THE PULSAR?

G291.0–0.1 is one of a class of supernova remnants which have Crab-like or plerionic properties (center brightened, flat spectrum, linearly polarized). It is commonly assumed that the driving force behind their emission is a pulsar, however, G291 joins a few other plerions that show no evidence of a pulsar, at either radio or high energy wavelengths.

Roger *et al.* [1] performed the last, thorough radio observations of this SNR; they confirmed its plerionic nature by finding a radio spectral index of $\alpha = 0.29$, and highly-ordered polarization: 7% at 5 GHz and 10% at 8 GHz. X-ray emission was detected at the same time by Wilson [2] using the Einstein Observatory. Its X-ray emission is positionally coincident with the radio SNR, center-brightened, and had a spectrum suggestive, but not certain of, non-thermal synchrotron emission.

Recently, Harrus *et al.* [3] have visited G291.0–0.1 and other plerions with the ASCA X-ray observatory, and confirmed a non-thermal component at the center, and a thermal emission component that extends over the entire remnant. Their spatial and spectral analyses imply the presence of a neutron star, however, searches in radio and X-ray have found no pulsed emission.

CP565, *Young Supernova Remnants: Eleventh Astrophysics Conf.*, edited by S. S. Holt and U. Hwang
© 2001 American Institute of Physics 0-7354-0001-6/01/$18.00

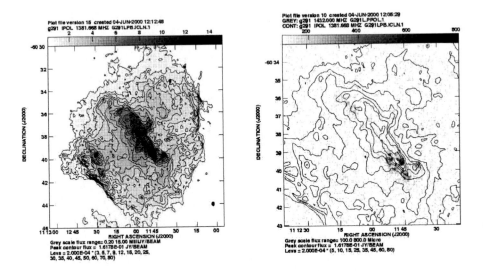

FIGURE 1. Left: contour and grayscale image of the total intensity at 1381 MHz (20 cm). The most prominent feature is an extended linear feature stretching diagonally across the center of the remnant. Right: linear-polarized intensity grayscale image at the same frequency superimposed on a contour map of the total intensity.

OBSERVATIONS

We performed radio continuum and HI spectral-line observations of G291.0–0.1 with four configurations of the Australia Telescope Compact Array. Data reduction was completed using standard procedures within the MIRIAD package. Deconvolved, total intensity images were formed at 13 and 20-cm wavelengths (left panels of Fig. 1 & 2). To search for compact emission, we also formed total intensity images using only the longest baselines of the interferometric data

Polarization images of Stokes Q and U were formed from individual continuum channels within each band and averaged to remove Faraday derotation effects in the polarized intensity images (right panels of Fig. 1 & 2). We found a Faraday rotation measure, RM = 390 rad cm^{-2}, consistent with that found by Roger *et al.*, by observing the rotation of the position angle within the 20-cm band channels and between 20 and 13 cm.

No previous HI observations have been conducted toward this remnant, so to better place constraints to the remnants distance, HI spectral-line data was acquired (Fig. 3) during two configuration observations using a correlator configuration of 1024 channels of width 4 kHz or approximately. 0.8 km/s of spectral resolution.

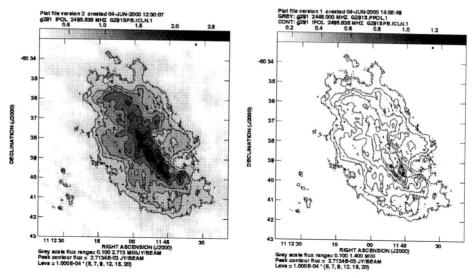

FIGURE 2. Left: contour and grayscale image of the total intensity at 2496 MHz (13 cm). Only the central extended feature remains because of radio spectral and interferometric spatial filtering effects. Right: linear-polarized intensity grayscale image at the same frequency superimposed on a contour map of the total intensity.

CONCLUSIONS

At our highest resolution and sensitivity (9″ and 100 mJy beam^{-1} at 20 cm, 3″ and 80 mJy beam^{-1} at 13 cm), we were unable to detect any compact source associated with a pulsar at the location of the peak X-ray emission in both ASCA and Einstein images. However, we must note that the X-ray peak is positionally coincident with the polarized radio emission seen at the SW edge of the extended linear feature in total intensity. The neutron star very likely inhabits this region of the remnant.

After correcting for Faraday rotation, magnetic fields in this remnant appear to be highly ordered and polarized (to greater than 10%). They are primarily oriented along the axis of the central linear feature, as was found by Roger *et al.* [1]. However, unlike their results, the detected polarized emission in our images is constrained to only the lower SW portion of the central linear feature (the plerionic core of the remnant).

The near distance of 3.5 kpc derived from the HI data is consistent with fits of the X-ray column density [3], and with the HI spectra of two neighboring HII regions [1]. Harrus *et al.* argue, that at a distance of 3.5 kpc, G291's synchrotron luminosity implies that a neutron star with a spin-down energy-loss rate of 2.5×10^{36} ergs s^{-1} inhabits the SNR. Assuming a surface magnetic field of $10^{12} - 10^{13}$ Gauss, and a braking index of 2.5 to 3, this implies an initial spin period in the range of 45 to 130 ms, and a current-epoch period in the range 46 to 160 ms.

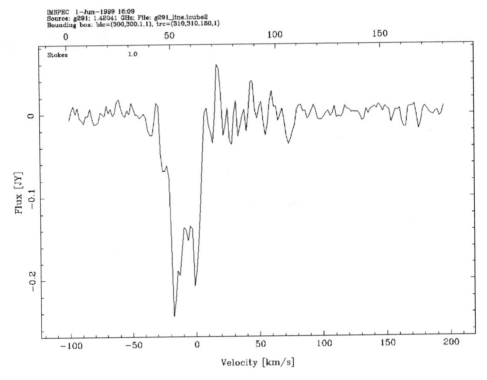

FIGURE 3. Non-weighted, neutral hydrogen-line spectrum integrated over a region covering the remnants center. Absorption is seen out to the tangent point ($\sim 20\,\mathrm{km\,s^{-1}}$), which yields a minimum distance (using the standard Galactic rotation model) to the remnant of 3.5 kpc.

ACKNOWLEDGEMENTS

D. Moffett acknowledges travel support from the Radio Astronomy Group of the University of Tasmania while a post-doc there. The Australia Telescope Compact Array is a facility of the Australia Telescope National Facility, operated by the CSIRO Division of Radiophysics.

REFERENCES

1. Roger *et al.* 1986, MNRAS 219, 815
2. Wilson 1986, ApJ, 302, 718
3. Harrus *et al.* 1998, ApJ, 499, 273

Chandra Observation of the Hard X-Ray Feature in IC 443

Christopher R. Clearfield[1], Charles M. Olbert[1], Nikolas E. Williams[1], Jonathan W. Keohane[1], and Dale A. Frail[2]

[1] *North Carolina School of Science and Mathematics, 1219 Broad St., Durham, NC 27705, USA*
[2] *National Radio Astronomy Observatory, P. O. Box O, Socorro, NM 87801, USA*

Abstract. We present new Chandra A01 data of the hard X-ray feature along the southern edge of the SNR IC443. With only 10 ks of observation time we were able to extract high resolution ($\sim 1''$) images that reveal a soft point source embedded in a hard comet-shaped nebula. The background-subtracted spectrum of the point source is best fit with an absorbed black body model of temperature and surface area $kT = 0.1 \pm 0.1$ keV and $A = 0.02^{+0.01}_{-0.02}(\frac{d}{1.5 \text{ kpc}})^2$ km^2, respectively.

INTRODUCTION

This paper focuses on the analysis and reduction of a short (10 ks) observation of IC 443 on April 10, 2000 during the first cycle of guest observations (A01). The observation examines the Hard X-ray Feature (HXF) on the Southern edge of IC 443 [1]. Companion papers, [2] and [3] discuss the physical interpretation and radio observations, respectively.

We use *Chandra Interactive Analysis of Observations (CIAO)* Software to produce our images and spectra. following the standard procedures outlined in Version 1.3 of the *CIAO* Beginner's Guide for *CIAO* Release 1.1. In addition, following procedures outlined in [4] we create a response function (RMF) and an ancillary response function (ARF). After reducing our data, we perform analysis using other image processing and spectral analysis tools (e.g. *AIPS* and *XSPEC*).

SPATIAL ANALYSIS

We extract high resolution ($\sim 1''$) images in both hard ($\epsilon > 2.1$ keV) and soft ($\epsilon < 2.1$ keV) spectral bands. Smoothed versions of these images are presented in Figure 1. In our companion paper [3] we also present a radio map for comparison.

The hard X-ray source is resolved by these high resolution Chandra observations (Figure 1) as a comet-shaped region of diffuse emission within which there is an

CP565, *Young Supernova Remnants: Eleventh Astrophysics Conf.*, edited by S. S. Holt and U. Hwang

unresolved point source, which we have designated *CXOU J061705.3+222127*. The nebula exhibits bow shock morphology with a width of $\simeq 35''$ and a minimum standoff distance of $r_s \simeq 8.5''$. The X-ray point source lies at the apex of the nebula, located at (epoch 2000) $\alpha = 06^h17^m05.31^s$, $\delta = 22°21'27''.3$, with errors of $\pm 2''$ in each coordinate. It is especially visible in the soft ($\epsilon < 2.1$ keV) X-ray band (Figure 1). On the other hand, the hard X-ray and radio [3] images are morphologically similar — both clearly showing the nebula but not the point source.

SPECTRAL ANALYSIS

A spectrum of CXOU J061705.3+222127 was extracted, as well as a spectrum of the surrounding hard cometary nebula, without the point source, (an annulus with an inner and outer radius of 4.3'' and 23.6'' respectively) which was used as background. These spectra were binned by a factor of 256, and the softest two or three channels were ignored, because of the ACIS-I poor response at low energies. Assuming that our channel 3 ($\epsilon \sim 1$ keV) is reliable, the spectrum is much better-fit by a black body than a power law model ($\Delta \chi^2 \sim 20$). On the other hand, channel 3 contains only 2 counts and makes either model fit poorly ($\chi_\nu^2 \sim 2.3$ or 3.9 respectively). Ignoring channel 3, on the other hand, yields mediocre, but acceptable, fits for either model ($\chi_\nu^2 \sim 1.1$), but qualitatively the black body model appears to fit better than the power law even ignoring channel 3 (Figure 2). The best-fit black body temperature and surface area are kT=0.1±0.1 keV and A=$0.02^{+0.01}_{-0.02}(\frac{d}{1.5 \text{ kpc}})^2$ km^2. The best-fit power law photon index is Γ=1.6±0.3. The 1-5 keV X-ray flux of the point source is $\sim 2 \times 10^{-13}$ erg s^{-1} cm^{-2}. All of these fits contained two free parameters and assumed K97's best-fit column density (N$_H$=1.3×10^{21}cm^{-2}).

CONCLUSIONS

These observations resolve the hard X-ray feature discovered by [1] as a comet-shaped region of diffuse emission within which there is an unresolved point source, which we have designated *CXOU J061705.3+222127*, at coordinates (epoch 2000) α=06^{h}17^{m}05.31^s, δ=22°21'27''.3 (with errors $\pm$ 2'' in each coordinate). The nebula exhibits bow-shock morphology with a width of $\simeq 35''$ and a minimum standoff distance of $r_s \simeq 8.5''$. The best fit of spectrum of the unresolved point source was a black body model with temperature and surface area are kT=0.1±0.1 keV and A=$0.02^{+0.01}_{-0.02}(\frac{d}{1.5 \text{ kpc}})^2$ km^2.

Discussion of the interpretation of these data can be found in [2] in these proceedings.

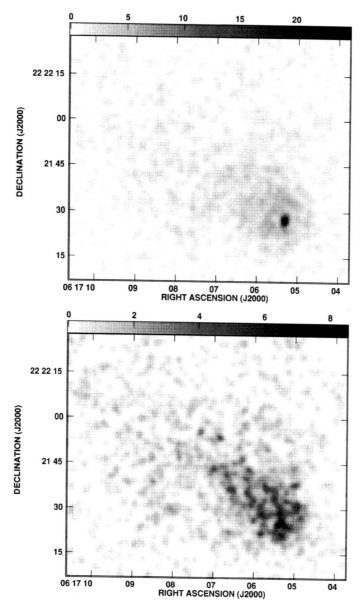

FIGURE 1. X-ray intensity images of the hard X-ray feature on the southern edge of IC 443. The image on the left is in the soft ($\epsilon < 2.1\,\mathrm{keV}$) photon energy band, scaled between 0 to 24.6 counts per beam. The image on the right is in the hard ($\epsilon > 2.1\,\mathrm{keV}$) photon energy band, scaled between 0 to 8.5 counts per beam. Both of these Chandra images have been smoothed with a 2″ Gaussian "beam." The soft point source stands out well against the surrounding nebula in the soft band image (left), while he cometary nebula is most apparent in the hard band image (right).

ACKNOWLEDGEMENTS

We thank G. E. Allen, K. A. Arnaud, J-H. Rho, N. Evans, and M. Johnson for help with data analysis. This research was funded by NASA grant NRA 97-0SS-14.

REFERENCES

1. Keohane, J. W., Petre, R., Gotthelf, E. V., Ozaki, M., and Koyama, K. 1997, ApJ, 484, 350. (K97)
2. Olbert, C. M., Clearfield, C. R., Williams, N. E., Keohane, J. W., and Frail, D. A. 2000, *These Proceedings*.
3. Williams, N. E., Olbert, C. M., Clearfield, C. R., Keohane, J. W., and Frail, D. A. 2000, *These Proceedings*.
4. Preswitch, A. CIAO Instrument and Calibration Guide 1999, CXC. Release 1.1

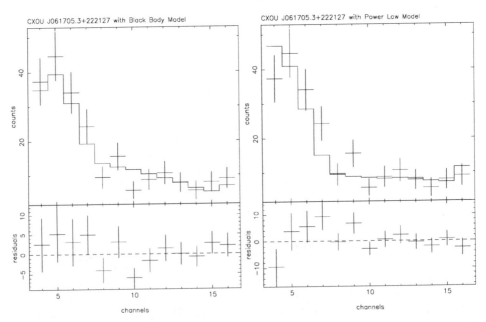

FIGURE 2. The Chandra spectrum of the soft point source, fit with black body (left) and power law (right). In both cases, there were two free parameters, and the column density was fixed to the ASCA value (K97).

Evidence of a Pulsar Wind Nebula in Supernova Remnant IC 443

Charles M. Olbert[1], Christopher R. Clearfield[1], Nikolas E. Williams[1], Jonathan W. Keohane[1], and Dale A. Frail[2]

[1] *North Carolina School of Science and Mathematics, 1219 Broad St., Durham, NC 27705, USA*
[2] *National Radio Astronomy Observatory, P. O. Box O, Socorro, NM 87801, USA*

Abstract. New *Chandra X-Ray Observatory* and *Very Large Array* observations of the hard X-ray feature along the southern edge of the supernova remnant IC 443 have revealed a comet-shaped nebula of hard emission with a soft X-ray point source at its apex. Based on the X-ray spectrum, X-ray and radio morphology, and the radio polarization properties, we argue that this object is a synchrotron nebula powered by the compact source. The derived parameters of the system favor an interpretation in which the central object is a young, energetic neutron star physically associated with IC 443. The cometary morphology of the nebula originates from the supersonic motion of the pulsar ($V_{PSR} \simeq 250 \pm 50$ km s^{-1}), which causes the relativistic wind of the pulsar to terminate in a bow shock and trail behind as a synchrotron tail. This velocity is consistent with an age of 30,000 years for the SNR and its associated pulsar.

INTRODUCTION

The mixed-morphology Galactic supernova remnant IC 443 (l,b=189.1°,+3.0°) has been the subject of extensive studies [1–4], and is well known for its clear interaction with surrounding molecular clouds [5,6] and shock chemistry ([3] and references therein). IC 443 is coincident with the unidentified EGRET source 2EG J0618+2234 (l,b=189.13°,+3.19°) [7,8], and thus has stimulated a good deal of theoretical work aiming to explain the production of GeV γ-rays by shell SNRs [9–13]. Also, an unresolved region of hard X-ray synchrotron emission [14] and flat radio spectral index (i.e $\alpha < 0.24$ vs $\alpha = 0.42$, where $F_\nu \propto \nu^{-\alpha}$) [15,16] had been observed and attributed to enhanced synchrotron emission from SNR/molecular cloud interaction [14,6].

An equally viable model, favoured by Chevalier [17] (hereafter C99), posits that the hard X-ray feature is synchrotron emission powered by an energetic neutron star, despite null results from previous timing analysis (i.e. [14,19]). In the absence of detectable pulsations, pulsars can be distinguished from other compact objects by the presence of an extended synchrotron nebula [18]. New data from the *Chandra*

CP565, *Young Supernova Remnants: Eleventh Astrophysics Conf.*, edited by S. S. Holt and U. Hwang
© 2001 American Institute of Physics 0-7354-0001-6/01/$18.00

X-ray Observatory and the *Very Large Array* indicate that this region does indeed contain such a cometary nebula that contains a compact object at its apex. These data suggest that this hard X-ray feature is a synchrotron nebula powered by an energetic neutron star.

The details of the reduction, analysis, spectral fitting, and many specific results of the *Chandra* data are discussed by Clearfield *et al.* [20]. Similarly, discussion of data from the *VLA* can be found in Williams *et al.* [21].

DISCUSSION

The X-ray point source in IC 443 (*CXOU J061705.3+222127*) is the newest in a growing number of young neutron star candidates associated with supernova remnants [22]. The observational properties of the cometary nebula and point source are concordant with expected characteristics [17,23]: flat spectrum radio emission (α_R=0.1-0.3), a steep X-ray spectrum (α_X=1.0-1.5) and a high degree of linear polarization (>5%). Morphologically, it resembles other pulsar wind nebulae (PWNe) detected towards SNRs W 44 and G 5.4-1.2 [24,25], both of which contain active pulsars. We assert that the X-ray point source in IC 443 is a young pulsar which has traveled from its birth place to its present location, near the edge of the decelerating SNR.

We proceed with inferring the salient physical parameters of the wind nebula and neutron star. The offset of the pulsar from its origin gives a measure of its transverse velocity V_{PSR}, which is made difficult owing to the uncertain location of IC 443's explosion center [26,4]. We use a shock radius r_s=7.4 pc and note that the pulsar has traveled a fraction $\beta{\simeq}0.9$ of this projected distance, giving $V_{PSR} = \beta \, r_s \, t_s^{-1}$. The shock velocity of the SNR can be expressed as $V_s = c_o r_s \, t_s^{-1}$, where c_o is a constant equal to $\frac{2}{5}$ for a remnant in the Sedov stage, while C99 argues for a value of $c_o = \frac{3}{10}$ for IC 443. Combining both equations and using the value of V_s=100 km s^{-1} (C99) gives $V_{PSR}=V_s \, \beta \, c_o^{-1}$=225-300 km s^{-1}, a result independent of distance and weakly dependent on the evolutionary state of the SNR (i.e. c_o).

Keohane et al. [14] fit the hard X-ray feature with an X-ray temperature kT$\simeq$1 keV, implying a sound speed $c_{local}{\simeq}100$ km s^{-1}. A bow shock angle of 90°-120° gives a minimum Mach number M$\simeq$1.1-1.5, and implies a velocity of $V_{PSR\,local}$=150 M$_{1.5}$ kT$_{1\,keV}^{0.5}$ km s^{-1}. Taking local expansion into account, this velocity could be as high as 250 km s^{-1}. Both methods of determining V_{PSR} are in good agreement with C99, indicating a remnant age on the order of 30,000 years.

To empirically compare this nebula with statistical studies of PWNe (e.g. [27]), we integrate the power-law model of [14] over the *Einstein* band (0.2-4 keV), deriving $L_x{\sim}5{\times}10^{33} \left(\frac{d}{1.5\,kpc}\right)^2$ erg s^{-1}. For this value of L_x, we derive $\dot{E}{\sim}2{\times}10^{36}$ erg s^{-1} [27]. $\dot{E}{\sim}10^4 L_R$ for known radio PWN [28,23], which yields a value of $\dot{E}{\sim}6{\times}10^{35}$, and we thus adopt $\dot{E}=10^{36}$ erg s^{-1} (i.e. $\dot{E}_{36}$=1). The spindown luminosity of a pulsar is related to its current period P and dipolar surface field B by $\dot{E} \propto B^2 P^{-4}$, while the characteristic age of the pulsar is expressed as $\tau_c \propto B^{-2}P^2$. It is straightforward to show that for a pulsar born spinning rapidly (i.e. $P_\circ \ll P$), which loses energy predominately via magnetic dipole radiation, $P = 145\,\mathrm{ms}\,(\tau_{30}\dot{E}_{36})^{-1/2}$ and $B = 3.3 \times 10^{12}\,\mathrm{G}\;\tau_{30}^{-1}\dot{E}_{36}^{-1/2}$, where we have chosen an age of 30,000 yrs (i.e. $\tau_{30} = 1$) for IC 443 used by C99 and confirmed above. This period and magnetic field are consonant with those of other young pulsars (e.g. PSR B1757$-$24) [25].

It is worth re-examining the nature of the unidentified EGRET source toward IC 443. Nel et al. (1996) [29] note that all known gamma-ray pulsars have a large (greater than 0.5) ratio of $\frac{\dot{E}_{33}}{d_{kpc}^2}$, where $\dot{E}_{33} = \frac{\dot{E}}{10^{33}\,\mathrm{erg\,s}^{-1}}$. For our derived $\dot{E} \sim 10^{36}\,\mathrm{erg\,s}^{-1}$ and $d \sim 1.5\,\mathrm{kpc}$, this ratio is over 400, similar to that of the gamma-ray emitting PSR B1706$-$44, strongly suggesting that the unidentified EGRET emission originates from the PSR or the PWN. However, our proposed PSR and PWN lie several arcminutes outside the formal 95% error radius of the EGRET source. Since the position of EGRET sources in the Galactic plane are subject to systematic error (e.g. [30]), it is entirely possible that the PSR and the PWN may be the source of this gamma-ray emission.

Despite the evidence in favor of a causal relationship between IC 443 and pulsar candidate, there are a few puzzles that remain to be explained. To first order, we expect the "tail" of the nebula to point back to the origin of the SNR and the pulsar. However, the geometric center of IC 443 lies almost due north of the pulsar, and not to the northeast as suggested by the direction of the cometary emission. Given the complex kinematics and distribution of the molecular gas in IC 443 [31,3] it may be naïve to ascribe the PSR/SNR birth place to the geometric center. A second problem concerns the absence of X-ray and radio pulsations from the point source [14,19]. In our view, these null results present no immediate problem for our hypothesis since pulsed X-rays from Vela-like pulsars have proven difficult to detect due to their soft spectra and contamination from the surrounding nebular emission (e.g. [32]). Furthermore, the radio beam may not intersect our line of sight [33].

The hard X-ray feature in IC 443 is best interpreted as a wind nebula, powered by a young pulsar. The PWN interpretation is suggested first on morphological grounds by the Chandra and VLA images, which show a comet-shaped nebula with a soft X-ray point source at its apex. The measurement of significant polarization and a flat radio spectrum further strengthens the argument that this is synchrotron emission from a PWN. The X-ray spectrum of the point source is best fit by a black body, suggesting that the emission is thermal in origin. The inferred physical properties of the nebula and point source (i.e. T, V_{PSR}, $\dot{E}$, P, B) support our hypothesis that IC 443 and the pulsar were produced approximately 30,000 years ago in a core collapse supernova event.

We thank J.H. Rho, D. J. Thompson, B. M. Gaensler, R. Petre, P. Slane, J. Kolena, and R. A. Chevalier for useful discussions. This research was funded by NASA grant NRA 97-0SS-14.

REFERENCES

1. Fesen, R. A. 1984, ApJ, 281, 658.
2. Braun, R. and Strom, R. G. 1986, A&A, 164, 193.
3. van Dishoeck, E. F., Jansen, D. J., and Phillips, T. G. 1993, A&A, 279, 541.
4. Asaoka, I. and Aschenbach, B. 1994, A&A, 284, 573.
5. Burton, M. G., Hollenbach, D. J., Haas, M. R., and Erickson, E. F. 1990, ApJ, 355, 197.
6. Bocchino, F. and Bykov, A. M. 2000, A&A, *in press* (astro-ph/0010157).
7. Sturner, S. J. and Dermer, C. D. 1995, A&A, 293, L17.
8. Esposito, J. A., Hunter, S. D., Kanbach, G., and Sreekumar, P. 1996, ApJ, 461, 820.
9. Sturner, S. J., Dermer, C. D., and Mattox, J. R. 1996, A&AS, 120, C445.
10. Sturner, S. J., Skibo, J. G., Dermer, C. D., and Mattox, J. R. 1997, ApJ, 490, 619.
11. Gaisser, T. K., Protheroe, R. J., and Stanev, T. 1998, ApJ, 492, 219.
12. Baring, M. G., Ellison, D. C., Reynolds, S. P., Grenier, I. A., and Goret, P. 1999, ApJ, 513, 311.
13. Bykov, A. M., Chevalier, R. A., Ellison, D. C., and Uvarov, Y. A. 2000, ApJ, 538, 203.
14. Keohane, J. W., Petre, R., Gotthelf, E. V., Ozaki, M., and Koyama, K. 1997, ApJ, 484, 350.
15. Green, D. A. 1986, MNRAS, 221, 473.
16. Kovalenko, A. V., Pynzar, A. V., and Udal'Tsov, V. A. 1994, Astronomy Reports, 38, 95.
17. Chevalier, R. A. 1999, ApJ, 511, 798 (C99).
18. Gaensler, B. M., Bock, D. C. ., and Stappers, B. W. 2000, ApJ, 537, L35.
19. Kaspi, V. M., Manchester, R. N., Johnston, S., Lyne, A. G., and D'Amico, N. 1996, AJ, 111, 2028.
20. Clearfield, C. R., Olbert, C. M., Williams, N. E., Keohane, J. W., and Frail, D. A. 2000, *these proceedings*.
21. Williams, N. E., Clearfield, C. R., Olbert, C. M., Keohane, J. W., and Frail, D. A. 2000, *these proceedings*
22. Helfand, D. J. 1998, Memorie della Societa Astronomica Italiana, 69, 791.
23. Gaensler, B. M., Stappers, B. W., Frail, D. A., Moffett, D. A., Johnston, S., and Chatterjee, S. 2000, MNRAS, 318, 58.
24. Frail, D. A., Giacani, E. B., Goss, W. M., and Dubner, G. 1996, ApJ, 464, L165.
25. Frail, D. A. and Kulkarni, S. R. 1991, Nature, 352, 785.
26. Claussen, M. J., Frail, D. A., Goss, W. M., and Gaume, R. A. 1997, ApJ, 489, 143.
27. Seward, F. D. and Wang, Z. 1988, ApJ, 332, 199.
28. Frail, D. A. and Scharringhausen, B. R. 1997, ApJ, 480, 364.
29. Nel, H. I. *et al.* 1996, A&AS, 120, C89.
30. Hunter, S. D. *et al.* 1997, ApJ, 481, 205.
31. Giovanelli, R. and Haynes, M. P. 1979, ApJ, 230, 404.
32. Becker, W. and Truemper, J. 1997, A&A, 326, 682.
33. Lorimer, D. R., Lyne, A. G., and Camilo, F. 1998, A&A, 331, 1002.

Very Large Array Data Pertaining to the Hard X-ray Source on the Southern Edge of Supernova Remnant IC 443

Nikolas E. Williams[1], Charles M. Olbert[1], Christopher R. Clearfield[1], Jonathan W. Keohane[1], and Dale A. Frail[2]

[1] North Carolina School of Science and Mathematics, 1219 Broad St., Durham, NC 27705, USA
[2] National Radio Astronomy Observatory, P. O. Box O, Socorro, NM 87801, USA

Abstract. We present multi-frequency ($\lambda 3.5$ cm, $\lambda 6$ cm, and $\lambda 20$ cm) *Very Large Array* data pertaining to the hard X-ray feature along the southern edge of the supernova remnant IC 443. They have revealed a cometary nebula blending with the shell of the remnant. Polarization images show an intriguing magnetic field structure which wraps around a point at the widest part of the cometary nebula and appears strongest where the tail blends with the shell of IC 443. The nebula has a mean flux density of 200 mJy and a spectral index much flatter than IC 443 as a whole.

INTRODUCTION

We present VLA observations of IC443, and in particular the hard X-ray feature on its southern edge. All radio observations were made on August 26 and December 31, 1997 with the *Very Large Array* (VLA[1]) in the C and D arrays, respectively. A log of the observations is summarized in Table 1. The data acquisition and calibration were standard, using J0632+103 as a phase calibrator and 3C 138 as both the flux density and polarization angle calibrator.

---

[1] The VLA is operated by the National Radio Astronomy Observatory (NRAO), which is a facility of the National Science Foundation (NSF), operated under cooperative agreement by Associated Universities, Inc. (AUI).

Wavelength (cm)	Time (min)	Beam ($'' \times ''$)	F_r (mJy)
3.5	13	8.6×7.6	195 ± 8
6	36	5.0×4.8	173 ± 11
20	53	15.5×14.5	229 ± 34

TABLE 1. VLA Observing Log

CP565, *Young Supernova Remnants: Eleventh Astrophysics Conf.*, edited by S. S. Holt and U. Hwang
© 2001 American Institute of Physics 0-7354-0001-6/01/$18.00

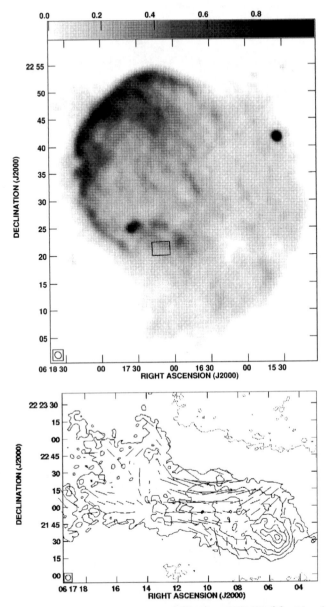

FIGURE 1. The top image is of the entirety of IC 443 at 327 MHz [1] with a box highlighting the area that the below image covers. The units of the grey scale key are Jy beam^{-1}, with the beam size shown in the corner of the plot . The bottom contoured radio image at $\lambda6$ cm is of the cometary nebula in IC 443 with contours at integral values of 10% of 2.8×10^{-3} Jy beam^{-1}. The polarized intesity (B) vectors are overlayed. The vector length scale is $50\,\mu$Jy beam^{-1} arcsec^{-1}.

The cometary morphology is most pronounced at radio wavelengths, extending some 2' to the southeast of the compact X-ray source [2]. Although we display only the λ3.5 cm image in Figure 1, similar structure is visible at λ6 cm and λ20 cm. The peak in the radio emission lies about 6" to the northeast of the compact X-ray source. There is no radio point source (> 2 mJy) coincident with the X-ray point source.

The distribution of linearly polarized emission (B) is shown in Figure 1. The B-vectors are generally parallel to the shock, as would be expected for evolved SNRs with circumferential magnetic fields, but the character of the magnetic field evidently changes substantially in the west of the image which is coincident with the area dominated by the X-ray emission [3]. Here, the B-vectors appear to wrap around the "head" of the bow cometary nebula, tracing a bow-shock morphology. The average degree of polarization in this region is 8%. This is likely a lower limit since, owing to the compact size of the nebula, there is likely some beam depolarization occurring. To the in the center of this image, where the field is more ordered, the percentage polarization is uncharacteristically strong, exceeding 25% in several locations.

The integrated flux density over the entire length of the cometary nebula is given for three frequencies in Table 1. An approximate value of 230±200 mJy was also obtained from the 327 MHz image presented by Claussen et al. (1997). The increased error in estimating the flux density at low frequencies is due to the uncertainty in subtracting out the background shell emission from the SNR. The cometary nebula has a radio spectral index which is considerably flatter than the

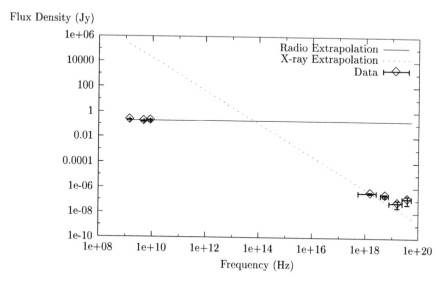

FIGURE 2. Above is a plot showing the relation between the radio and X-ray flux densities.

SNR as a whole (i.e $\alpha < 0.24$ vs $\alpha = 0.42$, where $F_\nu \propto \nu^{-\alpha}$) ([4]; [5]). Within the errors, the radio spectrum is well represented by a flat spectral index $\alpha_R \simeq 0.0$ and a mean flux density of 206 mJy. In contrast, [3] measured an X-ray spectral index of $\alpha_X \simeq 1.3 \pm 0.2$, which implies a break in the spectrum near 4×10^{13} Hz. This is illustrated in figure 2. The radio luminosity of the cometary nebula, determined by integrating the spectrum from 10 MHz to 100 GHz, is $L_R = 5.5 \times 10^{31} \left(\frac{d}{1.5\text{kpc}}\right)^2$ erg s^{-1}, or less than 1% of L_R for IC 443 as a whole.

ACKNOWLEDGEMENTS

We thank B. Koralesky and L. Rudnick for preparing the VLA observe files. This research was funded by NASA grant NRA 97-0SS-14.

REFERENCES

1. Claussen, M. J., Frail, D. A., Goss, W. M., and Gaume, R. A. 1997, ApJ, 489, 143.
2. Clearfield, C. R., Olbert, C. M., Williams, N. E., Keohane, J. W., and Frail, D. A. 2000, *these proceedings*.
3. Keohane, J. W., Petre, R., Gotthelf, E. V., Ozaki, M., and Koyama, K. 1997, ApJ, 484, 350.
4. Green, D. A. 1986, MNRAS, 221, 473.
5. Kovalenko, A. V., Pynzar, A. V., and Udal'Tsov, V. A. 1994, Astronomy Reports, 38, 95.

Cosmic Ray Acceleration and Pulsars

Pulsar Emission and Central Sources in Young Supernova Remnants

Alice K. Harding

NASA Goddard Space Flight Center, Greenbelt, MD 20771, USA

Abstract.
Recent high-energy detections of radio-quiet neutron stars are changing our view of central sources in young supernova remnants. There are at least several classes of sources that make up this radio-quiet population: rotation-powered pulsars with magnetic fields around 10^{12} G, non-rotation powered pulsars (possibly magnetars) and non-pulsating X-ray point sources. I will review pulsar emission mechanisms for both thermal and non-thermal emission in both rotation-powered and non-rotation powered sources. Radio quiescence in some cases may be caused by unfavorable alignment with the radio beams. In pulsars with magnetic fields exceeding 10^{14} G, radio quiescence may be due to the suppression of pair production by photon splitting.

INTRODUCTION

Not long ago, the Crab pulsar and its nebula were the paradigm for a central source inside a young supernova remnant. This led us to believe that all such sources would be radio-loud and powered by the rotational energy loss of the pulsar, which also lit up its plerionic remnant with a bright synchrotron glow. But over the last five years observations by X-ray telescopes such as RXTE, ASCA and Chandra have identified a new population of radio-quiet neutron stars in, or possibly associated with, young supernova remnants. There are at least four classes of such radio-quiet young neutron stars. The anomalous X-ray pulsars (AXPs), of which Kes 73 is a prototype [1], are steadily spinning down X-ray sources having periods in the range 6 - 12 s, luminosities $L_X \sim 10^{35}$ erg s^{-1}, show no sign of any companion, and have ages $\tau \lesssim 10^5$ years. The soft gamma-ray repeaters (SGRs) [2] are transient γ-ray burst sources that undergo repeated outbursts, but their quiescent emission properties are very similar to those of AXPs. Both AXPs and SGRs are observed to be spinning down with high period derivatives, implying surface magnetic fields in the range $10^{14} - 10^{15}$ G assuming dipole radiation torques. A third class of radio-quiet neutron stars are the young, rotation-powered X-ray pulsars discovered in N157B [3], RCW103 [4] and PuppisA [5] . Unlike the AXPs, these sources have dipole magnetic fields around 10^{12} G. Finally, there have been X-ray point

CP565, *Young Supernova Remnants: Eleventh Astrophysics Conf.*, edited by S. S. Holt and U. Hwang

sources detected in such very young remnants such as CasA [6], having no apparent pulsations or radio couterparts.

It is generally agreed that isolated radio pulsars are powered by the rotational energy loss of the neutron star. However, rotation cannot provide the power for all of the radio-quiet pulsars. AXPs and SGRs in particular have quiescent X-ray luminosities which exceed their spin-down power by at least an order of magnitude. If these sources are magnetars, neutron stars with fields larger than 10^{14} G, then magnetic field decay may provide their emission power [7]. It is also possible that their high fields may suppress the pair cascades which are thought to be necessary for the generation of coherent radio emission [8]. However, radio pulsars with magnetic fields around 10^{14} G have recently been discovered by the Parkes Multibeam survey [9]. One of the most intriguing questions then is what unmeasured property of the neutron star causes AXPs and high-field radio pulsars to have such different observational properties.

ROTATION-POWERED PULSARS

Nonthermal emission

Particle acceleration inside the pulsar magnetosphere gives rise to pulsed nonthermal radiation. Rotating, magnetized neutron stars are natural unipolar inductors, generating huge **vxB** electric fields. However, these fields are capable of pulling charges out of the star against the force of gravity [10] and it is believed that the resulting charge density that builds up in a neutron star magnetosphere is able to short out the electric field parallel to the magnetic field ($E_\parallel$)(thus allowing the field to corotate with the star) everywhere except at a few locations. These regions where $\mathbf{E} \cdot \mathbf{B} \neq 0$ are thought to occur above the surface at the polar caps and along the null charge surface, $\mathbf{\Omega} \cdot \mathbf{B} = 0$, where the corotation charge changes sign. These are the purported sites of particle acceleration and have given rise to the two classes of high energy emission models: polar cap models, where the acceleration and radiation occur close to the neutron star surface, and outer gap models, where these processes take place in the outer magnetosphere.

Polar cap models

Polar cap models for pulsar high energy emission date from the early work of [11], who proposed particle acceleration and radiation near the neutron star surface at the magnetic poles. There are two different types of polar cap accelerators possible. A space-charged-limited flow (SCLF) type accelerator forms when the neutron star surface temperature (measured in the range $T_s \sim 10^5 - 10^6$ K) exceeds the thermal emission temperature, T_e, and charged particles are freely extracted from star. A vacuum gap (V) type accelerator forms in the opposite case when $T_s < T_e$, charged particles cannot be extracted and a strong $E_{\parallel} \neq 0$ builds up right above the surface [12,13]. Although $E_{\parallel} = 0$ at the neutron star surface in SCLF accelerators, the space charge along open field lines above the surface falls short of the corotation charge, due to the curvature of the field [14] or to general relativistic inertial frame dragging [15]. The $E_{\parallel}$ generated by the charge deficit accelerates particles, which radiate inverse Compton (IC) photons by resonant scattering (when they reach energies $\gamma \sim 10^2 - 10^6$) and curvature (CR) photons (at energies $\gamma \lesssim 10^6$). Both IC and CR photons can produce $e^+ e^-$ pairs in the strong magnetic field. However, it is found [16] that in all but the very high-field pulsars, the IC pair formation fronts do not produce sufficient pairs to screen the $E_{\parallel}$ or are unstable, due to returning positrons which disrupt $E_{\parallel}$ near the surface. Harding & Muslimov [17] found that stable acceleration zones can form at 0.5 - 1 stellar radii above the surface, where the density of soft X-rays from the neutron star surface decreases and CR photons from both primary electrons and returning positrons produce stable pair formation fronts. The primary particle energies can then reach $\sim 10^7$ before pair production screens the field.

The type of polar cap cascade which produces high-energy radiation depends on the primary radiation mechanism, which in turn depends on which photons (IC or CR) control the screening of the accelerating field. In pulsars where IC-controlled acceleration zones are stable, particle energies are limited to Lorentz factors $10^5 - 10^6$ [17] and IC is both the dominant primary radiation mechanism and the initiator of the pair cascade [18]. In pulsars where IC photons either cannot screen the accelerating field or IC-controlled zones are unstable, the primary particles continue accelerating up to Lorentz factors $\sim 10^7$. CR is then the dominant primary radiation mechanism and initiates the pair cascade. In the original version of the CR-initiated polar cap pair cascade [19,20] the emergent cascade spectrum is dominated by synchrotron radiation from the pairs and has a very sharp high energy cutoff at several GeV due to pair production attenuation. Recently, Zhang & Harding [21] noted that the pairs produced in polar cap cascades may resonant-scatter the soft thermal photons from the neutron star surface, losing most of the remaining parallel energy they could not lose via synchrotron emission, and producing an additional component of the cascade spectrum at X-ray energies.

Outer gap models

The outer gap models for γ-ray pulsars are based on the existence of a vacuum gap in the outer magnetosphere which may develop between the last open field line and the null charge surface $(\mathbf{\Omega} \cdot \mathbf{B} = 0)$ in charge separated magnetospheres. The gaps arise because charges escaping through the light cylinder along open field lines above the null charge surface cannot be replenished from below. The first outer gap γ-ray pulsar models [22] assumed that emission is seen from gaps associated with both magnetic poles, but this picture, although successful in fitting the spectrum of the Crab and Vela pulsars, did not reproduce the observed pulsar light curves. More recent outer gap models assuming emission from one pole can more successfully reproduce the observed light curves [23]. Pairs from the polar cap cascades, which flow out along all the open field lines, will undoubtedly pollute the outer gaps to some extent, but this effect has yet to be investigated.

The electron-positron pairs needed to provide the current, and therefore allow particle acceleration, in the outer gaps are produced by photon-photon pair production. In young Crab-like pulsars, the pairs are produced by curvature photons from the primary particles interacting with non-thermal synchrotron X-rays from the same pairs. In older Vela-like pulsars, where non-thermal X-ray emission is much lower, the pairs were assumed to come from interaction of primary particle inverse Compton photons with infra-red photons. However, this original Vela-type model (CHR) predicted large fluxes of TeV emission, from inverse Compton scattering of the infra-red photons by primary electrons, which violates the observed upper limits [24] by several orders of magnitude. Cheng [25] revised the outer gap model for Vela-type pulsars by proposing another self-sustaining gap mechanism where thermal X-rays from the neutron star surface interact with primary particle radiation to produce pairs, replacing the infra-red radiation (which has also never been observed). Some of the accelerated pairs flow downward to heat the surface and maintain the required thermal X-ray emission. The modern outer gap Vela-type models [26,27] all adopt this picture.

As in polar cap models, it becomes more difficult for older pulsars to produce the pairs required to screen the field and "close the gap", so that young pulsars have thin gaps and old pulsars have thick gaps. However, unlike in polar cap (SCLF) models, pair production plays a critical role in production of the high energy emission: it allows the current to flow and particle acccleration to take place in the gap. Beyond a death line in period-magnetic field space, and well before the traditional radio-pulsar death line, pairs cannot close the outer gap and the pulsar cannot emit high energy radiation. This outer gap death line for γ-ray pulsars falls around $P = 0.3$ s for $B \sim 10^{12}$ G [28], putting Geminga just barely among the living. The observed non-thermal radiation in Crab-like pulsars is a combination of synchrotron emission and synchrotron self-Compton emission from pairs. In Vela-type pulsars, the non-thermal radiation is a combination of curvature and curvature self-Compton emission from the primaries at γ-ray energies, and synchrotron emission from the pairs at optical through X-ray energies. The high-energy spectra in both types of model have cutoffs around 10 GeV, due to the radiation-reaction cutoff in the primary particle spectrum, which are much less sharp than the attenuation cutoffs in polar cap model spectra.

Thermal Emission

Thermal X-rays may be produced at the stellar surface by cooling of the neutron star or by a variety of possible heating mechanisms that include the bombardment of the surface by backflowing particles and internal friction. Surface cooling will be the dominant source in young pulsars with ages $\tau \lesssim 10^5$ yr, and the X-ray emission of many young pulsars are consistent with standard cooling models [29]. For older pulsars, any residual cooling will be dominated by surface heating due to back-flowing positrons or internal friction [30]. Cooling emission from even the whole neutron star surface can produce pulsations because, due to the anisotropic conductivity in a strong magnetic, the poles will be slightly hotter than the equator [31]. However, gravitational bending of light near the neutron star tends to smear out the pulses [32], and the pulse fractions of observed thermal emission (in the range of $5\% - 25\%$) require either thermal hot spots or beamed emission due to the anisotropic magnetic scattering cross sections [33,34].

Predictions for polar cap heating, due to positrons from the pair cascade that are accelerated back toward the neutron star surface, vary widely among different models. Polar cap model vacuum-gap accelerators tend to produce X-ray luminosities from polar cap heating that are orders-of-magnitude higher than those observed because a large fraction of the positrons from the pair cascade return and heat the surface [12]. By contrast, in a polar cap SCLF accelerator the trapped positrons needed to screen the $E_{\parallel}$ are only a small fraction of the number of primary electrons because this field was created by a small charge imbalance in the first place [35]. The predicted X-ray luminosities due to heating by trapped positrons are much less than X-ray luminosities observed in young pulsars but become comparable to those observed in older pulsars, where they are a significant fraction of the spin-down luminosity [21,16]. SCLF models thus predict that many old pulsars, including millisecond pulsars, will have thermal X-ray components due to polar cap heating.

The pairs produced in the outer gaps that are accelerated back toward the neutron star surface radiate curvature photons that can produce pairs in the strong magnetic field near the surface, initiating pair cascades [36,37]. But the pairs still have enough residual energy to heat the surface at the footpoints of open field lines that thread the outer gaps, producing thermal emission at a temperature around 1 keV, which is much higher than the temperatures observed in thermal emission from pulsars. According to the model, this emission is not observed directly (except right along the poles) but only through the blanket of pairs produced by the downward-going particle cascades. The 1 keV photons are reflected back to the surface by the pair blanket through cyclotron resonance scattering [38], and are re-radiated from the entire surface at a temperature around 0.1 keV. Thus, these outer gap models predict three X-ray emission components from the return particles from the outer gaps: hard thermal emission from direct heating of the polar caps (seen only along the poles), soft thermal emission reflected from the pair blanket, and non-thermal emission from the downward pair cascades [39]. The components actually observed from a particular pulsar depend on inclination and viewing angle. The predicted X-ray luminosities for pulsars in the ROSAT band can account for the observed $L_x = 10^{-3} L_{sd}$ relation.

355

Radio quiescence

The lack of detected radio emission in otherwise normal rotation-powered pulsars could result from several possible factors. A narrow radio beam may be unfavorably aligned while the high-energy emission is visible due to either a larger beam or a beam pointed in a different direction. This would be quite possible in the case of thermal X-rays from the whole neutron star surface, while for non-thermal emission such a situation would be much for likely for outer gap models, where the high-energy and radio emission originate from opposite poles. A radio-quiet X-ray point source (e.g. source in CasA) could be thermal cooling emission that is too weakly beamed to produce detectable pulsations if the star has a low magnetic field. Lack of radio emission could also result for some sources in the galactic plane that are visible to X-ray detectors but difficult to detect at radio wavelengths due to scattering or dispersion effects at the sources (e.g. the SNR) or along the line of sight.

MAGNETARS

As pulsars spin down by magnetic dipole radiation torques, their supply of rotational energy, $E_{rot} = (1/2)I\Omega^2$, decreases in relation to their supply of magnetic energy, $E_B \sim B_0^2 R^3/6$, where B_0 is the surface magnetic field, I is the neutron star moment of inertia and Ω its rotation frequency. The magnetic energy of a pulsar will dominate over its rotational energy at an age, $\tau_B \sim 1.5 \times 10^5$ yr $(B_0/10^{14}\,\text{G})^{-4}$, where B_{14} is the surface field in units of 10^{14} G. All of the AXPs and SGRs are in this magnetically dominated regime, if in fact their high observed spin-down rates are due to magnetic dipole radiation. Although the issue of whether AXPs and SGRs are magnetars — sources that derive their primary power from their magnetic fields — is still somewhat in question, it is currently the leading theory of their origin and behavior.

Spin-down and field decay

Magnetars, magnetically-powered isolated neutron stars, spin down much faster than do neutron stars with fields in the normal pulsar range of $10^{11} - 10^{13}$ G. The spin-down timescale, or characteristic age,

$$\tau_{SD} = \frac{P}{2\dot{P}} = 1600 \left(\frac{P}{B_{14}}\right)^2 \text{ yr},\qquad(1)$$

from integrating the rotational energy loss rate for magnetic dipole radiation, $\dot{E}_D = I\Omega\dot{\Omega} \propto \Omega^4$, where P is the present period, predicts that magnetars will reach relatively long periods approaching 10 s in an age of several thousand years. Particle wind flows may also contribute to the rotational energy loss of some magnetars, affecting both the characteristic age and the derived surface magnetic field [40,41].

Magnetar fields can decay on relatively short timescales and this decay could be the mechanism by which the field energy is released to power the stellar activity [42]. Ohmic diffusion, in which the magnetic flux moves through the charged particle component, produces field decay on timescales too long to be useful for energy generation. However, ambipolar diffusion, in which the magnetic flux carries the charged particles with it as it moves out of the neutron star core, can operate on much shorter timescales in magnetar fields and are calculated to fall in the range of the derived characteristic ages of the SGRs and AXPs [43]. Field decay is thus a promising source of energy for these objects.

Burst and quiescent emission

The large magnetic flux diffusing out of the neutron star core, and eventually through the crust, creates imbalances and stresses. Thompson & Duncan [42] postulate that a large-scale rearrangement of the field, either through an interchange instability or both the internal and external field or shearing of the external field, can cause a massive reconnection event. These events could release a significant portion of the magnetic field energy, and may be responsible for the giant SGR flares observed from SGR 0526-66 and SGR 1900+14 having energies $\sim 10^{45}$ erg. The small SGR bursts having average energies of 10^{41} ergs may be due to cracking of the neutron star crust.

Steady, quiescent emission is observed from both SGRs and AXPs and the two types of sources have similar spectra. A dominant thermal component having temperature of $0.5 - 1$ keV is in most cases accompanied by a steep power law of photon index $2 - 3.5$ [44,45]. Even the luminosity of the quiescent emission exceeds the spin-down luminosity of the stars by at least an order of magnitude. According to Thompson & Duncan [43], this quiescent emission is also powered by the decaying field, through conduction of heat from the core, continually heating the crust to a temperature $T_{\mathrm{crust}} \simeq 1.3 \times 10^6 \, \mathrm{K} \, (T_{\mathrm{core}}/10^8 \, \mathrm{K})^{5/9}$, producing thermal emission of $L_x \sim 10^{35}$ erg. The transport of magnetic flux through the crust produces Hall turbulence, which causes small-scale fractures of the crust. The small-scale fractures may excite Alfven waves above the surface and accelerate particles [46] to radiate the power-law component.

Radio quiescence

One prominent feature of the AXPs and SGRs is that none of them have been firmly detected to have pulsed radio emission, making them a distinct category from the normal radio pulsars. Radio quiescence of magnetars has been attributed to the function of an exotic QED process, namely photon splitting, which, in a magnetic field in excess of roughly the critical field strength, $B_q = 4.4 \times 10^{13}$ G, acts as an efficient attenuator of gamma-rays before they are materialized to electron-positron pairs [8,47]. This argument, however, is not certain since complete pair-suppression requires that both photon polarizations split. This will be realized if all the three photon splitting modes permitted by charge-parity invariance in QED, i.e., $\perp\rightarrow\|\|$, $\perp\rightarrow\perp\perp$, and $\|\rightarrow\perp\|$, operate. However, under the weak linear vacuum dispersion limit (which may not be true in a magnetar environment) only one splitting mode ($\perp\rightarrow\|\|$) can fulfill the energy and momentum conservation requirements simultaneously [48]. Recently, three high magnetic field radio pulsars were detected to be above the "photon splitting deathline" for surface emission proposed by Baring & Harding [8], and one of them, PSR J1814-1744 resides very close to an AXP, 1E 2259+586, in $P-\dot{P}$ phase space, casting doubt on the effectiveness of photon splitting to suppress radio emission [9].

Zhang & Harding [49], after investigating the formation condition and the properties of the two types of polar cap accelerators (SCLF, or vacuum gap) in high field pulsars, conjectured that the different observational properties of PSR J1814-1744 and 1E 2259+586 are due to a geometric effect, i.e., different orientations of the magnetic pole with respect to the rotational pole. SCLF accelerators can form in higher fields for parallel rotators ($\mathbf{\Omega}\cdot\mathbf{B} > 0$), where the sign of the corotation charge above the polar caps, $\rho_c = -\mathbf{\Omega}\cdot\mathbf{B}/2ec$, is negative, because T_e for electrons is lower than T_e for ions. A SCLF accelerator has a much longer acceleration length, so that a pair formation front could form at a higher altitude when the local magnetic field is degraded to a low enough level, although pair production from the γ-rays produced in the near zone of the accelerators is strongly suppressed. Therefore, high-field radio pulsars may be parallel rotators while AXPs may be anti-parallel rotators.

SUMMARY

As high-energy observations discover more radio-quiet neutron stars, the radio pulsar surveys are no longer defining the nature of central sources in supernova remnants. In coming years, it is expected that the number of detected radio-quiet pulsars will greatly increase. Chandra and XMM have only recently begun their operations and have already made several important discoveries. The EGRET telescope on the Compton Gamma-Ray Observatory compiled a catalog of some 170 unidentified γ-ray sources above 100 MeV, many of which could be radio-quiet pulsars. The GLAST γ-ray satellite, set for launch in 2006, will be able to perform independent pulsation searches of all the EGRET sources and may be more sensitive than radio surveys in finding young pulsars (at least as point sources) inside supernova remnants.

REFERENCES

1. Vasisht & Gotthelf 1997, ApJ, **486**, L129.
2. Hurley, K. 2000, X-Ray '99 (Bologna) in press.
3. Marshall, F. E. et al. 1998, ApJ, 499, L179.

4. Torii, K. et al. 1998, ApJ, 494, L207.
5. Pavlov, G. G. et al. 1999, ApJ, 531, L53.
6. Tananbaum, H. 1999, IAU Circ. 7246.
7. Thompson, C. & Duncan, R. 1993, ApJ,
8. Baring, M. G. & Harding, A. K., 1998, ApJ, **507**, L55.
9. Camilo, F., et al. 2000, in Pulsar Astronomy: 2000 and Beyond - IAU Coll. **177**, 3.
10. Goldreich, P. & Julian, W. H. 1969, ApJ, **157**, 869.
11. Sturrock, P.A. 1971, ApJ, **164**, 529.
12. Ruderman, M.A. & Sutherland, P. G. 1975, ApJ, **196**, 51.
13. Usov, V. V. & Melrose, D. B. 1995, *Aust. J. Phys.*, **48**, 571.
14. Arons, J., ApJ 1983, **266**, 215.
15. Muslimov, A. G. & Tsygan, A. I. 1992, *MNRAS*, **255**, 61.
16. Harding, A. K. & Muslimov, A. 2001, in prep.
17. Harding, A. K. & Muslimov, A. G. 1998, ApJ, 508, 328
18. Sturner, S. J., Dermer, C. D. & Michel, F. C. 1995, ApJ, 445, 736.
19. Daugherty, J. K. & Harding, A. K. 1982, ApJ, **252**, 337.
20. Daugherty, J. K., & Harding A. K. 1996, Ap. J., **458**, 278.
21. Zhang, B. & Harding, A. K. 2000, ApJ, 532, 1150.
22. Cheng, K. S., Ho, C. & Ruderman, M. A. 1986, ApJ, **300**, 500.
23. Romani, R. W. & Yadigaroglu 1994, I.-A., ApJ, **438**, 314.
24. Nel, H. I. et al. 1993, ApJ, **418**, 836.
25. Cheng, K. S. 1994, *Proc. Toward a Major Atmospheric Cherenkov Detector*, ed. T. Kifune (Tokyo: Universal Academy), 25.
26. Romani, R. W. 1996, ApJ, **470**, 469.
27. Cheng, K. S. & Zhang, L. 1998, ApJ, **498**, 327.
28. Chen, K. & Ruderman, M. A. 1993, ApJ, **402**, 264.
29. Becker, W. & Trumper, J. 1997, A & A, 326, 682.
30. Shibazaki, N. & Lamb, F. K. 1989, ApJ, 346, 808.
31. Greenstein, G. & Hartke, G. J. 1983, ApJ, 271, 283.
32. Page, D. 1995, ApJ, 442, 273.
33. Pavlov, G. G., Shibanov, Yu. A., Ventura, J. & Zavlin, V. E. 1994, A & A, 289, 837.
34. Harding, A. K. & Muslimov, A. G. 1998, ApJ, 500, 862.
35. Arons, J. 1981, ApJ, 248, 1099.
36. Zhang, L. & Cheng, K. S. 1997, ApJ, **487**, 370.
37. Wang, F. Y.-H. et al. 1998, ApJ, 498, 373.
38. Halpern, J. & Ruderman 1993, ApJ, 415, 286.
39. Cheng, K. S. & Zhang, L. 1999, ApJ, 515, 337.
40. Harding, A. K., Contopoulos, I. & Kazanas, D. 1999, ApJ, 525, L125.
41. Thompson, C. & Blaes, O. 1998, Phys. Rev. D, **57**, 3219.
42. Thompson, C. & Duncan, R. C. 1995, MNRAS, **275**, 255.
43. Thompson, C. & Duncan R. C. 1996, ApJ, **473**, 322.
44. Mereghetti, S. & Stella, L. 1995, ApJ, **442**, L17.
45. Kouveliotou, C. et al. 1999, ApJ, **510**, L115.
46. Melia, F. & Fatuzzo, M. 1991, ApJ, **376**, 673.
47. Baring, M. G. & Harding, A. K. 2001, ApJ, in press.
48. Adler, S., 1971, Ann. Phy., **67**, 559.
49. Zhang, B. & Harding, A. K. 2000, ApJ, 535, L51.

Evidence for Cosmic Ray Acceleration in Supernova Remnants from X-ray Observations

R. Petre

NASA/GSFC, Greenbelt, MD 20771, USA

Abstract.
    Spatially-resolved X-ray spectroscopic observations over the past several years have led to the discovery of non-thermal X-ray emission arising in the shells of many young Galactic supernova remnants, most notably SN 1006 and Cas A. This emission is thought to be synchrotron emission from electrons shock acclerated to hundreds of TeV, and thus represents strong evidence that cosmic rays are accelerated in SNR shocks. The inferences made using the X-ray observations are corroborated by the detection of TeV γ-rays from three of these remnants. The recent launches of the *Chandra* and XMM-Newton Observatories provide powerful new probes for regions of shock acceleration in young supernova remnants. The status of the X-ray observations and their implications are reviewed, including some early *Chandra* results.

INTRODUCTION

For many years, it has been thought that cosmic rays with energy as high as the $\sim$3,000 TeV spectral turnover (referred to as the "knee") are produced by diffusive shock acceleration in Galactic supernova remnants (SNRs). The acceleration sites cannot be observed directly because the intervening Galactic magnetic fields deflect the trajectories of these energetic particles. Detection of synchrotron emission from SNR shells in the radio band verify that electrons with energies up to the GeV range are accelerated there. Until recently, however, it has not been possible to search for evidence of particles with energy closer to the knee. The dominant component in the X-ray band is the thermal emission from gas shock-heated to $\sim 10^7$ K. Only above $\sim$5 keV, where this thermal emission falls off, might it be possible to detect synchrotron emission in the form of a hard X-ray continuum component, but only if electrons are accelerated to sufficiently high energy, on the order of 100 TeV. However, even the total thermal flux is much lower than that predicted by extrapolating the slope of the radio spectrum through the X-ray band. Thus the synchrotron spectrum must have a break before the X-ray band

CP565, *Young Supernova Remnants: Eleventh Astrophysics Conf.*, edited by S. S. Holt and U. Hwang
2001 American Institute of Physics 0-7354-0001-6

which makes detecting the synchrotron component from very high energy electrons difficult to detect even above ~5 keV.

The broad bandpass, high sensitivity and moderate resolution spectral capabilities of the X-ray observatories launched in the mid 1990's, ASCA, RXTE and BeppoSAX, led to a breakthrough in our ability to detect evidence for highly relativistic particles in SNR shells. The first breakthrough came from the ASCA observations of the supernova remnant SN 1006. This remnant was known to have a featureless integrated spectrum above 1 keV. It was possible to construct satisfactory models of the spectrum by invoking extreme departure of shock heated gas of a particular composition from ionization equilibrium [1]. ASCA provided the first broad-band, spatially-resolved spectrum, which revealed that only the bright northeast and southwest limbs have a featureless spectrum, and that the emission from the remainder of the remnant is line-dominated thermal emission from gas enriched by supernova ejecta. The most plausible origin of the emission from the bright limbs is synchrotron radiation from electrons with energies up to ~100 TeV. As acceleration processes at these highly relativistic energies do not distinguish between positively and negatively charged particles, we infer that protons and nuclei are being accelerated to these energies as well, and therefore SN 1006 is the first identified cosmic ray source [2]. Theoretical models, both general [3,4] and specific [5], support this conclusion, or show that alternative interpretations fail [6]. Moreover, this conclusion has been dramatically confirmed by the subsequent detection of TeV γ-rays from SN 1006, which arise from Compton scattering of cosmic microwave background photons off the TeV electrons [7].

The SN 1006 discovery prompted a flurry of searches for synchrotron components in other remnants. Components have been found primarily via specific observations of very young remnants using RXTE [8] or BeppoSAX [9] and systematic searches through archival ASCA or RXTE data [10, 11]. Addtionally, some serendipitously discovered remnants also turned out to have continuum-dominated spectra [12, 13]. It is now generally the case that the analysis of the X-ray spectrum of a young SNR includes a search for a hard, nonthermal component. The presence of a hard X-ray continuum component does not automatically mean that another site of cosmic ray acceleration has been identified: other mechanisms, both thermal and nonthermal, can produce hard components [6, 14, 15], and care must be taken in each instance to rule out these. The most definitive way of eliminating other possible models is detecting TeV γ-ray emission; this has been done for only three remnants, SN 1006 [7], G347.5-0.5 [16], and Cas A [17].

CURRENT STATUS OF OBSERVATIONS

In Table 1 we list remnants whose shells have been reported to have a nonthermal emission component. Although they range in age from 350 yr to 15,000 yr, most are young (~10^3 yr). Those in bold face are the ones for which we are fairly certain the nonthermal component is contributed by synchrotron emission from TeV electrons.

TABLE 1. Galactic SNR's with Hard, Nonthermal X-Ray Spectral Components

Object	Age (ky)	R/X Match?	Low L_R	TeV γ-rays?	Type	Result	Ref.
Cas A	0.3	Y	N?	Y	II	Hard tail to ≥ 100 keV	8
Kepler	0.4	Y	N		II?	$\alpha \sim 2.5$	18
Tycho	0.4	N	N		Ia	Tail w/ $\alpha \sim 2$ to ≥ 20 keV	19
G266.3-1.2	~1	?	Y		II?	$\alpha \sim 2$; low ρ	13
SN 1006	1.0	Y	Y	Y	Ia	Obvious NT shell	2, 7
W49B	1-3	Y	N		?	Hard X-ray has $\alpha \sim \alpha_{1006}$	10
RCW 86	~2	N	N		II?	$\alpha \sim 2$ to ≥ 20 keV	20
G347.5-0.5	~4	?	Y	Y	II?	NT shell; $\alpha \sim 1.5$; low ρ	12
G156.2+5.7	15	Y	Y		II?	$\alpha \sim 2$; low ρ	11
IC 443	1-4	N	N		II	NT flux from e^- accel. in shock/cloud collision	21, 22
W50	5-10	N	N		II	NT flux from e^- accel. in jet/cloud collision	23

The list contains virtually all known shell-like Galactic SNRs younger than ~1,500 years.

In Figure 1 we show the spectra at energies above 5 keV for five of the remnants listed in Table 1, as observed by RXTE. In each case, the spectrum is represented by a power law or broken power law, with spectral index of $\alpha \sim 2$. This is steeper than the $\alpha \sim 1$ generally observed from synchrotron radiation associated with pulsars, and flatter than the $\alpha \geq 5$ that characterizes thermal emission that might be expected from these remnants in the 5-15 keV band. For three of these remnants, SN 1006, Cas A, and RCW 86, there is morphological evidence that this hard component is produced by shock-accelerated electrons.

In three of these remnants, SN 1006, G266.3-1.2, and G347.5-0.5, the non-thermal X-rays dominate. For G347.5-0.5 there is also a report of a TeV γ-ray detection [16]. These SNR as a group (possibly including G156.2+5.7) share another important property: they have low radio surface brightness. The low radio surface brightness and lack of strong thermal X-rays could both be the consequence of SNR expansion into a very low density ISM. Since the synchrotron emissivity scales as the ambient density n and the thermal emissivity as n^2, the nonthermal X-ray emission associated with the forward shock in a low density medium would be reduced substantially less than the thermal emission. Moreover, there would be insufficient interstellar material to produce a strong reverse shock, which in turn substantially reduces the thermal X-ray flux, as the X-ray emission from the reverse-shocked component is usually the dominant component. At the same time, the forward shock would decelerate more slowly, and as the maximum energy of diffusive shock accelerated particles depends strongly on the shock velocity, this enables the particles to attain higher energy, and thus emit higher energy X-ray synchrotron radiation.

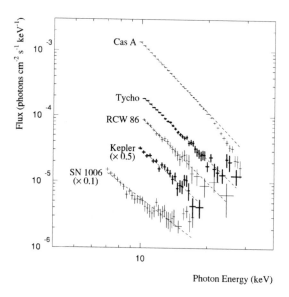

FIGURE 1. Spectra of hard X-ray components in five supernova remnants, as observed by RXTE. Each spectrum is characterized by a power law with spectral index $\alpha \sim 2$, and is probably produced by synchrotron emission by electrons shock-accelerated to TeV energy.

Chandra **RESULTS**

The recent launch of the *Chandra* and XMM/*Newton* observatories provide us with powerful new tools for identifying possible sites of cosmic-ray acceleration in SNR's. XMM/*Newton* offers broad band, high throughput imaging that will allow us to observe low surface brightness features and isolate nonthermal-emission regions at high energy. *Chandra's* superb angular resolution and modest-resolution spectroscopy allow us to isolate nonthermal emission regions that are intermixed with predominantly thermal emission [24]. *Chandra* has already yielded some results that offer a glimpse of how it will improve our understanding X-rays from shock-accelerated electrons.

In Cas A, *Chandra* observations have shown that there is no clean separation between thermal and nonthermal components (Figure 2), as suggested by lower-angular-resolution studies [25]. On the other hand, *Chandra* has revealed for the first time the precise locations of the forward and reverse shocks in Cas A [26]. The X-ray spectrum of the forward-shocked gas is very different from that of the dominant, reverse-shocked material. It requires the presence of a hard, featureless component absent from the reverse shock. Thus cosmic-ray acceleration appears to take place only near the forward shock. While this in itself is unsurprising, it points out that care must be taken in radio/X-ray comparisons, such as that carried out

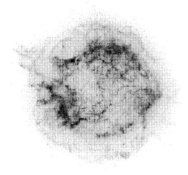 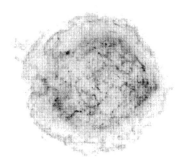

FIGURE 2. (Left) Broad band *Chandra* image of Cas A. (Right) *Chandra* image of Cas A at energies above 3 keV. The outer shock front is far more prominent than in the broad band image. A substantial fraction of the X-ray flux front from this outer shock is contributed by a hard, featureless component. This component contributes much less, and could be absent, in the bright X-ray ring.

by Keohane & Reynolds [27], to consider the radio emission only from those regions in which cosmic rays are accelerated.

The SNR 0540-693 in the Large Magellanic Cloud is a composite remnant whose plerion and pulsar have many properties in common with the Crab Nebula. The shell is dominated by thermal X-rays. A hard image (E higher than ~2 keV) reveals two arcs of hard emission diametrically opposite one another along the rim (Figure 3). It is unclear whether these arise as a result of interaction between jets from the pulsar and the shock-compressed shell material, as in the old Galactic SNR W50 [23], or from shock acceleration, as suggested by the morphological similarity to SN1006.

The *Chandra* data have also cast doubt on the association of at least one hard component with synchrotron emission from TeV electrons, in the irregularly-shaped, ejecta-dominated remnant W49B. The broad band ASCA image of W49B reveals an irregular, centrally filled morphology. The hard continuum ASCA image hinted at shell-like morphology more consistent with that in the radio [10]. The 3x4 arc minute size of W49B make ASCA imaging studies subject to the vagaries of image reconstruction; in addition, the continuum image has low signal to noise. In contrast, the *Chandra* images, which locate to high accuracy the origin of all detected photons, shows that the broad band image has an overall shape that

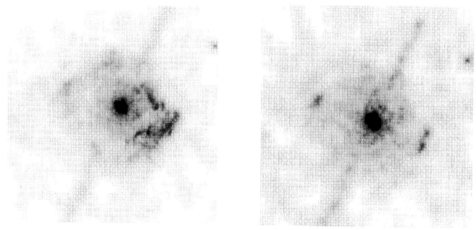

FIGURE 3. (Left) Broad band *Chandra* image of the Large Magellanic Cloud SNR 0540-693. (Right) *Chandra* image of 0540-693 at energies above 2 keV. The outer shock front is far more prominent than in the broad band image. A substantial fraction of the X-ray flux front from this outer shock is contributed by a hard, featureless component. This component contributes much less, and could be absent, in the bright X-ray ring. The bright linear feature running approximately from northwest to southeast is a detector artifact brought about by pulse pile-up from the bright pulsar.

resembles the radio more than previously thought, and that the hard continuum emission is, if anything, more centrally concentrated than the broad-band emission (Fig. 4). Thus the *Chandra* data suggest that the bulk of the hard continuum emission from W49B is thermal in origin.

In IC 443, perhaps more definitively, *Chandra* observations have shown that one of the concentrations of hard emission [21] can be attributed to an unresolved source, presumably a neutron star [22].

DISCUSSION

The considerable and ever-increasing number of detections of hard, nonthermal emission associated with the shocks of supernova remnants has allowed us to establish the following:

- The existence of nonthermal X-ray emission from the shocks of many supernova remnants. In at least some of these remnants the emission is synchrotron emission from electrons with energy around 100 TeV. This result is substantiated by the detection of TeV gamma rays from SN 1006, G347.5-0.5, and Cas A.

- *Chandra* has already shown that the situation in these remnants is more complicated than previously thought. In particular, there is not necessarily a

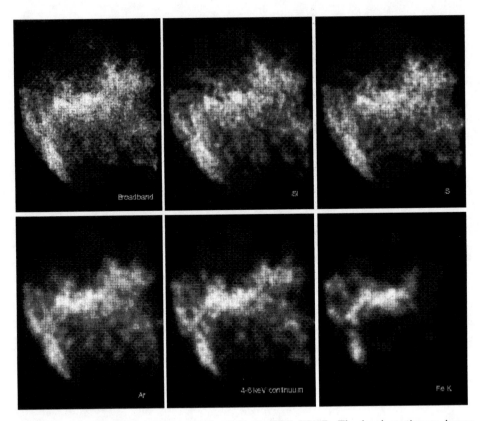

FIGURE 4. Narrow band images of the Galactic SNR W49B. The hard continuum image strongly resembles the images in the Ar and Fe K lines.

unique correspondence between the radio surface brightness and the nonthermal X-ray emission. We need to revise our notion about the relationship between radio- and X-ray-emission regions in these remnants, as only some of the regions in which GeV electrons are present also accelerate particles to TeV energy.

- There seems to exist an emerging class of low-radio-flux, shell-like remnants, whose X-ray emission is dominated by synchrotron radiation. SN 1006 is the class prototype. Their properties can be ascribed to expansion in a low density medium. Finding new members of this class is an observational challenge as these remnants' low radio surface brightness causes them to be overlooked or not detected in surveys.

In our discussions thus far we have only considered TeV electrons. Unfortunately, these observations to not allow us to test the relationship between these electrons and the protons and nuclei comprising the majority of cosmic rays. To understand that, we are forced rely on the inferences from shock-acceleration models (e.g., Ellision, Berezhko, & Baring [27]).

More importantly, these results do not resolve whether SNR's are responsible for accelerating most cosmic rays below the knee. The maximum energy of the electrons responsible for the X-ray emission is probably less than a few hundred TeV; the steepness of the X-ray synchrotron spectrum requires a turnover of the photon spectrum between the radio and X-ray bands, and thus of the electron spectrum around a few tens of TeV. Depending on the mechanism responsible for the turnover, there could be serious implications for the more massive, positively-charged particles [27]. If the turnover is due to the finite age of the remnant, then we will find no protons with higher energy. On the other hand, if the electron energy is governed by synchrotron losses or electron escape, then higher energy protons might be present.

One thing that seems likely is that shocks impart sufficient energy into particle acceleration to produce cosmic rays. Nonlinear-acceleration models suggest that 10 percent or more of the energy from the forward shock can be deposited in particles. Such models can successfully account for the emission from remnants like SN 1006 [28]. A recent *Chandra* result has provided the first observational support for these models. In the remnant E0102.2-7219 in the Small Magellanic Cloud, the very low ratio of the electron temperature to the shock velocity requires a second channel for substantial energy loss. The most likely such mechanism is particle acceleration [29], yet, curiously, the X-ray spectrum of this remnant shows no evidence for electron synchrotron radiation.

CONCLUSION

In the five years since SN 1006 was demonstrated to be the first detected cosmic ray source, we have made enormous progress using X-ray observations toward determining whether diffusive shock acceleration in supernova remnants is the primary mechanism for producing Galactic cosmic rays. While many issues remain to be settled before a definitive answer is obtained, with *Chandra* and XMM/*Newton* we have available the powerful observatories necessary for performing this task.

REFERENCES

1. Hamilton, A.J.S., Sarazin, C.L. & Szymkowiak, A.E., 1986, ApJ, 300, 698
2. Koyama, K., Petre, R., Gotthelf, E.V., Hwang, U., Matsuura, M., Ozaki, M., & Holt, S.S., 1995, Nature, 378, 255
3. Reynolds, S.P., 1996, ApJ, 459, L13
4. Reynolds, S.P., 1998, ApJ, 493, 375
5. Dyer, K.K., Reynolds, S.P., Borkowski, K.J., Petre, R., & Allen, G.E., 2001, ApJ, in press
6. Laming, J.M., 1998, ApJ, 499, L309
7. Tanimori, T., Hayami, Y., Kamei, S., Dazeley, S. A., Edwards, P. G., et al., 1998, ApJ, 497, L33
8. Allen, G.E., Keohane, J. W., Gotthelf, E. V., Petre, R., Jahoda, K., et al., 1997, ApJ, 487, L97
9. Vink, J., Kaastra, J.S., & Bleeker, J.A.M., 1997, A&A, 328, 628
10. Keohane, J.W., 1998, Ph.D. Thesis, University of Minnesota
11. Tomida, H., 2000, Ph.D. Thesis, Kyoto University
12. Koyama, K., et al., 1997, PASJ, 49, L7
13. Slane, P., Hughes, J.P., Edgar, R.J., & Plucinsky, P. 1999, BAAS, 195, 3306
14. Asvarov, A.I., Dogiel, V.A., Gusienov, O.H., & Kasumov, F.K., 1990, A&A, 229, 196
15. Tatischeff, V., Ramaty, R., & Kozlovsky, B., 1998, ApJ, 504, 874
16. Muraishi, H., Tanimori, T., Yanagita, S., Yoshida, T., Moriya, M., et al., 2000, A&A, 354, 57
17. Puelhofer, G., Voelk, H., Wiedner, C. A., et al., 1999, Proc. 26th International Cosmic Ray Conference, 3, 492
18. Decourchelle, A., Ellison, D.C., & Ballet, J., 2000, ApJ, 543, L57, 2000
19. Allen, G.E., Gotthelf, E.V., & Petre, R., 1999, Proc. 26th International Cosmic Ray Conference, 3, 480
20. Borkowski, K.J., Rho, J., Reynolds, S.P., & Dyer, K.K., 2001, ApJ, in press
21. Keohane, J.W., Petre, R., Gotthelf, E.V., Ozaki, M., & Koyama, K., 1997, ApJ, 484, 350
22. Keohane et al. 2001, these proceedings.
23. Safi-Harb, S., & Petre, R., 1999, ApJ, 512, 784
24. Hughes, J.P., Rakowski, C.E., Burrows, D.N., & Slane, P.O., 2000, ApJ, 528, L109
25. Holt, S.S., Gotthelf, E.V., Tsunemi, H., & Negoro, H., 199, PASJ, 46, L151
26. Gotthelf, E.V., Koralesky, B., Jones, T.W., Rudnick, L., Hwang, U., Petre, R., & Holt S.S., 2001, ApJ, in press
27. Reynolds, S.P., & Keohane, J.W., 1999, ApJ, 525, 368 Keohane & Reynolds (2000)
28. Ellison, D.C.; Berezhko, E.G., & Baring, M.G, 2000, ApJ, 540, 292
29. Hughes, J.P., Rakowski, C.E., & Decourchelle, A., 2000, ApJ, 543, L61

Particle Acceleration in Shock Waves of Young Supernova Remnants

Stephen P. Reynolds

North Carolina State University, Raleigh, NC 27695, USA

Abstract.
I shall briefly review the theory of diffusive shock acceleration, with particular emphasis on testable predictions for young supernova remnants. Evidence now exists for several predictions of nonlinear shock-acceleration theories: curvature in the radio spectrum of some young remnants, shock precursors of various kinds, and most importantly, acceleration of electrons, at least, to energies in excess of 10 TeV. As yet there is no definitive evidence for cosmic-ray ions. The best prospects for observational study of shock acceleration lie in optical observations of nonradiative shocks, to study dynamical precursors and compression ratios; X-ray observations of remnants with synchrotron components; and gamma-ray observations, both at MeV-GeV energies from satellites such as INTEGRAL and GLAST, and at TeV energies from ground-based air Cerenkov telescopes.

TEST-PARTICLE SHOCK ACCELERATION

The basic theory of diffusive shock acceleration of particles, also called first-order Fermi acceleration, was worked out in several roughly contemporaneous papers [1,2]. Many good reviews have been written on the process in the test-particle approximation, and some discuss nonlinear extensions as well [3,4,5]. Here I shall review very briefly the fundamental results of test-particle diffusive shock acceleration, and describe the evidence that the process is intrinsically nonlinear. I shall summarize general properties of nonlinear models, concentrating on results of particular interest for young supernova remnants. Space prohibits the referencing of all the efforts that have contributed to these results; I beg the forgiveness of uncited authors.

The basic geometry is shown in Figure 1. We assume a fully ionized, magnetized plasma enters from the left at constant velocity u_1. It is shocked and compressed at $x = 0$ and flows out toward $x = +\infty$ at postshock velocity u_2, so that the compression ratio is $r \equiv u_1/u_2 \equiv \rho_2/\rho_1$ where $\rho_1 (\rho_2)$ is the upstream (downstream) density. In the test-particle limit, cosmic rays are assumed to be a distinct population of energetically unimportant particles, scattering and diffusing in the system and satisfying a stationary diffusion equation in momentum space, with diffusion coefficient

CP565, *Young Supernova Remnants: Eleventh Astrophysics Conf.,* edited by S. S. Holt and U. Hwang
© 2001 American Institute of Physics 0-7354-0001-6/01/$18.00

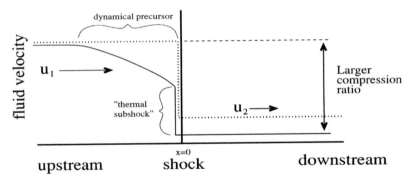

FIGURE 1. Shock velocity profiles. Dotted line: test-particle (unmodified) shock (discontinuous velocity jump). Solid line: shock profile modified by cosmic rays.

κ. The cosmic-ray distribution function $f(x,t)$ then is spatially constant behind the shock, dropping ahead of the shock as $\exp\left(-x/x_D\right)$ where $x_D(p) \equiv \kappa(p)/u_1$ is the diffusion length for particles of momentum p. The momentum distribution is a power-law, $f(x,p) \propto p^{-3r/(r-1)}$. This implies a power-law in energy, though with different slopes in the regimes $E \ll mc^2$ and $E \gg mc^2$. For $E \ll mc^2$, $f(E) \propto E^{-(2r+1)/(2r-2)}$. In the ultrarelativistic domain,

$$f(x, E) \propto E^{-(r+2)/(r-1)} \equiv E^{-s} \tag{1}$$

so that for a strong adiabatic shock with ratio of specific heats $\gamma = 5/3 \Rightarrow r = 4$, $f(E) \propto E^{-2}$. Electrons with this energy distribution would radiate synchrotron radiation with a power-law photon spectrum $S_\nu \propto \nu^{(1-s)/2} \equiv \nu^{-\alpha}$ with $\alpha = 0.5$ for $s = 2$. The rough agreement of the predicted photon spectrum with that observed in many synchrotron sources ($\alpha \sim 0.4 - 0.7$), and of the particle spectrum with the observed cosmic rays at Earth ($s \sim 2.75$, but steepened by propagation through the Galaxy), was the major factor in the almost immediate acceptance of this mechanism for the origin of energetic particle distributions in astrophysics. In the derivation of this result, nothing has been assumed about the energy dependence of κ, or about the nature of the scattering.

In general, the compression ratio of a shock with Mach number $\mathcal{M}$ and ratio of specific heats γ is given by

$$r^{-1} = \frac{\gamma - 1}{\gamma + 1} + \frac{2}{\gamma + 1}\frac{1}{\mathcal{M}^2} \tag{2}$$

(e.g., [6]). For $\gamma = 5/3$, this can be rewritten $\mathcal{M}^2 = (s+2)/(s-2)$; for $\gamma = 4/3$, for instance for a shock in a relativistic medium, $\mathcal{M}^2 = (2s+4)/(2s-3)$. So for nonrelativistic, adiabatic shocks, $s > 2 \Rightarrow \alpha > 0.5$, and for s significantly above 2, the Mach number must be quite low. For $\gamma = 4/3$, the maximum compression ratio is 7, implying $s = 1.5$. It should be stressed that the slope of the accelerated-particle distribution depends only on the compression ratio; that ratio can be much larger than 4 or even 7 for radiative shocks or shocks with other energy sinks, resulting in very hard spectra. In all cases, the spectrum has a flatter slope at nonrelativistic energies, steepening around mc^2. (See Fig. 2.)

The time to accelerate a particle from momentum p_i to p is given by [3,7]

$$\tau_{acc} = \frac{3}{u_1 - u_2} \int_{p_i}^{p} \left(\frac{\kappa_1}{u_1} + \frac{\kappa_2}{u_2} \right) \frac{dp'}{p'} \tag{3}$$

where the diffusion coefficient in general will have different values upstream and downstream.

To go further requires specifying some properties of the diffusion coefficient. We presume that the particle scattering is due to resonant scattering of particles by MHD waves. In the approximation known as "quasi-linear theory," in which particle scattering is treated as a small perturbation to motion in a constant magnetic field, the particle mean free path along the magnetic field is proportional to its gyroradius, $\lambda_{mfp} = \eta r_g$ with $r_g = E/e\langle B \rangle$ in the extreme-relativistic limit (and Gaussian units). The "gyrofactor" η is given by $\eta = (\delta B/\langle B \rangle)_{res}^{-2}$ where δB is the amplitude of the wave resonant with the particle (i.e., $r_g \sim \lambda_{MHD}$). The assumption of constant η is often made, but it should be pointed out that this corresponds to the assumption of a particular spectrum of MHD turbulence: constant energy per unit logarithmic bandwidth. The diffusion coefficient is given by $\kappa = \lambda_{mfp} v/3$.

For the first time the magnetic-field obliquity (characterized by θ_{Bn}, the angle between the upstream magnetic field and the shock normal) enters the discussion, since for oblique ($\theta_{Bn} > 0$) shocks, $B_2 > B_1$ and κ_2 differs from κ_1. For a parallel shock ($\theta_{Bn} = 0$), $\kappa_2 = \kappa_1 \propto p$ and we obtain from (3) (for $p \gg p_i$)

$$\tau_{acc}(p)(\text{parallel}) = \frac{3\kappa(p)}{u_1^2} \frac{r(r+1)}{r-1}. \tag{4}$$

Particle scattering due to MHD turbulence is likely to be anisotropic, with diffusion across field lines probably inhibited compared to diffusion along them. One traditional approximation is that particles scatter one gyroradius perpendicular to the field for every parallel scattering (e.g., [8]). This gives $\kappa_\perp = \kappa_\parallel/[1 + (\lambda_{mfp}/r_g)^2] \equiv \kappa_\parallel/(1 + \eta^2)$. Then, for shocks with non-zero obliquities, the diffusion coefficient along the shock normal is a combination of parallel and perpendicular diffusion, and

$$\kappa = \kappa_\parallel \cos^2 \theta_{Bn} + \kappa_\perp \sin^2 \theta_{Bn}. \tag{5}$$

Then the acceleration time given in Equation 4 is shortened for $\theta_{Bn} > 0$ by a substantial amount [8]. We can write $\tau_{acc}(\theta_{Bn}) = R_J \tau_{acc}(\theta_{Bn} = 0)$ as in [9], with $R_J \leq 1$, and dropping as η increases ($R_J \propto \eta^{-2}$ for large η). For a spherical shock encountering uniform magnetic field, all obliquities are achieved somewhere on the shock surface, and the results can be complicated [9].

371

The maximum energy to which particles can be accelerated by diffusive shock acceleration is clearly an important issue for cosmic ray origins, but also for young supernova remnants, as we shall see. Lagage & Cesarsky [10] first discussed this issue in the SNR context, for the limitation of finite acceleration time available. This age limitation is similar to one obtained by considering the finite size of a spherical remnant [11]. This was generalized to include radiative losses [12] (likely only important for electrons), and [9, 13] to include a change in the availability of MHD waves above some wavelength (i.e., an abrupt increase in the diffusion coefficient above the corresponding particle energy), resulting in particle escape upstream. In any of these cases, the form of the resulting distribution function should be well approximated as a power law with an exponential cutoff at some E_{max} [12, 14]. For each of the three possible limitations mentioned, one can obtain an expression for E_{max}: by equating the acceleration time to the lesser of the remnant age or the radiative-loss time, or, for the escape mechanism, by equating E_{max} to the energy of particles resonant with the maximum wavelength λ_{max} of MHD waves present. The results depend in detail on the shock compression ratio and obliquity [9], but are roughly

$$E_{max}(\text{age}) \sim 50 B_{\mu G} \left(\eta R_J \right)^{-1} \quad \text{TeV} \tag{6}$$

$$E_{max}(\text{loss}) \sim 100 u_8 \left(\eta R_J B_{\mu G} \right)^{-1/2} \quad \text{TeV} \tag{7}$$

$$E_{max}(\text{escape}) \sim 20 B_{\mu G} \lambda_{17} \quad \text{TeV} \tag{8}$$

where $u_8 \equiv u_1/10^8$ cms^{-1} and $\lambda_{17} \equiv \lambda_{max}/10^{17}$ cm. There is no R_J-dependence in the escape case. For a spherical remnant, maximum energies in the first two cases depend on obliquity through R_J as described in excruciating detail in [9]. But these results show that in general one ought to expect that in typical SNR conditions, electrons and ions can easily achieve energies in the TeV range. Electrons of energy $E_{100} \equiv E/100$ TeV radiate the peak of their synchrotron emission at $h\nu_{peak} \sim 2 (B/10 \ \mu G) E_{100}$ keV; even well above E_{max}, this can be significant since in the delta-function approximation in which each electron radiates all its energy at frequency ν_{peak}, the spectrum above ν_{max} drops only as $\exp -(\nu/\nu_{max})^{-1/2}$, and when calculated numerically with the exact single-particle emissivity, drops even slightly slower [9].

NONLINEAR MODIFICATIONS

It was realized very soon after the development of test-particle theories that feedback of the accelerated particles on the background thermal fluid could not be neglected in general [15]. If SNRs are to replenish the Galactic pool of cosmic rays on the observed timescale of about 2×10^7 yr [16], for a SN rate of two or three per century, efficiencies of order 10% are required, maybe more, so a significant fraction of SN energy must appear as cosmic rays, whose effects on the blast wave structure are considerable. As these particles diffuse ahead of the shock and are turned around and scattered back toward the shock, they react on the background fluid, causing it to begin to decelerate smoothly (in the shock frame) before a final "thermal subshock" is reached whose thickness is given by the thermal ion gyroradius, where the fluid velocity makes a last sudden drop (see Fig. 1). Furthermore, if particles at the high end of the energy spectrum are energetically important, the assumptions of steady-state diffusion with constant shock velocity break down. The particles may excite the waves that scatter them, producing another feedback loop, and the details of the shock structure on the scale of the thermal ion gyroradius may be influenced by accelerated particles, so that "injection" of particles into the suprathermal tail is self-regulated. If particles can escape the shock, their loss causes an effective increase in the ratio of specific heats, increasing the shock compression; if particles become relativistic, the mean fluid γ may also drop below 5/3.

These effects have been worked out in various ways by many individuals. Self-generated waves were a part of Bell's [2] original picture. Soon after Eichler's analytic work [15], Ellison [17] began simulating shock acceleration with a statistical, Monte Carlo approach in which the full distribution function of particles was calculated, without an artificial distinction between "thermal" and "nonthermal" particles. This simulation program has been a powerful tool for the study of analytically intractable situations. Most implementations of the Monte Carlo technique (see also [18]) have assumed a dynamical steady state, with the shock velocity profile being iterated until uniform fluxes of mass, momentum, and energy are achieved through the shock. The result is an extensive "dynamical precursor" extending a distance ahead of the shock of the order of the diffusion length of the highest energy cosmic rays, in which the density increases gradually, and the velocity decreases, from their far upstream values (see Fig. 1). A suprathermal tail automatically appears in the distribution function, as particles in the exponential tail of the Maxwellian "leak" into the acceleration process. Arbitrary momentum-dependence of the diffusion coefficient can be treated this way, though most work centers on the case of $\kappa \propto p$ (constant η).

An alternative approach describes the thermal and accelerated-particle distributions as two fluids, each satisfying conservation equations, with specified couplings between them (the "two-fluid" approach [19, 20]). These models normally require an a priori specification of particle injection (e.g., fixing the fraction of particle flux, or energy flux, through the shock that enters the cosmic-ray fluid). Furthermore, until recently computational demands restricted the momentum-dependence of κ that could be treated, and much of the early work uses a constant energy-independent average diffusion coefficient. However, in recent years more sophisticated versions of this approach have appeared, and broad agreement has been achieved on the most important nonlinear modifications of test-particle diffusive shock acceleration.

1. Precursors. The pre-deceleration (or, in the SNR frame, pre-acceleration) of preshock fluid by the advancing cosmic rays has already been mentioned (the "dynamical precursor."). Early models allowed a "cosmic-ray dominated" shock solution, in which the flow gradually decelerated to the post-shock value with no entropy generation in the thermal gas. Since this picture would not produce any heating, and therefore no X-ray emission, it seems to be unimportant for SNRs. Most calculations of shock structure now find a "thermal subshock" with compression ratio of around 3 persisting even in very efficient shocks. **2. Spectral curvature.** Most authors agree that the diffusion coefficient probably increases with energy in some fashion. If so, as particles gain energy, they diffuse further ahead of the shock where the velocity jump is larger, and receive higher fractional energy gains. This produces a concave-up (hardening) spectrum that was a feature of the first nonlinear calculations [15], and is a distinctive feature of all nonlinear models with κ rising with E. **3. High compression ratios.** If particles can escape the shock, carrying away significant energy, the shock is lossy, and large increases in compression ratio can be attained in principle [e.g., 20]. However, if mechanisms to move energy back from cosmic rays to the thermal gas are included, such as generation of Alfvén waves followed by damping, the compression ratios are not as high. But this mechanism pre-heats thermal gas (a "thermal precursor"), as well as providing higher-than-ambient MHD turbulence upstream (an "MHD-wave precursor"). **4. Lowering of mean temperature.** The overall manifestation of energy appearing in the tail of the distribution is the lowering of the mean temperature of the thermal peak below normally assumed Rankine-Hugoniot values, though unless the efficiencies are much larger than 50%, this effect is moderate in magnitude, and can easily be masked by incomplete electron heating.

RADIATIVE PROCESSES

The expected distribution of electrons from a strong SNR shock is sketched in Figure 2, with curvatures exaggerated. This entire electron distribution radiates bremsstrahlung, in which an electron with energy E radiates on average, photons with energy $h\nu \sim E/3$. Bremsstrahlung from the nonthermal tail of the distribution is referred to as "nonthermal," though it is simply the continuation of the thermal-bremsstrahlung peak. The electrons with energies below a few tens of keV, whether they are in the thermal peak or the nonthermal tail of the distribution, will excite line transitions in thermal gas. At relativistic energies, electrons can produce synchrotron radiation at all frequencies from radio through X-rays, depending on their maximum energy, and can also inverse-Compton upscatter photons of any ambient radiation field. Finally, the ion distribution, similar in general outline though different in detail from the electron distribution, can also produce radiation, though the only process likely to be significant and observable is the production of pions through inelastic collisions between protons of energy $E \gtrsim 1$ GeV and thermal ions, with the subsequent decay of the π^0's into gamma rays (the charged pions

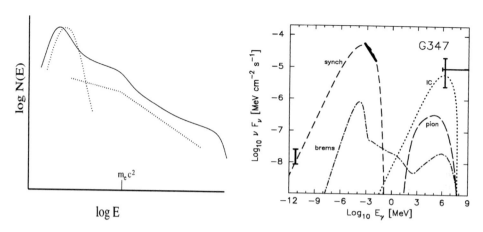

FIGURE 2. Left: Schematic electron distribution. Dotted line: test-particle result: a Maxwellian peak, and energetically unimportant power-law tail with arbitrary normalization. Solid line: nonlinear result. Right: Model νF_ν spectrum [Ellison, in preparation] showing contributions from various radiative processes. The model uses a broken power-law approximation to a nonlinear calculation of particle spectra.

decay into $e^\pm$ which are insignificant compared to locally accelerated electrons). The radio-to-TeV gamma-ray spectrum of SNRs has been modeled including these processes in [21, 22], where [22] uses a nonlinear Monte Carlo spectral calculation as input). Figure 3 shows a typical calculation.

Several features of these predictions need comment. Nonthermal bremsstrahlung should be present at some level in X-ray spectra above a few keV. It is distiguishable from synchrotron radiation by a straight or hardening spectrum; synchrotron radiation is always produced by the very highest energy electrons, where the distribution is steepening. Weak or nonexistent X-ray lines cannot be due to a nonthermal-bremsstrahlung continuum, since this would imply plenty of electrons with suitable energies to excite line emission. Synchrotron emission is produced by electrons with $E \sim 100$ TeV, whose cross-section for exciting atomic transitions is utterly negligible.

Inverse-Compton (IC) emission depends on the seed photon pool being scattered. The easiest to model is IC from cosmic-microwave-background photons (CMBIC); because of Klein-Nishina effects that reduce the cross-section for photons with energy larger than $m_e c^2$ in the electron rest-frame, the lowest-energy seed photons can produce the highest-energy scattered photons, in the TeV range accessible to ground-based air-Čerenkov detectors such as the Whipple Observatory, HEGRA, and CANGAROO. Other seed photon pools are present, but are less likely to produce distinguishable effects, though they will contribute to electron energy losses.

The IC spectrum will be quite hard, harder than either bremsstrahlung or π^0-decay gamma rays, with a slope equal to that of the radio synchrotron emission (except at the highest energies where the electron spectrum cuts off).

Ions can be detected only by the low-energy turnover below which pions cannot be produced, at photon energies around 100 MeV. For favorable circumstances, a detectable "bump" will be present in that vicinity. But it is also possible that the "bump" will be submerged in gamma rays from electron processes.

Observational evidence. At this time, reasonably good evidence exists for synchrotron emission at X-ray energies (see references in [23]). Bremsstrahlung from obviously non-Maxwellian populations has not been unambiguously detected. TeV gamma-ray emission has been seen from two remnants by the CANGAROO collaboration [24] and is reasonably well modeled as CMBIC, allowing the mean magnetic field strength in those remnants to be estimated [23, 25]. Clear evidence for hadronic processes is lacking, though hints were found in EGRET observations of unidentified Galactic-plane sources which have been attributed to SNRs [21].

PREDICTIONS

1. Precursors. MHD-wave precursors may have been inferred in the form of enhanced turbulence inferred to exist outside some SNRs with very sharply bounded radio emission, indicating short diffusion lengths for radio-emitting electrons ahead of the shock [26]. Of course, the limits are possible because any synchrotron precursor is too thin to observe. Since we expect electron diffusion lengths to increase with energy, a synchrotron X-ray halo should certainly be detectable outside remnants whose X-ray emission is dominantly synchrotron (and for which the shock can be unambiguously located). The cosmic-ray precursor is the most likely origin for the unexpectedly large line width in the narrow component of Hα emission seen in a few Balmer-dominated SNRs [27]. That component is presumably the upstream distribution of neutral H atoms, and it is too wide to be consistent with the population remaining neutral. Broadening by collisions with charged thermal particles heated by Alfvén waves, or directly with the lowest-energy nonthermal particles, is one explanation [28]. However, in a few remnants, one should be able to see the pre-acceleration of that neutral component (for a slit across the remnant face, i.e., where the shock velocity is mostly radial), and this is not seen [28].

2. Spectral curvature. The amount of curvature expected due to nonlinear effects is slight, amounting to only 0.1 or so in the synchrotron spectral index over the radio band [29]. However, it may have been seen in Tycho and Kepler [30].

3. Maximum energies. These predictions are not restricted to nonlinear models. Three remnants (SN 1006, G347.5–0.5, and G266.2+1.2) have essentially lineless X-ray spectra, implying synchrotron radiation and E_{max} values of order 10 – 100 TeV ([23], and references therein). Cas A has continuum emission out to 100 keV. In RCW 86, X-ray lines are too weak for any reasonable thermal explanation, but including a substantial synchrotron component produces sensible inferences. In all these cases, electrons are present with energies of order 100 TeV, but that part of the distribution is in the dying tail of the power-law. Samples of 14 Galactic and 11 LMC remnants [31] show that in no case can the electron spectrum preserve its low-energy power-law to beyond 100 TeV.

4. Higher compression ratios. Normally one infers post-shock quantities in SNRs, and deduces upstream ones assuming some compression ratio; if r is much larger, one just infers lower upstream values. Furthermore, derived shock velocities depend on the compression ratio; post-shock gas is moving at $(r-1)/r$ times the shock speed. However, some optical diagnostics give information on upstream gas, excited for instance by post-shock UV radiation. One cosmic-ray model for the Cygnus Loop found compression ratios of 4 to 6 [32].

5. Lower mean temperatures. A lower-than-expected ion temperature may be hard to distinguish from $T_e < T_i$ due to incomplete electron heating, as is expected (and observed; several Balmer-dominated SNRs show $T_e \lesssim 0.1 T_i$ [27]). However, in principle one could imagine using high-resolution X-ray spectroscopy to observe u_1 from Doppler shifts of line centroids; T_i from Doppler widths of lines (testable because of differences for different ions); and T_e from line ratios. One could then infer the energy going into cosmic rays, the amount of shock heating of electrons, and the compression ratio. The comparison between the predicted bremsstrahlung for the inferred T_e value and the observed continuum could give evidence for a synchrotron contribution to the continuum.

6. High acceleration efficiencies. The efficiency of electron acceleration can in principle be deduced if CMBIC radiation is observed, since that gives directly the electron population, and any synchrotron flux density then allows separate deduction of the magnetic-field strength. This has been done for SN 1006 [23, 24, 25] where mean magnetic fields of order 10 μgauss are found, implying several percent of the current postshock pressure $(3/4)\rho u_1^2$ going into relativistic electrons, and about 60 times as much energy in those electrons as in the magnetic field. More TeV detections can add to this database; TeV upper limits give lower limits on the magnetic field strength and upper limits on the efficiency.

7. Do SNRs accelerate ions at all? This question will require the next generation of MeV-GeV satellites, such as INTEGRAL and GLAST, to try to find the π^0 kinematic bump, with no guarantees of success.

CONCLUSIONS

Nonthermal particles are expected to represent a significant fraction of the energy budget of a young SNR. The process of diffusive shock acceleration can be studied by observing nonthermal emission from SNRs, but even basic shock properties will be affected by the presence of those particles. Making reliable inferences will demand simultaneous thermal and nonthermal analyses. No matter which is your signal and which is your noise, you will need to be aware of all these effects to interpret the high quality of information on young SNRs now becoming available.

REFERENCES

1. Axford, W.I., Leer, E., & Skadron, G. 1977, *Proc. 15th Int. Cosmic Ray Conf.*, 11, 132; Blandford, R.D., & Ostriker, J.P. 1978, ApJ, 221, L29
2. Bell, A.R. 1978, MNRAS, 182, 147
3. Drury, L.O'C. 1983, Rep. Prog. Phys., 46, 973
4. Blandford, R.D., & Eichler, D. 1987, Phys. Reports., 154, 1
5. Jones, F.C., & Ellison, D.C. 1991, SpSciRev, 58, 259
6. Spitzer, L. Physical Processes in the Interstellar Medium (New York: Wiley), 220
7. Forman, M.A., & Morfill, G. 1979, *Proc. 16th Int. Cosmic Ray Conf.*, 3, 467
8. Jokipii, J.R. 1987, ApJ, 313, 842
9. Reynolds, S.P. 1998, ApJ, 493, 375
10. Lagage, P.-O., & Cesarsky, C.J. 1983, A&A, 118, 223
11. Berezhko, E.G. 1996, Astropart.Phys., 5, 367
12. Webb, G.M., Drury, L.O'C., & Biermann, P.L. 1984, A&A, 137, 185
13. Reynolds, S.P. 1996, ApJ, 459, L13
14. Drury, L.O'C. 1991, MNRAS, 251, 340, and private communication
15. Eichler, D. 1979, ApJ, 229, 419
16. Garcia-Muñoz, M., Mason, G.M., & Simpson, J.A. 1977, ApJ, 217, 859
17. Ellison, Jones, & Eichler 1981, JGR 50, 110; Ellison & Eichler 1984, ApJ, 286, 691
18. Ostrowski, M. 1991, MNRAS, 249, 551.
19. Drury, L.O'C., & Völk, H. 1981, ApJ, 248, 344; Jones, T.W., & Kang, H. 1992, ApJ, 396, 575
20. Berezhko, E.G., & Völk, H.J. 1997, Astropart.Phys., 7, 183
21. Gaisser, T.K., Protheroe, R.J., & Stanev, T. 1998, ApJ, 492, 227; Sturner et al. 1997, ApJ, 490, 619
22. Baring et al. 1999, ApJ, 513, 311
23. Dyer, K.K. et al. 2001, ApJ, in press (astro-ph/ 0010424)
24. Tanimori, T. et al. 1998, ApJ, 497, L25; Muraishi, H., et al. 2000, A&A, 354, L57
25. Aharonian, F.A., & Atoyan, A.M. 1999, A&A, 351, 330
26. Achterberg, A., Blandford, R.D., & Reynolds, S.P. 1994, A&A, 281, 220
27. Smith, R.C. et al. 1991, ApJ, 375, 652; Ghavamian et al. 2000, ApJ, 535, 266
28. Smith, R.C., Raymond, J.C., & Laming, J.M. 1994, ApJ, 420, 286
29. Ellison, D.C., & Reynolds, S.P. 1991, ApJ, 382, 242
30. Reynolds, S.P., & Ellison, D.C. 1992, ApJ, 399, L75
31. Reynolds, S.P., & Keohane, J.W. 1999, ApJ, 525, 368 (Galaxy); Hendrick, S.P., Reynolds, S.P., & Borkowski, K.J. 2001, this volume; ApJ, in preparation (LMC)
32. Boulares, A., & Cox, D.P. 1988, ApJ, 333, 198

Nonthermal X-ray emission from Young Supernova Remnants

Eric van der Swaluw, Abraham Achterberg, and Yves A. Gallant

Astronomical Institute, Utrecht University, The Netherlands

Abstract. The cosmic-ray spectrum up to the knee ($E \sim 10^{15}$ eV) is attributed to acceleration processes taking place at the blastwaves which bound supernova remnants. Theoretical predictions give a similar estimate for the maximum energy which can be reached at supernova remnant shocks by particle acceleration. Electrons with energies of the order $\sim 10^{15}$ eV should give a nonthermal X-ray component in young supernova remnants. Recent observations of SN1006 and G347.3-0.5 confirm this prediction. We present a method which uses hydrodynamical simulations to describe the evolution of a young remnant. These results are combined with an algorithm which simultaneously calculates the associated particle acceleration. We use the test particle approximation, which means that the back-reaction on the dynamics of the remnant by the energetic particles is neglected. We present synchrotron maps in the X-ray domain, and present spectra of the energies of the electrons in the supernova remnant. Some of our results can be compared directly with earlier semi-analytical work on this subject by Reynolds [1].

INTRODUCTION

The number of supernova remnants (SNRs) with an observed nonthermal X-ray component is slowly increasing (Allen et al.[2]). This confirms the theory of diffusive shock acceleration (DSA), which predicts the production of relativistic particles at SNR shocks in the required energy range. Furthermore it strengthens the evidence that SNR shocks are indeed the sites where cosmic rays are accelerated up to the knee of the observed cosmic ray spectrum ($E \sim 10^{15}$ eV). It is expected that with the current observational facilities like Chandra and XMM, the details about SNRs with a nonthermal component and, hopefully, the number of SNRs identified as sources of nonthermal X-ray emission will increase. This motivates the extension of the current models of nonthermal X-ray emission of SNRs, in order to keep up with with the status of observational work on this subject. We present a model which consists of a hydrodynamics code calculating the flow of an evolving SNR coupled with an algorithm which simulates simultaneously the transport and acceleration of relativistic particles.

CP565, *Young Supernova Remnants: Eleventh Astrophysics Conf.,* edited by S. S. Holt and U. Hwang
© 2001 American Institute of Physics 0-7354-0001-6/01/$18.00

HYDRODYNAMICS

The evolution of a single supernova remnant (SNR) can be divided in four different stages (Woltjer [3]): the free expansion stage, the Sedov-Taylor stage, the pressure-driven snowplow stage and the momentum-conserving stage. We will only focus on the first two stages. In later stages, both synchrotron losses and the efficiency of the acceleration process, prevent particles to be accelerated to energies where they can emit X-rays by the synchrotron mechanism. In the free expansion stage the material of the SNR is dominated by expanding ejecta from the progenitor star, bounded by a shock which is driven into the interstellar medium (ISM). Due to deceleration by the ISM, a reverse shock accompanies the forward shock (McKee [4]). When the SNR has swept up a few times the ejected mass of the progenitor star, this reverse shock is driven back into the interior of the SNR. This marks the stage where the expansion of the SNR shock will make the transition to the Sedov-Taylor stage. We have performed hydrodynamical simulations for a young supernova remnant up to this transition. All the hydrodynamical simulations were performed with the Versatile Advection Code [1] (VAC, Tóth [5]). The calculations are performed on a spherically symmetric, 1D, uniform grid. As an initial condition we deposit thermal energy and mass in the first few grid cells. This leads to the formation of both the reverse shock and the forward shock, discussed above. The grid resolution is such that the forward and the reverse shock are both resolved, and give the right compression factor ($r = 4$) for a non-relativistic strong shock. We make the model of the SNR 2D axially symmetric by imposing a uniform magnetic field in the ISM, aligned with the symmetry axis, at the start of the simulation. By using the condition of a frozen-in magnetic field (ideal MHD) the magnetic field lines are dragged along with the fluid, determining the configuration of the magnetic field at later times.

PARTICLE ACCELERATION

In principle, the acceleration and propagation of particles in a flow can be investigated by solving a Fokker-Planck equation for the particle distribution in phase space (e.g. Skilling [6]). Instead, we employ a method which uses Itô stochastic differential equations (SDEs). The SDE method simulates the random walk trajectory of a test particle in a given flow. By considering many realizations of the SDE in the same flow one constructs the distribution of particles in phase space. This corresponds to the solution of the Fokker-Planck equation (Achterberg and Krülls [7]).

By performing hydrodynamical simulations of a young SNR, and by using the flow from these simulations as an input for the SDE method, we simultaneously describe *both* the particle acceleration and the hydrodynamical evolution. Losses due to synchrotron radiation and inverse Compton radiation are easily included in

---

[1] See http://www.phys.uu.nl/~toth/

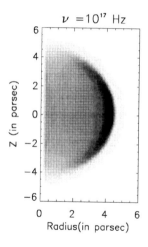

$\nu = 10^{17}$ Hz

Z (in parsec)

Radius(in parsec)

FIGURE 1. Synchrotron map for $\nu = 10^{17}$ Hz.

the SDE method. In this way, we get the electron distribution in phase space for a young SNR which realistically describes: (1) acceleration at the forward shock, (2) adiabatic losses due to the expansion of the SNR, (3) synchrotron losses due to the presence of magnetic fields and (4) inverse Compton losses due to the interaction of the electrons with the photons of the cosmic microwave background.

RESULTS

We present the results from a simulation of a SNR with a total mechanical energy of $E_0 \simeq 0.93 \times 10^{51}$ erg and an ejected mass of $M_{ej} = 5.6 M_\odot$. This expands into an uniform medium with density $\rho_0 = 10^{-24}$ g/cm^3 and an axially symmetric magnetic field with strength $B_0 = 10 \mu G$. We continuously inject particles at the forward shock of the SNR, starting at an age of $t = 100$ years up to the end of the simulation at an age of $t = 1000$ years. A total of $\sim 3.7 \times 10^6$ test particles were injected during the simulation. The injection was at a fixed momentum, and proportional with the area of the remnant, with a particle weight which takes into account the increase in the amount of material processed by the shock as it expands. At the end of the simulation we have the position and the momentum of each particle in the simulation box. Because the magnetic field strength throughout this box is also known, we can produce synchrotron maps at different frequencies. One example is shown in figure 1, which is reminiscent of the synchrotron maps presented by Reynolds [1]. Furthermore, figure 2 shows the spectrum of all the particles in the remnant. At low frequencies one can see the expected value of the spectral index for acceleration at a strong shock (compression ratio $r = 4$), i.e. $S_\nu \propto \nu^{-0.5}$. In the roll-off part of the spectrum, where the synchrotron and inverse Compton losses

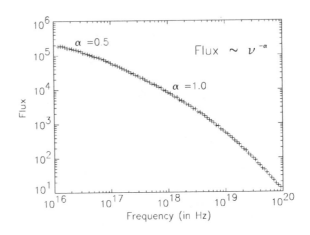

FIGURE 2. Spectrum of the total remnant.

start to compete with the energy gain due to the acceleration process, we get a power-law index of $S_\nu \propto \nu^{-1.0}$ at a frequency of 10^{18} Hz.

CONCLUSION

We have presented results from a method using a combination of hydrodynamical simulations and an algorithm simultaneously simulating particle acceleration in a SNR. The results are comparable with earlier work on this subject by Reynolds[1]. Future work will consider the evolution of a SNR in an ISM which is not uniform, like stellar wind cavities or molecular clouds. The combination of the SDE method and hydrodynamics is a strong tool to investigate the resulting morphology of the synchrotron emission.

REFERENCES

1. Reynolds S.P. 1998, ApJ, 493, 375
2. Allen G.E., Gotthelf E.V., Petre, R. 1999, "Evidence of 10-100 Tev Electrons in Supernova Remnants", in Proc. of the 26th ICRC, *astro-ph 9908209*
3. Woltjer L. 1972, ARA&A, 10, 129
4. McKee C.F. 1974, ApJ, 188, 335
5. Tóth G. 1996, Astrophys. Lett. Comm., 34, 245
6. Skilling J. 1975, M.N.R.A.S., 172, 557
7. Achterberg A. & Krülls W.M. 1992, A&A, 265, L13

Electron Acceleration in Young Supernova Remnants

William K. Rose

Department of Astronomy, University of Maryland, College Park, MD 20742 USA

Abstract.
There is evidence for production of relativistic electrons and amplification of magnetic fields in young supernova remnants. In a recent publication we have given a model for radio emission from SN 1993J [1]. In this paper we discuss theories of electron acceleration in supernova remnants such as SN 1993J. Similarities between particle acceleration processes in supernova remnants and radio galaxies undoubtedly exist and therefore observations of radio galaxies may provide crucial insight into interpreting supernova remnants.

INTRODUCTION

Radio emission by means of synchrotron radiation is a characteristic property of Cas A and other young supernova remnants. In 1960 Shklovskii [2] predicted that the radio brightness of such remnants should decrease approximately as the sixth power of their radii. His theory assumed that the injection of relativistic electrons ceases at the beginning of evolution and that the remnant consists of an adiabatically expanding cloud of relativistic electrons and other particles frozen into randomly oriented magnetic fields. Magnetic flux constancy was also assumed and so the magnetic field strength was predicted to decline as the square of the remnant radius. Subsequent observations of Cas A showed a decrease in radio emission. However, these observations as well as observations of other supernova remnants showed a substantially lower rate of decline than predicted by Shklovskii's theory. In order to attain agreement between theory and observations it became necessary to introduce physical mechanisms that explain magnetic field amplification and acceleration of relativistic particles. Turbulence can lead to the enhancement of existing magnetic fields. We have suggested that magnetic field amplification can result from plasma turbulence behind a collisionless shock front [1]. Three particle acceleration mechanisms namely type 1 Fermi acceleration, type 2 Fermi acceleration and acceleration by plasma waves have been introduced to explain electron acceleration.

CP565, *Young Supernova Remnants: Eleventh Astrophysics Conf.*, edited by S. S. Holt and U. Hwang

The correlation of SN II and SN Ib supernova with spiral arms and H II regions indicates that their progenitors have short lifetimes and consequently high mass (M $\gtrsim$ 9 M$\odot$). Supergiants, which are the progenitors of SN II supernovae, are known to have strong stellar winds. Moreover, the likely progenitors of SN Ib supernovae namely Wolf-Rayet stars are also observed to emit powerful winds. In recent decades radio emission has been observed from supernova ejecta interacting with the stellar winds of their progenitors. The first such radio observation occurred in 1971 when a bright radio source was detected in M101 at the position of SN 1970G. Approximately a decade later radio emission was observed from SN 1979C and SN 1980K. All three of these supernovae shared the property that the radio maximum was delayed relative to optical maximum. The radio light curves were interpreted as caused by synchrotron radiation with a low frequency free-free absorption cutoff that moved to lower frequency as the supernova shell expanded. Initially only weak radio emission was observed from SN 1987A because at the time of the outburst its progenitor was a blue supergiant with lower mass loss rate and higher outflow velocity than a red supergiant. Evidence that the progenitor of SN 1987A had previously been a red supergiant was obtained when enhanced radio emission was observed as the supernova ejecta caught up with the wind from the red supergiant [3]. Radio emission was also observed from SN 1993J, which occurred in the nearby galaxy M81 [4,5].

RESULTS

Fermi hypothesized that cosmic ray particles are accelerated to high energies as the result of scattering from magnetic fields associated with interstellar clouds. Such clouds have nonrelativistic velocities and are sufficiently massive that their velocities are unaffected by particle collisions. If a cloud is approaching the particle when scattering occurs then the particle is accelerated to higher velocity; however, the reverse occurs if the cloud is receding. Since there is a higher probability that randomly distributed clouds are approaching than receding the particle experiences a net acceleration. It can be shown that such acceleration is second order in velocity so that $\Delta E/E \; \alpha \; v^2/c^2$ where ΔE is the average gain in particle energy during a scattering. This acceleration process, which is known as second order Fermi acceleration, is probably too slow to make an important contribution to cosmic ray production. A more promising scenario for cosmic ray acceleration involves particle traversals of a supernova shock. This process is called first order Fermi acceleration because energy gain is first order in shock velocity. As an injected high velocity particle passes through the shock from upstream to downstream it experiences acceleration. Scattering in the downstream region randomizes the velocity direction of the injected particle without causing a change in energy. For this reason particles gain energy each time they cross the shock front [6, 7]. It can be shown that the resultant particle distribution function is $N(E) \; \alpha \; E^{-n}$ with n = 2, which is close to the cosmic ray spectrum of n = 2.5. This basic acceleration process is the widely accepted explanation for production of galactic cosmic ray protons and other ions with energies $\leq 10^{15}$ eV. Approximately 10% of supernova kinetic energy is required to explain cosmic ray acceleration. Only about 1% as much energy is necessary to account for production of relativistic electrons found in shell supernova remnants and it has been suggested that first order Fermi acceleration by shock front traversals also explains their acceleration. It is clear that shock acceleration is not as effective for electrons as ions [8]. It is difficult to inject electrons into a first order Fermi mechanism because the electron gyroradius is smaller than that of a proton of similar energy by a factor of $\sim .02$. Therefore, electron acceleration may take place by a different physical mechanism. In any case a separate explanation for initial relativistic electron injection is required.

It has been realized for some time that the supernova shock is collisionless. The relative velocity between supernova ejecta and ambient gas is approximately 10^4 kms^{-1}. It can be demonstrated that the stopping length of such a plasma cloud-cloud interaction due to the collisionless excitation of plasma waves by counter-streaming electrons is appreciably less than the collisional stopping length. When collisionless processes dominate the deceleration of a proton-electron plasma cloud incident protons are slowed by a polarization electric field and therefore their kinetic energy goes into the acceleration of target protons rather than into thermal energy. The dragging of electrons by protons in a collisionless shock increases the amount of kinetic energy available to accelerate electrons. Models of the radio source Cas A and a more recent model of SN 1993J [1] show that the relativistic electron energy is no greater than .1% of available kinetic energy.

Strong plasma turbulence is expected to lead to the formation of Langmuir solitons [9]. Such localized, oscillatory electric fields can accelerate electrons to relativistic energies. During plasma cloud-cloud interaction the level of plasma wave energy increases to the necessary threshold for plasma wave-wave instabilities such as the oscillating two stream instability to be excited. Subsequently the oscillating two stream instability and other parametric instabilities lead to the formation of Langmuir solitons [10, 11, 12]. The resultant electron acceleration and heating are so rapid that plasma wave energy levels are limited. The electric field of a Langmuir soliton is described approximately by the equation

$$E = E_0 \, cos \, \omega_p \, t \, sech \, (ax) \tag{1}$$

The above equation implies that the electric field oscillates at the plasma frequency ω_p and has spatial extent $\Delta x \sim 1/a$. The characteristic dimension of a soliton must be much larger than the Debye length λ_D. Ten Debye lengths is a reasonable estimate of its size. Because the electric field of a soliton is oscillatory only high speed electrons in the tail of the Maxwell-Boltzmann distribution experience substantial acceleration. When slower electrons traverse a soliton the sign of the electric field changes during transit. In a turbulent plasma the local electric field energy density can become comparable to the thermal energy density. If this condition is satisfied in a young supernova remnant such as SN 1993J then the soliton electric field will greatly exceed the Dreicer electric field and consequently runaway electrons are produced. For plausible plasma wave density levels and soliton dimensions it can readily be shown that electrons are rapidly accelerated to relativistic energies. It can also be shown that such acceleration continues after the electron becomes relativistic [12].

Radio emission was observed from SN 1993J shortly after the first peak in optical brightness. These observations showed that the strong radio emission came from an approximately spherically symmetric region of high ($\simeq$ 15,000 - 18,000 km s^{-1}) velocity [4, 5]. The stellar wind parameters of the red supergiant progenitor are estimated to be M = 10^{-4} M$\odot$/yr and v $\simeq$ 10 km s^{-1}. Recently we have given a model for the observed radio emission that assumes the relativistic electrons are accelerated by a collisionless shock front formed between the supernova ejecta and circumstellar gas [1]. Even if there is no pre-existing magnetic field the required magnetic field can be generated by plasma turbulence behind the shock front amplifying a seed magnetic field produced within the shock front. This model also interprets an observed high frequency break in the radio spectrum.

Previously we have argued that extragalactic jets emitted from massive black holes in AGN produce relativistic electrons in giant radio lobes as the result of collisionless interaction with ambient gas [10, 11]. If electron acceleration in these sources were caused by first order Fermi shock acceleration rather than by the plasma processes we have described then it should also be expected that relativistic proton energies in these sources should exceed those of electrons by a large ($\sim 10^2$) factor since protons are accelerated much more effectively by first order Fermi shock acceleration than electrons. Hopefully future observations will decide which of these alternatives is correct.

REFERENCES

1. Rose, W. K. 2000, in Cosmic Explosions, edited by S. S. Holt and W. W. Zhang (AIP: Woodbury).
2. Shklovskii, I. S. 1960, Sov. Astron., 4, 355
3. Lozinskaya, T. A. 1992, Supernovae and Stellar Wind in the Interstellar Medium (AIP: NY)
4. Bartel, N. et al. 1994, Nature, 368, 610
5. Marcaide, J. M. et al. 1995, Science, 270, 1475
6. Bell, A. R. 1978, MNRAS, 182, 147
7. Blandford, R. D. and Ostriker, J. P. 1978, ApJ, 221, L29
8. Blandford, R. D. 1988, in Supernova Remnants and the Interstellar Medium, edited by R. S. Roger and T. L. Landecker (Cambridge University Press: Cambridge)
9. Nicholson, D. R. 1983, Introduction to Plasma Theory (Wiley: N.Y.)
10. Rose, W. K., Guillory, J., Beall, J. H., Kainer, S. 1984, ApJ, 280, 550
11. Rose, W. K., Beall, J. H., Guillory, J., Kainer, S. 1987, ApJ, 314, 95
12. Rose, W. K. 1995 MNRAS, 276, 1191

Maximum Energies of Shock-Accelerated Electrons in Supernova Remnants in the Large Magellanic Cloud

S.P. Hendrick, S.P. Reynolds, & K.J. Borkowski

North Carolina State University, Raleigh, NC 27695, USA

Abstract.
Some supernova-remnant X-ray spectra show evidence for synchrotron emission from the extension of the electron spectrum producing radio synchrotron emission. For any remnant, if the extrapolated radio flux exceeds the observed X-ray flux (thermal or nonthermal), a rolloff of the relativistic-electron energy distribution must occur below X-ray emitting energies. We have studied the X-ray emission from a sample of 11 remnants in the Large Magellanic Cloud to constrain this rolloff energy. We assume that the electron distribution is a power law with an exponential cutoff at some E_{max} and radiates in a uniform magnetic field. If the radio flux and spectral index are known, this simple model for the synchrotron contribution depends on only one parameter which relates directly to E_{max}. Here we have modeled the X-ray spectra by adding a component for thermal radiation of a Sedov blast wave to the synchrotron model. For all 11 supernova remnants in this sample, the limits for E_{max} range between 10 and 70 TeV (for a mean magnetic field of 10 μG). We interpret E_{max} in the context of shock acceleration theories.

INTRODUCTION

Observations of the ion component of galactic cosmic rays show a straight power-law spectrum from a few GeV up to the "knee" at 1000 TeV. Shock acceleration of particles by supernova remnants is believed to create cosmic rays up to this energy. However, from radio synchrotron observations, we only learn about electron energies in the range 0.1 to 10 GeV. Both electrons and ions are accelerated by the shock, and are treated equally by the shock above 1 GeV, when all particles become relativistic. Between 5 GeV and 200 GeV, cosmic ray ions and electrons have power-law distributions that are consistent with the same slope at the cosmic ray source. Understanding the cosmic ray electrons should give us some information on the cosmic ray ions, which contain most of the energy in the cosmic ray spectrum.

The electron component of cosmic rays accelerated by a SNR shock can be cut off at high energies by any of several factors. Radiative (synchrotron and inverse-Compton) losses on electrons can cause the spectra to cut off. Finite shock age is

CP565, *Young Supernova Remnants: Eleventh Astrophysics Conf.*, edited by S. S. Holt and U. Hwang
© 2001 American Institute of Physics 0-7354-0001-6/01/$18.00

also a factor since the time required for a particle to be accelerated to a given energy by diffusive propagation back and forth across the shock is a sharply increasing function of particle energy. Finally, there can be a change in scattering due to the lack of long wavelength MHD waves around the shock. The first cutoff cause affects only the electron component, but finite age and change in scattering properties should affect ions and electrons equally.

X-ray observations have shown several SNRs with spectral evidence for nonthermal emission [1]. This nonthermal emission could be explained by extrapolating the radio synchrotron emission across the entire electromagnetic spectrum to X-rays. Comparing the X-ray flux to the extrapolated radio flux, we see that the electron spectrum must have steepened between the radio and X-ray regimes. We can determine at which frequency this steepening begins using a synchrotron cutoff model. The cutoff frequency is related to the maximum energy to which the shock can accelerate particles. Therefore, X-ray observations can put limits on the maximum energy, E_{max}, to which a SNR can accelerate electrons. The remaining question is: Do electron limits correspond to limits of ion acceleration?

DATA

The supernova remnants in this study (see Table 1) fall into two main categories: mature remnants, most likely in the Sedov evolutionary phase, and remnants from type Ia SNe, most of which exhibit Balmer lines in their optical spectrum, and which are probably in pre-Sedov phases. Remnants that appear to be in the Sedov phase include N23, N63A, N132D, and N49B [2]. Besides those objects, DEM 71 seems to be in a transition state from ejecta dominated spectra to a spectrum dominated by swept up ISM. 0534-69.9 has not been well studied, but its size and morphology (diffuse emission across the face of the shell) point toward Sedov phase. The other category is remnants of type Ia supernovae: 0453-68.5, N103B, DEM 71, 0509-67.5, 0519-69.0, and 0548-70.4, where the last four objects on this list also have an optical spectrum dominated by Balmer line emission [3]. 0453-68.5 and N103B are identified by characteristic lines in the X-ray spectra as type Ia. The youngest remnants in this survey are 0509-67.5, N103B, and 0519-69.0 [4]. Optical spectroscopy gives shock velocities which imply an age of 750-1500 years for 0519-69.0, and an upper limit of 1000 years for 0509-67.5 [3]. N132D has high velocity O-rich filaments within the shell, and both N103B and N63A contain optically bright knots.

Archival ACSA data were retrieved for each remnant, and processed with standard revision 2 criteria. The data for all four ASCA instruments, SIS0, SIS1, GIS2, and GIS3, were fit simultaneously. A complete discussion of the Sedov model used in this paper can be found in Borkowski, Lyerly, and Reynolds [5]. Details of the thermal fits will be presented in a later publication.

The possibility of synchrotron emission is modeled with a cutoff model (called *SRCUT* in XSPEC). This model assumes an exponential cutoff of the electron spectrum and is otherwise homogeneous. It should give the sharpest physically plausible cutoff in the synchrotron spectrum, allowing the most conservative upper limit on E_{max} and on the frequency ν_{cutoff} radiated by electrons with E_{max}. This rolloff frequency, along with the radio spectral index and flux at 1 GHz, are the only input parameters for the *SRCUT* model. Values for the spectral index and flux at 1 GHz have been documented in the literature [6], and were frozen at their values during fitting. In some cases, the flux at 1 GHz was extrapolated from the value of the flux at 408 MHz. The cutoff frequency ν_{cutoff} is related to E_{max} by

$$\nu_{cutoff} = 1.82 \times 10^{18} E_{max}^2 B \qquad (1)$$

DISCUSSION

The results of this study can be found in Table 1, where we have assumed a typical SNR magnetic field of 10 μG to convert ν_{cutoff} to E_{max}. None of the SNRs in this LMC sample is able to accelerate electrons in an unbroken power-law beyond 70 TeV. Observations of the spectrum of Galactic cosmic ray electrons indicate that the spectrum has already begun to fall below the cosmic ray ion spectra at 1 TeV. It is cosmic ray ions that extend out to the "knee" at 1000 TeV; we have no direct evidence of cosmic ray electrons near that energy. In this regard, our results that SNRs cannot accelerate cosmic ray electrons beyond 100 TeV without a break are consistent with Galactic cosmic-ray observations.

Theoretical models of shock acceleration predict that ions and electrons ought to have similar power law spectra in their putative source, SNR shocks. If the cutoffs we infer in the electron spectrum are due solely to radiative losses, then the ion spectrum will not be affected. However, the finite age of the shock, and a change in scattering, will affect both ions and electrons, so that the ion spectrum would not be able to extend beyond the electron spectrum. A cutoff of the electron spectrum below 100 TeV then casts into doubt the ability of supernova remnants to accelerate cosmic ray protons up to the "knee" at 1000 TeV.

Although there are no data on the cosmic ray spectrum of the LMC, the supernova remnant sample spans a wide range of evolutionary stages. Properties of these remnants should be similar to the properties of galactic SNRs. As opposed to a galactic survey, the oldest (and, presumably the largest) remnants in the LMC can still be observed in their entirety within a single ASCA observation. Young remnants have had a short time to accelerate particles, and they have a high limit for radiative losses. Older remnants, while having more time to accelerate particles, have much lower limits for radiative losses. These SNRs could become loss limited after a few thousand years. Therefore, older remnants could have an ion spectrum that extends beyond the maximum energies we have reported.

Object	$h\nu_{cutoff}$ (keV)	E_{max} (TeV)	Object	$h\nu_{cutoff}$ (keV)	E_{max} (TeV)
DEM 71	0.95	70	N63A	0.10	20
N49B	0.43	50	0534-69.9	0.06	20
N103B	0.26	40	0519-69.0	0.06	20
N23	0.22	30	0509-67.5	0.05	20
N132D	0.18	30	0453-68.5	0.02	10
0548-70.4	0.11	20			

TABLE 1. Cutoff frequency and maximum electron energy upper limits for supernova remnants in the LMC. The $h\nu_{cutoff}$ values are the 3σ upper limits of our fitted results. Note that while 10 μG was assumed for the magnetic field, actual SNR magnetic fields are quite uncertain.

CONCLUSION

This survey of LMC supernova remnants has found that the ability to accelerate electrons in the shock can be limited from the X-ray spectra. By determining the point at which the electron spectrum begins to steepen, we have found that 70 TeV is the upper limit for the maximum energy to which these LMC remnants can accelerate electrons in a straight power-law. With more detailed models and better X-ray data, the conservative upper limits on E_{max} found in this study will become lower. The question of particle acceleration by supernova remnants being able to accelerate particles up to the "knee" at 1000 TeV remains uncertain. The ion component could continue beyond the electron component, but finite shock age and change in scattering properties should affect both equally. It is unknown if there is a similar "knee" in the LMC cosmic ray spectrum, but the properties of LMC supernova remnants should be consistent with remnants in the galaxy. The fact that both the youngest and the oldest remnants in the LMC fail to accelerate electrons to within even an order of magnitude of the "knee" suggests that some revision may be needed in the picture of galactic cosmic ray origin.

REFERENCES

1. Koyama, K., et al. 1995, Nature, 378, 255; Koyama, K., et al. 1997, PASJ, 49, L7; Slane, P., et al. 1999, ApJ, 525, 357; The, L.-S., et al. 1996, AASupp, 120, C357
2. Hughes, J.P., Hayashi, I., & Koyama 1998, ApJ, 505, 732
3. Smith, R.C., Kirshner, R.P., Blair, W.P., & Winkler, P.F. 1991, ApJ, 375, 625
4. Hughes, J.P., et al 1995, ApJL, 444, L81
5. Borkowski, K.J, Lyerly, W.J., & Reynolds, S.P., 2000, ApJ, in press (astro-ph/0008066)
6. Mathewson, D.S, & Clarke, J.N., 1973, ApJ, 180, 725; Dickel, J.R., & Milne, D.K., 1995, AJ, 109, 200; Dickel, J.R., & Milne, D.K., 1998, AJ, 115, 1057; Filipovic, M.D., et al, 1998, A&A Supp, 127, 119

Spatially Varying X-ray Synchrotron Emission in SN 1006

Kristy K. Dyer, Stephen P. Reynolds, Kazik J. Borkowski[1], and Robert Petre[2]

[1] *North Carolina State University, Physics Department Box 8202, Raleigh, NC 27695, USA*
[2] *NASA's GSFC, LHEA Code 666, Greenbelt, MD 20771, USA*

Abstract.
    A growing number of both galactic and extragalactic supernova remnants show non-thermal (non-plerionic) emission in the X-ray band. New synchrotron models, realized as *SRESC* and *SRCUT* in XSPEC 11, which use the radio spectral index and flux as inputs and include the full single-particle emissivity, have demonstrated that synchrotron emission is capable of producing the spectra of dominantly non-thermal supernova remnants with interesting consequences for residual thermal abundances and acceleration of particles. In addition, these models deliver a much better-constrained separation between the thermal and non-thermal components, whereas combining an unconstrained powerlaw with modern thermal models can produce a range of acceptable fits. While synchrotron emission can be approximated by a powerlaw over small ranges of energy, the synchrotron spectrum is in fact steepening over the X-ray band. Having demonstrated that the integrated spectrum of SN 1006, a remnant dominated by non-thermal emission, is well described by synchrotron models I now turn to spatially resolved observations of this well studied remnant. The synchrotron models make both spectral and spatial predictions, describing how the non-thermal emission varies across the remnant. Armed with spatially resolved non-thermal models and new thermal models such as *VPSHOCK* we can now dissect the inner workings of SN 1006.

VERSIONS OF *SRESC* FOR SUBREGIONS OF SN 1006

    In Dyer et al. 2000 [1] we used new X-ray synchrotron models along with information from the radio synchrotron spectrum to determine with a new degree of accuracy the total amount of synchrotron emission in the spectrum of SN 1006. This analysis can be extended to regions of the supernova remnant (SNR). Here we separate the SNR into five regions – north and south limbs (northeast and southwest), north and south polar caps and center, as shown in Figure 1. Chandra observations of SNRs will be able to analyze much smaller regions, limited only by signal to noise, so these methods of spatial-spectral analysis of complex thermal and nonthermal emission, may be necessary for other SNRs.

CP565, *Young Supernova Remnants: Eleventh Astrophysics Conf.,* edited by S. S. Holt and U. Hwang
© 2001 American Institute of Physics 0-7354-0001-6/01/$18.00

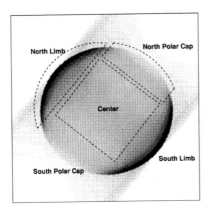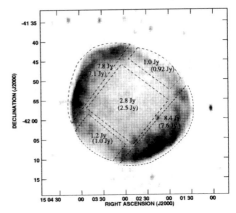

FIGURE 1. Left: Synchrotron emission for the simulated SNR. Right: The Molonglo + Parkes radio image with the subregions indicated. The value shown is the measured flux at 843 MHz, in parenthesis is the scaled 1 GHz value used in the model. Thanks to R.S. Roger for recovering the SN 1006 map.

The synchrotron models were developed [2,3] to describe the two dimensional distribution of synchrotron emission observed for a spherically symmetric SNR expanding in a uniform upstream magnetic field. The one-dimensional spectrum extracted for *SRESC* was was created by summing over the simulated SNR. Here we have developed subsets of the *SRESC* model: *LIMB*, *CENTER* and *CAP* (shown in Figure 1) by summing over discrete regions in the simulated SNR. The synchrotron emission from each region has a slightly different spectrum since it samples different populations of electrons interacting with the upstream magnetic field at different obliquity angles.

As in the *SRESC* model, these synchrotron models define the nonthermal X-ray emission with three parameters: the radio spectral index α, 1 GHz radio flux [Jy], and the rolloff [Hz]. The rolloff determines the curvature by specifying the frequency at which the model has dropped by a factor of ~ 6 from a powerlaw.

Many papers have investigated small changes in spectral index in radio synchrotron emission in SNRs [4]. However, flux uncertainties in interferometric measurements and nonlinear deconvolution processes have made unambiguous detections of these index changes difficult [5]. In any case, the expected spectral index variation ($\Delta\alpha \pm 0.01$) are expected to be small compared to the uncertainties in the spectral index ($\alpha \pm 0.2$). Here we assume the spectral index is constant across the SNR, a prediction that does agree with shock acceleration theory. For integrated spectral analysis the 1 GHz flux is readily available[1]. However, for spatially resolved

---

[1] Available for Galactic SNRs at Green's Catalogue
http://www.mrao.cam.ac.uk/surveys/snrs/index.html and for some Magellanic Cloud SNRs at
http://www.renaissoft.com/snrc/

TABLE 1. Predicted nonthermal X-rays in regions of SN 1006

Region	sub-models of SRESC[a]	Nonthermal Flux (% of total)
Center	CENTER	<16
North Polar Cap	CAP	13
South Polar Cap	CAP	8
North Limb	LIMB	37

[a] For $\alpha=0.60$ and rolloff (measured from the north limb) of 3×10^{17} Hz. See text for discussion.

spectral analysis the flux must be measured in each subregion. For large remnants such as SN 1006 this can be a problem. Maps from interferometers (such as the VLA) are unsuitable for measuring small scale fluxes, since interferometers are not sensitive to flux on angular scales beyond those corresponding to the smallest spacing. Correcting the image by adding a constant to bring it up to the single dish flux has not proven sufficiently accurate in small regions. The solution is to use images created with both interferometric and single dish data, which sample the flux at all spatial scales. Here we use a map made with the Molonglo interferometer plus the Parkes single dish telescope [6] to measure the flux in subregions. The curvature of the X-ray synchrotron is determined by *rolloff*. This factor was fixed by fitting a small region in the north limb expected to be dominated by synchrotron emission, and we have assumed, without evidence or theoretical expectation to the contrary, that this value is constant across the SNR.

RESULTS

These models reveal that there is significant nonthermal emission throughout SN 1006. Even the polar caps and center, which have prominent thermal lines, contain a synchrotron component (see Table 1). As might be expected, it is most difficult to determine the amount of synchrotron emission in regions where thermal emission dominates. Model fits of the center of SN 1006 will tolerate up to 16% of the flux in synchrotron. Differences in temperature and abundances between the fits may make it possible to argue for or against a synchrotron component for the center of the remnant.

Unfortunately there are no entirely satisfactory thermal models available to describe emission from Type Ia supernova remnants, such as SN 1006. Any attempts to model the thermal emission run into several complications: Young SNRs such as SN 1006 may require solutions intermediate between self-similar driven wave and Sedov [7]. Integrated fits to the full remnant showed high abundances [1] suggesting a reverse shock, so a two-shock description may be necessary. In light of this we use the plane shock model *VPSHOCK* as a starting point, making note of the fact it allows a range of ionization timescales, and can describe heavy element ejecta, but assumes a constant density and temperature.

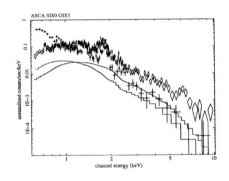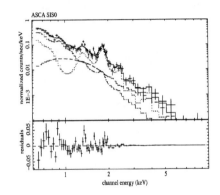

FIGURE 2. ASCA spectra of the South Polar Cap region of SN 1006. Left: GIS and SIS spectrum of regions of SN 1006 shown with the model predicted non-thermal emission Right: ASCA SIS with the *CAP* synchrotron model plus two plane parallel shocks (*VPSHOCK*), a low temperature with high abundances and a solar abundances with a higher temperature.

CONCLUSION

We have found that there is significant nonthermal emission in all regions of the SN 1006, not only the limbs. With these new models it is now possible to separate the thermal and nonthermal emission in spatially resolved regions (just in time for Chandra observations). This is an absolute prerequisite if the thermal emission is to be understood. This work allows us to address new questions, among them: TeV γ-rays have been observed from the north limb but not the south limb of SN 1006. Are there differences in the synchrotron emission from the two limbs? Preliminary results indicate that the curvature is different for synchrotron in the north and south limb. Hα is observed at the edge of the north polar cap region. Does the X-ray description of the thermal shock agree with optical observations?

REFERENCES

1. Dyer, K.K., Reynolds, S.P., Borkowski, K.J., Petre, R., Allen, G.E. 2001 *Separating Thermal and Nonthermal X-Rays in Supernova Remnants I: Total Fits to SN 1006 AD* **astro-ph/0010424** Accepted for the Astrophysical Journal March 10, 2001
2. Reynolds, S. P. 1996, ApJL, 459, L13
3. Reynolds, S. P. 1998, ApJ, 493, 375
4. Anderson. M. C., & Rudnick, L. 1993, ApJ, 408, 514
5. Dyer, K. K. & Reynolds, S. P. 1999, ApJ, 526, 365
6. Roger, R. S., Milne, D. K., Kesteven, M. J., Wellington, K. J. & Haynes, R. F. 1988, ApJ, 332, 940
7. Dwarkadas, V. V. & Chevalier, R. A. 1998, ApJ, 497, 807

The Non-Thermal X-Ray Emission of SN 1006 and the Implications for Cosmic Rays

G. E. Allen[1], R. Petre[2], and E. V. Gotthelf[3]

[1] MIT Center for Space Research, NE80-6029, Cambridge, MA 02139, USA
[2] Columbia Astrophysics Laboratory, Columbia University, Pupin Hall, 550 West 120th Street, New York, NY 10027, USA
[3] NASA/Goddard Space Flight Center, Laboratory for High Energy Astrophysics, Code 662, Greenbelt, MD 20771, USA

Abstract.

We present the results of a spectral analysis of *RXTE*, *ASCA*, and *ROSAT* data of SN 1006. These data were fit with several sets of thermal and non-thermal X-ray emission models to characterize the global spectral properties of the remnant. The present work represents the first attempt to model both the thermal and non-thermal X-ray emission over the entire X-ray energy band from 0.12–17 keV. The non-thermal X-ray spectrum is described by a broken power-law with low- and high-energy photon indices of 2.1 and 3.0, respectively. Since this spectrum steepens with increasing energy, our results support the claims that the emission is produced by synchrotron radiation from 100 TeV electrons. Using the radio and X-ray data, we estimate the parameters of the cosmic-ray electron, proton, and helium spectra. The results suggest that the ratio of the number densities of protons and electrons is 150 at 1 GeV and that the total energy in cosmic rays is 10^{50} erg. These results and the spectral index of the electrons at 1 GeV ($\Gamma_e = 2.14 \pm 0.12$) are consistent with the hypothesis that Galactic cosmic rays are predominantly accelerated in the shocks of supernova remnants. However, SN 1006 may or may not accelerate cosmic-ray protons to energies approaching the "knee" in the cosmic-ray spectrum.

INTRODUCTION

The search for evidence of the origin of Galactic cosmic rays has been an active area of research for many decades. While little evidence exists about the sites where very–high-energy nuclei are accelerated, the results of recent X-ray and gamma-ray observations indicate that at least some of the cosmic-ray electrons are accelerated in the shocks of supernova remnants [1–10]. For example, SN 1006 is one remnant for which there is evidence of the presence of cosmic-ray electrons that have been accelerated to energies as high as about 100 TeV [1,4]. In this paper, measurements

CP565, *Young Supernova Remnants: Eleventh Astrophysics Conf.*, edited by S. S. Holt and U. Hwang
© 2001 American Institute of Physics 0-7354-0001-6/01/$18.00

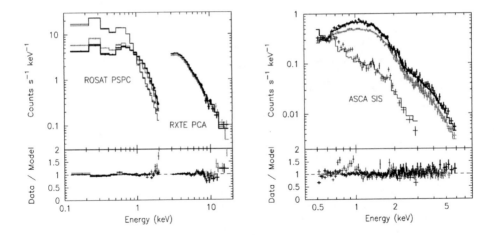

FIGURE 1. The *ROSAT* PSPC, *ASCA* SIS, and *RXTE* PCA data of SN 1006. The top panels include data for the south-western rim (thick black lines), north-eastern rim (thick gray lines), and center (thin black lines) of the remnant. The regions used for the PSPC and SIS data are mutually exclusive. The histograms through the data are the results of the best-fit model. The lower panels show the ratios of the data to the model. The model describes the entire 0.12–17 keV X-ray spectrum quite well.

of the X-ray and radio emission of the remnant are used to determine the parameters of the non-thermal cosmic-ray spectra.

DATA AND ANALYSIS

RXTE PCA, *ASCA* SIS, and *ROSAT* PSPC X-ray spectral data were simultaneously fit using several combinations of thermal and non-thermal emission components. The best-fit model includes broken–power-law and non-equilibrium–ionization components. The spectral data and the best-fit model are shown in figure 1. The fitted values of the interstellar column density, electron temperature, and ionization timescale are $n_H = 5.6 \times 10^{20}$ atoms cm^{-2}, $kT_e = 0.6$ keV, and $n_0t = 9 \times 10^9$ cm^{-3} s, respectively. The fitted relative abundance of silicon is 10–30 times larger than the solar abundance of silicon, which suggests that most of the X-ray–emitting silicon is reverse-shocked ejecta. The best fit values for the low-energy and high-energy photon indices of the broken power-law are $\Gamma_1 = 2.1 \pm 0.1$ and $\Gamma_2 = 3.0 \pm 0.2$, respectively. The difference between these two indices ≥ 0.7 at the 3 σ confidence level. Since this result shows that the non-thermal spectrum steepens with increasing energy, the result supports previous claims that the non-thermal X-ray emission from SN 1006 is produced by synchrotron radiation from 10-100 TeV electrons.

DISCUSSION AND CONCLUSIONS

The non-thermal X-ray spectrum and the available radio data of SN 1006 can be used to infer the parameters of the spectrum of the cosmic-ray electrons producing the synchrotron emission. For simplicity, the relativistic electron spectrum is assumed to have the form $dn_e/dE = A_e E^{-\Gamma_e} e^{-E/\epsilon_e}$. This form is the high-energy limit of Bell's expression [11]:

$$dn/dE = A g(E, m, \Gamma, \epsilon) \tag{1}$$
$$= A(E + mc^2)(E^2 + 2mc^2 E)^{-(\Gamma+1)/2} e^{-E/\epsilon}, \tag{2}$$

where $E\ (= (\gamma - 1)mc^2)$ is the kinetic energy of the particles and the exponential cutoff has been added to the formula of Bell. The parameters A, Γ, and ϵ have been estimated for cosmic-ray electrons, protons, and helium particles [12]. These parameters are listed in table 1 and the spectra are plotted in figure 2.

Table 1. Parameters of the Cosmic-Ray spectra.

Particle	A $(\mathrm{cm}^{-3}\ \mathrm{GeV}^{\Gamma-1})$	Γ	ϵ (TeV)
Electrons	2.4×10^{-9}	2.14 ± 0.12	10
Protons	1.0×10^{-6}	2.14	10
Helium	9.7×10^{-8}	2.14	20

Our assumptions about the shapes of the cosmic-ray spectra and about the relative numbers of non-thermal protons and electrons yield a ratio of the number of cosmic-ray protons to the number of cosmic-ray electrons $n_p^{cr}/n_e^{cr} = 150$ at 1 GeV (fig. 2). Within the rather large uncertainties of our estimates, this ratio is consistent with the ratio observed at Earth [13]. The energies in cosmic-ray electrons, protons, and helium nuclei can be obtained from equation 2 and the particle-spectra parameters described in table 1. The sum of these three energies $E_{cr} = 1 \times 10^{50}$ erg. These results and the differential spectral index of the electrons ($\Gamma_e = 2.14 \pm 0.12$) are consistent with the hypothesis that Galactic cosmic rays are accelerated predominantly in the shocks of supernova remnants. However, the inferred maximum energy of cosmic-ray electrons in SN 1006 ($\epsilon_e = 10$ TeV) is well below 100 TeV [14]. If the maximum energy of cosmic-ray protons $\epsilon_p = \epsilon_e$, the remnant does not accelerate particles to the energy of the "knee" at about 3000 TeV [15]. Yet, if the magnetic field strength is substantially larger than 10μG, the maximum energy of the electrons may be limited by synchrotron losses and the maximum energy of nuclei might be significantly larger than 10 TeV. Therefore, SN 1006 is clearly a significant source of Galactic cosmic rays (at least cosmic-ray electrons), but the remnant may or may not be capable of accelerating particles to energies as high as the energy of the knee. Similar conclusions are reported for other young Galactic shell-type remnants [3,5,15].

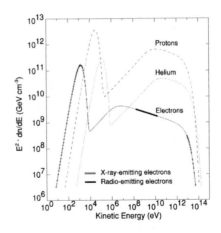

FIGURE 2. Estimates of the cosmic-ray electron, proton, and helium spectra of SN 1006. The low-energy and high-energy ends of the electron spectrum produce the observed thermal and non-thermal X-ray emission, respectively. The GeV electrons produce the observed radio emission. The ratio of the number densities of protons and electrons at 1 GeV is about 150, which is consistent with the ratio observed at Earth. The total cosmic-ray energy is dominated by the energy of the cosmic-ray protons.

We thank Stephen Reynolds for many thoughtful and informative discussions about particle acceleration in supernova remnants and about the photon emission processes of the accelerated particles.

REFERENCES

1. Koyama, K., et al. 1995, Nature, 378, 255
2. Koyama, K., et al. 1997, PASJ, 49, L7
3. Allen, G. E., et al. 1997, ApJ, 487, L97
4. Tanimori, T., et al. 1998, ApJ, 497, L25
5. Allen, G. E., Gotthelf, E. V., and Petre, R. 1999, in Proc. 26th Int. Cosmic Ray Conf. (Salt Lake City), 3, 480 (available at http://xxx.lanl.gov/abs/astro-ph/9908209)
6. Vink, J., et al. 1999, A&A, 344, 289
7. Slane, P., et al. 1999, ApJ, 525, 357
8. Muraishi, H., et al. 2000, A&A, 354, L57
9. Borkowski, K., et al. 2000, ApJ, submitted
10. Dyer, K. K., et al. 2000, in press
11. Bell, A. R. 1978, MNRAS, 182, 443
12. Allen, G. E., Petre, R., and Gotthelf, E. V. 2000, ApJ, submitted
13. Meyer, P. 1969, ARAA, 7, 1
14. Lagage, P. O., and Cesarsky, C. J. 1983, A&A, 125, 249
15. Reynolds, S. P., and Keohane, J. W. 1999, ApJ, 525, 368

X-ray Synchrotron in the Forward Shock of Cassiopeia A

F. Berendse[1,2], S.S. Holt[3], R. Petre[2], and U. Hwang[1,2]

[1] *University of Maryland College Park, College Park, MD 20742, USA*
[2] *NASA Goddard Space Flight Center, Greenbelt, MD 20771, USA*
[3] *F. W. Olin College of Engineering, Needham, MA 02492, USA*

Abstract.
We present high spatial resolution spectra of the forward shock of Cassiopeia A from a 50 ks observation obtained by the Advanced CCD Imaging Spectrometer on board the Chandra X-ray Observatory. We systematically compare the spectra of several regions of the forward shock and find that single-component non-equilibrium ionization models do not provide good fits to the spectra. Our preliminary results indicate that it is necessary to add a nonthermal, synchrotron component that dominates the continuum of the spectrum above 3 keV. We find that the shape of this nonthermal continuum changes from region to region which may indicate variable acceleration conditions within the remnant.

INTRODUCTION

Supernova remnants (SNRs) have long been believed to be a source of cosmic rays below the "knee" of the observed cosmic ray spectrum at 1000 TeV. The primary evidence for this has been the observed nonthermal radio spectrum of the shell of young remnants believed to originate from electrons accelerated by a first-order Fermi mechanism to GeV energies. However, evidence of the acceleration of cosmic rays to TeV energies in SNRs was lacking until the discovery of x-ray synchrotron radiation in the shell SN 1006 by Koyama et al. [1]

The first report of x-ray synchrotron emission in Cas A came from Allen et al. [2] based on their observations made with the Rossi X-ray Timing Explorer (RXTE). Their observations discovered a nonthermal power-law spectrum between 20-50 keV and concluded that this spectrum must be x-ray synchrotron — having ruled out other possible mechanisms such as inverse Compton and nonthermal bremsstrahlung.

RXTE does not have any imaging capability, so exactly where within Cas A the x-ray synchrotron emission is originating is still unknown. Studies with imaging x-ray observatories have concentrated on looking for a small nonthermal component within the bright, thermal reverse shock. A calibration observation of Cas A with

CP565, *Young Supernova Remnants: Eleventh Astrophysics Conf.*, edited by S. S. Holt and U. Hwang
© 2001 American Institute of Physics 0-7354-0001-6/01/$18.00

the Chandra X-ray Observatory brought forth evidence of a faint forward shock. [3] It is the primary goal of this work to determine if the synchrotron emission discovered by Allen et al. is originating in this recently-revealed forward shock.

DATA AND ANALYSIS

We observed Cas A with the back-side illuminated chip (S3) of the ACIS on board the Chandra X-ray Observatory on January 30-31, 2000. The effective integration time was 50 ks– ten times longer than the observation done during Chandra's calibration phase. The background subtraction was performed with a same-field background spectrum within close to the region of interest.

The thermal emission of the regions was modeled with the plane-parallel shock model of Borkowski et al. [4] This model assumes the thermal x-ray emission is coming from a plasma shocked by a plane-parallel shock. It is understood that this plasma has not yet reached ionization equilibrium, so in addition to a plasma temperature, it is also necessary to characterize the degree of ionization by introducing an ionization timescale $n_0 t$, where n_0 is the pre-shock density and t is the time since passing through the shock.

The x-ray synchrotron emission was modeled using a simplified version of Reynolds' synchrotron model. [5] This model assumes a power-law distribution with an exponential cutoff. Folding this distribution into the synchrotron emissivity of a single electron results in a power law distribution of photons below a break frequency, but a cutoff that is slower than exponential above the break frequency. The resulting photon spectrum can be parameterized by a radio spectral index α ($f_\nu = \nu^\alpha$), and a cutoff frequency.

THE CASE FOR A NONTHERMAL COMPONENT

To determine if a nonthermal component was necessary, we systematically fitted spectra from nine different regions to four different models. The regions chosen for this analysis are outlined in Figure 1. Three thermal-only models were used– one assuming equipartition of electrons and ions (i.e. $T_e = T_i$) and two assuming non-equipartition ($T_e = 0.5 T_i$ and $T_e = 0.1 T_i$). The fourth model was a thermal model in equipartition combined with x-ray synchrotron model. In the synchrotron component, the radio spectrum of Cas A was assumed to be $\alpha = 0.77$, taken from Green's Catalog of Galactic Supernova Remnants. [6]

In each region, it was necessary to have a synchrotron component in order to obtain a good fit to the data. The synchrotron model accounts for all the flux above 3 keV, except in the east and west regions, where an Fe K line at 6.7 keV can be seen. The synchrotron component also contributes significantly to the continuum emission below 3 keV. Table 1 shows the fitted ratio of nonthermal to thermal flux between 3 and 10 keV.

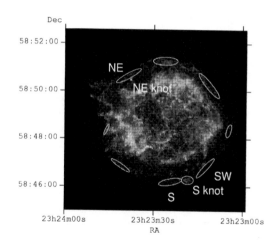

FIGURE 1. Regions of Cas A extracted for spectral analysis.

TABLE 1. Best fit parameters of nonthermal component.

	S	SE	SW	S knot	
nonthermal flux [a]	1.7×10^{-12}	1.7×10^{-12}	5.3×10^{-13}	1.5×10^{-12}	
thermal flux	2.5×10^{-14}	7.2×10^{-14}	2.0×10^{-14}	4.2×10^{-14}	
nonth./thermal flux	68	24	27	36	
break freq. $\times 10^{17}$ Hz	$3.6 \begin{pmatrix} 2.8 \\ 3.8 \end{pmatrix}$	$17 \begin{pmatrix} 10 \\ 29 \end{pmatrix}$	$1.4 \begin{pmatrix} 1.0 \\ 1.7 \end{pmatrix}$	$1.4 \begin{pmatrix} 1.2 \\ 1.8 \end{pmatrix}$	
	N	NE	NW	E	W
nonthermal flux	4.0×10^{-12}	3.0×10^{-12}	3.7×10^{-12}	6.3×10^{-13}	4.3×10^{-13}
thermal flux	1.2×10^{-13}	7.8×10^{-14}	8.8×10^{-14}	2.9×10^{-13}	1.3×10^{-13}
nonth./thermal	33	38	42	2.1	3.3
break freq. $\times 10^{17}$ Hz	1.7	$13.5 \begin{pmatrix} 9.5 \\ 19.5 \end{pmatrix}$	$2.1 \begin{pmatrix} 1.6 \\ 2.3 \end{pmatrix}$	0.86	$0.52 \begin{pmatrix} 0.38 \\ 0.76 \end{pmatrix}$

[a] Fluxes are in units of ergs/cm^2 s at 3-10 keV. Values in parentheses are 90% confidence intervals.

There appear to be some differences in the continuum shape from region to region. If the continuum above 3 keV is nonthermal as we suggest, then we may be seeing a variation in either the photon spectral index α or in the break frequency. If this is the case, then this would be the first discovery of a variation of nonthermal electron conditions within a SNR.

In light of this discovery, we found it necessary to allow the break frequency to vary as a free parameter. We present the best-fit break frequencies in Table 1 along with nonthermal and thermal fluxes derived from our model fits. We also replace the Reynolds model with a power law to give an idea of the spectral index variation between the regions in the table as well. We find that the break frequency is constant for most regions and is lower than the maximum break frequency found by Keohane & Reynolds. [7] However, two regions, the northeast and the southeast, have break frequencies an order of magnitude higher than the rest of the remnant.

FEATURELESS SPECTRUM OF THE NE KNOT

We extracted a tenth spectrum from an x-ray bright knot immediately behind the NE region. This featureless spectrum can be well-fitted ($\chi^2/\nu = 110/117$) by Reynolds' synchrotron model with a break frequency of 1.7 (+1.0/-0.5) $\times 10^{17}$ Hz assuming a radio spectral index of $\alpha = 0.77$. Hughes et al. also report a featureless spectrum in a western region of the reverse shock. These featureless spectra show that cosmic ray acceleration in Cas A is not likely isolated to one spot, but is taking place throughout the remnant and give additional proof that it is plausible for the forward shock to be dominated by nonthermal emission.

We also note that this featureless spectrum argues against a thermal-only interpretation of the forward shock spectra due to its location immediately behind the shock front. A featureless thermal spectrum would have a lower ionization timescale than a spectrum with Mg, Si, and S lines we see in the northeast shock front. Therefore, a thermal interpretation would have the featureless spectrum closer to the forward shock, but we observe the opposite.

CONCLUSIONS

During this preliminary analysis of the forward shock of Cas A, we have come to a few qualitative conclusions. The forward shock of Cas A appears to be dominated in the continuum above 3 keV by nonthermal emission. We have ruled out an thermal-only model for the spectrum of the forward shock. Based on the work done by Allen et al., we conclude the spectrum is dominated by x-ray synchrotron radiation from electrons accelerated by the forward shock. We have also discovered spatial variations in the continuum shape above 3 keV. This may be due to spatial variations in the conditions which affect the acceleration of electrons to TeV energies. We chose to interpret this as a variation in the break frequency and find it is constant for most regions, but two regions have break frequencies which vary by an order of magnitude.

REFERENCES

1. Koyama S., et al. 1995, Nature, 378, 255
2. Allen G. et al. 1997, ApJ, 487, L97
3. Gotthelf E. et al. 2000, BAAS, 195, 112.06
4. Borkowski K. et al. 2000, astro-ph/0008066
5. Reynolds S. 1998, ApJ, 493, 375
6. Green D. 2000, 'A Catalogue of Galactic Supernova Remnants (2000 August version)', Mullard Radio Astronomy Observatory, Cavendish Laboratory, Cambridge, UK
7. Keohane J. and Reynolds S. 1999, ApJ, 525, 368

Nonthermal X-Ray Emission from G266.2–1.2 (RX J0852.0–4622)

P. Slane[1], J. P. Hughes[2], R. J. Edgar[1], P. P. Plucinsky[1], E. Miyata[3], H. Tsunemi[3], and B. Aschenbach[4]

[1] Harvard-Smithsonian Center for Astrophysics, Cambridge, MA 02138, USA
[2] Rutgers University, Piscataway, NJ 08854-8019, USA
[3] Osaka University, Osaka 560-0043, Japan
[4] Max-Planck-Institut für extraterrestrische Physik, Garching, Germany

Abstract. The newly discovered supernova remnant G266.2–1.2 (RX J0852.0–4622), along the line of sight to the Vela SNR, was observed with ASCA for 120 ks. We find that the X-ray spectrum is featureless, and well described by a power law, extending to three the class of shell-type SNRs dominated by nonthermal X-ray emission. Although the presence of the Vela SNR compromises our ability to accurately determine the column density, the GIS data appear to indicate absorption considerably in excess of that for Vela itself, indicating that G266.2-1.2 may be several times more distant. An unresolved central source may be an associated neutron star, though difficulties with this interpretation persist.

INTRODUCTION

G266.2–1.2 was discovered by Aschenbach [1] using data from the ROSAT All-Sky Survey. Situated along the line of sight to the Vela SNR, the emission stands out above the soft thermal emission from Vela only at energies above ~ 1 keV. Iyudin et al. [2] reported the detection of ^{44}Ti from the source GRO J0852-4642, which was tentatively associated with the SNR. If correct, the very short ^{44}Ti lifetime ($\tau \approx 90$ y) would imply a very young SNR, and the large angular size would require that the remnant be very nearby as well. Estimates based on the X-ray diameter and γ-ray flux of ^{44}Ti indicate an age of ~ 680 y and a distance of ~ 200 pc [3]. The hard X-ray spectrum, if interpreted as high temperature emission from a fast shock, would seem to support this scenario. However, as we summarize here (see [4] for a detailed discussion), the X-ray emission is not from hot, shock-heated gas; it is nonthermal. Further, reanalysis of the COMPTEL data finds that the ^{44}Ti detection is only significant at the $2 - 4\sigma$ confidence level [5]. In the absence of such emission, and given that the X-ray emission is nonthermal, the nearby distance and young age may need to be reexamined.

CP565, *Young Supernova Remnants: Eleventh Astrophysics Conf.*, edited by S. S. Holt and U. Hwang
© 2001 American Institute of Physics 0-7354-0001-6/01/$18.00

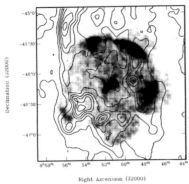 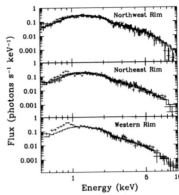

FIGURE 1. Left: ASCA GIS image of G266.2–1.2 ($E = 0.7 - 10$ keV). The image consists of a mosaic of 7 individual fields. Contours represent the outline of the Vela SNR as seen in ROSAT survey data with the PSPC. Right: ASCA spectra from both GIS detectors for regions of G266.2–1.2. The featureless spectrum is well described by a power law. Excess flux at low energies is presumably associated with soft thermal emission from the Vela SNR.

OBSERVATIONS AND ANALYSIS

We have carried out ASCA observations of G266.2–1.2 using 7 pointings, each of ~ 17 ks duration. The resulting image from the ASCA GIS is illustrated in Fig. 1 [6], where we also present GIS spectra from three regions along the rim of the remnant: 1) the bright NW rim; 2) the NE rim; and 3) the W rim. Unlike the line-dominated thermal emission typical of a young SNR, the spectra are featureless and well described by a power law of index ~ 2.6. G266.2–1.2 thus joins SN 1006 [7] and G347.3–0.5 [8,9] as a shell-type SNR dominated by nonthermal emission in X-rays. Weaker nonthermal components are observed for other SNRs as well, indicating the shock-acceleration of particles to energies of order $10 - 100$ TeV.

The best-fit spectral parameters for G266.2–1.2 yield $N_H \sim (1 - 4) \times 10^{21}$ cm^{-2}. We include a soft thermal component with the column density fixed at a low value appropriate for Vela [10]. The column density for the power law component is significantly higher than that for Vela. While simple scaling of the column density to estimate the distance to G266.2–1.2 is clearly rather uncertain, it would appear that the remnant is at least several times more distant than Vela; improved column density measurements are of considerable importance.

CO data [11] reveal a concentration of giant molecular clouds – the Vela Molecular Ridge – at a distance of $\sim 1 - 2$ kpc in the direction of Vela. The column density through the ridge is in excess of 10^{22} cm^{-2}, with the NE rim of G266.2–1.2 (nearest the Galactic Plane) falling along the steeply increasing column density region while the western rim lies along a line of much lower N_H (Fig. 2); the CO column density varies by more than a factor of 6 between these regions. The lack of a strong variation in N_H across G266.2–1.2 indicates that the remnant cannot be more distant than the Ridge. On the other hand, if N_H is much larger than for

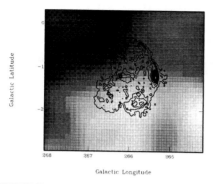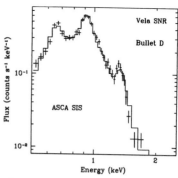

FIGURE 2. Left: CO emission ($V_{LSR} = -5 - +20$ km s^{-1}) in the direction of G266.2–1.2. Contours are GIS data. Note the Galactic coordinates. Right: ASCA SIS spectrum from Vela "Bullet D" showing thermal nature of emission, in contrast to nonthermal emission from G266.2–1.2.

the Vela SNR, as the GIS data reported here indicate, then G266.2–1.2 must be as distant as possible consistent with being in front of most of the Molecular Ridge gas.

In a recent study of optical emission in the vicinity of G266.2–1.2, Redman et al. [12] discuss a filamentary nebula lying directly along the edge of the so-called Vela "bullet D" region [13], just outside the eastern edge of Fig. 1 (in the direction of the extended contour), and argue that both structures represent a breakout region from G266.2–1.2. However, the ASCA spectrum of this region (Fig. 2) is clearly thermal, with large overabundances of O, Mg, and Si – quite in contrast to the nonthermal spectrum observed for G266.2–1.2. We conclude that an association between this region and G266.2–1.2 is unlikely.

The central region of the remnant contains a compact source, which we designate as AX J0851.9-4617.4, surrounded by diffuse emission that extends toward the northwest. The diffuse emission is well described by a power law of spectral index ~ 2.0, again accompanied by a soft thermal component that we associate with Vela. Given the nonthermal nature of the shell emission, however, it is quite possible that the diffuse central emission is associated with emission from the shell projected along the central line of sight. We note that the spectrum of this central emission appears harder than that from the rest of the remnant, perhaps suggesting a plerionic nature, but more sensitive observations are required to clarify this.

The spectrum of the compact source is rather sparse and can be adequately described by a variety of models. A power-law fit yields a spectral index that is rather steep compared with known pulsars. The luminosity, on the other hand, is quite reasonable for a young neutron star if the distance is indeed several kpc, and the observed column density is compatible with that for the SNR shell. A blackbody model leads to an inferred surface temperature of ~ 0.5 keV with an emitting region only $\sim 200 d_{kpc}$ m in radius which could be indicative of emission from compact polar cap regions of a neutron star, although this region is somewhat large for such a scenario. It is possible that the source is associated with a Be star within the position error circle [1], although the X-ray and optical properties appear inconsistent for such an interpretation [4].

Recent SAX observations [14] have revealed the presence of another source in the central region of G266.2–1.2. This source, designated SAX J0852.0-4615, has a harder spectrum than AX J0851.9-4617.4, but is considerably fainter. If at a distance of $\sim$ 3 kpc, its luminosity would be similar to the Vela pulsar, but this would place the source beyond the Vela Molecular Ridge. Further observations are required to investigate these central sources more completely.

CONCLUSIONS

The ASCA observations of G266.2–1.2 reveal that the soft X-ray emission from this SNR is dominated by nonthermal processes. This brings to three the number of SNRs in this class, and provides additional evidence for shock acceleration of cosmic rays in SNRs. Because of the bright and spatially varying background caused by the Vela SNR, limits on the thermal emission from G266.2–1.2 are difficult to establish. However, the ASCA data suggest a larger column density for this remnant than for Vela, indicating that G266.2–1.2 is at a larger distance, and perhaps associated with the star formation region located at a distance of $1 - 2$ kpc. The point source AX J0851.9-4617.4 could represent an associated neutron star, with the diffuse emission around the source being an associated synchrotron nebula, albeit a very faint one. Recent SAX studies reveal another possible candidate. Higher resolution X-ray observations of these sources are of considerable importance.

Acknowledgments. This work was supported in part by NASA through contract NAS8-39073 and grant NAG5-9106.

REFERENCES

1. Aschenbach, B. 1998, Nature 396, 141
2. Iyudin, A. F. et al. 1998, Nature 326, 142
3. Aschenbach, B., Iyudin, A. F., & Schönfelder, V. 1999, A&A 350, 997
4. Slane, P. et al. 2001 – to appear in ApJ (see astro-ph/0010510)
5. Schönfelder, V. et al. 2000, 5th Compton Symposium, Portsmouth, AIP Conf. Proc. 510, p. 54, eds. M.L. McConnell and J.M. Ryan
6. Tsunemi, H. et al. 2000, PASJ, in press (astro-ph/0005452)
7. Koyama, K. et al. 1995, Nature, 378, 255
8. Koyama, K. et al. 1997, PASJ 49, L7
9. Slane, P. et al. 1999, ApJ, 525, 357
10. Bocchino, F., Maggio, A., & Sciortino, S. 1999, A&A 342, 839
11. May, J., Murphy, D. C., & Thaddeus, P. 1988, A&ASS 73, 51
12. Redman, M. P., et al. 2000, ApJ, 543, L153
13. Aschenbach, B., Egger, R., Truemper, J. 1995, Nature, 373, 587
14. Mereghetti, S. 2001, ApJ - accepted (see astroph/0011554)

Some Special Examples

Circumstellar Nebulae in Young Supernova Remnants

You-Hua Chu

Astronomy Department, University of Illinois, 1002 W. Green Street, Urbana, IL 61801, USA

Abstract. Supernovae descendent from massive stars explode in media that have been modified by their progenitors' mass loss and UV radiation. The supernova ejecta will first interact with the *circumstellar* material shed by the progenitors at late evolutionary stages, and then interact with the *interstellar* material. Circumstellar nebulae in supernova remnants can be diagnosed by their small expansion velocities and high [N II]/Hα ratios. The presence of circumstellar nebulae appears ubiquitous among known young supernova remnants. These nebulae can be compared to those around evolved massive stars to assess the nature of their supernova progenitors. Three types of archeological artifacts of supernova progenitors have been observed in supernovae and/or young supernova remnants: (1) deathbed ejecta, (2) circumstellar nebulae, and (3) interstellar bubbles. Examples of these three types are given.

INTRODUCTION

Massive stars end their lives in supernova explosions, but it is not known exactly at which evolutionary stages massive stars explode. SN 1987A was the only supernova that had a known progenitor, Sk−69°202 [1], and its B3I spectral type was as shockingly different from theorists' expectation as could be. It was eventually recognized that Sk−69°202 was probably a binary, and the companion played a vital role in the stellar mass loss and stellar evolution before the supernova explosion [2]. At present, our knowledge of the transition from massive star to supernova is sketchy as best.

Massive stars lose mass throughout their lives in the forms of fast and slow winds. The mass loss rate heightens particularly toward the late evolutionary stages. The ejected stellar material has been detected as "circumstellar nebulae" around Wolf-Rayet (WR) stars, luminous blue variables (LBVs), and blue supergiants (BSGs) [3,21]. The circumstellar nebulae around these stars are visible because these stars have stellar winds to compress the circumstellar material into dense shells and UV fluxes to ionize the nebulae. It is observed that circumstellar nebulae around different types of stars have different sizes, expansion velocities, and abundances, depending on the history of the stellar mass loss [5].

CP565, *Young Supernova Remnants: Eleventh Astrophysics Conf.*, edited by S. S. Holt and U. Hwang

Circumstellar nebulae can also be detected in young SNRs before they are completely shredded beyond recognition by SNR shocks. The best known example of circumstellar material in a young SNR is the quasi-stationary flocculi (QSFs) in Cas A [6]. Noting a similarity in the kinematic properties and [N II]/Hα line ratios between these QSFs and the Galactic WR nebula NGC 6888, Kirshner and Chevalier [6] suggested that the progenitor of Cas A was a WN star. Similarly, if circumstellar nebulae are detected in other young SNRs, the physical properties of these nebulae can be compared to those of known nebulae around massive stars in order to determine the nature of the supernova progenitors.

In this paper, I will first give a brief review on the mass loss from massive stars and how the mass loss modifies the gaseous environment, then describe the different types of circumstellar nebulae and interstellar bubbles that young SNRs interact with, and finally give examples of young supernovae or SNRs that show such "archeological artifacts" of their massive progenitors.

EVOLUTION, MASS LOSS, AND BUBBLES OF MASSIVE STARS

The most massive stars, with $M_{ZAMS} > 40$ $M_\odot$, will evolve along the sequence O $\rightarrow$ Of $\rightarrow$ H-rich WN $\rightarrow$ LBV $\rightarrow$ H-poor WN $\rightarrow$ H-free WN $\rightarrow$ WC $\rightarrow$ supernova [7]. The less massive stars, on the other hand, evolve through the red supergiant (RSG) phase instead of the LBV phase [8]. Of these different evolutionary stages, fast stellar winds ($> 1,000$ km s^{-1}) are seen in main sequence O stars, WR stars, and BSGs; moderately fast winds ($< 1,000$ km s^{-1}) are seen in LBVs; and slow winds (~ 20 km s^{-1}) are seen in RSGs.

These various stellar winds interact with the ambient medium and form wind-blown bubbles [9]. The hydrodynamic evolution of these bubbles has been calculated for stars evolving along the sequence O star $\rightarrow$ LBV $\rightarrow$ WR star, e.g., a 60 $M_\odot$ star [10], and for stars evolving along the sequence O star $\rightarrow$ RSG $\rightarrow$ WR star, e.g., a 35 $M_\odot$ star [11].

The results of these calculations can be summarized as follows. Massive stars form "interstellar bubbles" during the main sequence stage, as the bubbles consist of mainly interstellar material. The copious mass loss during the LBV or RSG phase forms a small, slowly-expanding, circumstellar nebula within the cavity of the interstellar bubble. During the final WR phase, the fast stellar wind sweeps up and compresses the circumstellar material to form a "circumstellar bubble," which consists of mainly ejected stellar material.

Interstellar and circumstellar bubbles have both been observed around WR stars, LBVs, and BSGs [5]. Interstellar and circumstellar bubbles can be easily distinguished according to their patterns of expansion and abundances – circumstellar bubbles show regular expansion patterns and high N/O abundance ratios. In general, interstellar bubbles are large, up to a few tens of pc in radius. Circumstellar bubbles of WR stars are usually a few pc in radius and expand fast, $V_{exp} \geq 50$

km s^{-1}. Circumstellar bubbles of LBVs are small, ≤ 1 pc in radius; some expand slowly, $V_{exp} < 30$ km s^{-1}, while others expand fast with $V_{exp} \sim 50$–100 km s^{-1}. η Car is the most extreme but rare case [12].

DIAGNOSTICS OF CIRCUMSTELLAR NEBULAE

Circumstellar nebulae consist of material ejected by stars. Therefore, the nebular expansion is usually regular and shows point-symmetry with respect to the central star. Furthermore, the elemental abundances show enrichment in CNO products, i.e., high N/O ratios [13]. For normal nebular conditions, a high N abundance leads to high [N II]/Hα line ratios. It is thus easy to diagnose a circumstellar nebula using the expansion pattern and [N II]/Hα ratios.

Circumstellar nebulae in young SNRs can be detected using high dispersion spectroscopic observations of the Hα and [N II] λ6548 lines. The presence of a narrow emission component with anomalously high [N II]/Hα ratio would indicate the existence of a circumstellar nebula.

CIRCUMSTELLAR NEBULAE AND INTERSTELLAR BUBBLES IN YOUNG SNRS

Young SNRs contain a rich variety of archeological artifacts left behind by their massive progenitors. Three distinct types of circumstellar and interstellar nebulae have been observed: (1) deathbed ejecta, (2) circumstellar nebulae, and (3) interstellar bubbles. Descriptions and examples of these nebulae are given in the subsections below.

1. Deathbed Ejecta

Some supernovae, after the supernova light has faded, show narrow emission lines from circumstellar nebulae that are characterized by very high densities, $\gg 10^6$ H-atom cm^{-3}, and moderate expansion velocities, < 100 km s^{-1}. The best examples are SN 1987K [14], SN 1997ab [15], and SN 1997eg [16]. The Hα and [N II] lines of SN 1978K and SN 1997ab are shown in Figure 1.

The Hα+[N II] spectrum of SN 1978K shows a narrow, nebular Hα component superposed on the peak of the broad Hα component of the SN ejecta. The narrow nebular [N II]$\lambda\lambda$6548,6583 lines are also detected. The velocity profiles of the nebular components imply an expansion velocity < 50 km s^{-1}. The [N II]/Hα ratio is ~ 1. This nebular emission must originate from a circumstellar nebula. Low-dispersion spectra suggest a nebular density of a few $\times 10^5$ H-atom cm^{-3} [17].

The Hα spectra of SN 1997ab and SN 1997eg both have a narrow, nebular, P Cygni profile superposed on the broad profile of the SN ejecta [15,16]. The

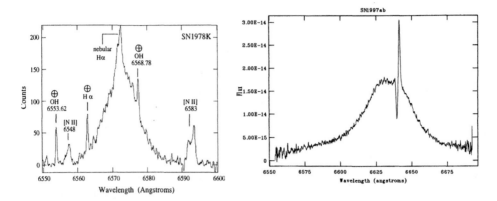

FIGURE 1. Left: High-dispersion Hα+[N II] spectrum SN 1987K, taken from [14]. **Right:** High-dispersion Hα spectrum of SN 1997ab, taken from [15], with permission from Blackwell Science Ltd.

[N II]/Hα ratio of the nebular component is high (Salamanca, private communication), and the expansion velocity of the nebular component is < 100 km s^{-1} [15,16], suggesting the existence of a circumstellar nebula. The P Cygni profile of the nebular component requires that the density is $\geq 10^7$ H-atom cm^{-3}.

The densities of the circumstellar nebulae around these three supernovae are orders of magnitude higher than the densities observed in circumstellar bubbles of WR stars [14]. The density of SN 1978K's circumstellar nebula is comparable to those of the densest LBV nebulae, but the densities of SN 1997ab's and SN 1997eg's circumstellar nebulae are higher than any known circumstellar nebulae. These circumstellar nebulae, therefore, must have a different origin from the known circumstellar nebulae around evolved massive stars. They may represent the stellar material ejected shortly prior to the supernova explosion, justifying the name "deathbed ejecta."

It would be of great interest to determine the elemental abundances of these circumstellar nebulae, and compare them to the nucleosynthesis yields of stellar evolution models. The results will shed light on the evolutionary stage of the SN progenitor.

2. Circumstellar Nebulae

The Case of SN 1987A

The first example given here is SN 1987A, where we are witnessing the collision between supernova ejecta and a circumstellar nebula. HST WFPC2 images of SN 1987A taken after the SN light had faded away revealed an inner ring and two outer rings (Figure 2). The [N II]/Hα ratios of these rings are 4.2 and 2.5 for

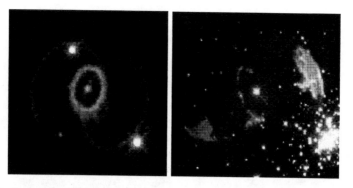

FIGURE 2. Left: SN 1987A and its circumstellar rings. **Right:** Sher 25 and its circumstellar rings. The progenitor of SN 1987A, Sk–69 202, was a B3 I star and Sher 25 is a B1.5 I star. Both inner rings are 0.4 pc in diameter!

the inner and outer rings, respectively [18]. These high [N II]/Hα ratios suggest anomalous nitrogen abundance, indicating that these rings are circumstellar, as opposed to interstellar, in origin. The expansion velocity of the inner ring is $\sim$ 10 km s^{-1}, indicating that it was not ejected during the SN explosion [19]. The morphology and kinematics of these circumstellar rings place constraints on the SN progenitor's mass loss history and lend support to its binarity [2].

It is interesting to note that circumstellar rings around B supergiants similar to those of SN 1987A's are probably rare. Only one case has been serendipitously found around the B1.5 I star Sher 25 in NGC 3603 (Figure 2, [20]). A search for circumstellar nebulae around >100 B supergiants in the Galaxy and the Magellanic Clouds has yielded null results [21]. The similarity in spectral type between Sher 25 and Sk–69 202 and the similarity in the size and physical properties of their circumstellar rings suggest that Sher 25 is a Galactic twin of Sk–69 202 and may very well be on the verge of a SN explosion [22].

The Case of SNR 0540–69.3

Next to SN 1987A, the second youngest SNR in the Large Magellanic Cloud (LMC) is SNR 0540–69.3 [23]. HST WFPC2 images of SNR 0540–69.3 in the Hα and [N II] lines show a [N II]-bright ring with a diameter of $\sim$5″, or 1.25 pc (see Figure 3). Spectroscopic observations of the [N II] lines show narrow velocity profiles indicating an expansion velocity $\leq$50 km s^{-1} [23]. This [N II]-bright ring is apparently a circumstellar bubble of the SN progenitor. The size of this circumstellar nebula is comparable to those of LBV nebulae, but smaller than WR bubbles [14]. It is possible that the SN progenitor exploded during or shortly after an LBV phase. The elemental abundances of this circumstellar nebula need to be measured in order to confirm the LBV hypothesis.

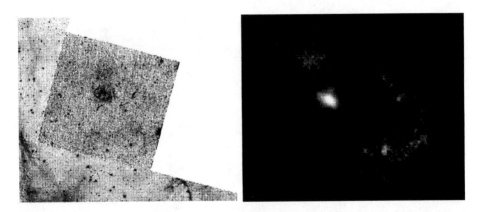

FIGURE 3. Left: HST WFPC2 image of SNR 0540–69.3 in the [N II]λ6583 line. The circumstellar nebula is visible near the center of the PC field. **Right:** Chandra HRC X-ray image of SNR 0540–69.3. The X-ray emission peaks at the region of circumstellar nebula. These two images have the same image scales.

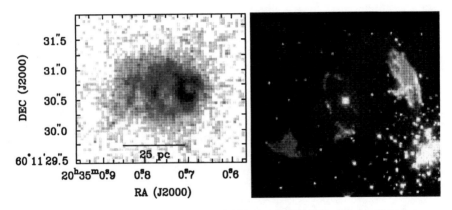

FIGURE 4. Left: HST WFPC2 Hα image of MF 16 in NGC 6946. **Right:** Echellogram and profile plot of the Hα+[N II] lines of MF 16 in NGC 6946. Both figures are taken from [28].

The Case of SNR MF16 in NGC 6946

The SNR MF 16 in NGC 6946 [24] at a distance of ~5 Mpc is known for its very high X-ray and optical luminosities, about 1,000 times as luminous as Cas A in X-rays [25,26]. HST WFPC2 images of MF 16 (Figure 4) show a multiple-loop morphology, which leads to the suggestion that the high luminosity of this remnant is caused by colliding SNRs. [27]. However. a different interpretation based on ground-based high-dispersion spectroscopic observations of MF 16 has been suggested [28]. The spectrum of MF 16 shows narrow Hα and [N II] lines superposed on the broad SNR components (Figure 4), and the [N II]/Hα ratio for the narrow component is high, ~1, indicating that the narrow components are emitted by N-enriched material. The expansion velocity implied by the velocity widths is <15 km s^{-1}. It is thus possible that the narrow components originate in a circumstellar bubble and the collision between this circumstellar bubble and the SNR causes the high X-ray and optical luminosities. HST STIS long-slit observations of MF 16 are needed to examine the location of regions with high [N II]/Hα ratios in order to determine the boundary of the circumstellar bubble. Ground-based, kinematically-resolved spectrophotometric observations similar to those of NGC 6888 [29] are needed to determine the abundances of this circumstellar material.

3. Interstellar Bubbles

Supernova ejecta in young SNRs, once having swept past the circumstellar material, can advance quickly to the inner walls of the interstellar bubble formed by the progenitor at the main sequence stage. The SNR N132D in the LMC is a good example [30]. The combination of high X-ray brightnesses and large linear sizes is usually indicative of such young SNRs. The detection/confirmation of an interstellar bubble is non-trivial, as illustrated in the example of N63A.

N63A is a SNR associated with the OB association LH83 in the LMC (Figure 5). Its optical image shows a three-lobed, clover-shaped nebulosity about 20″ (~ 5 pc) across. The two eastern lobes exhibit a high [S II]/Hα ratio, indicating a shock excitation, while the western lobe shows spectra characteristic of photoionization. The fact that N63A has such a high X-ray surface brightness and the fact that the optical size of the SNR N63A is much smaller than the extent of radio and X-ray emission (~70″) suggest that the SNR N63A is young and its SNR shock has just reached the inner walls of the progenitor's interstellar bubble.

HST WFPC2 images of N63A in Hα and [S II] resolved the filamentary structure in the two eastern lobes and diffuse emission in the western lobe, consistent with the excitation mechanisms diagnosed from their spectral properties [31]. Additionally, the WFPC2 images reveal a number of cloudlets as small as 0.1 pc across, within the X-ray-emitting regions of the SNR. The [S II]/Hα ratios and locations of these cloudlets suggest that these are shocked cloudlets lagging behind the shock front. These evaporating cloudlets are probably responsible for injecting mass into the

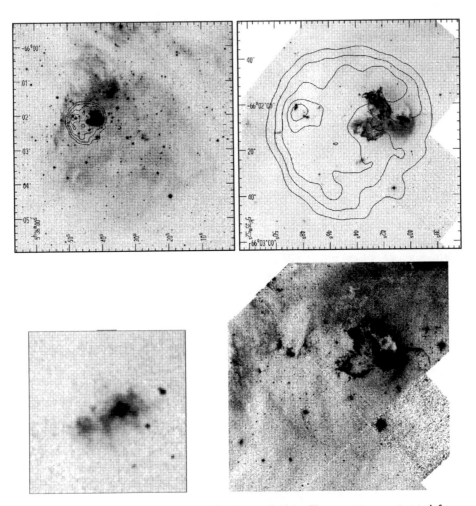

FIGURE 5. Upper Left: Hα image of N63, overlaid by X-ray contours extracted from a ROSAT HRI observation. This image shows that the SNR is associated with an OB association and embedded in an HII region. **Upper Right:** HST WFPC2 image of N63A in Hα. This image resolves not only the three bright lobes but also small cloudlets within the X-ray emission regions. **Lower Left:** Close-up of a small evaporating cloudlet. **Lower Right:** HST WFPC2 image of N63A in [O III]. The southern boundary of the SNR is delineated by a faint [O III] arc which is coincident with the X-ray boundary, suggesting that the [O III] arc marks the interstellar bubble blown by the supernova progenitor.

hot SNR interior to produce the high X-ray surface brightness. The Hα and [S II] images fail to show the location of the interstellar bubble.

A recent HST WFPC2 image of N63A in the [O III] λ5007 line finally shows a faint arc at the southern boundary of the SNR. The location of the arc coincides with the X-ray boundary. The long sought interstellar bubble of N63A is finally found.

SUMMARY AND CONCLUSION

Young SNRs offer the best laboratories to study the archeological artifacts of massive supernova progenitors. It is possible to detect the very latest stellar ejecta (the deathbed ejecta), circumstellar nebulae ejected at late evolutionary stages, and interstellar bubbles blown by the progenitors at main sequence stage. Detailed spectroscopic observations of these nebulae are necessary to determine their abundances in order to assess the evolutionary stages of the progenitors right before the supernova explosion.

This research is partially supported the the HST grant STScIGO-08110.01-97A.

REFERENCES

1. Walborn, N.R., Lasker, B.M., Laidler, V.G., & Chu, Y.-H., *ApJ* **321**, 41 (1987).
2. Podsiadlowski, P., *PASP* **679**, 717 (1992).
3. Chu, Y.-H. 1991, in IAU Symposium 143, Wolf-Rayet Stars and Interrelations with Other Massiv Stars in Galaxies, ed. K.A. van der Hucht and B. Hidayat (Dordrecht: Kluwer), p. 349
4. Chu, Y.-H. 1997, in CTIO/ESO/LCO Workshop "SN 1987A, Ten Years After,", in press
5. Chu, Y.-H., Weis, K., & Garnett, D.R., *AJ* **117**, 1433 (1999).
6. Kirschner, R.P., & Chevalier, R.A., *ApJ* **218**, 142 (1977).
7. Langer, N., Hamann, W.-R., Lennon, M., Najarro, F., Pauldrach, A.W.A., Puls, J., *A&A* **290**, 819 (1994).
8. Humphreys, R.M., & Davidson, K., *ApJ* **232**, 409 (1979).
9. Weaver, R., McCray, R., Castor, J., Shapiro, P., & Moore, R., *ApJ* **218**, 377 (1977).
10. García-Segura, G., Mac Low, M.-M., & Langer, N. *A&A* **305**, 229 (1996).
11. García-Segura, G., Langer, N., & mac Low, M.-M., *A&A* **316**, 133 (1996).
12. Morse, J.A., Davidson, K., Bally, J., Ebbets, D., Balick, B., & Frank, A., *AJ* **116**, 2443 (1998).
13. Esteban, C., Vílchez, J.M., Smith, L.J., & Clegg, R.E.S., *A&A* **259**, 629 (1992).
14. Chu, Y.-H., Caulet, A., Montes, M.J., Panagia, N., Van Dyk, S.D., & Weiler, K.W., *ApJ* **512**, L51 (1999).
15. Salamanca, I., Cid-Fernandes, R., Tenorio-Tagle, G., Telles, E., Terlevich, R.J., & Munoz-Tunon, C., *MNRAS* **300**, L17 (1998).

16. Salamanca, I. 2000, in the 1999 May Symposium of the STScI, "The Greatest Explosions Since the Big Bang: Supernovae and Gamma-ray Bursts," eds. M. Livio, N. Panagia, K. Sahu, in press

17. Ryder, S., et al., *ApJ* **416**, 167 (1993).

18. Burrows, C.J., et al., *ApJ* **452**, 680 (1995).

19. Crotts, A.P.S., & Kunkel, W.E., *ApJ* **366**, L73 (1991).

20. Brandner, W., Grebel, E.K., Chu, Y.-H., & Weis, K., *ApJ* **475**, L45 (1997).

21. Chu, Y.-H., Brandner, W., & Grebel, E.K., presented in CTIO/ESO/LCO Workshop "SN 1987A, Ten Years After," in press (1997).

22. Brandner, W., Chu, Y.-H., Einsenhauer, F., Grebel, E.K., & Points, S.D., *ApJ* **489**, L153 (1997).

23. Caraveo, P.A., Mignani, R., & Bignami, G.F., *Mem. Soc. Astron. Italiana* **69**, 1061 (1998).

24. Matonick, D.M., & Fesen, R.A., *ApJS* **112**, 49 (1997).

25. Schlegel, E.M., *ApJ* **424**, L99 (1994).

26. Blair, W.P., & Fesen, R.A., *ApJ* **424**, L103 (1994).

27. Blair, W.P., Fesen, R.A., & Schlegel, E.M., *AJ*, submitted (2000).

28. Dunne, B.C., Gruendl, R.A., & Chu, Y.-H., *AJ* **119**, 1172 (2000).

29. Esteban, C., & Vílchez, J.M., *ApJ* **390**, 536 (1992).

30. Morse, J.A., et al., *AJ* **112**, 509 (1996).

31. Chu, Y.-H., et al., in IAU Symposium 190, "New Views of the Magellanic Clouds," p. 143 (1999).

Young Supernova Remnants in the Magellanic Clouds

John P. Hughes

Rutgers University, Department of Physics and Astronomy, Piscataway, NJ 08854, USA

Abstract. There are a half-dozen or so young supernova remnants in the Magellanic Clouds that display one or more of the following characteristics: high velocity ($\gtrsim$1000 km s^{-1}) emission, enhanced metallicity, or a rapidly rotating pulsar. I summarize the current state of knowledge of these remnants and present some recent results mostly from the new X-ray astronomy satellites.

INTRODUCTION

The Magellanic Clouds (MCs) have been fertile hunting ground for supernova remnants (SNRs) for the last 40 years or so. Early work, focused on the optical [1] and radio [2] bands, resulted in the discovery of the first SNRs in the MCs, N132D, N49, and N63A. Since then deeper radio surveys as well as new X-ray surveys [3] in conjunction with optical spectroscopic follow-up have yielded a grand total of more than 50 confirmed remnants in the Clouds.

The common, well-known distances to the Clouds (50 kpc and 60 kpc for the large and small Clouds, respectively) allow for accurate estimates of such quantities as mass and energy, which have strong dependences on distance ($D^{5/2}$). The Clouds are close enough that angular resolution is not a major limiting factor, particularly now with the superb telescopes on the *Hubble Space Telescope* and *Chandra X-ray Observatory*. The generally low absorption to the Clouds makes possible UV and soft X-ray studies that are often precluded for Galactic remnants and indeed nearly all MC SNRs have been detected in the radio, optical, and X-ray bands. The sample of MC remnants is thus well-defined with reasonably clear selection criteria. With it one can sample the range of remnant types and SN progenitors, as well as probe the dynamical evolution of SNRs.

Since SNRs can exist as distinct entities for tens to hundreds of thousands of years, most of the remnants in the MCs are middle-aged or old. But chronological age alone is not the only determinant of youth; environment plays an equally important role. In a low density environment a remnant will expand to great size and the ejecta can dominate the dynamics and the emission properties for a considerable time. On the other hand, remnant evolution will proceed rapidly in the

CP565, *Young Supernova Remnants: Eleventh Astrophysics Conf.,* edited by S. S. Holt and U. Hwang

densest environments, passing quickly to the radiative phase. Thus a definition of what constitutes a young SNR is needed. Here I will use the following, largely observational, definition. A young remnant is one with a rapidly rotating, high spin-down-rate pulsar; high velocity oxygen-rich optical emission; other evidence for supernova (SN) ejecta; or a kinematic age of roughly 2000 yr or less. Note that I will not discuss SN1987A since it is the subject of an entire article elsewhere in this volume [4].

A SURVEY OF YOUNG MC REMNANTS

In this section I survey the properties and nature of the known young supernova remnants in the MCs. Objects are introduced in chronological order according to when each was identified as a young SNR, based on the criteria just described.

N132D

One of the original Large Magellanic Cloud (LMC) SNRs, N132D, can be considered to be young. Originally noted as an optical emission nebula [1] and a continuum radio source [2], N132D was confirmed as a remnant through optical spectroscopy [5]. In 1976 oxygen-rich optical filaments spanning a wide velocity range ($\sim$4000 km s^{-1}) were discovered in the center of N132D [6]. Although this is unquestionably emission from SN ejecta, N132D also shows an outer shell of normal swept-up interstellar medium (ISM) that in fact dominates the integrated X-ray emission [7] and appears to be associated with a giant molecular cloud [8]. N132D was almost surely a core-collapse SN, but no compact remnant has yet been detected. The age of the remnant estimated from the expansion and size of the oxygen-rich filaments is $\sim$3000 yrs, while an X-ray spectral analysis yields 2000–6000 yrs [9]. Preliminary results from *Chandra* HETG [10] and *XMM-Newton* [11] observations of N132D are discussed elsewhere in this volume. The EPIC MOS narrowband images of N132D show strong oxygen emission in the center, while at higher energies the most intense emission comes from the southern limb where the remnant is interacting with dense material presumably associated with the molecular cloud that lies there.

0540–69.3

It was not until the launch of the *Einstein Observatory* that additional examples of young remnants were discovered in the MCs. Three of the four new young remnants were discovered through their spatially-extended, soft X-ray emission and later identified as optical and radio emission nebulae. (More on them below.) The fourth remnant, SNR 0540–69.3 (also referred to as N158A), was known to be a nonthermal radio source [12] and had been suggested to be a SNR as early as 1973 [13]. High velocity oxygen-rich filaments with velocities spanning $\sim$3000 km s^{-1} were discovered [14] soon after the *Einstein* survey of the LMC showed SNR 0540–69.3 to be the third brightest extended soft X-ray source in the Cloud [15]. The X-ray emission is almost entirely nonthermal [16], like the Crab Nebula, and indeed SNR 0540–69.3 also harbors a rapidly spinning compact remnant, in this

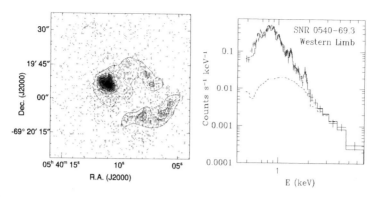

FIGURE 1. Left panel shows the HRC image of SNR 0540–69.3 as both raw counts per pixel (grayscale) and adaptively smoothed contours. Right panel shows the ACIS-S CCD spectrum of the faint western limb with thermal model and power law (dashed). Elemental abundances are consistent with swept-up ISM rather than SN ejecta.

case a 50-ms pulsar [17]. The spin-down age of the pulsar $\tau = P/2\dot{P}$ is roughly 1600 yr, while the kinematic age is ~800 yr [18]. This remnant is probably the second youngest SN in the LMC, only exceeded in its youth by SN1987A.

Within the last year *Chandra* HRC observations [19] have revealed the fine-scale structure of the remnant: a patchy incomplete shell surrounding an elliptically-shaped plerionic core that greatly resembles the Crab Nebula (Fig. 1). In the optical band the core, as imaged by *HST*, shows a beautiful filamentary structure with complex spectral variations with position [20]. The ACIS-S CCD spectral data (Fig. 1) show that the shell is swept-up ISM, as opposed to SN ejecta. A nonequilibrium ionization thermal model yields abundances for O, Ne, Mg, Si, and Fe of ~0.3 times solar, consistent with the low metallicity of the LMC. Under the plausible assumption that this emission represents the blast wave, then our fitted electron temperature of ~0.4 keV is remarkably low for such a young remnant. The X-ray spectrum also requires a hard power law tail with a photon index of $\alpha \sim 2.5$. Whether this emission is related to the pulsar itself or is an indication of nonthermal emission at the shock front, like in SN1006 and a few other Galactic remnants, remains to be determined. In any event, as the unique example of an oxygen-rich remnant that also contains a young pulsar, SNR 0540–69.3 is an important object for studying the link between pulsars and the massive stars that form them in core-collapse SNe.

1E0102.2–7219

Yet another oxygen-rich remnant was the next young one to be identified in the Clouds. This was also the result of an *Einstein* X-ray survey, this time of the Small Magellanic Cloud (SMC) [21]. The new remnant, 1E0102.2–7219, is the second brightest soft X-ray source in the SMC. Optical follow-up confirmed the identification as a SNR and revealed it to be extremely oxygen-rich [22] with [O

III] emission covering ~6500 km s^{-1} in velocity [23]. From this a kinematic age of ~1000 yr was estimated. *ASCA* showed the X-ray emission from 1E0102.2–7219 to be dominated by lines from He- and H-like species of O, Ne, Mg, and Si, although it was not possible to obtain a simple interpretation of the integrated spectrum [24]. No compact remnant, either pulsar or plerionic core, has been detected down to a flux limit of 10^{34} erg cm^{-2} s^{-1} [25]. However, the ACIS-S CCD data did reveal a faint shelf of X-ray emission presumably from the blast wave lying beyond the bright rim of oxygen-rich ejecta [25]. As for the ejecta, *Chandra* HETG line images from the past year reveal high velocity O- and Ne-rich X-ray emitting ejecta with different positions in the shell displaying different red- and blue-shifts [10]. The remnant's size depends on the emission line under study. In general the remnant appears larger in higher ionization species than it does in lower ones, an effect that is also seen in the lower spectral resolution ACIS-S CCD data [25]. One simple interpretation is that we are seeing the progressive increase in ionization caused by the reverse shock propagating inward through the ejecta. The outermost ejecta material was heated by the reverse shock first and hence has had the most time to ionize up to higher ionization species. The most recently shocked ejecta, however, lies interior to this and should show a lower ionization state. Although qualitatively consistent with the observations, this picture may be oversimplified. In addition to the temporal effect just described, the temperature and density in the ejecta may vary significantly from the contact discontinuity (outside) to the reverse shock (inside) and thereby cause spectral variations with position. The temperature and density variations contain information about the original distribution of matter in the ejecta and the circumstellar medium (CSM) (see, e.g., [26]). Hopefully, further study will allow us to disentangle the effects of time, density, and temperature and learn more about the ejecta and CSM in 1E0102.2–7219.

Very recently my collaborators and I used the *Chandra* ACIS-S CCD data on 1E0102.2–7219 to address some basic issues of shock physics [27]. The question we originally set out to investigate was "To what extent are electrons heated at high Mach number shocks in SNRs?" Do the electrons rapidly attain similar temperatures to the ions through anomalous heating processes driven by plasma instabilities at the shock front (see, for example, [28])? Or do the ion and electron temperatures initially differ by their mass ratio ($m_p/m_e = 1836$) with the electrons subsequently gaining heat slowly from the ions through Coulomb collisions?

Clearly the information needed for this study was the post-shock temperatures of both the electrons and ions. We determined the post-shock electron temperature from fits to the X-ray spectrum of the blast wave region that *Chandra* had revealed for the first time. Fig. 2 shows this spectrum and for comparison the spectrum of a portion of the bright rim of ejecta which displays a markedly different spectral character. The derived abundances from the blast-wave spectrum confirm its origin as swept-up ISM and, for a variety of nonequilibrium ionization spectral models, the derived electron temperature was constrained to be less than 1 keV. It is impossible at the present time to determine the post-shock ion temperature directly (since there is no H line emission from the remnant). We were, however, able to determine

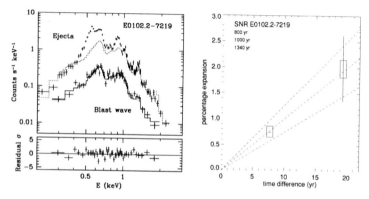

FIGURE 2. Left panel shows the ACIS-S CCD spectra of a portion of the outer blast wave and bright rim of ejecta in SNR 1E0102.2–7219. Right panel shows the expansion rate of this remnant from a comparison of the *Chandra* ACIS image with earlier *ROSAT* and *Einstein* images.

the velocity of the blast wave by measuring the angular expansion rate of the remnant. Previous images of 1E0102.2–7219 made by *ROSAT* and *Einstein* were compared to the *Chandra* image and it was found that a significant change in size had occurred (Fig. 2). From the expansion rate (0.100% ± 0.025% yr^{-1}) and known distance a blast wave velocity of ~6000 km s^{-1} was determined. If we assume that the electron and ion temperatures are fully equilibrated at the shock front, then the post-shock electron temperature we measure is about a factor of 25 lower than expected given this blast wave velocity. How low can the post-shock electron temperature be if the post-shock ion temperature is given by the Rankine-Hugoniot jump conditions? This is set by assuming that post-shock electrons and ions exchange energy solely through Coulomb interactions. In this case, we find that the minimum expected electron temperature is ~2.5 keV, still significantly higher than what we measure. Our explanation for this discrepancy is that efficient shock acceleration of cosmic rays has reduced the post-shock temperature of both the electrons and ions [29], or in other words, a large fraction of the shock energy has been diverted from the thermal particles and has instead gone into generating relativistic particles, which produce the remnant's radio emission [30]. Nonlinear models for efficient shock acceleration of cosmic rays [31] predict mean post-shock electron and ion temperatures of 1 keV for high Mach number shocks, i.e., 100–300, as appropriate to 1E0102.2–7219, lending strong quantitative support for this scenario.

E0509–67.5 and E0519–69.0

E0509–67.5 and E0519–69.0 are two apparently similar SNRs in the LMC that were discovered through their X-ray emission [3]. Their X-ray morphologies are strongly shell-like and reasonably symmetric. Optical follow-up [32] revealed them to be members of a rare class of faint SNRs, like Tycho and SN1006 in the Galaxy, that are dominated by Balmer line emission and show little or no forbidden line emission from, e.g., [O III] or [S II]. Like their Galactic cousins, these two LMC SNRs were originally believed to be young in part because they are among the smallest remnants in the LMC with diameters of 7 pc (E0509–67.5) and 8 pc (E0519–69.0). In addition for E0519–69.0, at least, there is evidence for a broad component to Hα with FWHM velocity widths ranging from 1300 km s^{-1} [33] to 2800 km s^{-1} [32]. This emission arises from charge exchange between fast protons and neutral hydrogen atoms in the immediate post-shock region [34] with the width of the line being related to the shock velocity and assumptions about the post-shock partition of energy into hot electron, ions, and cosmic rays [35]. The broad line in E0519–69.0 implies a probable shock velocity range of 1000–1900 km s^{-1} (although there are likely to be significant variations with position) and an age of 500–1500 yrs [33]. No broad Hα has yet been detected from E0509–67.5 which has been interpreted as implying a lower limit on the actual shock velocity in the remnant of >2000 km s^{-1} with a corresponding upper limit on the age of <1000 yr [33]. *ASCA* observations of these remnants [36] revealed the spectacular ejecta-dominated nature of their integrated X-ray emission. The spectra are dominated by a broad quasi-continuum emission of unresolved Fe L-shell lines around 0.7–0.9 keV as well as strong Kα lines of highly ionized Si, S, Ar, and Ca. The spectra were shown to be qualitatively consistent with the nucleosynthetic products expected from a Type Ia SN explosion, thereby significantly strengthening the connection between Balmer-dominated SNRs and the remnants of Type Ia SN.

Early *Chandra* results on E0519–69.0 [37] show an integrated spectrum quite similar to the *ASCA* one, but the lower effective background of the *Chandra* data allows for clear detection of the important Fe Kα line. There are spatial/spectral variations across the remnant with the most notable difference being a strong increase in emission in the 0.5–0.7 keV band at the rim.

The *Chandra* broad-band X-ray image of E0509–67.5 is shown in Fig. 3 along with the integrated spectrum. The shell is quite round, nearly complete, and brightest on the western side; all of which are quite similar to the Hα morphology [33]. The intensity fluctuations around the bright limb of the remnant are generally significant with peak to valley brightness variations of about a factor of two on spatial scales of just a few arcseconds. Whether this apparent clumping of the ejecta originated during the SN explosion process itself, arose later on through Rayleigh-Taylor instabilities from interaction with the ambient medium, or has some other origin remains to be investigated. There is no morphological evidence for a double shock structure: the X-ray emission is dominated by metal-rich ejecta and no outer blast wave component is detected. Again the spectrum is quite similar to the *ASCA* one [36], although here with *Chandra* we have detected Ar, Ca, and Fe Kα line emission at energies of 3.0 keV, 3.7 keV, and 6.4 keV, respectively. The Si and S Kα lines in the integrated spectrum of E0509–67.5 are quite strong (equivalent widths of 1.9 keV for Si and 1.2 keV for S), very broad (FWHM line widths of 100 eV and 120 eV), and appear at much lower energies than the He- and H-like Kα lines should (mean line energies are 1.804 keV and 2.377 keV). There is

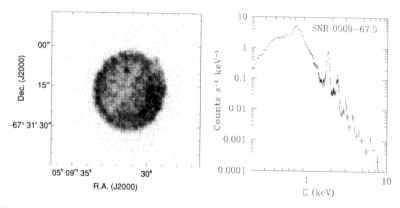

FIGURE 3. Left panel shows the ACIS-S CCD image of E0509–67.5 over the 0.2–7.0 keV band. Right panel shows the integrated spectrum of the remnant.

no change in the width or mean energy of the Si Kα line with radius, which argues against velocity effects (which would have to be enormous anyway) as the source of the line broadening. The most likely explanation for the broad Si line is that it is a blend of lines from a wide range of charge states, i.e., from Si^{12+} (He-like) to Si^{5+} or lower. A similarly broad range of charge states for S is required to explain that broad line. It seems likely that a range of temperatures or ionization timescales will be necessary to account for the broad lines; simple planar shock models tend to produce too narrow a distribution of charge states. The spectral analysis with radius does reveal one significant variation—the Si line equivalent width drops by a factor of 2 (to a value of $\sim$1 keV) right at the edge of the remnant—perhaps an indication of some dilution of the ejecta emission by a component of swept-up ISM.

N103B

First identified in 1973 as a SNR based on its radio and optical properties [13], N103B was detected as an extended, soft X-ray source by the *Einstein Observatory* [15]. It is the fourth brightest X-ray remnant in the LMC and is also one of the smallest LMC SNRs ($\sim$7 pc in diameter). Optically the remnant consists of several small, bright knots visible in Hα, [O III], and [S II]. Its proximity to the star cluster NGC 1850 and the H II region DEM 84 led to the suggestion that SNR N103B had a population I progenitor [38]. It was not until *ASCA* that N103B was revealed to be young based on its ejecta-dominated spectrum [36]. The surprising result was that the X-ray spectrum of N103B bears a remarkable similarity to the Balmer-dominated SNR E0519–69.0 and indeed both remnants are now believed to be remnants of Type Ia SN. The *Chandra* data [39] show a highly structured remnant that is much brighter on the western limb than toward the east. On large scales it resembles the radio remnant [40] but on finer spatial scales the bright radio and X-ray emission features tend not to overlap. There are complex spatial/spectral variations across the remnant including changes in the equivalent width of the He-

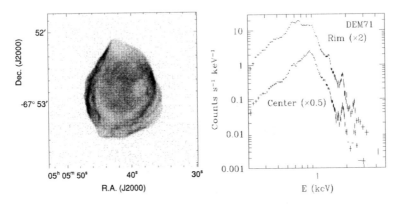

FIGURE 4. Left panel shows the ACIS-S CCD image of DEM71 over the 0.2–4.4 keV band. Right panel shows the spectrum of the remnant from the outer rim and the central region.

like Si Kα emission line which appears to increase radially outward. In contrast to E0509–67.5, the Si Kα line from N103B shows both a He- and H-like component indicating that N103B is at a higher mean ionization state.

N157B

Originally detected as a nonthermal radio source [12], and suggested as a possible SNR in 1973 [13], it was not until extended soft X-rays were detected by the *Einstein Observatory* [15] that N157B (also referred to as 30 Dor B) was confidently confirmed as a remnant. Its featureless powerlaw X-ray spectrum [16] and flat radio spectral index ($\alpha \sim -0.19$ [41] [42]) were taken as strong evidence that the remnant was Crab-like and should contain a rapidly rotating pulsar. The pulsar in N157B was discovered in 1998 [43] with a spin period of 16.1 ms making it the fastest known pulsar in a SNR. The spin-down rate of $\dot{P} = 5.126 \times 10^{-14}\,\mathrm{s\,s^{-1}}$ implies a characteristic age of $\tau = 5000$ yr which agrees well with other estimates of the remnant's age [44].

DEM71

A new example of a possible young SNR is DEM71. Initially noted as an optical nebulosity [45], it was detected as an extended soft X-ray source by *Einstein* and proposed as a SNR [3]. Optical follow-up revealed it to be a Balmer-dominated remnant [32] with broad Hα line emission [33]. Given its size (20 pc diameter) and shock velocity of $\sim$500 km s^{-1} (estimated from the broad Hα line), the remnant is $\sim$8000 yr old. *ASCA* spectroscopy revealed the first tantalizing hints of youth for this SNR in the form of an enhanced abundance of Fe [9].

The broadband image of DEM71 from *Chandra* (fig. 4) shows emission from an outer rim that matches nearly perfectly the optical Hα image, including the brightness of the eastern and western rims as well as the multiple filaments along the southern edge. The nearly circular, faint, diffuse emission that almost fills the interior has no counterpart in the optical band [41]. The ACIS-S CCD spectrum

426

of the center (Fig. 4) is dominated by a broad quasi-continuum of emission near 1 keV from Fe L-shell lines, plus Si and S Kα lines at higher energy. This spectrum is grossly different from that of the outer rim, which is generally softer and shows prominent O and Mg Kα lines. We propose that the rim represents the blast wave moving out into the ambient interstellar medium, while the interior emission is the reverse shock propagating through iron-rich ejecta. Detailed study of the *Chandra* data is now underway using the outer rim emission to study the basic physics of SN shocks and the central Fe-rich emission to investigate the evolution of metal-rich ejecta in SNRs.

CONCLUSIONS

I expect the list of "young" remnants in the MCs to grow as improvements in sensitivity make it easier to identify youthful aspects of older known remnants. It is likely that there are still a few pulsars left to be discovered in the MC SNRs either through direct measurement of pulsed radiation or from the identification of their associated pulsar-wind nebulae. If DEM71 can be considered to be a guide, then even remnants that are several thousand years old can show evidence for SN ejecta. N63A, N49, N23 in the LMC, all of which are smaller in size than N132D and DEM71, are in this age range and therefore have the potential for showing traits of youth. It will be worth keeping an eye on these remnants in the coming years.

Acknowledgments I gratefully acknowledge Dave Burrows, Anne Decourchelle, Gordon Garmire, Parviz Ghavamian, Karen Lewis, John Nousek, Cara Rakowski, and Pat Slane for their collaborations on various studies of MC SNRs. I would also like to thank Andy Rasmussan and Ehud Behar for sharing XMM results prior to publication. This work was partially supported by *Chandra* General Observer grant GO0-1035X.

REFERENCES

1. Henize, K.G. 1956, ApJS, 2, 315.
2. Mathewson, D.S., & Healey, J.R. 1964, The Galaxy and the Magellanic Clouds, ed., F. R. Kerr & A. W. Rodgers (Canberra: Australian Academy of Science) p. 283.
3. Long, K.S., Helfand, D.J., & Grabelsky, D.A. 1981, ApJ, 248, 925.
4. McCray, R., et al. 2001, this volume.
5. Westerlund, B. E., & Mathewson, D.S. 1966, MNRAS, 131, 371.
6. Danziger, I.J., & Dennefeld, M. 1976, ApJ, 207, 394.
7. Hughes, J.P. 1987, ApJ, 314, 103.
8. Banas, K.R., Hughes, J.P., Bronfman, L., & Nyman, L.-Å. 1997, ApJ, 480, 607.
9. Hughes, J.P., Hayashi, I., & Koyama, K. 1998, ApJ, 505, 732.
10. Canizares, C., et al. 2001, this volume.
11. Behar, E., et al. 2001, this volume.

12. Le Marne, A.E. 1968, MNRAS, 139, 461.

13. Mathewson, D.S., & Clarke, J.N. 1973, ApJ, 180, 725.

14. Mathewson, D.S., Dopita, M.A., Tuohy, I.R., & Ford, V.L. 1980, ApJ, 242, L73.

15. Long, K.S., & Helfand, D.J. 1979, ApJ, 234, L77.

16. Clark, D.H., Tuohy, I.R., Long, K.S., Szymkowiak, A.E., Dopita, M.A., Mathewson, D.S., & Culhane, J.L. 1982, ApJ, 255, 440.

17. Seward, F.D., Harnden, F.R., & Helfand, D.J. 1984, ApJ, 287, L19.

18. Kirshner, R.P., Morse, J.A., Winkler, P.F., & Blair, W.P. 1989, ApJ, 342, 260.

19. Gotthelf, E.V., & Wang, Q.D. 2000, ApJ, 532, L117.

20. Morse, J.A., et al. 2001, in prep. (shown at the conference by W.P. Blair).

21. Seward, F.D., & Mitchell, M. 1981, ApJ, 243, 736

22. Dopita, M.A., Tuohy, I.R., & Mathewson, D.S. 1981, ApJ, 248, L105.

23. Tuohy, I.R., & Dopita, M.A. 1983, ApJ, 268, L11.

24. Hayashi, I., Koyama, K., Ozaki, M., Miyata, E., Tsunemi, H., Hughes, J.P., & Petre, R. 1994, PASJ, 46, L121.

25. Gaetz, T.J., Butt, Y.M., Edgar, R.J., Eriksen, K.A., Plucinsky, P.P., Schlegel, E.M., & Smith, R.K. 2000, ApJ, 534, L47.

26. Chevalier, R.A. 1982, ApJ, 258, 790.

27. Hughes, J.P., Rakowski, C.E., & Decourchelle, A. 2000, ApJ, 543, L61.

28. Cargill, P.J., & Papadopoulos, K. 1988, ApJ, 329, L29.

29. Blanford, R.D., & Eichler, D. 1987, Phys. Rep., 154, 1.

30. Amy, S.W., & Ball, L. 1993, ApJ, 411, 812.

31. Ellison, D.C. 2000, in AIP Conf. Proc. 529, Acceleration and Transport of Energetic Particles Observed in the Heliosphere, ed. R.A. Mewaldt, et al. (New York AIP), 386.

32. Tuohy, I.R., Dopita, M.A., Mathewson, D.S., Long, K.S., & Helfand, D.J. 1982, ApJ, 261, 473.

33. Smith, R.C., Kirshner, R.P., Blair, W.P., & Winkler, P.F. 1991, 375, 652.

34. Chevalier, R.A., & Raymond, J.C. 1978, ApJ, 225, L27.

35. Chevalier, R.A., Kirshner, R.P., & Raymond, J.C. 1980, ApJ, 235, 186.

36. Hughes, J.P., et al. 1995, ApJ, 444, L81.

37. Williams, R.M., et al. 2001, this volume.

38. Chu, Y.-H., & Kennicutt, R.C. 1988, AJ, 96, 1874.

39. Lewis, K.T., Burrows, D.N., Nousek, J.A., Garmire, G.P., Slane, P., & Hughes, J.P. 2001, this volume.

40. Dickel, J., & Milne, D. 1995, AJ, 109,1.

41. Mathewson, D.S., Ford, V.L., Dopita, M.A., Tuohy, I.R., Long, K.S., & Helfand, D.J. 1983, ApJS, 51, 345.

42. Lazendic, J.S., Dickel, J.R., Haynes, R.F., Jones, P.A., & White, G.L. 2000, ApJ, 540, 808.

43. Marshall, F.E., Gotthelf, E.V., Zhang, W., Middleditch, J., & Wang, Q.D. 1998, ApJ, 499, L179.

44. Wang, Q.D., & Gotthelf, E.V. 1998, ApJ, 494, 623.

45. Davies, R.D., Elliott, K.H., & Meaburn, J. 1976, Mem. R.A.S., 81, 89.

An Investigation of SNRs in the Magellanic Clouds – SNR B0450-708 and SNR B0455-687

Miroslav D. Filipović[1,2], Paul A. Jones[1] and Graeme L. White[1]

[1] *University of Western Sydney Nepean, P.O. Box 10, Kingswood, NSW 2747 Australia*
[2] *Australia Telescope National Facility, CSIRO, P.O. Box 76, Epping, NSW 2121 Australia*

Abstract. We are studying the nature and the spatial distribution of $\sim$80 supernova remnants (SNRs) in the Magellanic Clouds (MCs). A multi-frequency comparison of these sources involves the latest radio, IR, optical and X-ray surveys with angular resolution of $\sim 1'$. Their morphology, birth rate and overall properties are investigated and compared with SNRs in our Galaxy and other galaxies. In this paper we are presenting initial results from two SNRs in the Large Magellanic Cloud (LMC).

INTRODUCTION

Statistical studies of SNRs in our Galaxy, despite the large number ($\sim$220) [1], have problems due to incompleteness and uncertain distances. SNRs in the MCs are approximately the same distance, however, they are still close enough to allow detailed spatially resolved investigations. The MCs contain all types of SNRs in various stages of evolution. Also, the MCs fortunately lie in a direction well out of the Galactic plane and the foreground densities of dust, gas and stars are small. SNRs in other galaxies such as M31, M33 or NGC 300 [2] appear about ten times smaller and one hundred times fainter then SNRs in the MCs.

We have made a low-resolution comparison between the Parkes radio and ROSAT All-Sky survey [3]. A source-intensity comparison of radio and X-ray source flux densities show very little correlation, but we note that the strongest SNRs at both frequencies are young SNRs from Population I.

About 60 SNRs and SNR candidates in the LMC and 20 in the SMC have been identified in our studies [4]. Of these, there are 23 SNR candidates in the LMC and 6 in the SMC which yet need to be confirmed as SNRs.

From the radio flux density of the SNRs we have estimated the SNR birth rate to be one every 100 ($\pm$20) yr and the star-formation rate (SFR) to be 0.7 ($\pm$0.2) $M_\odot \, yr^{-1}$. The SMC SNR birth rate was estimated to be one every 350 ($\pm$70) yr and the SFR was estimated to be 0.15 ($\pm$0.05) $M_\odot \, yr^{-1}$.

CP565, *Young Supernova Remnants: Eleventh Astrophysics Conf.*, edited by S. S. Holt and U. Hwang
© 2001 American Institute of Physics 0-7354-0001-6/01/$18.00

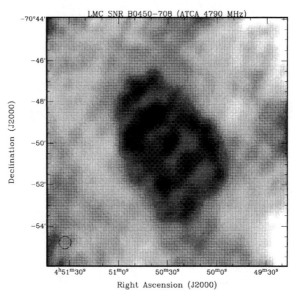

LMC SNR B0450-708 (ATCA 4790 MHz)

Right Ascension (J2000)

FIGURE 1. The ATCA 4790 MHz image of SNR B0450-708. The synthesized beam is 40″ × 40″ (lower left corner) and the r.m.s. noise (1σ) is 0.2 mJy.

ATCA OBSERVATIONS OF SNRS IN THE MCS

As a continuation of the Parkes radio-continuum investigations of SNRs in the MCs we use the Australia Telescope Compact Array (ATCA) mosaic surveys of both Clouds at 20/13 cm [5] and the MOST survey at 843 MHz [6,7].

In addition we have observed all SNRs and SNR candidates in both Clouds (~80) using the ATCA at 4790 MHz and 8640 MHz (λ=6 and 3 cm) with the 375 m configuration. The aim was to image all SNRs and SNR candidates in snapshot mode. The tradeoff of image complexity with uv coverage indicates that this compromise of resolution is justified. Observations were undertaken during 12 observing days from 1997 to 2000. These observations give us wide frequency coverage of these objects with sufficient resolution to resolve SNRs. Full ATCA 12-h synthesis observations of individual SNRs have been made by us and other groups [8].

LMC SNR B0450-708

We present here ATCA data for two SNRs in the LMC. The SNR B0450-708 (also known as MC 11) is the largest SNR in the MCs with size over 7′× 5′ (112×80 pc; D_{lmc}=55 kpc). The 4790 MHz image (Fig. 1) shows central radio emission with two distinctive and concentric shells. This SNR has a spectral index of α=−0.36±0.08

430

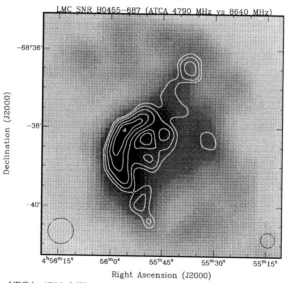

FIGURE 2. The ATCA 4790 MHz image of SNR B0455-687 (40″ × 40″ beam – lower left corner) overlaid with contours of the ATCA 8640 MHz image (22″ × 22″ beam – lower right corner). Contours are 1, 1.4, 2, 2.8, 3.4 and 4 mJy/beam

[4] and has been detected in Hα and MOST surveys [9]. From earlier optical observations it is classified as type 1 SNR [10]. Comparison with the ROSAT X-ray data (RASS survey) [4] did not show any X-ray emission from this SNR but further deeper X-ray study is required. Also, comprehensive comparison with the optical surveys such as [S II] and [O III] are underway.

LMC SNR B0455-687

The SNR B0455-687 (also known as N 86, DEM L33 and CAL 2) is located on the western edge of the LMC. In Fig. 2, we show the gray scale 4790 MHz image with contours from higher resolution 8640 MHz image to show shell-like but asymmetric structure which is a typical SNR characteristic. With a radio spectral index of α=−0.51±0.06 we confirm this source as a definite SNR.

FUTURE WORK

Our investigation of objects in the MCs are based on ATCA radio-continuum observations with the main aim of studying a complete sample of MC SNRs. The luminosity-diameter distribution will be used to study the evolution of SNRs in a statistical sense [7]. Also a comparison with the X-ray data (ROSAT, XMM and Chandra) will continue the radio – X-ray comparison [3]. Comprehensive comparison with the optical and IR surveys such as Hα, [S II], [O III], IRAS and DeNIS are planned as well.

In addition, individual SNRs will be studied in more detail. Some of their characteristics may be related to peculiar properties of the interstellar medium around the explosion site and/or different precursor stars. Theoretical models indicate that changes in density, clumpiness and other properties of the surrounding medium can have a significant effect upon the evolution of an SNR and its emission. However, the collective properties of the SNRs in the MCs studied to date are surprisingly consistent and quite similar to the population of Galactic SNRs [8], despite major differences between these galaxies.

REFERENCES

1. Green, D.A., 1998, 'A Catalogue of Galactic Supernova Remnants (1998 September version)', Mullard Radio Astronomy Observatory, Cambridge, United Kingdom, http://www.mrao.cam.ac.uk/surveys/snrs/
2. Pannuti, T., et al., 2001, this proceeding
3. Filipović, et al., 1998, A&AS **127**, 119
4. Filipović, Haynes, R.F., White, G.L., Jones, P.A., 1998, A&AS **130**, 421
5. Filipović, M.D., Staveley-Smith, L., 1998, In: Richtler, T., Braun, J.M., (eds.) The Magellanic Clouds and Other Dwarf Galaxies, Shaker Verlag, Aachen, p. 137
6. Ye T., 1988, PhD Thesis, Sydney University
7. Mills, B.Y., Turtle, A.J., Little, A.G., Durdin, M.J., 1984, *Aust. J. Phys.* **37**, 321
8. Dickel, J.R., Milne, D.K., 1994, *Proc. ASA* **11**, 99
9. Mathewson, D.S., Ford, V.L., Tuohy, I.R., et al., 1985, *ApJS* **58**, 197
10. Chu, Y.H., Kennicutt, R.C., 1988, *AJ* **95**, 1111

New Radio Supernova Remnants in the Core of M31

Lorant O. Sjouwerman[1] and John R. Dickel[2]

[1] Joint Institute for VLBI in Europe (JIVE), Dwingeloo, The Netherlands, sjouwerman@jive.nl
[2] University of Illinois at Urbana-Champaign, Urbana IL, USA and ASTRON, Dwingeloo, The Netherlands, johnd@astro.uiuc.edu

Abstract. Five 24-hour VLA observations at 8.4 GHz of the Andromeda galaxy have been combined into a very sensitive radio image of the central 500″ (1.7 kpc) of M31. The resulting image presented here has a resolution of 0.2″ and an rms noise level of about 4 μJy. These observations have detected three new shell supernova remnants within 250 pc of the center of M31, with sizes and luminosities of typical older remnants. In addition, a somewhat extended 27 μJy source with a full-width half maximum size of 0.73″×0.33″ (2.5×1.1 pc) is detected at $00^h42^m43\overset{s}{.}136$ and $41°16'05\overset{''}{.}06$ (J2000) which is within the positional uncertainty of the Type Ia supernova S And of 1885.

INTRODUCTION

The center of the Andromeda galaxy (M31) is an active area, containing an extended bright central radio core [1] with a central black hole candidate, M31* [2]. The central radio source M31* was found by Crane et al. [2], who were actually searching for the 100 year old radio remnant of the Type Ia supernova S Andromedae that appeared in 1885. Crane et al. did not detect any radio remnant of SN1885a, but instead found a 28 μJy compact radio source which they suggested to be the central black hole. Subsequent monitoring of this source revealed it was variable, supporting their view [3]. In the total monitoring program of M31*, five 24-hour VLA observations were taken over a time-span of 5.5 years at a frequency of 8.4 GHz, where only the extended A or B arrays of the VLA were utilized so that smooth emission was filtered out and making it possible to find compact sources that otherwise would have been blended into the background.

The image presented here consist of the five 24-hour monitor observations of the central region of M31 from Crane et al., which have been combined and mosaiced into a very sensitive and large high-resolution 8.4 GHz radio image of the center of that galaxy. The resulting image shown in Fig. 1 has a resolution of 0.2″ and an rms noise level of about 4 μJy. As well as a number of compact and faint μJy sources for which the identification is currently under investigation, at least three

CP565, *Young Supernova Remnants: Eleventh Astrophysics Conf.*, edited by S. S. Holt and U. Hwang
© 2001 American Institute of Physics 0-7354-0001-6/01/$18.00

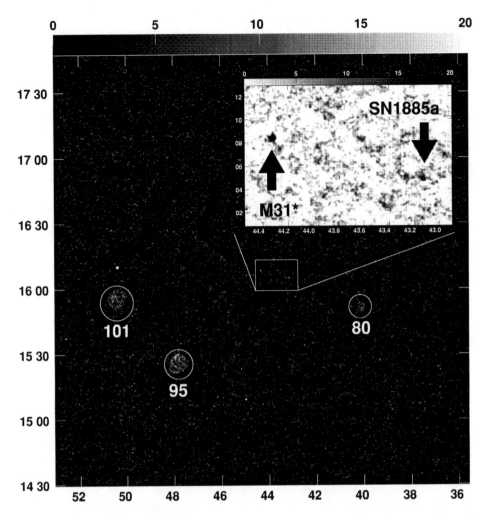

FIGURE 1. An 8.4 GHz VLA image of the central $200'' \times 200''$ of M31 (300 pc radius around M31* in the very center of the image) at $0.2''$ resolution. Right Ascension (J2000) is the horizontal coordinate, where the labels $00^h 42^m$ are omitted for clarity; vertical is Declination (J2000) without the $41°$. The noise in this image is $3.9\ \mu$Jy, and the grey scale ranges from 0 to $20\ \mu$Jy. The newly discovered, older supernova remnants are labeled with the source numbers from a catalog by Braun [4]. The inset shows a blow-up of the very central region with the radio sources associated with M31* and the remnant SN1885a of S And. Again, the labels $00^h 42^m$ and $41°16'$ are omitted for the RA and Dec, and the grey scale range is the same (but inverted).

new, older type and extended shell supernova remnants were detected within 250 pc of the center of M31. In addition, a somewhat extended 27 μJy radio source is detected within the positional uncertainty of S And, which we propose here to be the radio remnant of SN1885a.

THE NEW SUPERNOVA REMNANTS

Three newly detected supernova remnants lie within 250 pc of the center of M31, with sizes and luminosities of typical older remnants. They are summarized in Table 1, together with the proposed radio counterpart of SN1885a. S_{int} is the observed integrated flux density in mJy at 8.4 GHz, and L_{1GHz} is the 1 GHz luminosity (in units of 10^{22} erg sec^{-1} Hz^{-1}) scaled from the observed frequency of 8.4 GHz by the spectral index calculated by using our measurements and the 1.4 GHz measurements of Braun [4]. Because of low surface brightness and background confusion the flux densities of Braun's source "80" are uncertain and we adopt an approximate spectral index of -0.5 (average for supernova remnants [5]); the SN1885a remnant candidate was scaled by -0.6, more typical for young supernova remnants. The assumed distance to M31 is 700 kpc. Σ_{1GHz} is the 1 GHz surface brightness (in units of 10^{-20} W m^{-2} sr^{-1}). Three or four (or perhaps even more) supernova remnants within 1/8 kpc^3 indicates an active recent star formation. No estimate of the length of the star formation period has been attempted, because older, faint remnants remain undetected.

A RADIO COUNTERPART OF SN1885A?

A somewhat extended 27 μJy source with a full-width half maximum size of $0.73'' \times 0.33''$ (2.5×1.1 pc) is detected at $00^h42^m43\overset{s}{.}136$ and $41°16'05\overset{''}{.}06$ (J2000), which is within the positional uncertainty of the supernova S And of the year 1885. An HST iron absorption feature reported by Fesen et al. [6], and also attributed to the remnant of SN1885a, appears to lie at $00^h42^m43\overset{s}{.}005 \pm 0.002$ and $41°16'04\overset{''}{.}44 \pm 0.02$ (J2000) in our radio image. This position was determined by rotating the HST image to the proper J2000 alignment and placing peak P2 of the double optical nucleus at the position of M31* on the radio image; our radio-optical position, at $14.90''$ West and $3.97''$ South of P2/M31*, differs from the Fesen et al's published J2000 position of the supernova remnant ($00^h42^m42\overset{s}{.}89$ and $41°16'05\overset{''}{.}0$ [6]) for unknown reasons. If the remnant is 3-4 pc across, which is not unreasonable for a 100 year old supernova remnant, we could be viewing different parts of the same object in the different wavelength bands as the expanding shell interacts with different parts of the surrounding medium.

TABLE 1. Supernova remnants detected within 250 pc of M31*

R.A. and Dec. (J2000)		Diameter (FWHM) $''$	pc	PA deg	S_{int} mJy	Spec. Ind.	L_1	Σ_1	Ref.
$00^h42^m40\overset{s}{.}26$	$41°15'52\overset{''}{.}1$	8.0×8.0	28×28	-	0.53	[a]	90.	1.3	80
$00^h42^m43\overset{s}{.}136$	$41°16'05\overset{''}{.}06$	0.73×0.33	2.5×1.1	64	0.027	[b]	6.1	16.	[c]
$00^h42^m47\overset{s}{.}82$	$41°15'25\overset{''}{.}7$	9.6×8.0	33×28	137	1.41	-0.3	160.	1.9	95
$00^h42^m50\overset{s}{.}41$	$41°15'56\overset{''}{.}4$	11.6×8.4	40×30	143	1.29	-0.5	220.	2.1	101

S_{int} is the observed integrated flux in density in mJy at 8.4 GHz.
L_1 is the 1 GHz luminosity in units of 10^{22} erg sec^{-1} Hz^{-1}
Σ_1 is the 1 GHz surface brightness in units of 10^{-20} W m^{-2} sr^{-1}
References correspond to source numbers in Braun 1990 [4]
[a]: Adopted spectral index of -0.5 because flux densities are confused
[b]: Adopted spectral index of -0.6 as typical of young SNRs
[c]: Radio remnant of SN1885a?

SUMMARY

A high-sensitivity, high-resolution image of the central kpc of M31 has revealed at least three older type, previously unknown supernova remnants within 250 projected pc from the nucleus. Furthermore, a faint, extended radio source is located near the historical position of S Andromedae, the Type Ia supernova that appeared in 1885. These results indicate active recent star formation in the center of M31.

REFERENCES

1. Hjellming RM, Smarr LL, 1982, ApJ Lett. 257, L13
2. Crane PC, Dickel JR, Cowan JJ, 1992, ApJ Lett. 390, L9
3. Crane PC, Cowan JJ, Dickel JR, Roberts DA, 1993, ApJ Lett. 417, L61
4. Braun R, 1990, ApJ Sup. 72, 761
5. Dickel JR, 1991, in "Supernovae" (Santa Cruz Summer Workshop), ed. S. Woosley (Springer-Verlag), p. 675
6. Fesen RA, Gerardy CL, McLin KM, Hamilton AJS, 1999, ApJ 514, 195

Assessing the Nature of SNR Candidates in M 101

B. C. Dunne[1], R. A. Gruendl[1], Y.-H. Chu[1], and R. M. Williams[2]

[1] *Astronomy Department, University of Illinois, Urbana, IL 61801, USA*
[2] *NASA/GSFC, Greenbelt, MD 20771, USA*

Abstract.
We are conducting a study of SNR candidates in M 101 that have been identified in ground-based optical surveys. We have used archival *HST* WFPC2 Hα and continuum images, complemented by ground-based high-dispersion spectra. In this study, we determine the physical properties of the remnants, and use the continuum images to examine the stellar environment of the SNR candidates in order to determine whether the large candidates are superbubbles and to infer whether each remnant has a massive progenitor. Our study of the M 101 SNRs will aid our understanding of the distribution and environment of SNRs in a giant-spiral galaxy, and allow us to search for unusual SNRs in an enlarged sample. Here, we present preliminary results of this study.

INTRODUCTION

Supernova remnants (SNRs) play an important role in the structure and evolution of the interstellar medium of a galaxy. Surveys of extragalactic SNRs, besides the Magellanic Cloud SNRs, are limited by the poor resolution and sensitivity of X-ray and radio instruments and hence can only be carried out in the optical using the [SII]/Hα ratio. Of the SNR candidates identified at a distance of a few Mpc, ~50% are unresolved, ~20% are larger than 100 pc, and ~50% have only modest [SII]/Hα ratios. These properties suggest that many of the candidates are either not SNRs or are unusual remnants.

We have used archival *HST* WFPC2 Hα and continuum images, complemented by ground-based high-dispersion spectra, to study SNR candidates in M 101 (D=5.4 Mpc) that have been identified in ground-based optical surveys [1]. We determine the physical properties of the remnants from the Hα images and the high-dispersion spectra. Our study of the M 101 SNRs will aid our understanding of the distribution and environment of SNRs in a giant-spiral galaxy, and allow us to search for young and/or unusual SNRs in an enlarged sample. Here, we present preliminary results of 7 SNR candidates in M 101.

CP565, *Young Supernova Remnants: Eleventh Astrophysics Conf.*, edited by S. S. Holt and U. Hwang
© 2001 American Institute of Physics 0-7354-0001-6/01/$18.00

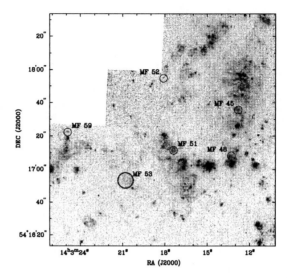

FIGURE 1. *HST* WFPC2 Hα image showing MF 45, MF 48, MF 51, MF 52, MF 53, MF 59, and their surrounding region in M 101.

OBSERVATIONS & ANALYSIS

Ninety-three SNR candidates were identified in M 101 [1]. In a search of the *HST* archives, 26 of the SNR candidates were found in archival WFPC2 Hα and continuum images[1]. We have obtained high-dispersion echelle spectra for 20 of these candidates with the Kitt Peak 4-m telescope in June 1999 and April 2000[2].

We have reduced and made a preliminary analysis of 7 of the observed SNR candidates: MF 25, MF 45, MF 48, MF 51, MF 52, MF 53, and MF 59. The WFPC2 Hα image containing 6 of these candidates is shown in Figure 1. The background-subtracted Hα-[NII] regions of the echelle spectra for the SNR candidates are presented in Figure 2. We have measured the Hα expansion velocity and [NII]/Hα flux ratios for each SNR candidate. As the SNR candidates are not spatially resolved, we have taken the expansion velocity to be the half-width at zero intensity. The spectral and physical properties of the candidates are presented in Table 1.

The lack of high expansion velocities (> 1000 km/s) indicates that none of these SNRs is very young. We are currently in the process of compiling a dataset on older LMC SNRs to compare with our M101 dataset. The LMC observations were made in December 2000.

[1] Based on observations with the NASA/ESA Hubble Space Telescope, obtained from the Data Archive at the Space Telescope Science Institute, which is operated by the Association of Universities for Research in Astronomy, Inc. under NASA contract No. NAS5-26555.

[2] High-dispersion spectra obtained at Kitt Peak National Observatory, National Optical Astronomy Observatories, which is operated by the Association of Universities for Research in Astronomy, Inc. (AURA) under cooperative agreement with the National Science Foundation.

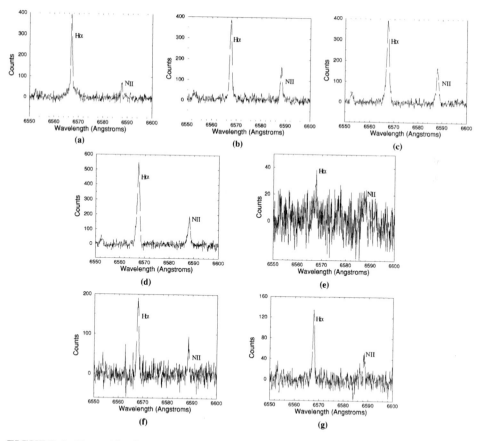

FIGURE 2. Hα and [NII] spectra for (a) MF 25, (b) MF 45, (c) MF 48, (d) MF 51, (e) MF 52, (f) MF 53, and (g) MF 59.

TABLE 1. Properties of the SNR Candidates

SNR	Diameter[1] (pc)	I(Hα)[1] (erg cm^{-2} sec^{-1})	Hα v_{exp} km sec^{-1}	[NII]/Hα
MF 25	30	7.5×10^{-15}	~230	0.18(0.36*)
MF 45	110 × 80	8.7×10^{-15}	~120	0.41
MF 48	stellar	1.7×10^{-14}	~120	0.41
MF 51	120	1.0×10^{-14}	~90	0.34
MF 52	9	9.4×10^{-16}	~150	0.51
MF 53	170	9.6×10^{-15}	~45	0.30
MF 59	40	1.7×10^{-15}	~225	0.48

* Narrow components only

DISCUSSION

Most of the SNR candidates exhibit an expansion velocity > 100 km/s, providing additional evidence that they are likely true SNRs. MF 53, however, is expanding at a significantly slower rate than 100 km/s. MF 53 is also the largest of the candidates at 170 pc across. Together, these suggest that MF 53 may be a superbubble rather than an SNR. Indeed, it has been shown that superbubbles can exhibit [SII]/Hα ratios > 0.5, the detection criteria for many optical SNR surveys [2]. MF 45 and MF 51 also exhibit large sizes (> 100 pc) and expansion velocities ~ 100 km/s. This suggests that MF 45 and MF 51 may be remnants of an unusual nature.

MF 25 shows a narrow Hα component superposed on a broad Hα component. This two-component feature is not detected in the [NII] line; the signal-to-noise level of the [NII] line, however, is much lower than that of the Hα line. The narrow component has an FHWM of ~ 30 km/s. MF 59 shows a similar structure, but the broad component is too close to the noise level for a definite detection. The other SNRs do not exhibit a two-component structure at detectable levels. MF 25 also exhibits a much lower [NII]/Hα ratio than the other SNRs. When the broad component of MF 25's Hα line is removed, however, the [NII]/Hα ratio becomes consistent with that of the other SNRs. This, along with the velocity structure of the Hα line, suggests that the SNR shell of MF 25 is interacting with circumstellar material around the remnant. The velocity structure is therefore a result of emission from both pre-shock and post-shock material. This is a similar conidition to the structure seen in the ultraluminous SNR of NGC 6946 [3].

CONCLUSIONS

We have presented preliminary results from our study of the SNR candidates in M 101. While the majority of the SNR candidates observed appear to be normal SNRs, several of the candidates exhibit unusual and interesting properties that call into question their status as SNRs. Once the study is completed, we hope that such interesting SNR candidates can be further studied to determine their true natures.

REFERENCES

1. Matonick, D.M., & Fesen R.A. 1997, ApJS, 112, 49
2. Chen, C.-H.R., Chu, Y.-H., Gruendl, R.A., & Points, S.D. 2000, AJ, 119, 1317
3. Dunne, B.C., Gruendl, R.A., & Chu, Y.-H. 2000, AJ, 119, 1172

Global VLBI Observations of M82

A. R. McDonald[1], T. W. B. Muxlow[1], A. Pedlar[1], M. A. Garrett[2],
K. A. Wills[3], S. T. Garrington[1], P. J. Diamond[1]

[1] *University of Manchester, Jodrell Bank Observatory, Macclesfield, UK*
[2] *Joint Institute for VLBI in Europe, Dwingeloo, The Netherlands*
[3] *University of Sheffield, UK.*

Abstract.
    Images are presented of the starburst galaxy M82 from a 20-station global very-long baseline interferometry (VLBI) experiment with an angular resolution of up to 3 mas. These images resolve several of the sources into a number of components and with future VLBI experiments the expansion of these components may be tracked on a $\sim$2 year timescale. Expansion of one of the brightest remnants in M82 has already been measured on a $\sim$10 year timescale using the European VLBI Network (EVN) alone and an expansion rate of 10,000 km s^{-1} calculated. Furthermore, limits on the deceleration of this remnant have been deduced. Further VLBI experiments will be able to better constrain the deceleration of the remnants on a $\sim$10 year timescale as they continue to interact with the interstellar medium in M82.

INTRODUCTION

M82 is generally considered to be the archetypal starburst galaxy. Far-infrared and radio luminosities imply a high star-formation rate and at a distance of 3.2 Mpc [1] it lends itself well to study with radio interferometry on parsec and sub-parsec scales. Radio images of M82 show a population of $\sim$50 compact (< 1 arcsecond) sources superimposed on a diffuse structure. Subsequent MERLIN observations [4] have resolved many of these sources and their shell-like structures combined with their non-thermal spectra [7] have led to their identification as young supernova remnants.

Early VLBI experiments concentrated on the brightest and most compact radio source in M82, 41.95+575. The European VLBI Network (EVN) 5 GHz map of Wilkinson & de Bruyn (1990) [6] showed structure which was interpreted as an elongated shell and limits were placed on its expansion speed of < 14000 km s^{-1}.

CP565, *Young Supernova Remnants: Eleventh Astrophysics Conf.*, edited by S. S. Holt and U. Hwang

RADIO SOURCES IN M82

The most compact source, 41.95+575

The new images of 41.95+575 show significant structure on smaller scales than had previously been observed with VLBI (figure 1). Structure is now seen which seems to link the well established NE and SW hotspots. This structure was previously suggested by the map of Wilkinson & de Bruyn (1990) [6] which these observations seem to confirm. In the 13.1 years which separates this EVN 5 GHz map and our global VLBI observations the source appears to have expanded along the major axis by 0.05 ± 0.02 pc. Given the uncertainties in identifying equivalent source structure between epochs and observing bands we prefer to express this as an upper limit to the radial expansion rate of 2000 km s^{-1}.

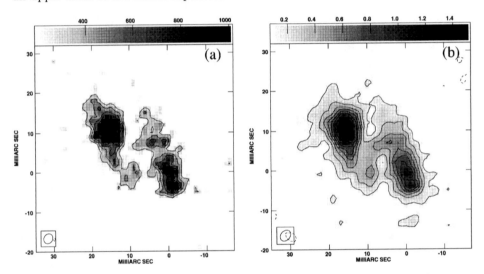

FIGURE 1. The compact radio source, 41.95+575. (a) shows a map made with purely uniform weighting ($\sim$ 2.5 mas resolution) and (b) shows a map made with an intermediate weighting scheme ($\sim$ 3 mas resolution).

An adequate supernova-related model which fully explains both the flux density decay ($\sim$8 percent per year) and the unusual radio morphology of this source has yet to be constructed. Hence, we pose the question as to whether this source is really supernova-related.

The shell-like remnant, 43.31+592

The global VLBI images reported in this paper (figure 2) represent the third epoch for which a map has been produced of sufficient resolution to image the structure of this source. Observations carried out with the EVN at λ18 cm in 1986 and 1997 provided the first two epochs [5]. A disadvantage associated with the study of the SNR in M82 is that we do not know the explosion dates of the supernovae, since any optical monitoring is severely hampered by dust extinction. This clearly adds an extra free parameter to any models which we may want to derive.

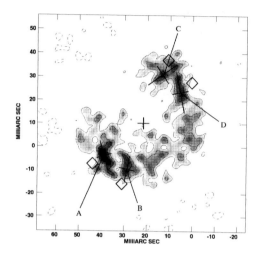

FIGURE 2. The shell-like remnant, 43.31+592. Labels A to D mark the positions of the most significant 'knots' of radio emission and the diamonds show their predicted positions in 10 years.

Using the nomenclature of Huang et al. (1994) [2] we parameterise the size evolution of the shell as $D_{pc} = kT_{yr}^{\delta}$. As well as the size of the remnant at epochs 1986, 1997 and 1998 we also know that the remnant existed in 1972 since it appears to be present in the 8.1 GHz map of Kronberg & Wilkinson (1975) [3]. This allows us to place a limit on the overall 'deceleration parameter' between 1972 and 1998 of $\delta > 0.73 \pm 0.11$ and a corresponding age of between 28 and 35 years. Figure 3 shows some example size evolution curves which fit the data for various values of the deceleration parameter. This implies that the remnant hasn't been in the adiabatic Sedov phase of its evolution for much of that time, although clearly the size evolution is heavily under-sampled.

CONCLUSION

We have imaged the most compact radio sources in M82 with a global VLBI array at λ18 cm. The brightest and most compact source, 41.95+575, shows an unusual morphology which has yet to be fully explained by a supernova-related model and we have placed a limit on the expansion speed of the source. The shell-like remnant 43.31+592 has been imaged with a linear scale comparable to the VLA

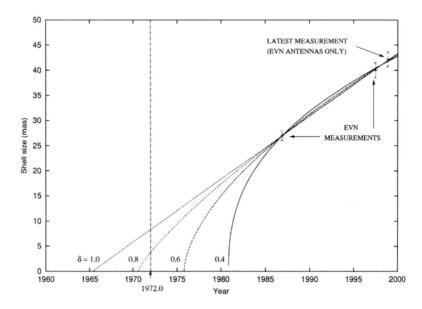

FIGURE 3. Example size evolution curves for the shell-like SNR, 43.31+592.

images of the galactic supernova remnant, Cassiopeia A. A limit has been placed on the power-law deceleration of this SNR. With future global VLBI experiments is should be possible to measure the expansion of this remnant on a ∼2 year timescale and its deceleration on a ∼10 year timescale.

The young SNR in M82 fill a gap in parameter space between the more evolved SNR of both the Milky Way and the Magellanic Clouds and young radio supernovae such as SN1993J.

REFERENCES

1. Burbidge E. M., Burbidge G. R., and Rubin V. C. 1964, ApJ, 140, 942
2. Huang Z. P., Thuan T. X., Chevalier R. A., Condon J. J. and Yin Q. F. 1994, ApJ, 424, 114
3. Kronberg P. P. and Wilkinson P. N. 1975, ApJ, 300, 430
4. Muxlow T. W. B., Pedlar A., Wilkinson P. N., Axon D. J., Sanders E. M. and de Bruyn A. G. 1994, MNRAS, 266, 455
5. Pedlar A., Muxlow T. W. B., Garrett M. A., Diamond P. J., Wills K. A., Wilkinson P. N. and Alef W. 1999, MNRAS, 307, 761
6. Wilkinson P. N. and de Bruyn A. G. 1990, MNRAS, 242, 529
7. Wills K. A., Pedlar A., Muxlow T. W. B. and Wilkinson P. N. 1997, MNRAS, 291, 517

X-ray and Radio Observations of Supernova Remnants in NGC 300

Thomas Pannuti[1,2], Miroslav D. Filipović[3,4], Nebojsa Duric[2],
Wolfgang Pietsch[4] and Andrew Read[4]

[1] *MIT Center for Space Research, NE80-6015, 70 Vassar Street, Cambridge, MA 02139*
[2] *Department of Physics & Astronomy, University of New Mexico, Albuquerque, NM 87131*
[3] *University of Western Sydney-Nepean, Kingswood, NSW 2747, Australia*
[4] *Max-Planck-Institut für Extraterrestrische Physik, D-85748-Garching, Germany*

Abstract. We present the results of our search for candidate X-ray and radio supernova remnants (SNRs) in the nearby Sculptor Group Galaxy NGC 300. We have made radio observations of this galaxy using the Very Large Array at 6 cm and 20 cm and supplemented this data with archived ROSAT PSPC observations. We find that the majority of the optically-identified SNRs in this galaxy are not detected at the 3-sigma level in either the X-ray or the radio, possibly indicating a bias effect in optical surveys that favors the detection of SNRs outside of HII regions. We have also prepared samples of candidate radio and X-ray SNRs in this galaxy. Our work has revealed sixteen new candidate X-ray and radio SNRs, complementing the 28 SNRs previously detected in this galaxy using optical methods and raising the total number of known SNRs in this galaxy to 44.

INTRODUCTION

We are currently conducting a multi-wavelength campaign to search for supernova remnants (SNRs) in nearby galaxies, using a combination of X-ray, optical and radio observations. Traditionally, Galactic SNRs have been the best-studied SNRs, but a thorough understanding of Galactic SNRs is prevented by several obstacles, such as massive extinction along lines of sight within the Galaxy and significant uncertainties in the distances to these sources. In contrast, the study of samples of SNRs in nearby galaxies offers the opportunity to examine a sample of sources where the effects of these obstacles can be minimized. By choosing galaxies located at high Galactic latitudes with nearly face-on orientations, the effects of both Galactic and internal extinction are decreased considerably. In addition, uncertainties in the distances to extra-galactic SNRs are reduced to uncertainties in the distance to the target galaxy. Therefore, the creation of samples of extra-galactic SNRs will allow a thorough examination of SNRs that is nearly free of the limitations which have hampered the study of Galactic SNRs.

CP565, *Young Supernova Remnants: Eleventh Astrophysics Conf.*, edited by S. S. Holt and U. Hwang

TABLE 1. X-ray and Radio Properties of Optically-Identified SNRs in NGC 300

	X-ray Emission[a]			Radio Emission		
SNR	log L$_{Total}$ (ergs s^{-1}) (0.1-2.0 keV)	Column (10^{20} cm^{-2})	kT (keV)	S$_6$ (mJy)	S$_{20}$ (mJy)	α (S$_\nu \propto \nu^{-\alpha}$)
N300-S10	36.35	$20.3^{+33.4}_{-20.3}$	$0.10^{+0.03}_{-0.10}$	0.14±0.04	0.29±0.07	+0.6±0.3
N300-S11	–	–	–	0.39±0.06	0.89±0.15	+0.7±0.2
N300-S26	36.50	$6.98^{+17.6}_{-3.2}$	$0.51^{+0.45}_{-0.29}$	–±0.10	0.22±0.07	>0.7

[a] Based on thermal bremsstrahlung fits made by Read and others [3].

We present the results of our search for radio and X-ray SNRs in the nearby Sculptor Group galaxy NGC 300. This galaxy has already been the subject of two optical surveys for SNRs, first by D'Odorico and others [1] and more recently by Blair and Long [2]. Both of these works searched for SNRs in NGC 300 by identifying emission nebulae in [S II] and Hα images of this galaxy that possessed flux ratios of [S II]/Hα ≥ 0.4, and these works yielded a total of 28 optically-identified SNRs. We have searched for X-ray and radio counterparts of these optically-identified SNRs as well as for new candidate X-ray and radio SNRs in this galaxy.

OBSERVATIONS

NGC 300 was observed using the Very Large Array at the wavelength of 6 cm in the CnB configuration and at the wavelength of 20 cm in the BnA configuration. The Hα image was made using the 0.6 m Curtis/Schmidt telescope at the Cerro Tololo Inter-American Observatory. Finally, observations made of NGC 300 using the ROSAT PSPC for a total integration time of 46306 seconds were obtained from the Goddard Space Flight Center Archives.

Candidate radio SNRs were identified as non-thermal radio sources that were coincident with regions of Hα emission in NGC 300. To identify candidate X-ray SNRs, we considered the 15 X-ray sources in this galaxy that were detected and analyzed by Read and others [3] in the ROSAT PSPC observations. Those authors attempted model fits of the spectra of these sources, and found that some of them were best fit by thermal bremsstrahlung models with energy distributions that peaked at $kT \approx 1$ keV or less. Those particular X-ray sources which were also associated with regions of Hα emission in NGC 300 were identified as candidate X-ray SNRs.

RESULTS

Surprisingly, of the 28 previously known optically-identified SNRs in this galaxy, only three – NGC 300-S10, NGC 300-S11 and NGC 300-S26 – were also detected at another wavelength. Two of these SNRs, NGC 300-S10 and NGC 300-S26, possess

TABLE 2. New Candidate X-ray Supernova Remnants in NGC 300

SNR[a]	RA (J2000.0)	Dec (J2000.0)	log L_{Total} (ergs s^{-1}) (0.1-2.0 keV)	Column (10^{20} cm^{-2})	kT (keV)
#1	00 54 31.73	-37 38 16.4	36.17	$16.9^{+47.4}_{-16.9}$	$0.58^{+1.77}_{-0.48}$
#2	00 54 37.67	-37 42 42.7	36.35	$20.3^{+33.4}_{-20.3}$	$1.29^{+8.58}_{-1.29}$
#5	00 54 45.07	-37 41 32.5	36.56	$70.1^{+28.5}_{-36.2}$	$0.13^{+0.08}_{-0.05}$
#11	00 55 20.60	-37 48 10.3	36.57	$1.33^{+0.92}_{-0.75}$	$0.57^{+0.27}_{-0.16}$

[a] Identifications and spectral fits made by Read and others [3].

counterparts in the X-ray (sources #4 and #10, respectively, from the list of Read and others [3]) and the radio, while NGC 300-S11 possesses a counterpart in the radio but not in the X-ray. The X-ray and radio properties of these SNRs are listed in Table 1, and multi-wavelength images of NGC 300-S10 are given in Figure 1.

Our separate searches for new candidate SNRs yielded fourteen candidate radio SNRs and four candidate X-ray SNRs. Two of the candidate radio SNRs (NGC 300-R2 and NGC 300-R5) are coincident with two of the candidate X-ray radio SNRs (#2 and #5, respectively), and when these redundancies are taken into account, a total of 16 new candidate SNRs have been found by our survey to complement the 28 SNRs previously known through optical searches. Properties of these new SNRs are listed in Tables 2 and 3.

CONCLUSIONS

The failure to detect X-ray and radio emission from the vast majority of optically-identified SNRs in this galaxy is a remarkable result of this work. In addition, by including new SNRs detected at X-ray and radio wavelengths we have increased the number of SNRs known in this galaxy by 16 to a total of 44. We hypothesize that optical searches for SNRs are biased toward the detection of sources in regions of low density and low optical confusion: such surveys would therefore select against SNRs located in regions of massive star formation, and thus may preferentially select the SNRs parented by low mass Population II stars (namely, white dwarfs). In contrast, X-ray and radio searches favor the detection of SNRs created in regions of high density (which are conducive for the production of X-ray and radio emission by SNRs), and may therefore preferentially select the SNRs parented by high mass Population I stars.

REFERENCES

1. D'Odorico, S. et al. 1980, A&AS, **40**, 67.
2. Blair, W. P. and Long, K. S. 1997, ApJSS, **108**, 261.
3. Read, A. M. et al. 1997, MNRAS, **286**, 626.

TABLE 3. New Candidate Radio Supernova Remnants in NGC 300

SNR	RA (J2000.0)	Dec (J2000.0)	S_6 (mJy)	S_{20} (mJy)	α ($S_\nu \propto \nu^{-\alpha}$)
NGC 300-R1	00 54 38.2	−37 41 47	−±0.10	0.36±0.07	>1.1
NGC 300-R2	00 54 38.4	−37 42 42	−±0.10	0.60±0.14	>1.5
NGC 300-R3	00 54 43.4	−37 43 11	0.27±0.04	0.57±0.10	+0.6±0.2
NGC 300-R4	00 54 44.9	−37 41 10	−±0.10	0.24±0.07	>0.7
NGC 300-R5	00 54 45.1	−37 41 49	−±0.10	0.23±0.07	>0.7
NGC 300-R6	00 54 50.3	−37 40 31	0.14±0.04	0.30±0.10	+0.6±0.4
NGC 300-R7	00 54 51.1	−37 40 59	−±0.10	0.22±0.07	>0.7
NGC 300-R8	00 54 51.1	−37 41 45	−±0.10	0.30±0.10	>0.9
NGC 300-R9	00 54 51.3	−37 46 22	−±0.10	0.24±0.07	>0.7
NGC 300-R10	00 54 51.8	−37 39 39	0.28±0.07	0.48±0.09	+0.4±0.3
NGC 300-R11	00 55 03.6	−37 42 49	0.28±0.04	0.35±0.11	+0.2±0.3
NGC 300-R12	00 55 03.7	−37 43 21	0.13±0.04	0.42±0.10	+1.0±0.6
NGC 300-R13	00 55 12.6	−37 41 38	0.30±0.06	0.50±0.14	+0.4±0.3
NGC 300-R14	00 55 30.1	−37 39 20	−±0.10	0.28±0.07	>0.9

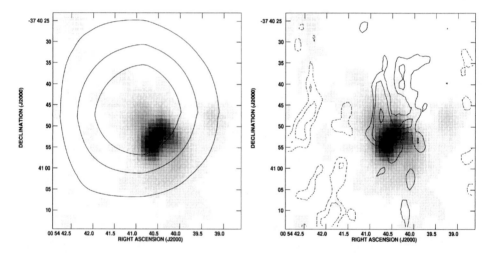

FIGURE 1. Multi-wavelength images of the optically-identified SNR NGC 300-S10. Hα emission is depicted in greyscale in both images, while X-ray and 6 cm emission is depicted in contours in the left and right images, respectively. The SNR is located on the northern edge of the bright HII region.

X-ray Fluxes of Radio-Selected Remnants

E.A. Miller[1] and L. Rudnick[2]

[1] *University of Maryland, College Park, MD 20742, USA*
[2] *University of Minnesota, Minneapolis, MN 55455, USA*

Abstract.
    We report the initial measurements of X-ray fluxes from the RASS for SNRs from the Green catalog. There is a significant correlation between radio and X-ray fluxes, but with approximately two orders of magnitude scatter in X-ray flux at a given radio flux level. Surprisingly, there is no apparent difference in the distributions for shell and for non-shell remnants. Some of the correlation is due to the largest sources being the strongest. Work continues on these measurements, with a goal of obtaining X-ray fluxes or limits on all Green Catalog sources with diameters less than 30 arcminutes.

INTRODUCTION

A variety of important physical questions regarding SNRs rest on their relative luminosities at different bands. To our knowledge, there is no unbiased compilation of X-ray flux densities, accumulated in a uniform way, which can be compared with radio fluxes. The closest approach to this was the early Seward catalog [1], but it contains only X-ray bright sources, and no upper limits.

Towards this end, we have begun a small program of measuring RASS [2] fluxes for SNRs in the Green catalog [3], to compare with the radio values. Although there are many non-detections at the RASS sensitivity – which could be supplemented with pointed observations where available – it offers the advantage of a reasonably uniform sensitivity limit for radio- or other non-X-ray selected samples.

I   MEASUREMENTS

The region around each SNR was downloaded from the ROSAT Data Browser [4]. Only results from the 1-3 keV band are shown here. We first determined the approximate centroid of the x-ray emission, using the radio position if there was insufficient X-ray signal. The incremental and cumulative fluxes in rings surrounding this centroid were then plotted, and used to identify a local background region, free from the target SNR or nearby sources. A least squares fit of the cumulative

CP565, *Young Supernova Remnants: Eleventh Astrophysics Conf.,* edited by S. S. Holt and U. Hwang

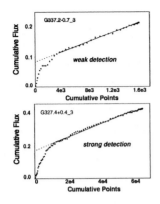 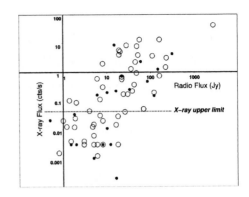

FIGURE 1. Left: Plots of cumulative flux vs. cumulative number of pixels in concentric rings around X-ray centroid for two remnants. The lines show the least square fits to the source-free regions around the remants. The extrapolation to zero points gives an estimate of the total remnant flux. Right: Plot of X-ray count rate vs. radio flux as described in the text, for all sources measured to date. This is a biased sample, including sources selected to be X-ray bright. Open symbols are shell-remnants; filled symbols represent all others.

flux as a function of the cumulative number of pixels was then used to calculate the total X-ray flux and errors. Cumulative flux plots of two sources are shown in Figure 1. At low count rates, the total X-ray flux became increasingly sensitive to the choice of backround range; robust values were obtained above a count rate of ≈ 0.05 cts/s in the 1-3 keV band.

II A BIASED SAMPLE

Figure 1 (right) shows the initial results from this study, plotting X-ray flux (total cts/sec in the 1-3 keV band) vs. radio flux (Jy) from the Green Catalog, and normalized to a frequency of 1.0 GHz. It contains both sources selected from the Green catalog (unbiased with regards to X-ray flux), supplemented by strong X-ray sources taken from the Seward catalog. X-ray fluxes below 0.05 counts/sec should all be considered upper limits.

There is clearly a significant correlation between X-ray and radio fluxes, some of which is due to the size of the source, as discussed further below. At each radio flux, however, there is a spread of at least 2 orders of magnitude in the X-ray fluxes. It is not clear what variable regulates this large range. Curiously, there is no clear difference in the distribution of shell (open-circle) and non-shell (filled-circle)

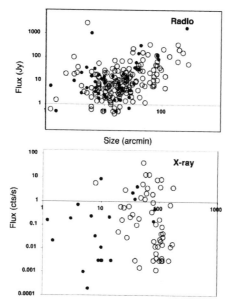

FIGURE 2. Flux dependence on size for both X-ray fluxes as measured here, and radio fluxes for all SNRs in the Green catalog. Symbols as in previous figure.

SNRS.

III SIZE DEPENDENCE

Figure 2 explores the dependence of X-ray and radio fluxes on the SNR diameter, defined as the geometric mean between the major and minor diameters quoted by Green. The radio sources represent the entire Green catalog. The largest sources are indeed the strongest, although there is very large scatter, and no apparent *flux vs. size* dependence below 30 arcmin diameter. There is no obvious difference between the shell (open symbols) vs. non-shell (closed symbol) remnants.

Turning to the X-ray fluxes (as measured here), we again see a tendency for the largest sources to be the brightest, but with a very large scatter and many faint, large sources. The paucity of large, non-shell X-ray sources may be due in part to a selection bias, and should not yet be taken seriously.

IV AN UNBIASED SAMPLE

We selected all remnants with diameters less than 15 armin in a restricted latitude range from Green's catalog. Figure 3 shows the X-ray, radio flux relationship for that subgroup, which is therefore unbiased with regards to X-ray flux. The observed

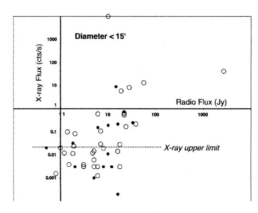

FIGURE 3. Plot of X-ray count rate vs. radio flux as described in the text, for a radio-selected, unbiased X-ray sample with diameters less than 15 arcminutes. Symbols as in previous figure.

distribution mimics the correlation seen in the biased sample, again with no obvious difference between shell and non-shell sources.

We intend to extend these measurements to all sources in Green's catalog with diameters < 30′. Initial scientific objectives will be to quantify the dependence on size, the relationship between X-ray and radio emissivities, and to explore the missing parameter(s) which regulate the large spread in their ratios. A remaining curiousity is the similarity between shell- and non-shell remnant distributions. We will carefully examine selection effects which could influence this result.

ACKNOWLEDGEMENTS

We are pleased to acknowledge support for SNR research at Minnesota from NSF grant AST96-19438 and REU support for EAM through NSF grant PHY97-32157 to the Univ. of Minn. We are grateful for the extensive help of Jakob Englhauser in our early use of the ROSAT Data Browser, and for D. Green's continuing toil in maintaining the SNR catalog. R. Petre provided the inspiration for this work and expert advice at some critical moments.

REFERENCES

1. Seward, F. D. 1990, ApJS, 73, 781
2. Voges, W., 1992, in Proc. of Satellite Symp. 3, ed. T.D. Guyenne & J.J. Hunt, ESA ISY-3 (Noordwijk:ESA), 9
3. Green D.A., 2000, 'A Catalogue of Galactic Supernova Remnants (2000 August version)', Mullard Radio Astronomy Observatory, Cavendish Laboratory, Cambridge, United Kingdom (http://www.mrao.cam.ac.uk/surveys/snrs/)
4. ROSAT Data Browser, (http://wave.xray.mpe.mpg.de/rosat/data-browser)

Hypernovae

Supernovae, Hypernovae and Gamma Ray Bursts

Arnon Dar

Physics Department and Space Research Institute, Technion, Haifa 32000, Israel

Abstract. Recent observations suggest that gamma ray bursts (GRBs) and their afterglows are produced by highly relativistic jets emitted in core collapse supernova explosions (SNe). The result of the event, probably, is not just a compact object plus a spherical ejecta: within a day, a fraction of the parent star falls back to produce a thick accretion disk around the compact object. Instabilities in the disk induce a sudden collapse with ejection of jets of highly relativistic "cannonballs" of plasma in opposite directions, similar to those ejected by microquasars. The jet of cannonballs exit the supernova shell/ejecta reheated by their collision with it, emitting highly forward-collimated radiation which is Doppler shifted to γ-ray energy. Each cannonball corresponds to an individual pulse in a GRB. They decelerate by sweeping up the ionised interstellar matter in front of them, part of which is accelerated to cosmic-ray energies and emits synchrotron radiation: the afterglow. The Cannonball Model cannot predict the timing sequence of these pulses, but it fares very well in describing the total energy, energy spectrum, and time-dependence of the individual γ-ray pulses and afterglows. It also predicts that GRB pulses are accompanied by detectable short pulses of TeV neutrinos and sub TeV γ-rays, that are much more energetic and begin and peak a little earlier.

INTRODUCTION

I was asked to discuss hypernovae — hypothetical spherical fireballs that are generated by gravitational collapse of very massive stars and generate GRBs [1,2]. But "I come to bury Caesar not to praise him" (Shakespeare, "Julius Caesar" Act. III Sc.II): Various observations suggest that **most** GRBs are produced by highly collimated ultrarelativistic jets from stellar collapse [3,4], probably, from supernova explosions [5–11] and not in spherical explosions that convert kinetic energy to GRBs with total γ-ray energy in excess of 10^{54} erg.

In my talk I shall review briefly the evidence that GRBs are associated with SNe. Then I will review the recently proposed Cannonball (CB) Model of GRBs that was recently proposed by Dar and De Rújula [10,11] and explains how GRBs are produced in SNe. Its success in describing the total energy, energy spectrum, the time-dependence of the individual γ-ray pulses in GRBs and the GRB afterglows is demonstrated in an expanded version of this talk [12].

CP565, *Young Supernova Remnants: Eleventh Astrophysics Conf.*, edited by S. S. Holt and U. Hwang

THE GRB–SNE ASSOCIATION

There is mounting evidence for an association of supernova explosions (SNe) and GRBs. The first example was GRB 980425 [13,14] within whose error circle SN1998bw was soon detected optically [15] and at radio frequencies [16]. The chance probability for a spatial and temporal coincidence is less than 10^{-4} (e.g. [15]), or much smaller if the revised BeppoSAX position (e.g., [17]) is used in the estimate. The unusual radio [16,18] and optical [15,19] properties of SN1998bw, which may have been blended with the afterglow of GRB 980425, support this association. The exceptionally small fluence and redshift of GRB 980425 make this event very peculiar, a fact that we discuss in detail in section 7.

Evidence for a SN1998bw-like contribution to a GRB afterglow [20] was first found by Bloom et al. [21] for GRB 980326, but the unknown redshift prevented a quantitative analysis. The afterglow of GRB 970228 (located at redshift z = 0.695) appears to be overtaken by a light curve akin to that of SN1998bw (located at $z_{bw} = 0.0085$), when properly scaled by their differing redshifts [22]. Let the energy flux density of SN1998bw be $F_{bw}[\nu, t]$. For a similar supernova located at z:

$$F[\nu, t] = \frac{1+z}{1+z_{bw}} \frac{D_L^2(z_{bw})}{D_L^2(z)} \times$$
$$F_{bw}\left[\nu \frac{1+z}{1+z_{bw}}, t \frac{1+z_{bw}}{1+z}\right] A(\nu, z), \qquad (1)$$

where $A(\nu, z)$ is the extinction along the line of sight. The SN–GRB association in the case of GRB 970228 was reconfirmed by Reichart [23] and by Galama et al. [24]. Evidence of similar associations is found for GRB 990712 [25,26], GRB 980703 [27] and GRB 000418: an example that we show in Fig. 5. In the case of GRB 990510 the observational evidence [28] is marginal. For the remaining cases in Table I the observational data preclude a conclusion, for one or more reasons: the late afterglow is not measured; $F_{bw}[\nu']$ is not known for large $\nu' \simeq \nu(1+z)$; the GRB's afterglow or the host galaxy are much brighter than the SN. The case of GRB 970508, for which the afterglow in the R band is brighter than a SN contribution given by Eq. (1), is shown in Fig. 6.

All in all, it is quite possible that a good fraction of GRBs are associated with SNe, perhaps even *all* of the most frequent, long-duration GRBs. The converse statement —that most SNe of certain types are associated with GRBs— appears at first sight to be untenable. The rate of Type Ib/Ic/II SNe has been estimated from their observed rate in the local Universe (e.g. [29]) and the star formation rate as function of redshift, to be 10 s^{-1} in the observable Universe [30]. The observed rate of GRBs is a mere 1000 y^{-1}.

Thus, very few of these SNe produce *visible* GRBs. But, if the SN-associated GRBs were beamed within an angle $\theta \sim 3.6 \times 10^{-3}$, only a fraction $\pi \theta^2 / 4\pi \sim 3 \times 10^{-6}$ would be visible, making the observed rates compatible and making possible a rough one-to-one SN–GRB association (or a ten-to-one association for $\theta \sim 1 \times 10^{-2}$).

THE CANNONBALL MODEL OF GRBS

The ejection of matter in a supernova (SN) explosion is not fully understood. The known mechanisms for imparting the required kinetic energy to the ejecta are inefficient: the theoretical understanding of core-collapse SN events is still unsatisfying. It has been proposed [31] that the result of a SN event is not just a compact object plus a spherical ejecta: a fraction of the parent star may fall back onto the newly born compact object. The infalling material with non-vanishing angular momentum settles into an accretion disk or a thick torus around the compact object. As observed in other cases of significant accretion onto a compact object, such as microquasars (e.g. [32–35] and active galactic nuclei in which the infalling material is processed in a series of "catastrophic" accretions, jets of relativistic CBs are ejected [5–7,10,11] and references therein).

The cannonball (CB) model of GRB [10,11] assumes that a sequence of highly relativistic CBs is emitted during a core-collapse SN. These cannonballs may be emitted during or right after the initial collapse [5–9] and hit a massive shell at a typical radius of $R_S \sim 3 \times 10^{15}$ cm formed by strong wind emission in the Red Supergiant or variable Blue Giant presupernova phase (for evidence see e.g., [36,37]). They may be emitted in a second collapse [31] at a time t_{fall} of $\mathcal{O}(1)$ day after a SN core-collapse. By this time the SN outer shell, traveling at a velocity $v_S \sim c/10$ (see, e.g. [38]) has moved to a distance: $R_S = 2.6 \times 10^{14}$ cm . As it hits a massive shell, the CB slows down and heats up. Its radiation is obscured by the shell up to a distance of order one radiation length from the shell's outer surface. As this point is reached the emitted radiation from the CB which continues to travel, expand and cool down, becomes visible. This radiation, which is boosted and collimated by the ultrarelativistic motion of the CB arrives as a short GRB pulse. The duration of the pulse is its radiative cooling time after it becomes visible. The total duration of a GRB with many pulses is the total emission time of CBs by the central engine. As the mechanism producing relativistic jets in accretion is not well understood, the CB model is unable to predict the timing sequence of the successive GRB pulses but, the CB model is quite successful in describing the total energy, energy spectrum and time-dependence *within single GRB pulses* [11].

Relativistic aberration, boosting and collimation

Let $\gamma = 1/\sqrt{1 - \beta^2} = E_{CB}/(M_{CB}c^2)$ be the Lorentz factor of a CB, that diminishes with time as the CB hits the SN shell and as it subsequently plows through the interstellar medium. Let t_{SN} be the local time in the SN rest system, t_{CB} the time in the CB's rest system and t the time of a stationary observer viewing the CB at redshift z from an angle θ away from its direction of motion. Let x be the distance traveled by the CB in the SN rest system. The relations between the above timings are:

$$dt = \frac{(1+z)\, dt_{CB}}{\delta} = \frac{(1+z)\, dt_{SN}}{\delta\,\gamma} = \frac{(1+z)\, dx}{\beta\,\gamma\,\delta\, c} \tag{2}$$

where the Doppler factor δ is:

$$\delta \equiv \frac{1}{\gamma\,(1 - \beta\cos\theta)} \simeq \frac{2\,\gamma}{(1 + \theta^2\gamma^2)}\,, \tag{3}$$

and its approximate expression is valid for $\theta \ll 1$ and $\gamma \gg 1$, the domain of interest here. In what follows we will set t=0 at the moment when the CB hits the shell. Notice that for large γ and $\theta\gamma \sim 1$, there is an enormous "relativistic aberration": $dt \sim dt_{SN}/\gamma\delta$ and the observer sees a long CB story as a film in extremely fast motion.

The energy of the photons radiated by a CB in its rest system, E^{γ}_{CB} and the photon energy, E, measured by a cosmologically distant observer are related by:

$$E = \frac{\delta\, E^{\gamma}_{CB}}{1 + z}\,, \tag{4}$$

with δ as in Eq.(3). If E^{rest}_{pulse} is the CB total emitted radiation in its rest frame, an observer at a luminosity distance $D_L(z)$ from the CB that is viewing it at an angle $\theta \ll 1$ from its direction of motion would measure a "total" (time- and energy-integrated) fluence per unit area:

$$\frac{dF}{d\Omega} \simeq \frac{(1+z)\, E^{rest}_{pulse}}{4\,\pi\, D^2_L}\, \delta^3\,. \tag{5}$$

Only if traveling with a large Doppler factor ($\gamma \gg 1$ and angle $\theta \sim 1/\gamma$ relative to the line of sight), will a CB at a cosmological distance be visible.

The making of the GRB

The general properties of GRB pulses in the CB model are not sensitive to the complex details of the CB's collision with the shell. They can be estimated from the overall energetics and approximate treatment of the cooling of the CB by radiation and expansion as it reaches the transparent outskirts of the shell.

A CB, in its rest system, is subject to a flux of high energy nuclei and electrons with energy $m\,c^2$, i.e., ~ 1 TeV per nucleon. Roughly 2/3 of the baryonic energy escapes through $\pi^{\pm}$ decay into muons and neutrinos and 1/3 is converted into γ-rays in the hadronic cascages (through $\pi^0 \to \gamma\gamma$ decay) within $X_p \approx m_p/\sigma_{in}(pp) \approx 50\,\mathrm{g\,cm^{-2}}$, where $\sigma_{in}(pp)$ is the nucleon-nucleon inelastic cross section. These high energy photons initiate electromagnetic cascades by $e^+ e^-$ pair production that convert their energy to thermal energy within the CB with a radiation length of $X_{\gamma e} \simeq 63$ g cm^{-2}.

While the electrons are being thermalized, they contribute a nonthermal high-energy tail of photons emitted via the "free-free" process. Such a power-law tail in an otherwise approximately-thermal emission is observed from young supernova remnants (see, e.g. [39]) and clusters of galaxies (e.g. [40–42], both of which are systems wherein a dilute plasma at a temperature of $\mathcal{O}(1\,\text{keV})$ is exposed to a flux of high energy cosmic rays. Thus, the CB emission can be modeled as a black body spectrum with a nonthermal power-law tail.

The density profile of the outer layers of an SN shell as a function of the distance x to the SN center as inferred from the photometry, spectroscopy and evolution of the SN emissions (e.g. [38]), can be fit by a power law, x^{-n}, with $n \sim 4$ to 8. The radiation length of thermalized photons in a hydrogenic plasma is $X_\gamma^{\text{ion}} \approx m_p/\sigma_T \approx 2.6\,\text{g cm}^{-2}$. If the quasi-thermal emission rate from the CB, within X_γ^{ion} from its surface, is in dynamical equilibrium with the fraction of energy deposited by the CB's collision with the shell in that outer layer, then the temperature of the CB's front is roughly given by:

$$T(x) \simeq \left[\frac{(n-1)\, X_\gamma\, m_p\, c^3\, \gamma^2\, \sigma_{\text{in}}(\text{pp})}{6\, \sigma\, x_{tp}\, X_{\gamma e}\, \sigma_T^2} \right]^{\frac{1}{4}} \left[\frac{x}{x_{tp}} \right]^{-\frac{n}{4}}, \tag{6}$$

where x_{tp} (t_{tp}) is the distance (time) where (when) the remaining column density of the SN shell is 1 radiation length. Remarkably, only the Lorentz factor of the CBs when they exit the shell, but neither their mass nor their energy, appear in the above expression, except for the fact that, for the result to be correct, they must be large enough for the CB to pierce the shell and remain relativistic.

The CB temperature at t_{tp} is not sensitive to the exact value of n, unlike its time dependence. For $n = 8$ and $M_s = 10\,\text{M}$. the value of x_{tp} is $\approx 3\,R_S$, and for t close to t_{tp} or later:

$$T(t) \simeq 0.16\,\text{keV} \left[\frac{t_{tp}}{t} \right]^2 \left[\frac{\gamma(t)}{10^3} \right]^{\frac{1}{2}}. \tag{7}$$

This estimate is valid as long as the surface temperature of the optically thick CB is higher than the internal temperature of the CB that under the assumption of isentropic expansion decreases with time only like $T_{\text{CB}} \sim 1/R_{\text{CB}} \sim 1/t$.

The total radiated energy, in the CB rest frame, is roughly the thermal energy deposition within one radiation length from its front surface. After attenuation in the SN shell, it reduces to:

$$E_{\text{pulse}}^{\text{rest}} \approx \frac{\sigma_{\text{in}}(\text{pp})\, \pi\, [R_{\text{CB}}^{tp}]^2\, \bar{X}_\gamma\, m_p\, c^2\, \gamma(t)}{3\, X_{\gamma e}\, \sigma_T^2}, \tag{8}$$

where $\bar{X}_\gamma$ is the radiation length in the obscuring shell averaged over the black body spectrum. For a typical γ-ray peak energy of $E_p \sim 1\,\text{MeV}$ in the SN rest frame,

$\bar{X}_\gamma \simeq 10\,\mathrm{g\,cm^{-2}}$. Consequently, the CB's radius at transparency is $R_{CB}^{tp} = 4 \times 10^{11}$ cm and $E_{pulse}^{rest} \sim 3 \times 10^{45}$ erg, for $\gamma(t) \sim 10^3$.

This is consistent with the estimated GRB energies in Table I provided that the typical Doppler factors of GRBs are in the range $300 \leq \delta \leq 1000$.

If, for the sake simplicity, the energy spectrum of the surface radiation from a CB is approximated by a thermal black body radiation then, the observed energy and time dependence of the photon intensity (photon number per unit area, N) of a single pulse in a GRB at an angle θ relative to the CB's motion is predicted to be:

$$\frac{dN}{dE\,dt} \equiv \frac{1+z}{4\,\pi\,D_L^2}\,\delta^2\,\frac{dn_\gamma}{dE\,dt}\,, \tag{9}$$

$$\frac{dn_\gamma}{dE\,dt} \simeq \frac{2\,\pi\,\sigma}{\zeta(3)}\,\frac{[R_{CB}[t]\,E\,(1+z)/\delta]^2\;Abs(E,t)}{Exp\,\{E\,(1+z)/(\delta\,T[t])\}-1}\,, \tag{10}$$

with $R_{CB}[t] \simeq c\,t/\sqrt{3}\gamma$ and $T[t]$ as in Eq.(6), and where

$$Abs(E,t) = Exp\left[-\frac{X_S(x[t])}{X_\gamma(E\,(1+z))}\right] \tag{11}$$

is the attenuation of the flux in the shell.

The mean photon energy of a black body radiation is approximately 2.7 T. Thus, around peak energy flux, the mean energy of the observed photons is

$$< E_\gamma > \simeq \frac{0.45}{1+z}\left[\frac{\delta}{10^3}\right]\,\mathrm{MeV}\,. \tag{12}$$

It yields $< E_\gamma > \simeq 0.23\,\mathrm{MeV}$ for $\delta \simeq 10^3$ and z=1 –the mean redshift of the GRBs listed in Table I–, in good agreement with that measured in GRBs [43].

For n = 4 the temperature decrease approximately as $1/t$. For n > 4 it diminishes faster than $1/t$ and for n = 8 it decreases faster than $1/t^2$, the "faster" being due, in both cases, to the effect of a decreasing $\gamma(t)$. For $T \sim 1/t^2$ behaviour, the pulse width narrows with time like $\sim E^{-0.5}$ in agreement with the analysis of Fenimore et al., [44] who found, from a large sample of GRB pulses, that it narrows like $t \propto E^{-0.46}$.

SOME SIMPLIFICATIONS AND APPROXIMATE PREDICTIONS

Eqs. (10–11) yield approximately,

$$\frac{dN}{dE\,dt} \propto \frac{(E\,t)^2}{Exp\{E\,t/H\}-1}\,Exp\left\{-[t_{tp}/t]^{n-1}\right\}\,\Theta[t]\,. \tag{13}$$

The total photon intensity and energy flux are, in this approximation:

$$\frac{dN}{dt} \propto \Theta[t] \, \frac{t_{tp}}{t} \, \text{Exp} \left\{ - [t_{tp}/t]^{n-1} \right\} , \tag{14}$$

$$F_E(t) \propto \Theta[t] \, \left[\frac{t_{tp}}{t} \right]^2 \, \text{Exp} \left\{ - [t_{tp}/t]^{n-1} \right\} . \tag{15}$$

Let the peak γ-ray energy at a fixed time during a GRB pulse be defined as $E_p^\gamma(t) \equiv \max [E^2 \, dI_\gamma/dE \, dt]$. Its value is $E_p^\gamma(t) \simeq 3.92 \, \delta \, T[t]/(1+z)$, so that, for t near or after t_{tp}:

$$E_p^\gamma(t) \simeq E_p^\gamma(t_{tp}) \, \Theta[t] \, \frac{t_{tp}}{t} . \tag{16}$$

The total "isotropic" energy of a GRB pulse – inferred from its observed fluence assuming an isotropic emission– can be deduced from Eq. (5), to be:

$$E_{iso} = \frac{4\pi D_L^2 \, F}{1+z} \simeq E_{pulse}^{rest} \, \delta^3 . \tag{17}$$

If CBs were "standard candles" with fixed mass, energy and velocity of expansion, and if all SN shells had the same mass, radius and density distribution, all differences between GRB pulses would result from their different distances and angles of observation. For such standard candles it follows from Eqs.(2-4,14, 15) that the observed durations (half widths at half maximum) of the photon intensity and of the energy flux density (Δt_I and Δt_F), their peak values (N_p and F_p), and the peak energy (E_p^γ) in a single GRB pulse are roughly correlated to the total "observed" isotropic energy (E_{iso}) as follows:

$$\Delta t_I \propto (1+z) \, [E_{iso}]^{-1/3} , \tag{18}$$

$$\Delta t_F \propto (1+z) \, [E_{iso}]^{-1/3} , \tag{19}$$

$$F_p \propto [E_{iso}]^{4/3} \, (1+z)^{-1} , \tag{20}$$

$$E_p^\gamma \propto [E_{iso}]^{1/3} \, (1+z)^{-1} . \tag{21}$$

$$N_p \propto E_{iso}, \tag{22}$$

These approximate correlations can be tested using the sample of 15 GRBs with known redshifts. Because of the strong dependence of the CB pulses on the Doppler factor and their much weaker dependence on the other parameters, they may be approximately satisfied (see, e.g. [45]) in spite of the fact that CBs and SN shells are likely to be sufficiently varied not to result in standard candles.

All of the above approximate predictions of the CB model [11] are in good qualitative agreement with observations. The success of the 'exact' CB model prediction in explaining the temporal structure of GRBs, the spectrum of individual pulses and their temporal evolution is demonstrated in [10–12].

461

GRB AFTERGLOWS

Far from their parent SNe, the CBs are slowed down by the interstellar medium (ISM) they sweep, which has been previously ionized by the forward-beamed CB radiation (traveling essentially at $v = c$, the CB is "catching up" with this radiation, so that the ISM has no time to recombine). As in the jets and lobes of quasars, a fraction of the swept-up ionized particles are "Fermi accelerated" to cosmic-ray energies and confined to the CB by its turbulent magnetic field, maintained by the same confined cosmic rays [5]. The bremsstrahlung and synchrotron emissions from the high energy electrons in the CB, boosted by its relativistic bulk motion, produce afterglows in all bands between radio and X-rays, collimated within an angle $\sim 1/\gamma(t)$, that widens as the ISM decelerates the CB.

A CB of roughly constant cross section, moving in a previously ionised ISM of roughly constant density, slows down according to $d\gamma/dx = -\gamma^2/x_0$, with $x_0 = M_{CB}/(\pi R_{CB}^2 n m_p)$ and n the number density along the CB trajectory. For $\gamma^2 \gg 1$, the relation between the length of travel dx and the (red-shifted, relativistically aberrant) time of an observer at a small angle θ is $dx = [2 c \gamma^2/(1 + \theta^2 \gamma^2)] [dt/(1 + z)]$. Inserting this into $d\gamma/dx$ and integrating, one obtains:

$$\frac{1 + 3\,\theta^2\gamma^2}{3\,\gamma^3} = \frac{1 + 3\,\theta^2\gamma_0^2}{3\,\gamma_0^3} + \frac{2\,c\,t}{(1 + z)\,x_0} \,, \tag{23}$$

where γ_0 is the Lorentz factor of the CB as it exits the SN shell. The real root $\gamma = \gamma(t)$ of the cubic Eq. (23) describes the CB slowdown with observer's time.

The radiation emitted by a CB in its rest system (bremsstrahlung, synchrotron, Compton-boosted self-synchrotron), is boosted and collimated by the CB's motion, and its time-dependence is modified by the observer's time flowing $(1 + z)/\delta$ times faster than in the CB's rest system. For $\gamma \gg 1$, an observer at small θ sees an energy flux density:

$$F[\nu] \sim \delta^3 \, F_0(\nu\,[1 + z]/\delta) \, A(\nu, z)\,, \tag{24}$$

where $F_0(\nu_0)$ is the CB emission in its rest frame, $\delta(t)$ is given by Eq. (3) with $\gamma = \gamma(t)$ as in Eqs. (23), and $A(\nu, z)$ an eventual absorption dimming.

Neglecting energy deposition through collision of the CB with the ISM during the afterglow regime and energy losses due to expansion and radiation, the CB's afterglow is dominated by a steady electron synchrotron radiation from the magnetic field in the CB (valid only for slowly expanding/cooling CBs in a low density ISM and for frequencies where the attenuation length due to electron free–free transitions exceeds its radius). The spectral shape of synchrotron emission in the CB rest frame is $F_0 \sim \nu_0^{-\alpha}$, with $\alpha = (p - 1)/2$ and p the spectral index of the electrons. For equilibrium between Fermi acceleration and synchrotron and Compton cooling, $p \approx 3.2$ and $\alpha \approx 1.1$, while for small cooling rates, $p \approx 2.2$ and $\alpha \approx 0.6$ [47] or $p \simeq 1.2$ and $\alpha \simeq 0.1$ if Coulomb losses dominate. At very low radio frequencies self-absorption becomes important and $\alpha \approx -1/3$ (2.1) for optically thin (thick) CBs.

Since $\gamma(t)$, as in Eq. (23), is a decreasing function of time, the afterglow described by Eq. (24) may have a very interesting behaviour. An observer may initially be outside the beaming cone: $\theta^2 \gamma^2 > 1$, as we shall argue to be the case for GRB 980425, for which we estimate $\gamma_0^2 \theta^2 \sim 200$ (other relatively dim GRBs in Table I, such as 970228 and 970508, may also be of this type). The observed afterglow would then initially rise with time. As γ decreases, the cone broadens, and around $t = t_p$ when $\gamma\theta \sim 1$ (when the observer enters the 'beaming cone' of the CB) the afterglow peaks and then begins to decline. Beyond the peak, when $\gamma^2\theta^2 \gg 1$ and where $\gamma \sim t^{-1/3}$, the afterglow declines like

$$F[\nu] \sim F_0[\nu, t_p] \left[\frac{(t_p/t)^{1/3}}{1 + (t_p/t)^{2/3}} \right]^{3+\alpha+\beta} A(\nu, z), \qquad (25)$$

where $F_0 \sim \nu^{-\alpha} t^{-\beta}$ in the CB rest frame. For steady emission, i.e., for small emission and energy deposition rates, $\beta = 0$. For an emission rate which is in equilibrium with the energy deposition rate in the CB rest frame by the incident ISM particles (that is proportional to δ^2), $\beta = 2$. Eq. (25) with $0 \le \beta \le 2$ describes well the late afterglows of most GRBs as demonstrated in [10–12]. Their radio afterglows must be corrected for absorption in the CB and along the line of sight.

When the CB enters the Sedov–Taylor phase its radius increases as $t^{2/5}$. The Lorentz factor of electrons decreases like $t^{-6/5}$. In equipartition, the magnetic field decreases like $t^{-3/5}$. Wijers et al. [48] have shown that these facts lead to an afterglow decline $F_\nu \sim t^{-(15\alpha-3)/5} \sim t^{-2.7}$ for $\alpha \sim 1.1$ CB at rest, as was observed for the late-time afterglows of some GRBs. Note, however, that if the CB enters the Sedov-Taylor phase in flight, Eq. (24) predicts $F_\nu \sim t^{-(50\alpha+6)/15}$, i.e., $F_\nu \sim t^{-2.4}$ for $\alpha \sim 0.6$ which changes to $F_\nu \sim t^{-4}$ for $\alpha \sim 1.1$.

GRB 980425: A SPECIAL CASE

In the list of Table I, GRB 980425 stands out in two —apparently contradictory— ways: it is, by far, the closest ($z = 0.0085$, $D_L = 39$ Mpc) and it has, by far, the smallest implied spherical energy: 8.1×10^{47} erg, 4 to 6 orders of magnitude smaller than that of other GRBs. However, if GRB 980425 was not abnormal, Eq. (17) tells us that the peculiarities of of GRB 980425 can be understood if its source was "fired" from SN1998bw with a bulk-motion Lorentz factor $\gamma \sim 1000$, at an angle $\theta \approx 15/\gamma$ relative to our line of sight. Then, for $\theta \ll 1$ and $\gamma \gg 1$, its projected sky velocity

$$v_T^+ \approx \frac{2\gamma^2\theta}{(1 + \gamma^2\theta^2)} c, \qquad (26)$$

yields superluminal velocities [46] $v_T \approx 2\,c/\theta$ for $\gamma^2\theta^2 \gg 1$, and $v_T \approx 2\,c\,\gamma^2\,\theta$, for $\gamma^2\theta^2 \ll 1$, provided $2\,\gamma^2\,\theta > 1$. Its transverse superluminal displacement, D_T, from the SN position can be obtained by time-integrating v_T, as in Eq. (26), using $\gamma(t)$ as in Eq. (23). The result can be reproduced, to better than 10% accuracy, by using the approximation $v_T \sim 2\gamma^2\,\theta\,c$, valid for $\gamma < 1/\theta$ (or $t > t_p$, the afterglow's peaktime):

$$D_T \simeq \frac{2\,c\,t_p}{\theta}\left[\frac{t}{t_p}\right]^{1/3}. \tag{27}$$

At present ($t \sim 850\,\mathrm{d}$), the displacement of the GRB from the initial SN/GRB position is $D_T \sim 20\,(\gamma_0/10^3)$ pc, corresponding to an angular displacement $\Delta\alpha \sim 100\,(\gamma_0/10^3)$ mas, for $z = 0.0085$, $t_p \sim 3.5 \times 10^6$ s.

The two sources, now a few tens of mas away, may still be resolved by HST. In Fig. 7 we show our prediction for the late-time V-band light curve of SN1998bw/GRB 980425. The SN curve is a fit by Sollerman et al. [50] for energy deposition by ^{56}Co decay in an optically thin SN shell. The GRB light curve is our predicted afterglow for GRB 9980425, as given by Eqs. (23,24), and constrained to peak at the position of the second peak in the radio observations [16,49]. The fitted normalization is approximately that of the mean afterglow of the GRBs in Table I, suppressed by the same factor as its γ-ray fluence relative to the mean γ-ray fluence in Table I. The joint system has at present (day ~ 850) an extrapolated magnitude $V \sim 26$ [51]. It can still be resolved from its host galaxy ESO-184-G82 by HST and, perhaps, by VLT in good seeing conditions. An extrapolation of the V-band late time curve of SNR1998bw [50] suggests that the present magnitude of the SN is $V \sim 28$, which is near the detection limit of HST and is dimming much faster than ^{56}Co-decay would imply.

A GRB as close as GRB 980425 ($z = 0.0085$) should occur only once every ~ 10 years and its associated supernova may be only occasionally observed. For typical GRBs ($z \sim 1$) there is no hope of resolving them with HST into two separate SN and GRB images. Resolving them with VLBI would also be arduous. For these reasons, we exhort interested observers to consider immediate high-resolution optical (STIS) and radio (ATCA and VLBI) follow-up observations of SN1998bw and the afterglow of GRB980425.

X-RAY LINES IN GRB AFTERGLOWS

Lines in the X-ray afterglow of GRBs have been detected in four GRBs (GRB 970508 [52]; GRB 970828 [53]; GRB 991216 [54]; GRB 000214 [55]). Their energies are listed in Table II. They were interpreted as iron lines emitted from a large mass of iron that was photoionized by the GRB. However, this interpretation raises many questions (e.g. [56]). The CB model offers an alternative interpretation for the origin of the X-ray lines in the "early" GRB afterglow - hydrogen recombination lines which are Doppler shifted to X-ray energies by the CB relativistic motion:

As long as the CB is opaque to its internal radiation, it expands with the relativistic speed of sound, $c\sqrt{3}$ in its rest frame, and cools with $T_{CB} \sim 1/R_{CB} \sim 1/t$. When $R_{CB} \simeq [3\,M_{CB}\,\sigma_T/(4\,\pi\,m_p)]^{1/2}$, where $\sigma_T \simeq 0.65 \times 10^{-24}$ cm^2 is the Thomson cross section, it becomes optically thin and its internal radiation escapes. This end to the CB's γ-ray pulse takes place at $t \simeq R_{CB}/\sqrt{3}\,c\,\delta \simeq 2 \times (10^3/\delta)$ s in the observer frame. Due to the escape of its internal radiation, its internal pressure drops and its expansion rate is slowed down by sweeping up the ISM. During this phase it cools mainly by emission of bremsstrahlung and synchrotron radiation. Its ionization state is described by the Saha The exponential term in the Saha equation confines this recombination phase of the CB to a temperature around 4500K (for CBs with free electron density 10^5 cm$^{-3} < n_e < 10^6$ cm^{-3}, as we shall estimate later). The recombination produces strong emission of Ly$_\alpha$ line (and, perhaps, a recombination edge above the Ly$_\infty$ line) which is Doppler shifted by the CB motion to X-ray energy in the observer frame:

$$E_\alpha \simeq \frac{10.2}{(1+z)}\left[\frac{10^3}{\delta}\right] \text{ keV} , \tag{28}$$

$$E_{edge} \simeq \frac{13.6}{(1+z)}\left[\frac{10^3}{\delta}\right] \text{ keV} . \tag{29}$$

The total number of these recombination photons is approximately equal to the baryonic number of the CB, $N_b \simeq E_{CB}/m_p\,c^2\,\gamma(0) \simeq 6.7 \times 10^{51}\,(E_{CB}/10^{52}\,\text{erg})\,(10^3/\gamma(0))$. Thus, the photon fluence of these lines (line-fluence) at a luminosity distance D_L is

$$N_{lines} \simeq N_b\,\frac{(1+z)^2\,\delta^2}{4\,\pi\,D_L^2} . \tag{30}$$

The measured energies of the observed X-ray lines in the above 4 GRBs and the Doppler factors implied by their interpretation as hydrogen recombination features are listed in Table II. The inferred Doppler factors are consistent with those needed in the CB model ($\delta \sim 10^3$) to explain the intensity and duration of the GRB pulses.

Note that the 4.4 keV line in the afterglow of GRB 991216 [54] may be either a hydrogen recombination edge from the same CB that produces the 3.49 keV line, or a Ly $-\alpha$ line from another CB.

There are various independent tests of the interpretation of the X-ray lines as hydrogen recombination features that are Doppler shifted to X-ray energies by the CB motion:

a. Time and duration of line emission: The mean time for radiative recombination in hydrogenic plasma is $r_{rec} \approx 3 \times 10^{10}\,T^{-1/2}\,n_e$ s^{-1}. In the observer frame this recombination time is

$$\Delta t_L \approx \frac{3 \times 10^{10}\,(1+z)\,T^{1/2}}{n_e\,\delta} \text{ s} . \tag{31}$$

465

The emission rate of bremsstrahlung by an hydrogenic plasma is $L \simeq 1.43 \times 10^{-27} \, n_e^2 \, T^{1/2} \, \mathrm{erg \, cm^{-3} \, s^{-1}}$, and the cooling time of the CB to temperature T in the observer frame is

$$t_{\mathrm{brem}} \simeq \frac{2.9 \times 10^{11} \, (1+z) \, T^{1/2}}{n_e \, \delta} \, \mathrm{s} \,. \tag{32}$$

Thus, the ratio $\Delta t_L / t_{\mathrm{brem}} \simeq 0.10$, where the dependence on Doppler factor, electron density, temperature and redshift of the CB has been canceled out, is a universal ratio for CB afterglows, independent of their detailed properties. This prediction is consistent with the observations of the X-ray line in GRB 980828 where $\Delta t_L / t_{\mathrm{brem}} \simeq 0.05$ with a large ($\sim$ factor 2) uncertainty [53].

However, the above estimate is valid only for a single CB (originally ejected, or one that was formed by overtaking and merger of separately ejected CBs). Generally, the afterglows of different CBs cannot be resolved either spatially or temporally. Their individual afterglows are blended into a single afterglow. Because of their different Lorentz and Doppler factors, X-ray line emission may extend over a much longer time, $\Delta t_L \simeq t_{\mathrm{brem}}$. This may be the case in the observations of X-ray line emission during the afterglow of GRB 970508 and GRB 000214 by BeppoSAX [52,54,55] and of GRB 991216 by Chandra [54] which were too short both in time and of statistics to measure accurately enough the duration of the line emission.

b. **Photon fluences** : Table II also reports the total photon fluence in X-ray lines during the observation time and the total baryon number required to produce the lines as predicted by Eq. (30). These baryon numbers are within the range expected for SNe CBs/jets ($N_b \simeq 6.5 \times 10^{51 \pm 1}$). However, because the observation times may not have extended over the full time of line emission, these baryon numbers may underestimate the baryon number of the CB/jet.

c. **Line width** : Thermal broadening of both the recombination lines and the recombination edge are rather small. However, the Doppler factors of the CBs decrease with time during the line emission because of the decrease of their Lorentz factors due to their deceleration by the ISM. At late time, $\gamma \sim t^{-1/3}$, for a single CB and as a result the line energy shifts by $\Delta E_L \simeq (\Delta t_L / 3 \, t_{\mathrm{berm}}) \, E_L \simeq 0.04 \, E_L$ during the line emission. The decrease in the line energy with time during the line emission is a clear fingerprint of the origin of the lines. When integrated over time (in order to increase statistics) the line shift will appear as a line broadering. The above estimate of this line broadening is consistent with the reported widths of the X-ray lines.

d. **The CB radius during line emission:** The electron density in the CB during the line emission can be inferred from the observed time and duration of the line emission, using Eqs. (31),(32). Then, the CB radius can be estimated from the total baryon number which is inferred from the measured line-fluence. This radius estimate is not sensitive to the accuracy of the observations since it depends on the third root of N_b / n_e. The three GRBs with measured redshifts yield $R_{\mathrm{CB}} \sim (1-2) \times 10^{15} \, \mathrm{cm}$ for $t \sim 1-2 \, \mathrm{days}$ in the observer frame. If a CB continues to expand within the first few weeks with the same mean speed of expansion as in the first day or two, its radius after a month reaches $\simeq (3-6) \times 10^{16}$ cm (only then the swept up ISM mass begins to be comparable to the CB mass). Indeed, from VLA observations of scintillations [57] in the radio afterglow of GRB 970508 and their disappearance after a month it was inferred [58] that the linear size of its source a month after burst was $\approx 10^{17}$ cm , i.e., corresponding to $R_{CB} \sim 5 \times 10^{16}$ cm .

CONCLUSIONS

GRBs and their afterglows may be produced by jets of extremely relativistic cannonballs in SN explosions. The cannonballs which exit the supernova shell/ejecta reheated by their collision with it, emit highly forward-collimated radiation which is Doppler shifted to γ-ray energy. Each cannonball corresponds to an individual pulse in a GRB. They decelerate by sweeping up the ionised interstellar matter in front of them, part of which is accelerated to cosmic-ray energies and emits synchrotron radiation: the afterglow. When the cannonballs cool below 4500K, electron-proton recombination to hydrogen produces Ly-α emission which is Doppler shifted to X-ray energy.

The Cannonball Model cannot predict the timing sequence of the GRB pulses, it fares very well in describing the total energy, energy spectrum, and time-dependence of the individual γ-ray pulses and their afterglow and explains the X-ray lines observed in some GRB afterglows.

The Cannonball model predicts that GRB pulses are accompanied by short pulses of TeV neutrinos and sub TeV γ-rays, which begin and peaka little earlier and are much more energetic than the GRB pulses. These high energy emission should be visible in ground based (sub TeV photon) and underground (neutrino) telescopes.

There are other events in which a variety of GRBs could be produced by the CB mechanism: large mass accretion episodes in binaries including a compact object, mergers of neutron stars with neutron stars or black holes [59,60], transitions of neutron stars to hyperon- or quark-stars [20,61], etc. In each case, the ejected cannonballs would make GRBs by hitting stellar winds or envelopes, circumstellar mass or light. I discussed only core-collapse SN explosions, as the GRBs they would produce, although relatively "standard", satisfactorily reproduce the general properties of the heterogeneous ensemble of GRBs, their afterglows and even their X-ray line emission.

Acknowledgements: The material presented in this talk is basesd on work done in collaboration with Alvaro De Rújula. The research was supported in part by the Fund for Promotion of Research at the Technion and by the Hellen Asher Fund for Space Research.

Table I - Gamma ray bursts of known redshift z

GRB	z	D_L^a	F_γ^b	E_γ^c	M^d
970228	0.695	4.55	0.17	0.025	25.2
970508	0.835	5.70	0.31	0.066	25.7
970828	0.957	6.74	7.4	2.06	—
971214	3.418	32.0	1.1	3.06	25.6
980425	0.0085	0.039	0.44	8.14 E-6	14.3
980613	1.096	7.98	0.17	0.061	24.5
980703	0.966	6.82	3.7	1.05	22.8
990123	1.600	12.7	26.5	19.8	24.4
990510	1.619	12.9	2.3	1.75	28.5
990712	0.430	2.55	—	—	21.8
991208	0.706	4.64	10.0	1.51	>25
991216	1.020	7.30	25.6	8.07	24.5
000131	4.51		3.51	11.0	
000301c	2.040	17.2	2.0	2.32	27.8
000418	1.119	8.18	1.3	0.49	23.9
000926	2.066	17.5	2.5	2.98	24

Comments: a: Luminosity distance in Gpc (for $\Omega_m = 0.3$, $\Omega_\Lambda = 0.7$ and $H_0 = 65\,\mathrm{km\,s^{-1}\,Mpc^{-1}}$. b: BATSE γ–ray fluences in units of $10^{-5}\,\mathrm{erg\,cm^{-2}}$. c: (Spherical) energy in units of 10^{53} ergs. d: R-magnitude of the host galaxy, except for GRB 990510, for which the V-magnitude is given.

Table II - GRB afterglows with X-ray lines

GRB	z	E_{line}	F_{line}	Δt_{obs}	δ_{line}	N_b
970508	0.835	3.4	1.5	30	612	3×10^{51}
970828	0.957	5.04	0.45	24	967	$> 7 \times 10^{50}$
991216	1.020	3.49	0.39	12	691	$> 1.3 \times 10^{51}$
991216	1.020	4.4	0.47	12	800	$> 1.3 \times 10^{51}$
000214		4.7	1.0	104	≥ 460	

Comments: Line energies in keV. Observation times in ks. All lines were assumed to be a Doppler shifted Ly_α lines. The line fluences (in $\mathrm{cm^{-2}}$) and the baryon numbers are lower limits because of partial observation times.

REFERENCES

1. Paczynski, B., 1998, ApJ, 494, L45
2. Woosley, S. E. 1993, ApJ, 405, 273
3. Shaviv, N. J. & Dar, A. 1995, ApJ, 447, 863
4. Dar A. 1998, ApJ, 500, L93
5. Dar, A. & Plaga, R. 1999, A&A, 349, 259
6. Cen, R. 1999, ApJ, 524, 51
7. Woosley, S. E. & MacFadyen, A. I. 1999, A&AS, 138, 499

8. MacFadyen, A. I. & Woosley, S. E. 1999, ApJ, 524, 168

9. MacFadyen, A. I., Woosley, S.E. & Heger, A. 1999, astro-ph/9910034

10. Dar, A. & De Rújula, A. 2000a, astro-ph/0008474 (A&A)

11. Dar, A. & De Rújula, A. 2000b, astro-ph/0012227 (A&A)

12. Dar A. 2000, astro-ph/0101007

13. Soffitta, P., et al. 1998, IAU Circ. No. 6884

14. Kippen, R. M., et al. 1998, GCN 67

15. Galama, T. J., et al. 1998, Nature, 395, 670

16. Kulkarni, S. R., et al. 1998a, Nature, 395, 663

17. Pian, E., et al. 1999, A&A, 138(3), 463

18. Wieringa, M. H., et al. 1999, A&AS, 138, 467

19. Iwamoto, K., et al. 1998, Nature, 395, 672

20. Dar, A. 1999a, A&AS, 138, 505

21. Bloom, J. S., et al. 1999, Nature, 401, 452

22. Dar, A. 1999b, GCN Report No. 346

23. Reichart, D. E., 1999, ApJ 521, L111

24. Galama, T. J., et al. 2000, ApJ, 536 185

25. Hjorth, J., et al., 2000, ApJ, 534, 147L

26. Sahu, K. C., et al. 2000, ApJ, 540, 74

27. Holland, S., et al. 2000, submitted

28. Sokolov, V. V., et al. 2000, to be published

29. van den Bergh, S. & Tammann, G. A. 1991, ARA&A, 29, 363

30. Madau, P. 1998, astro-ph/9801005

31. De Rújula, A. 1987, Phys. Lett. 193, 514

32. Mirabel, I. F. & Rodriguez, L. F. 1994, Nature, 371, 46

33. Belloni T., et al., 1997, ApJ, 479, 145

34. Mirabel, I. F. & Rodriguez, L. F. 1999a, ARA&A, 37, 409

35. Mirabel, I. F. & Rodriguez, L. F. 1999b, astro-ph/9902062

36. Salamanca, I., et al. 1998, MNRAS, 300L, 17

37. Fassia, A., et al. 2000, astro-ph/0011340

38. Nakamura, T. et al. 2000, astro-ph/0007010

39. Dyer, K. K., et al. 2000, astro-ph/0011578

40. Fusco-Femiano, R., et al. 1999, ApJ, 513, L21

41. Rephaeli, Y., et al. 1999, ApJ, 511 L21

42. Fusco-Femiano, R., et al. 2000, astro-ph/0003141

43. Preece, R. D., et al., 2000, ApJS, 126, 19

44. Fenimore, E. E., et al. 1995, ApJ, 448, L101

45. Plaga, R. 2000, astro-ph/001206

46. Rees, M. J. 1966, Nature, 211, 468

47. Dar, A., & De Rújula, A. 2000c, astro-ph/0005080 (MNRAS in press)

48. Wijers, R. A. M. J., Rees M.J. & Meszaros, P. 1997, MNRAS, 288, L5

49. Frail D. A., et al. 1999, http://www.narrabri.atnf.csiro.au/public/grb980425/

50. Sollerman, J., et al. 2000, astro-ph/0006406

51. Fynbo, J. U., et al. 2000, astro-ph/0009014

52. Piro, L., et al. 1999, ApJ, 514, L73

53. Yoshida, A. et al. 1999, A&A, 138S, 433
54. Piro, L. et al. 2000, Science, 290, 955
55. Antonelli, L. A., et al. 2000, astro-ph/0010221to be published
56. Vietri, M. 2000, astro-ph/0011580
57. Goodman, J. 1997, NA, 2, 449
58. Taylor, G. J., et al. 1997, Nature, 389, 263
59. Paczynski, B. 1986, ApJ, 308, L43
60. Goodman, J., Dar, A. & Nussinov, S. 1987, ApJ, 314, L7
61. Dar, A. De Rújula, A. 2000d, astro-ph/0002014 (MNRAS)
62. Klose, S., et al. 2000, astro-ph/0007201
63. Fruchter, A. S., et al. 2000, GCN 627
64. Fruchter, A. S., et al. 1999, astro-ph/9903236

Hypernovae: Observational Aspects

I.J. Danziger

Osservatorio Astronomico di Trieste, Trieste, 34131 Italy

Abstract. Following the discovery of a supernova SN 1998bw very probably associated with the the γ-ray burst source GRB980425 the term "hypernova" has been re-defined to embrace SNe of high energy associated with GRB's. Here we discuss in some detail the observed and inferred properties of SN 1998bw and its relation to other previously observed SNe to which it may be related. A brief account is given of other relatively nearby SNe where a possible association with GRB's has been suggested ex post facto. There are also cases where less direct but suggestive evidence for a SN/GRB association has been adduced for well observed GRB events and their associated afterglow. These are also referred to. Our conclusion is that observationally we are still near the beginning of understanding which types of SNe give rise to GRB's and how frequently they occur. Nevertheless objects such as SN1998bw surely give rise to young supernova remnants with high velocity oxygen-enriched filaments.

INTRODUCTION

The name "hypernova" was first given [1] to a hypothetical model for a GRB based on an earlier model [2] of the catastrophic collapse of a massive star (possibly a single WR star), whereby the inner core of $10 M_\odot$ forms a black hole while the outer part forms a massive disk (microquasar). A beamed pair fireball oriented along the rotation axis of the accreting black hole creates the GRB. Hence it has been called a "failed" supernova in the conventional sense. Nevertheless the subsequent visual luminosity was supposed to exceed magnitude -25. It is not clear whether we have observed such an event yet.

The coincidence in position and time of SN1998bw with GRB980425 has focussed attention on the possibility that SNe produce γ-ray bursts. Because the observations of SN1998bw coupled with modelling showed that the explosion energy of this SN exceeded 10^{52} ergs, resulting in very large expansion velocities, this type of SN was called hypernova [3]. Unlike the classification system for other SNe which depends almost exclusively on spectroscopy, this definition requires modelling to elucidate its nature since, as we shall see, not even the light curve differs significantly in shape or luminosity from some other more mundane objects. The similarity of the light curve to that of the Type Ia SN1991T is quite striking even

CP565, *Young Supernova Remnants: Eleventh Astrophysics Conf.*, edited by S. S. Holt and U. Hwang
© 2001 American Institute of Physics 0-7354-0001-6/01/$18.00

though the progenitors and type of explosion were quite different. This new classification is therefore not very useful for early on-line observations of SNe.

The Beppo-Sax early observations of GRB980425 allowed the discovery and location of the associated SN1998bw in the low luminosity, almost face-on spiral galaxy ESO 184-G82, very probably a galaxy of metallicity lower than solar. The low resolution images of the galaxy suggest that SN1998bw lies superimposed on a spiral arm, while the beautiful high resolution HST images show that it lies well within a complex of young stars and gas [4].

The xray emission associated with SN1998bw has been difficult to interpret [5] because although Beppo-Sax has detected xrays in that direction, it is not yet clear what proportion are associated with the SN itself and what proportion with its host galaxy. The more credible xray source decreased in brightness over a 6-month period, but the spectra do not allow one to distinguish between, for example, a synchrotron or an inverse-Compton process. If we ascribe the 40-700keV flux to SN1998bw alone and assume isotropic emission, we see that the absolute flux is at least 4 orders of magnitude lower than that of GRB's at redshifts of 0.5 and beyond.

The radio emission from SN1998bw was both very intense and prompt (peaking 10 days after outburst at 6cm), the combination being unusual [6]. This calls for relativistic or sub-relativistic shock propagation but nevertheless a very high expansion rate of the envelope, which in fact has been observed optically. In spite of this somewhat exceptional behaviour which helped to reinforce the association of the supernova with the GRB, the subsequent evolution of the radio emission was not unlike that observed in other Type Ib,c SNe even if the maximum flux density exceeded that from previously observed Type Ib,c by a large factor.

OBSERVATIONS AT OPTICAL WAVELENGTHS

Polarization.

Spectrophotometric measurements of linear polarization were made at 2 epochs, 9 and 25 days after GRB [7]. The average polarization in the range 4000-7000 Å was 0.8 and 0.4 percent with some suggestion of a change in wavelength dependence between the two. A correction for Galactic interstellar polarization is somewhat uncertain because of differences in published values in this region of the sky, but net polarization remains. This is confirmed by independent measurements [8] which included a star very close to the line-of-sight to SN1998bw. The fact that some change seems to occur, if not in the average amount but in the wavelength dependence, suggests that it is intrinsic to the supernova. However the polarization does not by itself indicate the degree of asymmetry, if any, of the expanding envelope, since one cannot know the angle of inclination between the line-of-sight and the axis of symmetry of a non-spherical figure. Indeed it is possible that large scale brightness changes or clumping giving an uneven brightness distribution in

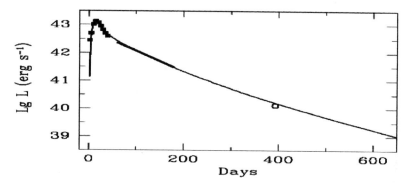

FIGURE 1. Bolometric light curve comparing model (full line) with observations (filled squares) [10]. Last point (open circle) added by author assuming V observations scale with bolometric magnitude as for earlier points.

various quadrants could result in net polarization. The percentage polarization in SN1998bw is similar to that observed in Type Ib,c SNe.

Light Curve.

The visual light curve has been measured to beyond 400 days. It had by day 60 reached an exponential decay which is steeper than that of the ^{56}Co decay line. The light curve around maximum light has been modelled [3], [9], [10] with somewhat different models. The first named authors found a series of symmetric models with different combinations of explosion energy and ejecta mass that equally well matched the early light curve. To resolve this degeneracy they could use a separate relation between kinetic energy, mass and velocity of the ejecta. Since available spectroscopy provided velocities of the envelope, this allowed a unique solution for mass of the ejecta and kinetic energy of the explosion. By using a $13.8 M_\odot$ CO core, the remnant of a more massive star, one arrives at an ejecta mass of $10.9 M_\odot$ and an explosion energy of 3.10^{52}ergs, and a mass of radioactive nickel of $0.7 M_\odot$ from the luminosity at maximum. It was conjectured that the original mass of the star may have been $30\text{-}40 M_\odot$ which before explosion had become a WR star. However after 60 days this model does not fit the observed light curve, falling below the observed luminosities before returning to a better fit after 320 days. It has been speculated that this mis-match might be caused by the fact that explosion was asymmetric. Subsequent modelling [11], [12] with well mixed symmetric models has given good to excellent fits to both early and late light curves. The results of the latter modelling are illustrated in Figure 1.

Early Spectra.

The early spectra were peculiar with very broad features indicative of a high velocity of expansion. In fact these peculiarities were mainly responsible for much of the confusion in typing this SN in the early phases. The model discussed above [3] seemed to be vindicated by the good match of the spectrum synthesized with its parameters and the velocity measured from the SiII6355 absorption lines. The first velocity measured 8 days after explosion still indicated a high velocity of 28000 kms^{-1}, and an extrapolation backwards in time guided by the expectation of the best-fit model would suggest that velocities in the outermost layers at outburst could have been well in excess of 60000 kms^{-1}. A velocity from the apparent IR CaII triplet absorption is complicated by the possible presence of OI7775 absorption. We note here that the broad emission-like features are not due to discrete emission lines but result from the absence of significant line absorption allowing redward shifting photons in the expanding envelope to finally escape.

The IR spectra taken at 3 epochs namely 8, 33, and 51 days after the GRB are of particular interest because of the question of helium. It transpired, as will be seen later, that of all possible types SN 1998bw is most similar to a Type Ic albeit with much higher expansion velocities in the photospheric phases. But Type Ic SNe have been traditionally distiguished from Type Ib because of a lack of helium absorption lines in the optical region. They are also not unambiguously obvious in SN 1998bw. However there is a conspicuous PCygni profile in all 3 spectra at 1.083μ which corresponds to the well known triplet series line of HeI at a velocity of 18000 kms^{-1}. There is another PCygni profile, less conspicuous but certainly present, near 2.058μ which corresponds to the transition in the singlet series 2^1S-2^1P. Both features become more conspicuous relative to the continuum as time progresses and remain constant in velocity. Both characteristics can be understood qualitatively in the context of an expanding envelope surrounding a continuum source (photosphere) where the helium would be confined to a relatively thin shell with narrow velocity range and either optically thick or thin. That the velocity is 18000 kms^{-1} at a time when the velocity of the SiII6355 absorption is 13000 kms^{-1} or less suggests that the helium layer lies on the very outer edge of the envelope. There may be other weaker absorption features contaminating HeI1.083μ and some become more prominent at day 51. Since the colour temperature of SN1998bw is too low to excite HeI lines thermally or radiatively, one must assume that non-thermal electrons resulting from capture of γ-rays [13] are responsible. The fact that HeI lines are not unambiguously detected in the optical spectra could result from non-thermal excitation altering conventional line ratios. The high velocities and density of lines may result in considerably more smearing of these lines into invisibility. If helium is present even in the outer layers it raises the question whether revised models are necessary and what exactly constituted the exploding progenitor.

Late Spectra.

The nebular spectrum is seen to develop at about day 94 with the appearance of [OI]6300, MgI]4571 and [CaII]7300 in emission. Other features near 5200Åwith less certain identifications also appear. The profile of [OI]6300 as well as that of MgI]4571 somewhat later are sharply peaked at zero velocity and this tells us immediately that there is a concentration of OI at low velocities, since an evenly

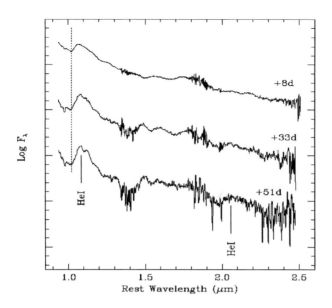

FIGURE 2. The IR spectra of SN1998bw at 3 epochs with the PCygni lines of HeI1.083μ and HeI2.058μ marked. Note the constant velocity of absorption and the increase in contrast of emission to continuum with time.

filled emitting envelope would produce a parabolic shaped profile. It is remarkable that this type of profile is observed in other Type Ib,c SNe [14]. In addition to the sharp peak there is structure seen in the blue wing of [OI]6300 displaced by about 2400 kms^{-1}. Careful inspection of spectra of other Type Ib,c SNe reveals that it is a common feature suggesting some structural features in the oxygen emitting region. The asymmetrical profile of MgI]4571 may be explained with the counterpart of this structure on the blue side, again a common feature of Type Ib,c SNe. This [OI]6300 profile has been modelled with a bimodal distribution of oxygen heated by the absorption of γ-rays from the decay of ^{56}Co in a symmetrically expanding envelope [15]. The model also successfully reproduces the temporal evolution of the [OI]6300 luminosity. It requires the central concentration of OI as foreseen by the sharply peaked profile as well as a broader distribution to accomodate the observed blue component.

The broad emission feature near 5200Å is certainly a blend of lines with possible contributions from [FeI], [FeII], [FeIII], FeII and MgI. Multiplets of these species are known to exist in astronomical environments which may have conditions somewhat similar to those prevailing in the envelope of SN1998bw. We note in particular the example of the Type II SN1987F in whose spectra allowed transitions of FeII have been identified [16]. Therefore a comparison of profiles of this feature with that of [OI]6300 or MgI]4571 must await a detailed accounting of all possible contributors.

475

A study of the temporal behaviour of the 5200Å feature may help because one begins to see in the latest spectra evidence of narrowing of this line. What is also of interest is whether the emission, if due to ions of Fe, reflects the bulk of the newly synthesized iron. All of these observational aspects have been elaborated in more detail elsewheres [7].

OTHER HYPERNOVAE?

Nearby Supernovae

There are several other SNe which, because of there abnormal characteristics followed by a possibly identified GRB association, have been suggested as other examples of hypernovae. It should be carefully noted however that in the cases discussed below the associations do not compare in quality with that of SN1998bw. This is because the positional coincidence is not nearly so good and the temporal coincidence is arguably even worse because the time of outburst of these SNe are only estimates that could be uncertain by at least several or more days.

One such case is SN1997ef whose early photospheric spectrum resembles that of SN1998bw more than any other SNe. The light curve and spectrum have been modelled in the same way as for SN1998bw [17]. It is intrinsically fainter than SN1998bw and the indicated photospheric velocities are lower and more uncertain. This is reflected in a less good fit to the model and a synthesized spectrum which is less impressive than that for SN1998bw. The best fit symmetric model indicates the explosion of a $10M_\odot$ CO core with an explosion energy of 8.10^{52}ergs, and producing $0.15M_\odot$ of ^{56}Ni. Thus it is a less extreme example of the SN1998bw type, where a possible association with GRB971115 has been suggested.

Another suggested possible GRB/SN association is that of SN1997cy and GRB970514 [18], [19], [20]. This SN is of Type II and one of the brightest recorded, exceeding the luminosity at maximum of SN1998bw. The $H\alpha$ line has 3 components one of which indicates very high velocities. Without ejecta-wind interaction the light curve on the exponential tail indicates that $2.6M_\odot$ of ^{56}Co were produced. However an ejecta-wind model in which the mass of the ejecta is $5M_\odot$ and the kinetic energy is 3.10^{52} ergs fits the light curve, and allows a production of as much as $0.7M_\odot$ of ^{56}Co. Since the original mass of this star was supposed to be approximately $20M_\odot$, one might question why there has been no sign of emission by oxygen which should have been produced in copious amounts or even exist at detectable levels if its abundance were solar in the $5M_\odot$ ejecta.

To summarize — at the time of writing the following GRB/SN associations have been suggested: SN1998bw/GRB980425; SN1997ef/GRB971115; SN1997cy/GRB970514; SN1999E/GRB980910; SN1999eb/GRB991002. Of these the first 2 SNe are of Type Ib,c and the last 3 of Type II. The positional coincidences lay within 2-5° , and the temporal coincidence is uncertain to at least 5 days.

Supernovae and Distant GRB Events

There are 2 well observed GRB events where there is a significant deviation from the power law decay normally expected and observed in the optical afterglow. In fact the slowing of the light curve for GRB980326 was modelled [21] by adding to the expected afterglow a component from a SN matching the shape and luminosity of that of SN1998bw. This led to a redshift of approximately 0.9 for this object with the above assumption.

In the case of GRB970228 where the redshift is known (z=0.695), a good fit to the light curve was obtained again using the known characteristics of SN1998bw but also with the constraint of the redshift [22].

CONCLUSION

The case of SN1998bw and GRB980425 demonstrates that some classes of GRB's are produced by supernovae. Although such SNe as well as others may explode and expand in an asymmetric way, the observations so far cannot convincingly demonstrate that significant asymmetries exist. This could simply be due to a lack of appropriate diagnostic criteria. It is also clear that SN1998bw bears a striking resemblance to other Type Ib,c SNe particularly in the nebular phase where one sees a concurrence in relative line strengths and velocities of expansion. This remains something of a mystery since SN1998bw seemed to explode with abnormally high velocities, a property which sets it apart from other Type Ib,c SNe. Furthermore the seemingly high explosion energies compel one to ask whether in SN1998bw we have seen a distinctly new kind of SN, or just the high energy end of a sample containing all Type Ib,c SNe. The case of SN1993J which started out as a Type II and developed into something resembling a Type Ib,c object only adds to the complexity. The various ways in which evolution in binary systems could affect the outer layers of these exploding stars is a subject in its infancy if in these various cases the existence of only a little hydrogen, or no hydrogen and a little helium, or no hydrogen and no helium are all that create variations in the basic explosion of a CO core of varying mass. It is clear that only more demonstrated coincidences of GRB's with SNe will clarify what is the full range of SN types which can produce γ-ray bursts. If GRB's are beamed this would surely mask a significant proportion of GRB/SN associations.

Another interesting result emerging from modelling observed light curves is the relation between progenitor mass and the mass of ^{56}Ni produced in SNe. Currently there are about 6 SNe for which one has results including the 3 objects called hypernovae. The progenitor masses range from 10-40$M_\odot$ and the mass of ^{56}Ni varies monotonically from 0.002 to 0.7$M_\odot$. These massive stars evolve rapidly and whether they play a significant part in chemical enrichment of galaxies at early phases is a matter that may be clarified by measuring such rates as a function of redshift.

In the context of this conference it is worth asking what type of young supernova remnant would evolve from an object such as SN1998bw. On the basis of the nebular spectrum one could fairly safely predict that it would be one with high velocity oxygen-enriched filaments. On the basis of the modelling so far one might expect to see evidence of substantial amounts of iron, one order of magnitude greater than that expected from SN1987A. But we cannot yet safely predict whether ring or annular structures would develop in this enriched material as is possibly seen in N132D, Cas A, and 1E0102.2-7219 in the Small Magellanic Cloud . As mentioned earlier the oxygen profiles can be modelled without resort to significant asymmetry. The distribution of oxygen and iron in such remnants coming from Chandra observations promises to be important. Another Type Ib,c SN 1996aq, with rather pronounced double-peaked [OI]6300 and MgI]4571 profiles may be the most obvious candidate for ejection of enriched material with a ring-like geometry. As far as is known this SN is not associated with a GRB.

REFERENCES

1. Paczyński B.,*ApJ* **494**, L45 (1998)
2. Woosley S.E.,*ApJ* **405**, 273 (1993)
3. Iwamoto K. et al.,*Nature* **395**, 672 (1998)
4. Fynbo J.U. et al.,*ApJ* **542**, L89 (2000)
5. Pian E. et al.,*ApJ* **536**, 778 (2000)
6. Weiler K.W. et al.,*The Largest Explosions Since the Big Bang: Supernovae and Gamma Ray Bursts, Proceedings of the 1999 STSCI Symposium*, in press
7. Patat F., et al.,*ApJ* **in press**
8. Kay L.E., et al.,*IAU Circ.* **n. 6969**, (1998)
9. Woosley S.E., et al.,*ApJ* **516**, 788 (1999)
10. Hoflich P., et al.,*ApJ* **521**, 179 (1999)
11. Sollerman J. et al.,*ApJ* **537**, L127 (2000)
12. Chugai N.N.,*Astr. Letters* **26**, 797 (2000)
13. Lucy L.B.,*ApJ* **383**, L308 (1991)
14. Filippenko A.V. et al.,*AJ* **100**, 1575 (1990)
15. Chugai N.N. and Danziger I.J.,*to be published*
16. Filippenko A.V.,*AJ* **97**, 726 (1989)
17. Iwamoto K., et al.,*ApJ* **534**, 660 (2000)
18. Woosley S.E., et al.,*ApJ* **516**, 788 (1998)
19. Germany L.M., et al.,*ApJ* **533**, 320 (2000)
20. Turatto M., et al.,*ApJ* **534**, L57 (2000
21. Bloom J.S., et al.,*Nature* **401**, 453 (1999)
22. Galama T.J., et al.,*ApJ* **536**, 185 (2000)

Conference Summary

Reflections on Young SNRs: Rapporteur Summary of the 11th Maryland Astrophysics Conference

P. Frank Winkler[1] and Knox S. Long[2]

[1] *Middlebury College, Middlebury, Vermont 05753, USA*
[2] *Space Telescope Science Institute, Baltimore, Maryland 21218, USA*

Abstract. At this conference we have seen a wealth of new results on young supernova remnants, the supernovae that produce them, the environments that surround them, and the pulsars or more exotic compact objects that may be left behind. Great progress has been made through the combination of observations at wavelengths from radio through gamma rays—both from the ground and from space—with creative theoretical efforts and powerful computational models. Especially impressive are the many spectacular early results of X-ray observations from the *Chandra* and *XMM-Newton* missions. In this paper, we give a personal view of some of the highlights of the conference, colored by the perspective of almost six decades (three for each of us) of work on SNRs.

INTRODUCTION

As scientists who began our astronomical careers working on X-rays from SNRs in the early 1970s, we have to pause to marvel at the leaps in capabilities for the X-ray observations made possible by *Chandra* and *XMM-Newton* and reported in numerous oral and poster papers at this meeting.

When we began, sounding rockets equipped with slat collimators were being used to produce one-dimensional scans of the surface brightness distributions of the Cygnus Loop [1] and Vela [2] with an angular resolution of ~ 0.5 degree. Most, if not all of the SNRs discussed at this conference were unresolved. As a result, we all share some of the awe that Galileo and his friends must have felt in seeing the wonders of the optical sky through a telescope for the first time. The exquisite images of young SNRs shown at this conference, with resolutions approaching $1''$, represent an improvement in angular scale of order 1000 compared with what was possible in the 1970's—an even greater improvement than Galileo's telescope made over the naked eye.

What passed for spectroscopy in our first experience, with the MIT experiment on the OSO-7 satellite, was color ratios among a few very broad energy bands. The

CP565, *Young Supernova Remnants: Eleventh Astrophysics Conf.*, edited by S. S. Holt and U. Hwang
© 2001 American Institute of Physics 0-7354-0001-6/01/$18.00

crystal spectrometer on the Einstein observatory launched in 1978 represented an unprecedented increase in spectral resolution—but with a throughput so low that every detected photon received almost as much attention as one of our children. The combination of both spatial and spectral resolution with the high sensitivity of today's new instruments, and the scientific advances—many reported at this meeting—that they have made, are truly cause to celebrate. X-ray astronomy has at last come of age.

SUCCESSES

The advances in X-ray astronomy, and the corresponding advances in other branches of astrophysics, have enabled many successes. We'd like to note some that have impressed at this meeting—with apologies to the many presenters whom the limitations of space and time do not permit us to mention individually.

Supernova Classification. First is the convergence of supernova classification into a few principal types based on their spectra, as discussed here by David Branch and Alex Filippenko. The vast majority of SNe can be classified as either Type Ia, or as core collapse SNe—a group that comprises Types II, Ib, and Ic. The unusual SN events that fit none of these classes are relatively few, though as many as 5% of outbursts classified as SN II—ones like 1997bs, for example—may be impostors, perhaps extragalactic analogs to η Car. This is a significant advance from a decade ago, when each new meeting on SNe resulted in at least one new class of event. It has been made possible by systematic searches which are discovering approximately a hundred new SNe each year, most of which yield high quality spectra that can be compared with sophisticated synthetic spectra based on model-atmosphere calculations. The large sample and good models for comparison has made it possible for patterns to emerge. Furthermore, for the heterogeneous category of core-collapse SNe we are beginning to understand why spectra from different types evolve as they do, to distinguish between those features that arise from the progenitor and those that are produced by primarily by the environment, and to get good statistics for the fraction of SNe in each category. At last we are getting close to matching SNe of different types with their likely progenitors.

SN Ia as Standard Candles for Cosmology. A second success, though tangential to this meeting, has been the narrowing of the dispersion in peak luminosity of SN Ia, making them invaluable as standard candles for cosmology. The dispersion is intrinsically narrow because SN Ia are "standard bombs"—white dwarfs that accrete matter from a companion until they exceed the Chandrasekhar limit, leading to sudden collapse and runaway nuclear burning (deflagration) from the inside out. But, as David Branch has described, the already low dispersion can be further narrowed by using an empirically based correlation of peak luminosity with decline rate [3]. Branch has given a qualitative explanation for standardization procedure, but a more rigorous understanding would engender greater confidence that we are avoiding systematic effects that may affect the use of SN Ia for measuring cosmological distances.

SN1987A. Since the time when SN1987A began to fade optically, Dick Mc-Cray has been telling us that it would resurrect itself and that the results of that resurrection would be spectacular to behold. That prediction is now coming to pass. Radio emission and X-rays from this young SNR began to increase about 1200 days after the explosion. Radio images described by Dick Manchester show that the radio emission is arising from inside the optical ring being monitored with *HST*, and that the spectral index has flattened from α of -0.97 in 1993 to -0.88 in 1999. In Lyα, STIS spectroscopy reveals that the optical ring has broad wings with velocities from $-15,000$ to $+10,000$ km s^{-1}, acquired as neutral H in the ejecta passes through the reverse shock. As this line structure changes with time, we will develop a 3-dimensional map of the reverse shock and ejecta. Today, there are also about five bright spots around the ring, arising from material with densities of order 10^5 cm^{-3} and evolving on time scales of order 1 year. These developing spots probably indicate the initial encounters of the SN blast wave with the equatorial ring of circumstellar material. New *Chandra* images show general agreement with those obtained at radio wavelengths, including the existence of a pronounced E-W asymmetry. Grating spectra of SN1987a obtained with *Chandra* indicate electron temperatures of $\sim$4 keV and ion temperatures of $\sim$100 keV, indicating that electrons and ions are not being brought into equilibrium behind the shock. Furthermore, Dick McCray assures us that the best is yet to come: SN1987A is brightening at all wavelengths, and could be as much as 100 times brighter in about seven years.

Radio SNe and their Evolution to SNRs. While SN1987A will surely be the SNR whose early evolution will be explored in most detail over the coming decades, Kurt Weiler and his collaborators have amassed radio lightcurves of several dozen young radio SNRs resulting from massive star explosions. In addition, several young extragalactic SNRs, including 1986J and 1993J, have been resolved with VLBI. The radio emission is believed to stem from the interaction of the outermost ejecta with circumstellar material shed as winds from the progenitor. Since the expansion velocities of SN ejecta can reach 10,000 km s^{-1}, about 1000 times larger then the velocities of pre-supernova winds in massive stars, in ten years the light curve can probe 10,000 years of the progenitor's mass loss history. While some of these light curves resemble predictions for a simple, uniformly expanding $1/r^2$ wind, the light curves of others show complex structure that must reflect variations in their winds. The prospect of continued monitoring of these SNe lends hope that we may ultimately make the connection between these extragalactic SNe and the young SNRs in our own Galaxy. The fact that no Type Ia SNe have been detected at radio wavelengths places a limit on pre-supernova mass loss rates of $\lesssim 10^{-8}$ $M_\odot$ yr^{-1}.

Ejecta-Rich SNRs. Nowhere are the new *Chandra* and *XMM-Newton* data making more of an impact than for the ejecta-rich remnants. In Cas A and E0102-

7219 especially we've seen dramatic evidence for the long believed but seldom apparent picture of a blast wave propagating outward rapidly through interstellar and/or circumstellar material, heating it to keV temperatures, while a reverse shock propagating inward through denser ejecta leads to far brighter, but lower temperature, emission. This model has been widely utilized since its development by Gull [4] [5] and McKee [6], but nowhere has the two-shock structure been so apparent as in the *Chandra* image of the Cas A, shown and discussed by both Una Hwang and Larry Rudnick, which shows not only the familiar bright inner ring, but also a crisply defined hard X-ray blast wave outlining the more diffuse post shock emission seen in both radio and X-rays.

The reverse shock reveals itself even more dramatically, especially in the spatially resolved grating spectra of E0102, discussed by Kathy Flanagan and Claude Canizares. Here we see not simply a bright ring of shocked ejecta, but variations in the size of the ring which show the progressive ionization behind the reverse shock. Emission lines from less ionized species appear on the inside, at the leading edge of the reverse shock, with more highly ionized species appearing at larger radii where the ionization is more fully developed. It is inescapable that the shock is working its way inward through the ejecta.

Another spectacular success from *Chandra* has been the analysis of the first-light image of Cas A by Hughes *et al.* [7] and discussed here by Una Hwang. Individual knots as small as 3″ show a variety of different abundance patterns indicative of explosive oxygen burning, incomplete Si burning, and incomplete Si burning with an admixture of Fe. Abundance variations in knots should not be a surprise, of course, since large variations in abundances in the optical knots have been seen in optical spectra for decades, but the X-ray emission presumably probes a much larger fraction of the ejecta with lines from more of the key species that probe nucleosynthesis. Furthermore, the Fe-rich knots are farthest from the center, which must indicate a large-scale inversion of material originating in the progenitor core.

SNR Expansion and Kinematics. Una Hwang also discussed recent X-ray measurements of proper motions that have been carried out for Cas A [8] [9], Tycho [10], and Kepler [11]. In all cases the X-ray measurements give expansion rates almost twice as high as the corresponding radio measurements. Exploring the cause of these differences is one of the directions future research will have to follow.

For E0102, Jack Hughes, as well as Anne Decourchelle and Rob Petre, discussed X-ray proper motion studies, carried out over a 20-year baseline by comparing *Einstein, ROSAT*, and *Chandra* data [12]. The measured expansion rate of $\sim$ 0.1% yr^{-1} indicates that the blast-wave velocity must be $\sim$ 6000 km s^{-1}, but the temperature of the post-shock X-ray emission is far lower than expected from straightforward shock heating. This leads to the suggestion that much of the blast-wave energy must go into diffuse particle acceleration, and thus provides another argument for the that SN shocks play a key role in producing cosmic rays. In the spirit of our opening remarks, to measure proper motions for any SNR in X-rays is a significant achievement; to do so for one in the Magellanic Clouds is stunning!

Also for E0102, Kris Eriksen presented beautiful optical Fabry-Perot velocity maps, which show us the distribution of emitting material in radial velocity space. Since the ejecta have presumably been expanding at a constant velocity since the explosion, radical velocity is equivalent to position along the line of sight, giving us a true three-dimensional picture. Observational studies like this may provide a rigorous arena for testing the detailed 3-dimensional models like those discussed by John Blondin.

Infrared Observations. Although Eli Dwek [13] and others have long emphasized the importance of dust emission in SNRs, and although about one third of galactic supernova remnants were actually detected with IRAS [14], most of us would probably agree that IR studies of supernova remnants have not had a huge impact on the field. This may have been due to the lack of good observing opportunities. That situation has begun to change, however, with the launch of ISO, the production of the 2MASS survey, and prospects for SOFIA and, especially, SIRTF. ISO has provided the first high quality spectra of the other well-known historical SNe—Cas A, Kepler, Tycho and the Crab, as well as RCW103 and IC443, which are bright due to their interactions with molecular clouds, which can be directly observed in the IR. And as Richard Arendt described, the IR emission lines provide important diagnostics in a temperature and density regime not accessible at other wavelengths. Furthermore, studies of emission from dust are beginning to provide information about dust minerology and composition, as well as dust formation and destruction mechanisms in supernova remnants. Cas A is actually much brighter in the IR than at X-ray wavelengths, and 2MASS imagery resembles the radio more than the X-ray morphology. Contour maps of silicates in Cas A tend to be anti-correlated with emission from IR Ne lines.

Diffuse Particle Acceleration, Cosmic Rays, and SN 1006. The remnant of the SN of AD 1006, which eluded even basic understanding for many years, is proving to be a Rosetta Stone for connecting SN shocks and cosmic rays. From ASCA spectroscopy [15], we now understand that the bright X-ray shell is synchrotron-dominated. From the detection of TeV γ-rays from SN 1006 [16], we have confirmation of the existence of electrons with energies up to 10^{14} eV, as expected from diffuse particle acceleration models for cosmic rays. As discussed by Rob Petre, the existence of high-energy electrons does not directly prove that the TeV *protons* we observe in cosmic rays are produced in SNRs, but all of the processes theorists have explored accelerate protons as well as electrons. Furthermore, now that the SN 1006 spectra are understood, detailed spectral analyses of X-ray data from *ASCA*, *Chandra* and other X-ray satellites are providing strong evidence of non-thermal tails in other SNRs, including Cas A, Kepler, and Tycho. Steve Reynolds described how these particle acceleration mechanisms will work, noting that if SNRs do produce the bulk of cosmic rays, then this process requires 10-20% of their energies. Furthermore, for observers like ourselves, Steve pointed out that tests of diffusive particle acceleration are now possible through observations of spectral curvature, through measurement of the highest energy electrons in hard X-rays, and through imaging observations of precursors and synchrotron halos.

Crab and other Plerions. As Virginia Trimble noted in her opening talk, our picture of the Crab Nebula has evolved from the time Bevis first observed it in 1731 through the period of Lord Ross, who produced a drawing more closely resembling a pineapple than a crab. At this conference, Jeff Hester showed us beautiful movies of the expanding wisps inside the axi-symmetric synchrotron nebula surrounding the Crab pulsar obtained with HST, a preview of a coordinated *HST/Chandra* series that will follow over the next few years. The sound-crossing time, as well as the X-ray synchrotron energy-loss time scale, is less than a week in these wisps. He also described the Rayleigh-Taylor instabilities observed in the thermal plasma surrounding the synchrotron nebula. It is quite clear that the Crab will remain a laboratory par excellance for studying the interaction of synchrotron and thermally-dominated plasmas for the indefinite future. Perhaps more exciting, however, are prospects for fully understanding the location of the freely expanding supernova ejecta which may have been observed in CIV in absorption, reaching a blueward velocity of $2500 \ km \ s^{-1}$ [17]. If this is correct, then we may be on our way to fully understanding a nebula first observed in 1731.

Even if we succeed with the Crab, there are other plerions to study. Brian Gaensler gave us an overview of the current census which includes 9 plerions and 23 composites, or about 10% of all Galactic SNRs. Of these 5 contain compact powering objects, although the pulsar is not always at the peak of the X-ray emission. The morphologies of these SNRs are heterogeneous, which is not surprising given that environment plays a large role in all classes of SNRs. They are distinguished from center-filled thermal remnants by the facts that the plerions and composites have center-filled radio morphologies and flat radio spectral indices. The combination of spectral and spatial resolution available with *Chandra* and *XMM-Newton* suggests not only that such remnants will continue to be discovered over the next several years, but that progress in our understanding will also occur because of our new found ability to separate the various emission components.

Pulsars. Another dramatic development, from historical perspective, is the growing number of associations between SNRs and pulsars. In a review at a 1981 meeting on supernovae in Cambridge, Lyne [18] noted that while we were virtually certain that SNe make both SNRs and pulsars, we knew (then) of ~ 140 SNRs, and ~ 330 pulsars in the Galaxy—yet only two were associated: the Crab and Vela. By contrast, Dick Manchester reported here that compact objects—pulsed or otherwise—are now known in 17 SNRs, including a substantial fraction of the young SNRs. This represents really significant progress.

Another recent observational development, summarized at this meeting by Alice Harding, has been the emergence of one or more classes of pulsars, including the Anomalous X-ray Pulsars and Soft Gamma Ray Repeaters, whose energy appears to be dominated by magnetic fields rather than by rotation. Further observation

486

of such objects, together with critical models for interpreting the data, is likely to be a fertile field in the near future.

QUESTIONS AND CHALLENGES

Although the wealth of new data available at various wavelengths has increased our understanding of young SNRs and validated some of the conceptions under-pinning the analysis of the earlier data, many questions, some quite old and others quite new, remain. Our list, stimulated in part by Chris McKee's introductory talk on issues and prospects, is as follows:

Can we connect SNRs with their progenitors? What type stars and ex-plosions lead to which classes of remnants? In a few specific cases there is enough detailed information to make well founded inferences—most notably for Cas A, where Rob Fesen's observations that the very fastest ejecta show only nitrogen lines leads to the reasonable conclusion that the progenitor was a Wolf-Rayet star with a thin layer of nitrogen on its surface. But others remain enigmatic, especially, as Anne Decourchelle described, Kepler's SNR. In that case, we are not really sure whether it was a Type Ia or a core-collapse SN.

Jack Hughes has outlined somewhat more speculative connections that have now been made for a number of young SNRs in the Magellanic Clouds based on abun-dances measured in their X-ray plasma. You-Hua Chu showed us another inter-esting approach to the problem of identifying the progenitor star for a young SNR through observing its circumstellar environment. From the density profile and composition of circumstellar layers that are revealed as the blast wave works it way outward, it may be possible to reconstruct the mass-loss history of the pre-SN star, and to pinpoint the identity of the suspect progenitor.

Kurt Weiler has showed how the radio development of a few SNe into SNRs has now been traced for as long as 30 years. But if we wish to make connections with remnants with ages of centuries or millennia, we must develop either extraordinary patience or new techniques.

Core-collapse SNe explode near where they were born, and that environment should not only be rich with stars of similar age, but also with molecular clouds. Roger Chevalier described the interaction of a young SNR inside a molecular cloud, but we have few if any really good examples of such remnants.

Finally, most of the SNRs that we have discussed at this meeting are at least an order of magnitude younger than the large numbers of new SNRs that have been discovered in the past decade in nearby galaxies like M33, M31, NGC300, NGC2403, and NGC6946 using combinations of narrow-band optical and radio imagery. Even though the ejecta from these supernova remnants will have been diluted to non-detection, these samples contain significant information about the nature of SNe and especially their role in creating the ISM, if only we can interpret the selection effects.

Where is the iron from SN Ia's? Ever since Stirling Colgate's early work [19], we've believed that SN Ia produce $\sim 0.5 M_\odot$ of ^{56}Ni whose decay ^{56}Ni $\rightarrow^{56}$ Co $\rightarrow^{56}$ Fe leaves a residue of $\sim 0.5 M_\odot$ of iron, a scenario reviewed by David Branch at this meeting. Yet in not one SNR has this much iron been observed. Jack Hughes has presented evidence from *Chandra* data for iron-enriched plasma in the interior of 0509-67.5 and DEM71, but a leap of faith from the observed tip of highly ionized iron to a much larger iceberg of cold iron is necessary to approach $0.5 M_\odot$. The closest we've come is in SN 1006, where the fortuitous placement of the UV-bright Schweizer-Middleditch star behind the center of the remnant has made possible the observation of cold iron via absorption lines [20]. But plausible extrapolation from the single core sample through SN 1006 still falls short of the expected amount by a factor of 3.

Where are the runaway companions from SN Ia's? A second problem of SN Ia—of somewhat shorter history—arises from the widely accepted models that SN Ia result from WD accretion in binary systems. When the parasitic white dwarf explodes, its host companion would be released, and remain visible. Yet we know of exactly zero examples of runaway companions associated with remnants that are believed to have originated in SN Ia's. Where are they?

Where are all the young SNRs? Where are all the young SNRs in the Galaxy? in M31, M33? In prolific SN producers like M83 or NGC6946? Many speakers have highlighted this problem. Chris McKee has suggested that many Galactic SNe occur in regions already swept clean by earlier SNe from stars in the same association, so with nothing to run into they produce no bright remnant. Some may be hidden within the highly obscured inner disk of the Galaxy. But surely there should be the odd Cas A or N132D in all of a galaxy like M31. We should find them! Let's hope that for some future meeting on supernova remnants—perhaps in 2006 to celebrate the millennial anniversary of SN 1006—we will hear reports on expanded population of young remnants and see patterns emerge which today we can only guess at.

How are SNe and Gamma-Ray Bursts Related? Finally, we would all like to know what makes a γ-ray burst. There are now at least 15 γ-ray bursts with known redshifts. The association of SN1998bw with GRB980425 makes it quite likely that at least some GRBs are related to SNe, but this was not an extremely energetic event. However, as Chris McKee noted, some events require energies of 5×10^{54} ergs if the event is isotropic, and there are no good models for such explosions.

How do we make compare the data with the models quantitatively? With the rapid advances of computational power, hydrodynamical simulations of SNe and SNRs have become increasingly sophisticated, incorporating much more of the physics we believe to operate in these objects. As discussed by John Blondin, we are no longer limited to simple comparisons with the predictions of Sedov and other one-dimensional models. Qualitatively, the "pseudo-images" that result from the hydrodynamical simulations resemble what we observe. In practice, however, observers and modelers are going to need to develop techniques to compare the simulations with the data. This is not going to be easy. X-ray observatories are expensive, and supernova remnants that can be observed in detail are few compared to computers and the number of models which will be generated. Efforts to use the simulations to model properties that can be widely applied to real data—emission measure vs temperature plots, for example—are going to be fundamental for the confrontation of theory with fact.

John Raymond described the physical processes that need to be included in order to model X-ray spectra with high-resolution, along with some of the uncertainties in the atomic physics. But every one of us who has tried to understand spectra at any wavelength has learned that quantitative analysis is difficult, especially in the case of supernova remnants, where variations in physical conditions along the line of sight produce lines from one ion in one region and lines from another somewhere else. All is not lost by any means, but the low resolution of proportional-counter spectra—and to a somewhat lesser extent even CCD spectra—have covered a multitude of sins. All but a few of us have been cloaked in innocence that we will now have to lose if we are to make progress. Along the way, we will need to be mindful of "local minima in our knowledge space," and Virginia Trimble's warning that band-wagon effects have infested our field.

CONCLUDING REMARKS

This has been a wonderful conference. The spectacular results we have seen at this meeting—especially the X-ray data which have so impressed the two of us who started our careers in X-ray astronomy—have profound implications for our understanding of young SNRs, as well as other areas of astrophysics. We leave with a sense of excitement, not only about what has been presented, but also about the promise of the next several years. We are grateful to the organizers for a meeting which ran extraordinarily smoothly, and to the speakers and poster presenters for presenting so many interesting scientific results while holding tightly to the schedule. Special thanks are also due to NASA, ESA, the taxpayers of several nations, and especially to the dedicated scientists and engineers who have worked so hard to make missions like *Chandra* and *XMM/Newton* such resounding successes.

Acknowledgements. PFW's SNR research is supported by NSF Grant AST-9618465, and by NASA Grant NAG5-8020. KSL's work on SNRs with *Chandra* is supported by NASA grant GO0-1120X.

REFERENCES

1. Rappaport, S., Doxsey, R., Solinger, A., Borken, R., *Ap.J.* **194**, 329 (1974).
2. Moore, W. E, & Garmire, G.P. *Ap.J.* **206**, 247 (1976).
3. Phillips, M.M., *Ap.J.* **413**, L105 (1993).
4. Gull, S.F., *M.N.R.A.S.* **162**, 135 (1973).
5. Gull, S.F., *M.N.R.A.S.* **171**, 263 (1975).
6. McKee, C.F., *Ap.J.* **188**, 335 (1974).
7. Hughes, J.P., Rakowski, C.E., Burrows, D.N., & Slane, P.O., *Ap.J.* **528**, L109 (2000).
8. Koralesky, B., Rudnick, L., Gotthelf, E.V., &Keohane, J.W., *Ap.J.* **505**, L27 (1998).
9. Vink, J., Bloemen, H., Kaastra, J.S., &Bleeker, J.A.M., *A.&Ap.*, **339**, 201 (1998).
10. Hughes, J.P., *Ap.J.* (*in press*).
11. Hughes, J.P., *Ap.J.* **527**, 298 (1999).
12. Hughes, J.P., Rakowski, C.E., & Decourchelle, A. *Ap.J.* **543**, L61 (2000).
13. Dwek, E., *Ap.J.*, **247**, 614 (1981).
14. Arendt, R. G. , *Ap.J.Supp.*, **70**, 181 (1989).
15. Koyama, K., Petre, R., Gotthelf, E. V., Hwang, U., Matsura, M., Osaki, M., & Holt, S. S., *Nature* **378**, 255 (1995).
16. Tanimori, it et al., *Ap.J.* **497**, L25 (1998).
17. Sollerman, J., Lundqvist, P., Lindler, D., Chevalier, R. A., Fransson, C., Gull, T. R., Pun, C. S. J., Sonneborn, G., *Ap.J.* **537**, 861 (2000).
18. Lyne, A.G., in *Supernovae: A Survey of Current Research, Proceedings of the NATO Institute held at Cambridge, U.K.*, M.J. Rees & R.J. Stoneham, eds., pp. 405-417 (1982).
19. Colgate, S.A., & McKee, C., *Ap. J.*, **157**, 623 (1969).
20. Hamilton, A.J.S., Fesen, R.A., Wu, C.-C., Crenshaw, JD.M., & Sarazin, C.L., *Ap.J.*, **481**, 838 (1997).

Appendices

Conference Program
16 – 18 October, 2000

Monday, October 16, 2000

8:30 **Session #1: Introduction**
Chair: Steve Holt

Virginia Trimble	Pineapples and Crabs: When Young SNRs Were Even Younger
Chris McKee	Issues and Prospects

10:00 Coffee

10:30 **Session #2: Stage–Setting**
Chair: Jim Stone

David Branch	Type Ia Supernovae
Alex Filippenko	Type Ib/c and II Supernovae
John Blondin	Hydrodynamics

Discussion and Poster Presentations
12:30 Lunch

2:00 **Session #3: SNe to SNR**
Chair: George Sonneborn

Dick McCray	SN1987A: The Birth of a Supernova Remnant
Roger Chevalier	Molecular Clouds
Rob Fesen	The SN–SNR Connection

Discussion and Poster Presentations
4:00 Tea

4:30 **Session #4: Young SNRs – I**
Chair: Rob Petre

Una Hwang	X-ray Observations
Bill Blair	UV/Optical Observations
Rick Arendt	IR Observations

Discussion and Poster Presentations
6:30 Poster Session

493

Tuesday, October 17, 2000

8:30 **Session #5: Young SNRs – II**
Chair: Andy Szymkowiak

John Raymond   Emission Processes
Claude Canizares High Resolution X-ray Spectroscopy

Discussion and Poster Presentations
10:00   Coffee

10:30 **Session #6: Young SNRs – III**
Chair: John Dickel

Kurt Weiler   Radio Observations
Larry Rudnick   Cas A
Anne Decourchelle Kepler

Discussion and Poster Presentations
12:30 Lunch

2:00 **Session #7: Crab and Related Remnants**
Chair: Cole Miller

Jeff Hester   The Crab Nebula: The Gift that Keeps on Giving
Bryan Gaensler   Plerions and Pulsar Wind Nebulae
Dick Manchester Pulsars, Magnetars, SGRs

Discussion and Poster Presentations
4:00   Tea

4:30 **Session #8: Cosmic Ray Acceleration and Pulsars**
Chair: Nicholas White

Alice Harding   Pulsar Emission Mechanisms and the Nature of
Central Sources in Young Supernova Remnants
Rob Petre   Evidence for Cosmic Ray Acceleration in
Supernova Remnants from X-ray Observations
Steve Reynolds   Particle Acceleration in Shock Waves of
Young Supernova Remnants

Discussion and Poster Presentations
7:00 Banquet
Nicholas White

Wednesday, October 18, 2000

8:30 **Session #9: Some Special Examples**
     Chair: Kathryn Flanagan

    You-Hua Chu   Circumstellar Nebulae in Young SNRs
    Jack Hughes   LMC Sample

    Discussion and Poster Presentations
10:00   Coffee

10:30 **Session #10: Hypernovae**
      Chair: You-Hua Chu

    Arnon Dar   What Makes Them Different?
    John Danziger   Observational Evidence

    Discussion

12:00   Rapporteurs

    Knox Long & Frank Winkler

12:40 Lunch
 2:00   End of Conference

List of Attendees

Name	Affiliation	Email
Allen, G.	MIT	gea@space.mit.edu
Arendt, R.	RITSS/NASA/GSFC	arendt@stars.gsfc.nasa.gov
Arnaud, K.	NASA/GSFC, UMD	kaa@lheamail.gsfc.nasa.gov
Balsara, D.	U. of Illinois	dbalsara@ncsa.uiuc.edu
Bandiera, R.	Osserv. Astr., Arcetri	bandiera@arcetri.astro.it
Behar, E.	Columbia U.	behar@astro.columbia.edu
Benjamin, R. A.	U. of Wisconsin	benjamin@wisp.physics.wisc.edu
Bennett, C.	NASA/GSFC	charles.l.bennett.1@gsfc.nasa.gov
Berendse, F.	UMD, NASA/GSFC	berendse@astro.umd.edu
Blair, W.	Johns Hopkins U.	wpb@pha.jhu.edu
Blondin, J.	NCSU	john_blondin@ncsu.edu
Boffi, F.	STScI	boffi@stsci.edu
Boldt, E. A.	NASA/GSFC	Elihu.A.Boldt.1@gsfc.nasa.gov
Borkowski, K.	NCSU	kborkow@unity.ncsu.edu
Boyce, K.	NASA/GSFC	Kevin.R.Boyce.1@gsfc.nasa.gov
Branch, D.	U. of Oklahoma	branch@mail.nhn.ou.edu
Canizares, C.	MIT	crc@space.mit.edu
Cheng, E.	NASA/GSFC	Edward.S.Cheng.1@gsfc.nasa.gov
Cheung, C.	NASA/GSFC	ccheung@pop600.gsfc.nasa.gov
Chevalier, R. A.	U. of Virginia	rac5x@virginia.edu
Chin, G.	NASA/GSFC	Gordon.Chin.1@gsfc.nasa.gov
Chu, Y.-H.	U. of Illinois	chu@astro.uiuc.edu
Clearfield, C.	NCSSM	clearfieldc@ncssm.edu
Cline, T. L.	NASA/GSFC	Thomas.L.Cline.1@gsfc.nasa.gov
Crannell, C. J.	NASA/GSFC	crannell@gsfc.nasa.gov
Crotts, A.	Columbia U.	arlin@astro.columbia.edu
Danziger, J.	Osserv. Astr., Trieste	danziger@ts.astro.it
Dar, A.	Technion	arnon@physics.technion.ac.il
Davis, D.	MIT/CXC	dsd@space.mit.edu
Decourchelle, A.	Service d'Astrophysique	adecourchelle@cea.fr
DeLaney, T.	U. of Minnesota	tdelaney@astro.umn.edu
Dickel, J.	U. of Illinois	johnd@astro.uiuc.edu
Dolan, J.	NASA/GSFC	Joseph.F.Dolan.1@gsfc.nasa.gov
Drachman, R.	NASA/GSFC	drachman@stars.gsfc.nasa.gov
Drake, P.	U. of Michigan	rpdrake@umich.edu
Dunne, B.	U. of Illinois	carolan@astro.uiuc.edu
Dwek, E.	NASA/GSFC	eli.dwek@gsfc.nasa.gov
Dyer, K.	NCSU	Kristy_Dyer@ncsu.edu

Eriksen, K.	CfA	keriksen@cfa.harvard.edu
Fahey, D.	NASA/GSFC	fahey@stars.gsfc.nasa.gov
Felten, J. E.	NASA/GSFC	felten@stars.gsfc.nasa.gov
Fesen, R.	Dartmouth College	fesen@snr.dartmouth.edu
Filipovic, M. D.	U. of W. Sydney	m.filipovic@uws.edu.au
Filippenko, A.	UC–Berkeley	alex@astro.berkeley.ed
Fisher, R.	NASA/GSFC	Richard.R.Fisher.1@gsfc.nasa.gov
Flanagan, K.	MIT	kaf@space.mit.edu
Gaensler, B.	MIT	bmg@space.mit.edu
Gehrels, N.	NASA/GSFC	Cornelis.A.Gehrels.1@gsfc.nasa.gov
Gerardy, C.	Dartmouth College	gerardy@dartmouth.edu
Ghavamian, P.	Rutgers U.	parviz@physics.rutgers.edu
Graber, J.		jgra@loc.gov
Greenhouse, M.	NASA/GSFC	matt@stars.gsfc.nasa.gov
Gull, T.	NASA/GSFC	Theodore.R.Gull.1@gsfc.nasa.gov
Gursky, H.	NRL	herbert.gursky@nrl.navy.mil
Harding, A. K.	NASA/GSFC	Alice.K.Harding.1@gsfc.nasa.gov
Harrington, P.	U. of Maryland	jph@astro.umd.edu
Harrus, I.	NASA/GSFC	imh@lecanard.gsfc.nasa.gov
Hendrick, S.	NCSU	sphendri@unity.ncsu.edu
Hester, J.	Arizona State U.	jhester@asu.edu
Hinkle, C.	NCSU	clhinkle@unity.ncsu.edu
Holt, S.	Olin College	stephen.holt@olin.edu
Hughes, J.	Rutgers U.	jph@physics.rutgers.edu
Hwang, U.	NASA/GSFC	hwang@orfeo.gsfc.nasa.gov
Iping, R.	NASA/GSFC	rosina@taotaomona.gsfc.nasa.gov
Jahoda, K. M.	NASA/GSFC	Keith.M.Jahoda.1@gsfc.nasa.gov
Jones, F. C.	NASA/GSFC	Frank.C.Jones.1@gsfc.nasa.gov
Kazanas, D.	NASA/GSFC	kazanas@milkway.gsfc.nasa.gov
Keilty, K.	Rice U.	keilty@spacsun.rice.edu
Keohane, J.	NCSSM	keohane@ncssm.edu
Kothes, R.	NRC of Canada	roland.kothes@nrc.ca
Kundu, M.	U. of Maryland	kundu@astro.umd.edu
Laming, M.	NRL	jlaming@ssd5.nrl.navy.mil
Lawrence, S.	Columbia U.	lawrence@astro.columbia.edu
Leventhal, M.	U. of Maryland	ml@astro.umd.edu
Lewis, K.	Penn State U.	lewis@astro.psu.edu
Long, K.	STScI	long@stsci.edu
Maeda, Y.	Penn State U.	maeda@astro.psu.edu
Manchester, R.	Australia Tel.–CSIRO	Dick.Manchester@atnf.csiro.au
Maran, S.	NASA/GSFC	hrsmaran@eclair.gsfc.nasa.gov
Marshall, F. E.	NASA/GSFC	marshall@rosserv.gsfc.nasa.gov

McCray, R.   U. of Colorado–JILA   dick@jila.colorado.edu
McDonald, A.   Jodrell Bank Obs.   amd@jb.man.ac.uk
McKee, C.   UC–Berkeley   cmckee@astro.berkeley.edu
Michael, E.   U. of Colorado   michaele@colorado.edu
Miller, C.   U. of Maryland   miller@astro.umd.edu
Milne, P.   NRL/NRC   milne@gamma.nrl.navy.mil
Mitchell, J. W.   NASA/GSFC   mitchell@lheavx.gsfc.nasa.gov
Moffett, D.   Furman U.   David.Moffett@furman.edu
Morse, J.   U. of Colorado   morsey@casa.colorado.edu
Mushotzky, R. F.   NASA/GSFC   mushotzky@lheavx.gsfc.nasa.gov
Olbert, C.   NCSSM   olbertc@ncssm.edu
Ormes, J. F.   NASA/GSFC   ormes@lheamail.gsfc.nasa.gov
Panagia, N.   ESA/STScI   panagia@stsci.edu
Pannuti, T.   MIT–CSR   tpannuti@space.mit.edu
Patnaude, D.   Dartmouth College   daniel.patnaude@dartmouth.edu
Petre, R.   NASA/GSFC   Robert.Petre.1@gsfc.nasa.gov
Polidan, R.   NASA/GSFC   rpolidan@pop600.gsfc.nasa.gov
Porter, F. S.   NASA/GSFC   porter@milkyway.gsfc.nasa.gov
Ramaty, R. R.   NASA/GSFC   ramaty@lheamail.gsfc.nasa.gov
Rasmussen, A.   Columbia U.   arasmus@astro.columbia.edu
Raymond, J.   CfA   jraymond@cfa.harvard.edu
Reynolds, S.   NCSU   steve_reynolds@ncsu.edu
Rho, J.   IPAC/Caltech   rho@ipac.caltech.edu
Ritz, S. M.   NASA/GSFC   Steven.M.Ritz.1@gsfc.nasa.gov
Rose, W. K.   U. of Maryland   wrose@astro.umd.edu
Rudnick, L.   U. of Minnesota   larry@tc.umn.edu
Safi-Harb, S.   U. of Manitoba   samar@physics.umanitoba.ca
Sankrit, R.   Johns Hopkins U.   ravi@pha.jhu.edu
Schlegel, E.   CfA   eschlegel@cfa.harvard.edu
Senda, A.   Kyoto U.   senda@cr.scphys.kyoto-u.ac.jp
Serlemitsos, P.   NASA/GSFC   pjs@astron.gsfc.nasa.gov
Seward, F.   SAO   fds@cfa.harvard.edu
Shafer, R.   NASA/GSFC   shafer@achamp.gsfc.nasa.gov
Shapiro, M.   U. of Maryland   shapiro@sigmanet.net
Silverberg, B.   NASA/GSFC   silverberg@stars.gsfc.nasa.gov
Sjouwerman, L.   JIVE   sjouwerm@jive.nl
Slane, P.   CfA   slane@cfa.harvard.edu
Smith, R.   CfA   rsmith@cfa.harvard.edu
Smither, R.   Argonne   rks@aps.anl.gov
Sonneborn, G.   NASA/GSFC   george.sonneborn@gsfc.nasa.gov
Sramek, R.   NRAO   dsramek@nrao.edu
Stahle, C. K.   NASA/GSFC   ckstahle@lheapop.gsfc.nasa.gov

Stecher, T.	GSFC/NASA	Stecher@UIT.gsfc.nasa.gov
Stecker, F. W.	NASA/GSFC	Floyd.W.Stecker.1@gsfc.nasa.gov
Stiller, B.		bstiller@capaccess.org
Stockdale, C.	U. of Oklahoma	stockdale@phyast.nhn.ou.edu
Stollberg, M. T.	USNO	stollberg.mark@usno.navy.mil
Stone, J.	U. of Maryland	jstone@astro.umd.edu
Streitmatter, R. E.	NASA/GSFC	streitmatter@lheavx.gsfc.nasa.gov
Sturner, S.	NASA/GSFC	Steven.J.Sturner.1@gsfc.nasa.gov
Sugerman, B.	Columbia U.	ben@astro.columbia.edu
Szymkowiak, A. E.	NASA/GSFC	aes@milkyway.gsfc.nasa.gov
Temkin, A.	NASA/GSFC	Aaron.Temkin.1@gsfc.nasa.gov
Trasco, J.	U. of Maryland	jtrasco@astro.umd.edu
Trimble, V.	U. of Maryland	vtrimble@astro.umd.edu
Valinia, A.	NASA/GSFC	valinia@milkyway.gsfc.nasa.gov
van der Swaluw, E.	Astr. Inst., Utrecht	E.vanderSwaluw@astro.uu.nl
Vink, J.	Columbia U.	jvink@astro.columbia.edu
Weiler, K.	NRL	weiler@rsd.nrl.navy.mil
White, N. E.	NASA/GSFC	nwhite@lheapop.gsfc.nasa.gov
Williams, N.	NCSSM	nik@astro.ncssm.edu
Williams, R.	NASA/GSFC	rosanina@lhea1.gsfc.nasa.gov
Winkler, P. F.	Middlebury College	winkler@middlebury.edu
Woltjer, L.	OHP, Arcetri	woltjer@obs-hp.fr
Woodgate, B.	NASA/GSFC	woodgate@stars.gsfc.nasa.gov
Yoshida, T.	Ibaraki U.	yoshidat@mito.ipc.ibaraki.ac.jp
Zhang, W.	NASA/GSFC	zhang@xancus10.gsfc.nasa.gov

AUTHOR INDEX

A

Achterberg, A., 379
Allen, G. E., 230, 395
Arendt, R. G., 163
Aschenbach, B., 267, 403

B

Baganoff, F. K., 177
Ballmoos, P. V., 317
Balsara, D., 89
Bandiera, R., 329
Bautz, M. W., 177
Benjamin, R. A., 89, 222
Benson, B. A., 222
Berendse, F., 399
Blair, W. P., 153, 189
Blondin, J. M., 59, 81
Boffi, F. R., 73
Borkowski, K. J., 387, 391
Bouchet, P., 129
Branch, D., 31
Brandt, W. N., 177
Burrows, D. N., 177, 181

C

Cambresy, L., 197
Canizares, C. R., 213, 226, 230
Cesaroni, R., 329
Chevalier, R. A., 109
Chu, Y.-H., 77, 409, 437
Clearfield, C. R., 337, 341, 345
Cowan, J. J., 77
Cox, D. P., 89, 222
Crotts, A., 129, 133, 137

D

Danziger, I. J., 471
Dar, A., 455
Davis, D. S., 213, 226, 230
Decourchelle, A., 257

DeLaney, T., 279
Dewey, D., 213, 226, 230
Diamond, P. J., 441
Dickel, J. R., 279, 325, 433
Ditmire, T., 85
Doty, J. P., 177
Dunne, B. C., 437
Duric, N., 445
Dyer, K. K., 391

E

Edgar, R. J., 403
Eriksen, K. A., 193

F

Feigelson, E. D., 177
Fesen, R. A., 119
Filipović, M. D., 267, 429, 445
Filippenko, A. V., 40
Flanagan, K. A., 213, 226, 230
Frail, D. A., 337, 341, 345

G

Gaensler, B. M., 295, 333
Gallant, Y. A., 379
Garmire, G. P., 177, 181
Garrett, M. A., 441
Garrington, S. T., 441
Ghavamian, P., 189
Gotthelf, E. V., 395
Graber, J. S., 321
Green, A., 333
Gruendl, R. A., 437
Gull, T. R., 315

H

Harding, A. K., 351
Hendrick, S. P., 387
Hester, J. J., 285

Hinkle, C. L., 81
Holt, S. S., 185, 399
Houck, J. C., 213, 226, 230
Hughes, J. P., 181, 403, 419
Hwang, U., 143, 399

J

Jones, P. A., 267, 429

K

Keilty, K. A., 85
Keohane, J. W., 337, 341, 345
Kirshner, R. P., 193
Koo, B.-C., 197
Koralesky, B., 275, 279
Kothes, R., 271
Kunkel, W., 137

L

Lacey, C. K., 237
Lawrence, S., 129, 133, 137
Lewis, K. T., 181
Liang, E. P., 85
Lindler, D., 315
Long, K. S., 481
Lundqvist, P., 315

M

Maeda, Y., 177
Manchester, R. N., 305
McCray, R., 95
McDonald, A. R., 441
McKee, C. F., 17
Miller, E. A., 449
Milne, P. A., 173
Miyata, E., 403
Moffett, D., 333
Montes, M. J., 237
Morris, M., 177
Morse, J. A., 193
Muxlow, T. W. B., 441

N

Neri, R., 329
Nousek, J. A., 181

O

Olbert, C. M., 337, 341, 345

P

Panagia, N., 73, 237
Pannuti, T., 445
Pedlar, A., 441
Petre, R., 185, 360, 391, 395, 399
Pietsch, W., 445
Plucinsky, P. P., 403
Pravdo, S. H., 177

R

Raymond, J. C., 189, 203
Reach, W. T., 197
Read, A., 445
Reich, W., 271
Remington, B. A., 85
Reynolds, S. P., 369, 387, 391
Rho, J., 197
Ricker, G. R., 177
Rose, W. K., 383
Rubenchik, A. M., 85
Rudnick, L., 247, 275, 279, 449
Rupen, M. P., 77

S

Schattenburg, M. L., 213, 226, 230
Schlegel, E. M., 69
Schulz, N. S., 230
Shigemori, K., 85
Sjouwerman, L. O., 433
Skinner, G., 317
Slane, P., 181, 403
Smither, R., 317
Sollerman, J., 315
Sonneborn, G., 315

Sramek, R. A., 237
Stockdale, C. J., 77
Sugerman, B., 129, 133, 137

T

Townsley, L. K., 177
Trimble, V., 3
Tsunemi, H., 403

V

van der Swaluw, E., 379
Van Dyk, S. D., 237

W

Weiler, K. W., 237
White, G. L., 429
Williams, N. E., 337, 341, 345
Williams, R. M., 185, 437
Wills, K. A., 441
Winkler, P. F., 193, 481

Source Index

η Carinae, 47, 78

0453-68.5, 388, 390

0509-67.5, 121, 388, 390, 488

0519-69.0, 185, 186, 188, 388

0534-69.9, 388, 390

0540-69.3, 120, 121, 291

0548-70.4, 388, 390

1E 161348-5055, 197

1E 1841-045, 310, 311

1RSX J1708-4009, 311

2EG J0618+2234, 114

3C 10 (also Tycho and SN 1572), 5

3C 358 (also Kepler and SN 1604), 257

3C 391, 116

3C 58, 5, 9, 121, 296, 298, 300, 332

AX J0851.9-4617.4, 405, 406

Cassiopeia A, 4, 5, 12–14, 17, 21–23, 27, 62, 65, 79, 89, 91, 92, 96, 120–122, 124, 125, 127, 144, 146–151, 153–155, 165–168, 170, 174, 176, 193, 195, 204, 209, 214, 220, 247, 257, 259, 263, 264, 272, 275, 296, 352, 356, 360–364, 383, 385, 399, 410, 415, 478, 483–485, 487, 488

Crab (also SN 1054), 5, 7–9, 11, 12, 16, 79, 119, 120, 127, 154, 155, 164–168, 170, 176, 204, 208, 285, 296–302, 305, 315, 317, 320, 326, 329–332, 351, 353, 364, 420, 421, 486

CTA 1, 298

CTB 37AB, 4

CTB 80, 4, 298, 327

CTB 87, 332

Cygnus Loop, 13, 21, 111, 153–155, 158, 165, 189, 200, 208, 296, 377

Cygnus X, 13

DEM 71, 388, 426, 427, 488

DEM 84, 425

E0102-72, 153, 156–158, 193, 213, 226, 230, 263, 484, 485

ESO 184-G82, 472

G10.0-0.3, 312

G11.2-0.3, 5, 271

G21.5-0.9, 274, 298, 301, 326, 327

G266.2-1.2, 403

G266.3-1.2, 362

G291.0-0.1, 333, 334

G309.2-0.6, 264

G327.1-1.1, 297

G332.4-0.4, 197

G337.0-0.1, 312

G347.3-0.5, 263, 379

G4.5+6.8 (also 3C 358, Kepler and SN 1604), 257

Geminga, 301, 306, 310, 354

GRB 000418, 456

GRB 970228, 456, 463, 468, 477

GRB 970508, 456, 463, 464, 466–468

GRB 970514, 53, 476

GRB 970828, 464, 468
GRB 971115, 476
GRB 980326, 456, 477
GRB 980425, 23, 26, 52, 456, 463,
    464, 468, 471, 472, 476, 477,
    488
GRB 980703, 456
GRB 980828, 466
GRB 980910, 53, 476
GRB 990510, 26, 456, 468
GRB 990712, 456
GRB 991002, 476
GRO J0852-4642, 403
Guitar Nebula, 300

IC 443, 109, 111–116, 165, 170, 185,
    337, 341, 365

J0437-4715, 301
J1846-0258, 309, 313

Kepler (also 3C 358 and SN 1604),
    5, 12, 14, 62, 121, 155, 158,
    165, 168, 169, 176, 257, 272,
    279, 376, 485, 487
Kes 25, 4
Kes 75, 4, 5, 309, 310

Large Magellanic Cloud, 37, 120,
    121, 123, 153, 154, 156, 179,
    181, 185, 186, 188, 213, 217,
    291, 296, 297, 310, 312, 327,
    364, 365, 377, 387, 389, 390,
    413, 415, 420, 423–425, 429,
    438
LH83, 415

M100, 122

M101, 384, 437
M31, 9, 14, 119, 123, 124, 127, 154,
    160, 429, 433, 487, 488
M33, 127, 154, 429
M81, 49, 384
M82, 441
M83, 3, 154, 488
MCG +03-28-022, 122
MF 16, 414, 415
Milky Way, 3
MSH 11-54, 4
MSH 11-62, 333

N103B, 121, 181, 388, 390, 425, 426,
    444
N132D, 121, 153, 154, 156–158, 195,
    388, 390, 415, 419, 420, 427,
    478, 488
N157B, 298, 301, 302, 309, 310, 313,
    326, 327, 351, 426
N158A, 420
N23, 388, 390, 427
N49, 312, 313, 419, 427
N49B, 388, 390
N63A, 388, 390, 415–417, 419, 427
NGC 1058, 47, 77
NGC 1850, 425
NGC 1952, 6, 9
NGC 2403, 47, 78, 487
NGC 300, 445, 487
NGC 3031, 49
NGC 3603, 413
NGC 4321, 44
NGC 4449, 121, 127
NGC 4496A, 33
NGC 4527, 33
NGC 4536, 33

NGC 5377, 43
NGC 6888, 410, 415
NGC 6946, 154, 415, 487
Nova Aql 1918, 12, 13
Nova Ophiuchi, 12, 257
Nova Per 1901, 12

PKS 1459-41, 5
PSR B1046–58, 301
PSR B1055–52, 301
PSR B1610–50, 301
PSR B1757–24, 298
PSR J0537-6910, 308, 313
PSR J1119-6127, 308, 309
PSR J1435-6100, 308
PSR J1740-3052, 308
PSR J1808-3658, 310
PSR J1814-1744, 311, 358
PSR J1846-0258, 309, 313
Puppis A, 13, 121, 154, 155, 165,
    176, 351

RCW 103, 4, 165, 169, 170, 197–
    200, 351, 485
RCW 86, 21, 176, 362
RX J0852.0-4622, 174

S Andromedae *(also SN 1885A)* 9,
    14, 433, 435, 436
SAX J0852.0-4615, 406
SGR 0526-66, 312, 313, 357
SGR 1806-20, 312, 313
Sgr A East, 177
Sgr A West, 177, 178, 180
Sher 25, 413
Small Magellanic Cloud, 21, 153,
    193, 213, 226, 230, 263, 421,
    429, 478

SN 1006, 5, 13, 36, 37, 120, 121,
    148, 154, 155, 158, 160, 165,
    176, 206, 208, 209, 263, 268–
    270, 272, 360–362, 365, 367,
    377, 379, 391, 395, 399, 404,
    421, 423, 485, 488
SN 1054 *(also Crab)* 5, 6, 8, 9, 11,
    13, 285, 290
SN 1523, 4, 5
SN 1572 *(also Tycho)* 5, 12, 36, 37,
    120
SN 1604 *(also Kepler)* 5, 12, 257
SN 1885A *(also S Andromedae)* 3,
    9, 14, 160, 433–435
SN 1895B, 4, 12
SN 1923A, 4
SN 1937C, 4
SN 1950B, 78, 79
SN 1954A, 4
SN 1954J, 47
SN 1957D, 3, 78, 79, 121
SN 1959D, 79
SN 1961V, 3, 47
SN 1970G, 78, 79, 121, 384
SN 1972E, 4
SN 1978K, 4, 121, 122, 243, 244,
    411, 412
SN 1979C, 44, 45, 79, 121, 122,
    240, 242, 384
SN 1980K, 45, 78, 79, 121, 122,
    240, 242, 384
SN 1983N, 241
SN 1984A, 35
SN 1984L, 49
SN 1985L, 121
SN 1986J, 77, 79, 243
SN(R) 1987A, 3, 4, 20, 21, 42, 48,

49, 51, 56, 65, 69, 70, 95,
119–121, 129, 133, 137, 155,
176, 244, 245, 262, 384, 409,
412, 413, 420, 421, 478, 483

SN 1987F, 475

SN 1987K, 48

SN 1988Z, 70, 122, 243

SN 1990B, 240, 241

SN 1991bg, 33

SN 1991T, 32, 33, 35, 471

SN 1992H, 43

SN 1993J, 49–51, 243–245, 383–386,
444, 477, 483

SN 1994ak, 46

SN 1994I, 243

SN 1994W, 46

SN 1994Y, 45

SN 1995N, 122, 123

SN 1997ab, 411, 412

SN 1997bs, 47

SN 1997cy, 53, 69, 70, 72, 476

SN 1997ef, 476

SN 1997eg, 411, 412

SN 1998bw, 24–26, 52, 456, 463,
464, 471–477

SN 1998S, 51, 53

SN 1999aa, 33

SN 1999E, 53, 70, 476

SN 1999eb, 476

SN 369, 13

SN 837, 4, 5

SN1961V, 77

SS 433 *(also W50)* 4, 5, 120, 327

Tycho *(also SN 1572)* 8, 12–14, 21,
37, 148, 154, 158, 160, 165,
176, 185, 257, 259, 260, 264,

272, 376, 423, 485

Vela, 154, 155, 174, 176, 267–269,
301, 302, 305, 327, 353, 403–
406

Vela Z, 267

Virgo, 33

W28, 113, 116

W44, 115, 116, 176, 327

W50 *(also SS 433)* 120, 327, 364

W7, 31

Z Cen, 12

Subject Index

absorption, 3, 8, 13, 35, 40, 41, 44–46, 48, 51, 76, 86, 111, 114, 123, 160, 182, 189, 191, 192, 238, 240, 242–244, 271, 292, 315, 333, 384, 403, 419, 435, 462, 463, 474, 475

absorption profile, 123, 191, 292

absorption, free-free, 114, 240, 243, 384

absorption, HI, 271, 333

absorption, interstellar, 155, 231

absorption, synchrotron self, 238, 240

accretion, 34, 36, 254, 305, 308, 310, 311, 323, 455, 457, 467, 488

adiabatic losses, 298, 299, 381

adiabatic phase (also Sedov-Taylor phase), 85, 86

Balmer-dominated emission, 123, 185, 189, 259, 376, 388, 424–426

binary systems, 31, 73, 75, 76, 97, 122, 238, 243, 305–308, 310, 327, 328, 467, 477, 488

binary systems, double-degenerate, 32

binary systems, single-degenerate, 31

black hole, 13, 26, 42, 177, 254, 308, 321–324, 386, 433, 467, 471

blue giant, 97, 457

blue supergiant, 18, 138, 384, 409

bremsstrahlung, 11, 114, 115, 204, 211, 374, 375, 377, 399, 446, 462, 465, 466

bremsstrahlung, inverse, 204

bremsstrahlung, nonthermal, 204

bubbles, iron-nickel, 65–67, 125, 145, 193, 195

cannonball, 455, 457, 467

Cepheid distance, 33, 78

Chandrasekhar mass, 31, 32, 289, 482

charge transfer, 207

Chevalier-Nadyozhin solution, 19

circumstellar material, 23, 32, 34, 35, 40, 45, 50, 51, 76, 79, 95–99, 102, 105, 107, 139, 185, 220, 237, 248, 250, 386, 440

circumstellar nebulae, 409

circumstellar rings, 98, 99, 101, 102, 105–107, 129

clumping, of ambient material, 239, 240, 243

clumping, of ejecta, 21, 65, 123, 424

clumping, of wind material, 123, 243

collisional (Coulomb) heating, 105, 210, 422

collisionless heating, 21, 189

composite supernova remnants, 296, 325–327

contact discontinuity, 19, 81, 82, 137, 138, 262, 263, 422

cooling, radiative, 22, 85–87, 199,
204, 210, 222, 224, 225, 457
core collapse, 4, 10, 18, 20–22, 26,
65, 72, 120, 138, 153, 154,
173, 181, 193, 248, 268, 321,
343, 420, 421, 455, 457, 467,
482, 487
cosmic ray acceleration *(also electron and particle acceleration)*, 19, 21, 149, 360, 395,
399, 402, 406, 423
cosmic ray heating, 160
cosmic ray spectrum, 113, 379, 384,
387, 389, 390, 399
cosmic ray spectrum, knee of, 360,
367, 379, 387, 389, 390, 395,
397, 399

deflagration, 31, 33, 35, 36, 482
density distribution, of ejecta, 17–
19, 216, 229
density profile, exponential, 60–62
density profile, power-law, 66, 81,
262
detonation, delayed, 35, 36
detonation, supersonic, 35
Doppler shifts, 41, 71, 97, 100–102,
107, 149–151, 194–196, 213,
216, 220, 227, 254, 290, 315,
377, 455, 458, 460, 461, 464–
468
dust, composition of, 125, 163–165,
248
dust, destruction of, 20, 107, 169
dust, emission from , 20, 163–165,
168, 209, 260, 485
dust, excitation temperature, 164

dust, formation of, 20, 163, 248,
485
dust, heating of, 209
dust, sputtering of, 20

ejecta, asymmetric, 40, 101, 107
ejecta, deathbed, 409, 411
ejecta, kinematics of, 173
ejecta, mixing of, 65
ejecta, oxygen-rich, 120, 153, 154,
156–158, 187, 193, 210, 213,
226, 230, 388, 420–422, 471,
478
ejecta-dominated, 17–19, 59, 62, 148,
179, 188, 249, 272, 285, 289,
291, 364, 388, 424, 425
electron acceleration, 98, 115, 164,
263, 325, 383, 387, 399, 402
electron distribution, rolloff in, 381,
387, 389, 392, 393
electron heating, 21, 210, 377, 422
entropy, 82
equilibration, electron-ion, 105, 158,
178, 189, 190
excitation, by protons, 206
expansion velocity, 77, 98, 123, 125,
185, 195, 251, 259, 267, 275,
277, 290, 409, 411–413, 415,
438, 440, 461, 471, 473, 474,
483
explosion, asymmetric, 105, 120, 250,
477
explosive burning, of O, 125
explosive burning, of Si, 146, 484

Fermi acceleration, 113, 114, 263,
399, 462

fireball, 471
fractal structure, 112
free expansion, 23, 100, 101, 119, 123, 193, 195, 196, 225, 260, 275, 285, 289–293, 380, 486
free-bound emission *(also bremsstrahlung)*, 133, 135
free-free emission *(also recombination)*, 22, 133, 135, 164, 240, 331

Galactic halo, 89
galaxy, dwarf, 3
galaxy, elliptical, 42
galaxy, spiral, 42, 384, 437, 472
galaxy, starburst, 441
gamma ray bursts, 17, 23–26, 40, 52–54, 72, 351, 455–458, 460, 461, 463–468, 471, 472, 474, 476, 477, 488
gamma ray bursts, afterglows from, 23, 26, 455, 456, 462–468, 471, 477
gravitational collapse, 110, 455
gravitational radiation, 26, 32

Herbig-Haro objects, 200, 204
HII region, 42, 178, 181, 244, 416, 445, 448
hypernovae, 4, 14, 26, 72, 455, 471

initial mass function, 22, 23
instability, convective, 62–64, 81
instability, cooling, 86, 88, 288
instability, gravitational, 109
instability, hydrodynamic, 59, 62, 263
instability, Kelvin-Helmholtz, 288

instability, Rayleigh-Taylor, 50, 62, 63, 81–83, 260, 278, 285, 289, 290, 292, 293, 424, 486
instability, Richtmeyer-Meshkov, 62, 63
instability, thermal, 103
interstellar bubble, 138, 409–411, 415–417
inverse Compton losses, 381
inverse Compton radiation, 45, 114, 211, 250, 252, 353, 354, 375, 380, 381, 399
ionization equilibrium, 178, 179, 187, 203, 204, 207, 215, 222, 232, 361, 400

jet, 22, 52, 72, 86, 87, 124, 125, 244, 250, 264, 286–288, 297, 301, 326–328, 364, 386, 455, 457, 462, 467

kinematic age, 153, 154, 193, 195, 196, 214, 420–422
Kompaneets approximation, 18

light curve, 9, 10, 25, 26, 32, 33, 35, 37, 41, 42, 44, 45, 47, 49, 65, 69, 73, 76, 77, 79, 97, 105, 120, 129, 130, 140, 238, 244, 245, 257, 354, 384, 456, 464, 471, 473, 476, 477, 483
light echo, 33, 97, 107, 137–139
line profile, 45, 51, 72, 102, 315
line profile, P Cygni, 40, 41, 43, 44, 51, 69, 411, 412, 474, 475
luminous blue variables, 40, 78–80, 409–413

511

Mach number, 17, 21, 189, 190, 211, 282, 342, 370, 422, 423

magnetar, 310, 313, 327, 351, 352, 356–358

magnetic field amplification, 84, 260, 277, 383, 386

magnetic field decay, 299, 352, 357

main sequence wind, 137

masers, 109, 113, 115, 116, 197, 200

mass loss, 25, 34, 45, 46, 62, 73, 75, 110, 120, 129, 137, 221, 238, 240, 242–244, 384, 409, 410, 413, 483, 487

mass transfer, 34, 42, 48, 49

mixed morphology remnants, 179, 341

molecular clouds, 89, 109, 197, 341, 404, 420

morphology, barrel-shaped , 89, 170, 197, 267, 269

morphology, shell, 177, 178, 197, 198, 209, 210, 230, 267, 271, 296, 299, 302, 325, 362, 364, 367, 397, 403, 404, 423, 431, 441–444

nebula, bipolar, 99, 137–140

nebula, comet-shaped, 298, 301, 337–339, 341–343, 345–348

neutron star, 9, 10, 13, 34, 42, 166, 254, 268, 305, 306, 309–311, 313, 321, 322, 324, 325, 333, 335, 341, 342, 352–357, 365, 403, 405, 406

neutron star, cooling of, 8, 354, 355

nonequilibrium ionization, 187, 228, 230–232, 421, 422

nucleosynthesis, 17, 20, 26, 107, 146, 156, 158, 185, 193, 412, 424

outer gap, 352–356

pair cascades, 352, 353, 355

pair formation (inverse Compton), 353, 358

particle acceleration, 17, 107, 148, 149, 210, 250, 252, 254, 257, 263, 264, 279, 325, 352, 354, 369, 379, 380, 382, 404, 484, 485

particle acceleration, nonlinear, 369, 373, 375, 423

particle injection, 113–115, 326, 329, 373, 381, 384

photoionization, 110, 156, 204, 210, 290, 292, 415

planetary nebula, 165

plerion, 23, 120, 168, 274, 295, 313, 325–327, 329, 332, 333, 335, 351, 364, 405, 421, 422, 486

polar caps, 352–355, 358, 391, 393, 394, 405

polarization, 8, 11, 40, 51, 52, 72, 91, 110, 115, 244, 249, 251, 264, 271, 272, 287, 296, 300, 306, 326, 333–335, 341–343, 345–347, 358, 385, 472, 473

precursor, dynamic, 374

precursor, magnetic, 160, 197, 200

precursor, MHD-wave, 374

precursor, radiative, 197, 200

pressure, interclump, 110

pressure, internal, 465

pressure, magnetic, 6, 89, 91, 110

512

pressure, postshock, 113, 377

pressure, ram , 61, 113, 137, 138, 290

pressure, thermal, 90, 91, 110

Primakoff blast wave, 18

proper motion, 4, 8, 14, 149, 195, 214, 250, 258, 259, 264, 275–277, 279, 301, 302, 484

pulsar, 4, 9, 11, 17, 19, 27, 62, 63, 115, 120, 153, 154, 199, 273, 285–289, 291–293, 295–302, 305, 315, 317, 321, 324, 326–328, 331, 333, 335, 341–343, 351–358, 362, 364, 405, 421, 422, 426, 427, 481, 486

pulsar braking index, 298, 300, 335

pulsar glitch, 305, 311

pulsar nebula, 19, 63, 115, 273, 295, 296, 300, 302, 427

pulsar spin period, 295, 301, 335, 426

pulsar wind, 115, 285–287, 289, 292, 293, 295, 296, 302, 331, 342, 427

pulsar, characteristic age of, 306, 308, 309, 343, 356, 357, 426

pulsar, quiescent emission from, 351, 357

pulsar, spin-down age of, 154, 421

pulsar, spin-down luminosity of, 285, 295, 298, 299, 310, 342, 355, 357

pulsar, spin-down rate of, 305, 306, 310, 312, 335, 356, 426

pulsars, anomalous X-ray, 302, 305, 351, 352, 356–358

pulsars, millisecond, 305–307, 310,

355

radiative cooling, 22, 85–87, 199, 204, 210, 222, 224, 225, 457

radiative phase *(also snowplow phase)*, 85, 86, 111, 420

radioactive decay, 10, 96, 125, 173, 209

radioactive heating, 65, 145, 195

recombination, 43, 99, 204–208, 210, 224, 291, 464–467

red giant, 34, 36, 97

red supergiant, 73, 74, 137, 384, 386, 410, 457

red supergiant wind, 74, 137, 242

reflection nebulae, 137

relativistic wind, 285, 295, 341

resonant scattering, 41, 189, 192, 353, 371

Reynolds number, 17

Sedov-Taylor phase *(also adiabatic phase)*, 17, 19, 380, 388, 443, 463

Sedov-Taylor solution, 61, 87, 90

shock heating, 156, 325, 377

shock velocity, 21, 66, 113–115, 132, 149, 155, 189, 191, 200, 229, 282, 290, 342, 367, 373, 376, 424, 426

shock, adiabatic, 115, 370

shock, blast, 11, 18, 19, 24, 59–63, 66, 67, 82, 85–88, 90–92, 95, 96, 98, 99, 101–105, 107, 110, 130, 132, 143, 148, 149, 160, 194, 207, 214, 224, 230, 238–241, 244, 245, 249,

250, 263, 282, 362, 363, 367,
373, 379–381, 387, 399, 400,
402, 421–424, 427, 483, 484,
487

shock, bow, 261, 299–301, 338, 341,
342, 347

shock, compression by, 113, 114,
369, 370, 373, 374, 377, 380,
381

shock, nonradiative, 21, 158, 189,
259

shock, oblique, 67, 102, 103, 282

shock, plane-parallel, 148, 232, 393,
400

shock, radiative, 85, 88, 102, 103,
109, 115, 259, 371

shock, refracted, 130

shock, reverse, 18–20, 50, 61–65,
67, 95, 96, 99–101, 104, 105,
107, 138, 143, 145, 148, 160,
186, 193, 195, 200, 207, 213–
216, 220, 226, 230, 232, 248–
250, 257, 262, 263, 277, 282,
299, 362, 363, 380, 393, 396,
399, 402, 422, 427, 483, 484

shock, termination, 301

shock, wind, 211, 288

snowplow phase *(also radiative phase)*,
85, 380

soft gamma ray repeaters, 305, 351,
352, 356, 357

star formation, 10, 17, 22, 23, 110,
116, 163, 200, 406, 435, 436,
447, 456

stellar atmosphere, 101

stellar winds, 25, 48, 64, 69, 70, 73,
97, 99, 241, 243, 244, 261,

262, 382, 384, 386, 409, 410,
467

subshock, thermal, 373, 374

superbubbles, 437, 440

superluminal motion, 288

supernovae, birth rate of, 22, 54

supernovae, jet-induced, 52, 72

supernovae, subluminous, 33, 37,
120, 123, 124, 127

supernovae, Type Ia, 13, 18, 31, 40,
41, 48, 60, 120, 121, 123,
124, 127, 145, 153, 154, 158,
160, 169, 170, 174–176, 181,
183, 185, 237, 259, 262, 321,
324, 362, 388, 393, 424, 425,
433, 436, 471, 482, 483, 487,
488

supernovae, Type Ib/c, 40, 120, 121,
125, 158, 165, 214, 221, 237,
238, 240, 241, 243, 245, 384,
472–478, 482

supernovae, Type II, 40, 69–73, 77–
79, 111, 120–122, 127, 165,
168–170, 174, 175, 237, 238,
240, 242, 243, 245, 475–477

supernovae, Type II pec, 70, 120,
121

supernovae, Type II-L, 40, 42, 44,
45, 69, 120–123, 127

supernovae, Type II-P, 40, 42–45,
51, 120–122

supernovae, Type IIn, 40, 45–47,
51

synchrotron flickering, 89, 91, 92

synchrotron halos, 376

synchrotron losses, 289, 299, 300,
326, 327, 329–332, 367, 380,

381, 397

synchrotron nebula, 154, 271, 273,
      285–287, 289–293, 341, 342,
      406

TeV energy particles, 115, 252, 263,
      354, 360–365, 367, 369, 372,
      376, 377, 387, 389, 390, 394,
      399, 402, 455, 458, 467, 485
turbulence, 65, 66, 89, 91, 92, 110,
      114, 190, 211, 371, 376, 383,
      385, 386
turbulence, Langmuir, 211
turbulence, MHD, 91, 92, 110, 114,
      190, 371
turbulent energy, 66, 82, 84
turbulent medium, 89, 90, 114
turbulent mixing, 22, 207
two-photon emission, 133, 135

white dwarf, 10, 12, 31–36, 43, 49,
      120, 154, 305, 308, 488
Wolf-Rayet stars, 25, 70, 120, 125,
      146, 156, 193, 194, 384, 409,
      410, 412, 413, 417, 471, 473,
      487

TABLE OF PHYSICAL CONSTANTS

CONSTANT	SYMBOL	MKS	CGS	OTHER
speed of light	c	$3.00 \cdot 10^8$ m/s	$3.00 \cdot 10^{10}$ cm/s	(2.997925)
electron charge	e	$1.60 \cdot 10^{-19}$ coul	$4.80 \cdot 10^{-10}$ esu	
Planck constant	h	$6.63 \cdot 10^{-34}$ J•s	$6.63 \cdot 10^{-27}$ erg•s	
	$\hbar$	$1.05 \cdot 10^{-34}$ J•s	$1.05 \cdot 10^{-27}$ erg•s	
	hc	$1.99 \cdot 10^{-25}$ J•m	$1.99 \cdot 10^{-16}$ erg•cm	200 MeV•fm
	$\hbar c$	$3.15 \cdot 10^{-26}$ J•m	$3.15 \cdot 10^{-17}$ erg•cm	
Boltzmann constant	k	$1.38 \cdot 10^{-23}$ J/K	$1.38 \cdot 10^{-16}$ erg/K	$8.6 \cdot 10^{-5}$ eV/K
	k/h	$2.08 \cdot 10^{10}$ s^{-1}/K	$2.08 \cdot 10^{10}$ s^{-1}/K	
	k/hc	69.5 m^{-1}/K	0.695 cm^{-1}/K	
Gravitational constant	G	$6.67 \cdot 10^{-11}$ N•m^2/kg^2	$6.67 \cdot 10^{-8}$ dy•cm^2/gm^2	
Gas constant	R	8.314 J/K•mole	$8.31 \cdot 10^7$ erg/K•mole	
Avogadro's number (= R/k)	N	$6.02 \cdot 10^{26}$ amu/kg	$6.02 \cdot 10^{23}$ amu/kg	$6 \cdot 10^{23}$ molecules/mole
electron mass	m_e	$9.11 \cdot 10^{-31}$ kg	$9.11 \cdot 10^{-28}$ gm	0.51 MeV
proton mass	M_p	$1.67 \cdot 10^{-27}$ kg	$1.67 \cdot 10^{-24}$ gm	938 MeV
neutron mass	M_n	$1.67 \cdot 10^{-27}$ kg	$1.67 \cdot 10^{-24}$ gm	939 MeV
pion mass (=270•m$_e$)	m_π	$2.46 \cdot 10^{-28}$ kg	$2.46 \cdot 10^{-25}$ gm	140 MeV
muon mass (=207•m$_e$)	m_μ	$1.89 \cdot 10^{-28}$ kg	$1.89 \cdot 10^{-25}$ gm	106 MeV
classical elect radius (=e^2/mc^2)	r_c	$2.82 \cdot 10^{-15}$ m	$2.82 \cdot 10^{-13}$ cm	
Compton wavelength (=h/mc)	λ_c	$2.43 \cdot 10^{-12}$ m	$2.43 \cdot 10^{-10}$ cm	0.02 Å
Thomson cross-section	σ_T	$6.65 \cdot 10^{-29}$ m^2	$6.65 \cdot 10^{-25}$ cm^2	
Planck length $(=\sqrt{\hbar G/c^3})$	l_{Pl}	$1.61 \cdot 10^{-35}$ m	$1.61 \cdot 10^{-33}$ cm	
Planck time $(=\sqrt{\hbar G/c^5})$	t_{Pl}	$5.39 \cdot 10^{-44}$ s	$5.39 \cdot 10^{-44}$ s	
Planck density (=c^5/ℏ G^2)	ρ_{Pl}	$5.16 \cdot 10^{96}$ kg/m^3	$5.16 \cdot 10^{93}$ gm/cm^3	
Bohr radius (=ℏ ^{2}/me^2)	r_B	$0.53 \cdot 10^{-10}$ m	$0.53 \cdot 10^{-8}$ cm	0.5 Å
Fine structure constant (=e^2/ℏ c)	α	$7.30 \cdot 10^{-3}$	$7.30 \cdot 10^{-3}$	1/137
Bohr magneton (=eℏ /2m$_e$c)	μ_B	$9.27 \cdot 10^{-24}$ J/T	$9.27 \cdot 10^{-21}$ erg/gauss	
Nuclear magneton (=eℏ /2M$_p$c)	μ_N	$5.05 \cdot 10^{-27}$ J/T	$5.05 \cdot 10^{-24}$ erg/gauss	
Permittivity of vacuum	ε_o	$8.85 \cdot 10^{-12}$ fd/m		$1/4\pi\varepsilon_o = 9.0 \cdot 10^9$